W9-ADV-693

INTERDISCIPLINARY MATHEMATICS

BY ROBERT HERMANN

1. General Algebraic Ideas.
2. Linear and Tensor Algebra.
3. Algebraic Topics in Systems Theory.
4. Energy Momentum Tensors.
5. Topics in General Relativity.
6. Topics in the Mathematics of Quantum Mechanics.
7. Spinors, Clifford and Cayley Algebras.
8. Linear Systems Theory and Introductory Algebraic Geometry.
9. Geometric Structure of Systems -- Control Theory and Physics, Part A.
10. Gauge Fields and Cartan-Ehresmann Connections, Part A.
11. Geometric Structure of Systems-Control Theory, Part B.
12. Geometric Theory of Nonlinear Differential Equations, Bäcklund Transformations, and Solitons, Part A.
13. Algebro-Geometric and Lie Theoretic Techniques in Systems Theory, Part A, by R. Hermann and C. Martin.
14. Geometric Theory of Nonlinear Differential Equations, Bäcklund Transformations, and Solitons, Part B.
15. Toda Lattices, Cosymplectic Manifolds, Bäcklund Transformations and Kinks, Part A.
16. Quantum and Fermion Differential Geometry, Part A.
17. Differential Geometry and the Calculus of Variations, 2nd Ed.
18. Toda Lattices, Cosymplectic Manifolds, Bäcklund Transformations and Kinks, Part B.
19. Yang-Mills, Kaluza-Klein, and the Einstein Program.
20. Cartanian Geometry, Nonlinear Waves, and Control Theory, Part A.
21. Cartanian Geometry, Nonlinear Waves, and Control Theory, Part B.
22. Topics in the Geometric Theory of Linear Systems.
23. Topics in the Geometric Theory of Integrable Mechanical Systems.

LIE GROUPS: HISTORY, FRONTIERS AND APPLICATIONS

Note: This series has outgrown its original plan, hence it will now bifurcate. SERIES A will continue the series of translations of the classics.

SERIES A

1. Sophus Lie's 1880 Transformation Group Paper. Translation by M. Ackerman, Comments by R. Hermann.
2. Ricci and Levi Civita's Tensor Analysis Paper, Translation and Comments by R. Hermann.
3. Sophus Lie's 1884 Differential Invariants Paper, Translation by M. Ackerman, Comments by R. Hermann.
4. Smooth Compactification of Locally Symmetric Varieties, by A. Ash, D. Mumford, M. Rapoport and Y. Tai.
5. Symplectic Geometry and Fourier Analysis, by N. Wallach.
6. The 1976 Ames Research Center (NASA) Conference on the Geometric Theory of Nonlinear Waves.
7. The 1976 Ames Research Center (NASA) Conference on Geometric Control Theory.
8. Hilbert's Invariant Theory Papers. Translation by M. Ackerman, Comments by R. Hermann.
9. Development of Mathematics in the 19th Century, by Felix Klein, Translated by M. Ackerman, Appendix "Kleinian Mathematics from an Advanced Standpoint", by R. Hermann.
10. Quantum Statistical Mechanics and Lie Group Harmonic Analysis, Part A, by N. Hurt and R. Hermann.
11. First Workshop on Grand Unification, by P. Frampton, S. Glashow, and A. Yildiz.
12. Inverse Scattering Papers: 1955-1963, by I. Kay and H.E. Moses.
13. Geometry of Riemannian Spaces, by Elie Cartan. Translated by J. Glazebrook, Commentary by R. Hermann.

SERIES B: SYSTEMS INFORMATION AND CONTROL

1. Geometry and Identification, P.E. Caines and R. Hermann, Eds.
2. Berkeley-Ames Conference on Nonlinear Problems in Control and Fluid Mechanics.

LIE GROUPS: HISTORY, FRONTIERS AND APPLICATIONS

VOLUME XIV

The Geometric Theory of Ordinary Differential Equations and Algebraic Functions

By Georges Valiron

TRANSLATED BY JAMES GLAZEBROOK

MATH SCI PRESS

53 JORDAN ROAD
BROOKLINE, MASSACHUSETTS 02146

Translation of

Equations Fonctionelles, Applications

(Vol. 2 of *Cours D'Analyse Mathematique*),
Chapters I-XI

Second Edition, 1950

Permission to translate given by Masson et Cie
Publisher of French Edition

Library of Congress Cataloging in Publication Data

Valiron, Georges, 1884-
The geometric theory of ordinary differential equations and algebraic functions.

(Lie groups ; v. 14)
Translation of: Equations fonctionnelles, applications, Chapter 1-11 (v. 2 of Cours d'analyse mathématique, 2nd ed., 1950)
Bibliography: p.
1. Differential equations. 2. Functions, Algebraic. I. Title. II. Series.
QA372.V334 1984 515.3'5 84-21314
ISBN 0-915692-38-4

MATH SCI PRESS
53 JORDAN ROAD
BROOKLINE, MASSACHUSETTS 02146

Printed in the United States of America

PREFACE

by Robert Hermann

Since this translation of the first part of the second volume of Valiron's treatise is meant as an historical accompaniment to the work in the *Interdisciplinary* series (Volumes 23 and 24) on Integrable Systems, I will not write Commentaries, as I have done for the other translations in the *Lie Group* series.

In recent years, a branch of mathematics called *geometric analysis* has crystallized from a diverse background. It has been very fruitful and stimulating to think of analytical problems from a geometric point of view, and to develop analytic techniques to tackle geometrically posed problems. Developing the mathematical background of physics, mechanics, and control theory has also been a major stimulant, and conversely, the mathematical tools developed in response should be increasingly useful in these applied disciplines.

While there have been (to my knowledge) no attempts at describing the historical background to this development of a hybrid subject between geometry and analysis, it is my opinion that the French *Cours d'Analyse* tradition has played a major role. As a student of Don Spencer (who is the prototypical "geometric analyst"!) in the 1950's, I tried to understand the classical background and examples for what I saw and heard around me: I discovered that these treatises were full of valuable information. Later on, as I became interested in the mathematical background of physics, mechanics, and control theory, I also found much that was of value in these applications. In the mathematical training of the late 19th and early 20th century, the sense of unity between algebra, analysis, and geometry had not yet been lost -- as Mike Hazewinkel once put it to me, "Nobody told them that you couldn't study algebra, analysis, and geometry at the same time." As an additional bonus, there was still some sense of common culture between "pure" and "applied" mathematics.

In recent years, there has been a turning in even the most avante garde areas of "pure" mathematics toward themes that were common in the 19th century. This makes the *Cours d'Analyse* even more valuable and relevant for today's students of pure and applied mathematics, physics, and engineering.

I have chosen this piece of Valiron's *Cours d'Analyse* to translate, even though it is not the most famous or historically important treatise in this traditional form. (I would have chosen Picard, if these were the criteria. Goursat's version also is very significant, but, of course, it has long been available in an English translation.) However, because it is virtually the last "Cours d'Analyse" (the first edition was published in 1945), I have found it to be most valuable as a bridge between the "classical" and "modern" points of view, and as a source of insight into many topics of 19th century mathematics, which I want to understand from a modern point of view.

The first part of Valiron's *Cours* (which I am not planning to translate) is subtitled *Theories des Fonctions*, and is an excellent exposition (with the classical flavor) of early 20th century real and complex analysis, with many fascinating explanations of the 19th century background. For example, it has an excellent treatment of elliptic functions, which is background to the treatment of Abelian functions (≡ algebraic curves ≡ function fields of one variable ≡ meromorphic functions on fields on a Riemann surface) given in this volume. The original of the material presented here forms the contents of the first eleven chapters of the second volume of the *Cours d'Analyse*, subtitled *Equations Fonctionelles, Applications*. I have chosen to publish these separately because they form a superb exposition of the classical background of what one might call the "geometric" theory of ordinary differential equations. (Hille's treatises are the best contemporary versions.) This material is valuable background to work in differential geometry, Lie group theory, and global analysis/differential topology. It also contains much important to the contemporary theory of integrable dynamical systems, especially the Painlevé theory of ordinary differential equations. Chapters I and II contain a valuable exposition of the 19th century theory of *algebraic functions of one complex variable*, and the auxiliary theory of *compact Riemann surfaces*. Chapters 12-16, which I plan to publish later, treats the classical differential geometry of curves and surfaces, and related material in partial differential equations.

Again, I would like to thank James Glazebrook for his translation. Joyce Martin has also been of great help in typing and editorial work, and I also thank her.

PREFACE

The main purpose of this volume is to expose the student to the material of a course in differential and integral calculus, which was not covered in the volume entitled *Theorie des functions*, namely: the elements of the theory of analytic functions of several variables, the theory of differential equations and first order partial differential equations, the calculus of variations and applications of analysis to geometry. But, as in the first volume, I have made some inclusions to the theories studied, occasionally with some detail. As certain results relating to the properties of solutions of analytic differential equations necessitate some acquaintance with the theory of algebraic functions, I have devoted two chapters to this theory and that of abelian integrals. This has allowed me to prove some theorems relating to algebraic plane curves, which are no longer taught in specialised mathematics courses, and to present, once more, the unity of mathematics.

Three types of equations defining functions are studied in this book: algebraic equations in two variables, (ordinary) differential equations and partial differential equations. All of these equations are functional equations, hence the justification in calling the first part of the title: *Equations functionelles*. As for *applications*, these are the applications of analysis to geometry and include some applications of Abel's theorem on abelian integrals.

The first two chapters are devoted to algebraic functions of one variable and to abelian integrals; the theorem of Picard on uniformisation is established with the help of a recent theorem due to Bloch. The elements of the theory of analytic functions of several variables and the method, due to Cauchy, of the majorant functions, are discussed in the third chapter: the theorems of existence of the solutions of analytic differential equations are established along with the essential inclusion due to Weierstrass and Poincaré, concerning the way in which the solutions depend on the initial conditions and how the parameters arise in these equations. There follows a chapter on the theory of contact and (curve) envelopes; for the latter, I have strived to present an accurate account. This chapter immediately serves to produce an application of the theory of analytic functions, and a pathway to the study of singular solutions of differential equations.

The theory of differential equations accounts for six chapters. The elementary theory of these equations, a lot of which should be familiar to the student, is covered in Chapter V. The general account of linear differential equations in the complex domain, presented in Chapter VI, contains some of the results of Poincaré on equations with rational coefficients. This is followed, in Chapter VII, by a more prolonged study of the equations of Gauss, Legendre and Bessel, and the associated polynomials and series expansions. In Chapter VIII, having introduced the more elementary theorems of Briot and Bouquet on the singularities of first order analytic equations, the theorem of Painlevé on first order equations, the results of Briot and Bouquet and Hermite on first order algebraic equations with uniform solutions, and the more recent results of Boutroux and Malmquist, I discuss the fundamental works of Painlevé on equations with fixed critical points, and restrict matters to the most elementary equation of this theory. Chapter IX is concerned with differential equations over the real; the method of successive approximation due to Picard is presented together with the same additions as in the method of the majorant functions. The general theorem on the existence of solutions of a solvable first order equation is established (following Montel) by the method of polygonal approximation due to Cauchy. It is followed by some observations on the practical methods of approximation. The more elementary theorems of Poincaré on the form of real integral curves of rational equations of the first order, and those of Sturm and Liouville on linear equations of the second order, are outlined in Chapter X.

On the calculus of variations, in Chapter XI, I discuss the more elementary examples and those are supplemented by the method of Weierstrass and Hilbert.

The applications of analysis to the theory of curves and surfaces are the subjects of Chapters XII, XIII and XIV. In the theory of surfaces, I include, for the sake of completion, the theorem of Bonnet on total curvature, the theorem of Gauss on geodesic polygons and some observations on minimal surfaces.

The last two chapters deal with partial differential equations. The first order equations are studied in Chapter XV. The existence theorems of the solutions are here deduced from theorems relating to differential equations without assuming the analyticity of those given. Chapter XVI is concerned with several classical results relating to second order equations in two variables: equations of Monge-Ampère and linear equations; the telegraphic equation is integrated by Riemann's method, the Laplace equation is discussed from the point of view of the Dirichlet problem, and the heat equation of Fourier, a short study of which completes this volume, leads to the study of the infinitely differentiable, non-analytic functions studied by Holmgren.

Table of Contents

Chapter I

ALGEBRAIC FUNCTIONS OF ONE VARIABLE

The simplest multiform functions, defined by a functional relation, are the algebraic functions. The polynomial $P(z,u)$ in z and u being irreducible and of degree m greater than 1 with respect to u, the equation $P(z,u) = 0$ defines u as a function of z. This is a function with m branches, known as an algebraic function. The case where m equals 2, provides a function with two branches, the explicit expression of u as a function of z is then easily obtained. We could effectively write in the same way, u as a function of z when $m = 3$, by reducing the equation in u to the canonical form and applying the formula of Cardano. For the fourth degree, we would have just as explicit a formula. However, we intend studying the general properties of algebraic functions, their singularities on the Riemann surfaces that they define, and the connection of these surfaces.

It is clear that the study of an algebraic function is tied up with that of the algebraic curve to which it corresponds, by considering as belonging to this curve, not only the real points corresponding to z and u, but also the imaginary points.

The theory of algebraic curves originates from Descartes (*La Geometrie*, 1637) and Newton [the definition of order (1636) and verification of its invariance under co-ordinate transformations]. It was developed towards and during the 18th century [by L'Hopital (1696), Newton; Gua; Malves (the study of double points, 1740)] and took flight when Euler, interested in the intersection of two curves, introduced imaginary points (1748): Cramer in 1750 gave the first results relating to the determination of a curve of degree nn by $1/2\, n(n+3)$ points. Bezout, in 1766, proved completely the theorem on the number of common points of two curves. But the building up of a complete theory has been the work of the 19th century. When Poncelet (1822), Plücker (1830), Steiner (1832) gave the first examples of quadratic transformations, as Poncelet and Plücker studied the influence of singularities on the class (Plücker formulae), Abel, in 1826, produced his theorem on the functions defined by integrals of rational fractions in u and z, evaluated on an algebraic curve $P(z,u) = 0$ and Jacobi, in 1834, posed the problem of parametric representation of groups of points of the curve by means of sums of such integrals. Solved in a particular case by Göpel and Rozenhain (1844-1847), this problem of Jacobi inversion was developed in the general case by Riemann and Weierstrass. In the meantime, Puiseaux published in 1854, an important paper on algebraic functions. Riemann's fundamental paper (*Theorie der Abelachen Funktionen* (1857)), in which were introduced the notions of genus, class of curves, a curve corresponding to a Riemann surface, etc.), was the origin of all further works due mainly to Cremona (1863), Roch (1864), Clebsch and Gordon (1866), Cayley (1865), Neumann (1865), Schwarz (1870), Brill and Nöther (1873) and Bertini (1877). Along with these we could justifiably associate the famous work of Poincaré on uniformisation.

I. DEFINITIONS--CIRCULAR SYSTEMS

1 - Mondromy theorem

We know that given a Taylor series

$$\sum_{0}^{\infty} a_n (z - z_0)^n \tag{1}$$

with finite positive radius of convergence R_0, it defines an analytic function $f(z)$ which is its analytic continuation and whose values could be obtained by a sequence of radial continuation.

The radial continuation of the series (1) takes effect following (I, 199). Given a half-line D originating from the point z_0, we consider the point z_1 of intersection of D and the circumference $|z - z_0| = R_0$. If z_1 is not a singular point of $f(z)$, there exists a Taylor series

$$\sum_{0}^{\infty} b_a (z - z_1)^n \tag{2}$$

whose radius of convergence R_1 is positive and finite, and whose sum is equal to the sum of (1) at every point z common to the circles $|z - z_0| < R_0$ and $|z - z_1| < R_1$. We repeat the same operation starting from the point z_2, The intersection of D and $|z - z_1| = R_1$ (z_2 being outside the segment z_0, z_1). If z_2 is not singular, we extend $f(z)$ in a circle $|z \quad z_2| < R_2$, etc. Two cases are possible: either the continuation is possible right along D, or else a singular point is met at the end of a finite or infinite number of operations (in the second case, the radii $R_p \to 0$ as $p \to \infty$). In all cases, if z' is a point of D attained by this extension, z' belongs to a circle $|z - z_p| < R_p$, $|z' - z_0|$ being greater than or equal to $|z_p - z_0|$, and we could find a number ε sufficiently small such that, for $|z'' - z'| < \varepsilon$, each point of the segment z_0, z'' belongs to one of the circles $|z - z_q| < R_q$, $0 \le q \le p$. The value of $f(z'')$ obtained by means of the Taylor series at $2 - 3p$, is therefore the same as that which would be obtained by continuing the series (1) along the segment z_0, z''. It follows that the set of points z that could be attained by radial continuation, by considering all the half-lines D from z_0, forms a domain and that defined on each point z of this domain, is a unique value $f(z)$, the function $f(z)$ thus obtained being holomorphic in this domain which is the star-set of holomorphy with centre z_o (I, 199).

It is clear that if we carry out the analytic continuation of the series in (1) arbitrarily by means of the Taylor elements along a curve which belongs to the domain of holomorphy of (1), the value obtained at each point z, will be the value of the holomorphic function defined by radial continuation. Generally, it is possible to extend the series (1) along a continuous curve originating from z_o. We consider the first point z_1 of intersection of this curve with the circumference $|z - z_o| = R_o$ and, if z_1 is not singular, we consider the series (2) giving the first continuation, then the point z_2 of intersection of the circumference $|z - z_1| = R_1$ with the curve, and so on.

Monodromy Theorem. *Let* Δ *be a simply connected domain and let a function element be given a Taylor series (1),* z_o *belonging to* Δ. *If it is possible to continue analytically this element along every continuous curve originating from* z_o *and completely inside* Δ, *the branch of the analytic function is holomorphic in* Δ.

PRELIMINARY REMARKS

I - We shall assume that Δ does not contain the point at infinity. If Δ does contain the point at infinity, we would make a homographic transformation sending one of the boundary points of Δ to infinity. (If Δ contains all of the plane, together with the point at infinity, the radial continuation of (1) would show that the radius of convergence of this series would be infinite, the series would define an entire holomorphic function at infinity and hence a constant.)

II - The statement implies that the radius of convergence of each element deduced from (1) by continuation in Δ, is at least equal to the shortest distance from the centre of this element to the boundary of Δ.

III - The term monodromy arises from the word monodrome that is sometimes used to characterise a function holomorphic in a domain. To prove the theorem, we can proceed as in the construction of the star-set holomorphy. We introduce a family of simple continuous curves originating from z_o and possessing the following properties:

1° Every point of Δ other than z_o lies on one and only one of these curves.

2° If z' and z are two points of Δ, the portion of the curve of the family joining z' to z_o when z' tends towards z.

Under these conditions, if we define at each point z of Δ, a func- $f(z)$ equal to the value obtained by continuing along the curve of the family joining z to z_o and by replacing the circle of convergence of each element by the concentric circle "tangent" to the boundary of Δ, the function $f(z)$

is uniform and the second condition imposed on the curves, shows that it is holomorphic.

Everything reverts to proving the existence of the family of curves in question. We could manage by replacing Δ by a sequence of polygons whose interiors tend towards Δ. But it suffices to note that if we make a conformal representation of Δ on the circle $|Z| < 1$ in such a way that z_o becomes $Z = 0$, the curves of Δ which correspond to the radii of the circle $|Z| < 1$, settles matters.

This conformal representation gives another proof of the theorem. To the analytic function obtained by continuation of (1) in Δ, there corresponds the function deduced by extension in $|Z| < 1$ of the element of centred $Z = 0$ deduced from (1) by the inverse transformation, $z = Q(Z)$. The continuation of this element in $|Z| < 1$ is possible along every path, hence along the radii. The radius of convergence of this element is at least 1; the function transformed by $f(z)$ is holomorphic for $|Z| < 1$, therefore $f(z)$ is holomorphic in Δ.

2 - Algebraic functions and corresponding Riemann surfaces

We have seen, quite succinctly (I, 203) how the theorem of implicit functions allows us to define the algebraic functions. We are going to take up this matter again as another means of proving the continuity of roots of an equation. This theorem on the continuity had already been given for sequences of functions (a consequence of Rouche's theorem, I, 193). Generally, we have:

Theorem. *Assume the function* $f(z,u,v,w)$ *is continuous in the domain* $|z - z_o| < \alpha$, $|u - u_o| < \beta$, $|v - v_o| < \beta$, $|w - w_o| < \beta$ *and holomorphic with respect to* z, *and that the equation* $f(z,u_o,v_o,w_o) = 0$ *admits the number* z_o *as a root of order* p. *Under these hypotheses, the equation* $f(z,u,v,w) = 0$ *also admits* p *roots in* z *in the domain* $|z - z_o| < \alpha'$, $|u - u_o| < \beta'$, $|v - v_o| < \beta'$, α' *being arbitrary, but sufficiently small, and* β' *depending on* α'.

Let us write

$$f(z,u,v,w) \equiv f(z,u_o,v_o,w_o) - [f(z,u_o,v_o,w_o) - f(z,u,v,w)] = 0 \quad .$$

As the function $f(z,u_o,v_o,w_o)$ is holomorphic and zero for $z = z_o$, we can find a circle $|z = z_o| \leq \alpha' < \alpha$, in which the function is non-zero for $z \neq z_o$, and on the circumference of which its modulus will be greater than a positive number γ. On account of the continuity which induces uniform continuity, we have

$$|f(z,u_o,v_o,w_o) - f(z,u,v,w)| < \gamma \cdot$$

providing $|z-z_o| \leq \alpha'$, $|u-u_o| < \beta'$, $|v-v_o| < \beta'$, $|w-w_o| < \beta'$, where β' is less than β and sufficiently small. Following Rouche's, the equations $f(z,u,v,w) = 0$ and $f(z,u_o,v_o,w_o) = 0$, will have the same number of roots, which proves the theorem.

APPLICATION TO ALGEBRAIC FUNCTIONS

Let $P(z,u)$ be an irreducible polynomial in z and u, i.e., cannot be reduced to a product of two polynomials in z and u, and let m be the degree of $P(z,u)$ with respect to u for an arbitrary value of z. We have

$$P(z,u) = a_n(z)u^n + \dots + a_o(Z) \quad , \tag{3}$$

where $a_n(z)$ are polynomials. For a value of z, the roots of the equation $P(z,u) = 0$, where u is the unknown, are distinct in general. Two roots are indistinct if and only if the equations in

$$P(z,u) = 0 \quad , \quad \frac{\partial}{\partial z} P(z,u) = 0 \tag{4}$$

have a common solution. On eliminating u from these two equations (4), we obtain the discriminant of (3), $\Phi(z)$. The equation

$$\Phi(z) = 0 \tag{5}$$

gives the values of z for which the two zeros of (3) are indistinct. The polynomial $\Phi(z)$ which could be obtained by the method of divisions, could only be identically zero if the first members of equation (4) have a common divisor, accordingly if $P(z,u)$ were reducible.

When z is distinct the roots ζ of equation (5), all the roots of $P(z,u) = 0$ are simple. For $z = \zeta$ a root of (5), $P(z,u)$ has at least one multiple zero (possibly infinite), but could even have simple zeros. Let us consider a pair of finite values (z_o,u_o) such that the equation $P(z_o,u) = 0$ admits u_o as a simple root. Following the theorem on continuity, the equation in u

$$P(z,u) = 0 \tag{6}$$

admits a root $u(z)$, unique in the domain $|z-z_o| < \alpha'$, $|u-u_o| < \beta'$, where β' is arbitrary but sufficiently small, and α' depends on β'. $u(z)$ then tends to u_o when $z \to z_o$. By applying this result to a point z and with value $u(z)$, we see that the function $u(z)$ is continuous in the circle $|z-z_o| < \alpha'$. This function, both uniform and continuous, is holomorphic. To show this, it suffices to show that it is derivable. If we give to z an increment Δz and if Δu is the corresponding increment to u, we have $P(z+ \ z, u+\Delta u) = 0$, $P(z,u) = 0$, hence

$$P(z+\Delta z,\ u+\Delta u) - P(z,u) = 0$$

or

$$[P(z+\Delta z,\ u+\Delta u) - P(z,u+\Delta u)] + P[(z,u+\Delta u) - P(z,u)] = 0$$

which can also be written, since the first bracket is the increment of a holomorphic function of z, and the second a holomorphic function of u, as

$$\Delta z\left[\frac{\partial P}{\partial z}(z,u) + \eta\right] + \Delta u\left[\frac{\partial P}{\partial u}(z,u) + \eta'\right] = 0$$

where η and η' tend to zero when Δz and Δu tend to zero (I, 187). Now when $\Delta z \to 0$, Δu tends to zero, and $\frac{\partial P}{\partial u}$ is non-zero for the values z and $u(z)$, since $u(z)$ is a simple root of (6). The ratio of Δu to Δz hence has a limit; $u'(z)$ exists and we have

$$u'(z) = -\frac{\frac{\partial P}{\partial z}((z,u(z))}{\frac{\partial P}{\partial u}((z,u(z))}$$

U(z) is *holomorphic.*

Having proved this point, let us denote by $\zeta_1, \zeta_2, \ldots, \zeta_4$ the points in the plane of z, that are the images of roots of equation (5) and also the roots ζ_4' of $a_n(z) = 0$ distinct from these; they provide the points for which (3) has an infinite simple zero. If z_o is distinct from these points ζ and ζ', the m roots of (6) are finite and simple. If u_o is one of them, there corresponds to it, following that which preceded, a solution $u(z)$ of equation (6), which is holomorphic in a certain circle $|z - z_o| < \rho$. This holomorphic function admits a Taylor series expansion of the form (1) about the point z_o. If we replace u by this expansion in P(z,u), we obtain an entire series in $zz - z_o$, that is identically zero for $|z - z_o| < \rho$, therefore identically zero in its circle of convergence, which is that of the series in (1). If we analytically continue this solution, each new element so obtained, in turn coincides with the one proceeding it in the common region of the circles of convergence, hence verifying (6) in this region and consequently throughout its circle of convergence. The analytic function so defined, is therefore a solution of (6); moreover, its singularities at a finite distance could only be the points ζ_j and ζ_k' since, at every point distinct from these, the solutions of (6) are holomorphic.

The solutions of (6) are therefore branches of one of several analytic functions whose singular points are the points ζ_j, ζ_k' *or certain or these points.*

If z tends towards a point ζ_k', a branch of $u(z)$ could tend towards a finite value u' which is a simple root of (6) for $z = \zeta_k'$. Then the branch $u(z)$ coincides about ζ_k' with the holomorphic solution defined from $(\ _k',u')$, this branch is still holomorphic at the point ζ_k'. If the branch $u(z)$ tends towards infinity when $z \to \zeta_k'$, this signifies that $1/u(z)$ tends to the zero root of the equation in $1/u$

$$a_n(z) + a_{n-1}(z)\frac{1}{u} + \ldots + a_o(z)\frac{1}{u^n} = 0 \quad ,$$

which is a simple root; we see that $1/u(z)$ is again holomorphic about ζ_k' and admits this point as a zero. In this case, the point ζ_k' is a pole for the branch envisaged. *Every point* ζ_k' *is therefore a pole for a branch.*

The study of $u(z)$ when z moves away indefinitely, is made by introducing the transformation $1/z$. If we put $Z = 1/z$ and if the equation in u so obtained has all its roots simple at the origin $Z = 0$, one possibly infinite, the branches $u(z)$ are all holomorphic at infinity, save at most one which could admit $z = 0$ as a pole. If there are multiple roots for $Z = 0$, the point $z = \infty$ must be considered as a point ζ_j. If z tends towards a point ζ_j, certain branches could tend towards a simple root of (6); these branches are again holomorphic at this point, with the exception of at most one branch which could admit this point as a pole.

Let us assume that when z tends towards ζ_j, $u(z)$ tends towards a multiple finite root u_j of $P(\zeta_j,u) = 0$; let p be the order of multiplicity of this root. Following the theorem on continuity, the equation $P(\ ,\ ) = 0$ has p and only p roots which tend towards u_j when $z \to \zeta_j$. Let us trace out a circle $|z - \zeta_j| < \rho$, ρ being less than the shortest distance from the point ζ_j to the other possible singular points, and let us consider the domain Δ obtained by fixing one of the radii D of this circle. Inside Δ which is simple connected and which contains neither the point ζ_q nor the point ζ_k', the m branches $u(z)$ are holomorphic following the monodromy theorem. Let us consider one of these which takes the value u_j at the point ζ_j. If z describes a closed contour about ζ_j, lying with Δ and starting from a point z' of D and to return there (Fig. 1), $u(z)$ starts with a value $u_1(z')$ and takes a value $u_2(z')$ either distinct or not from the first, after this rotation. If $u_2(z') = u(z')$, this equality continues along the radius D, since along D the m solutions of (6) are all distinct. *In this case the branch* $u(z)$ *is uniform throughout the circle* $|z - \zeta_j| < \rho$ *and tends toward a finite limit as* $z \to \zeta_j$; *it is still holomorphic at the point* ζ_j.

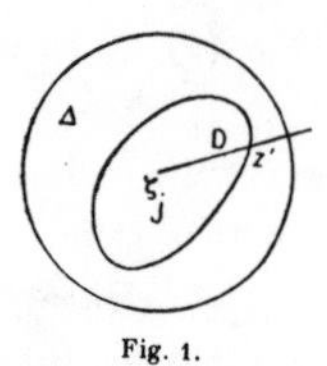

Fig. 1.

Let us now assume that $u_2(z') \neq u_1(z')$. If we start from z' with the value of $u_2(z')$, the corresponding branch $u(z)$ will give, after the same rotation, a value $u_3(z')$, distinct from $u_2(z')$ [unless on making a circuit in the opposite sense one does not pass from $u_2(z')$ to $u_1(z')$]. If $u_3(z') = u_1(z')$, the same equality holds right along D and by putting $z - \zeta_j = Z^2$, the branch $u(z)$ in question, becomes a uniform function of Z in the circle $|Z^2| < \rho$. As this function of Z remains bounded at the point 0, it is holomorphic there. If $u_3(z') \neq u_1(z')$, a new circuit will be made which leads to a value $u_4(z')$. If $u_4(z') = u_1(z')$, $u(z)$ will become uniform by setting $z - \zeta_j = Z^3$ and it will then be a holomorphic function of Z for $Z = 0$. Otherwise, it will be recommenced. This being true for all the branches tending towards u_j when $z \to \zeta_j$, we see that these branches could be divided into one or more groups: *branches which are holomorphic for* $z = \zeta_j$; *groups of* s *branches which permute between themselves under rotation about* $z = \zeta_j$, *for such a group* $u(z) - u_j$ *is a holomorphic function of* Z *for* $|Z| < \rho^{1/s}$ *if we have* $z - \zeta_j = Z^{ss}$, *and is zero for* $Z = 0$; *moreover,* s *is the smallest integer having this property.* Hence we have, in this case,

$$u(z) = u_j + \sum_1^{\infty} c_\mu Z^\mu = u_j + \sum_1^{\infty} c_\mu (z - \zeta_j)^{\mu/s} \qquad |z - \zeta_j| < \rho \tag{7}$$

where the values of μ for which $c_\mu \neq 0$ have their greatest common division with s. Formula (7) gives the values of the s branches which permute between themselves about ζ_j. This point ζ_j is *an algebraic critical point of order* s for those branches, or *a point of ramification of order* s. We also say that those s branches form a circular system of order s.

AN IMPORTANT REMARK

In what follows the notation used in (7) $(z - \zeta_j)^{\mu/s}$ *signifies* $[(z - \zeta_j)^{1/s}]^{\mu}$, *the chosen branch for* $(z - \zeta_j)^{1/s}$ *being the same for all terms.*

If the equation $P(\zeta_j, u) = 0$ admits an infinite multiple root, we consider $u(z)$ by putting $u = 1/\nu$. For solutions of the equation in ν which tends towards zero when $z \to \zeta_j$, we will have the same situation as above. Certain branches will admit the point ζ_j as a pole, others might group themselves in a circular system, but we will have

$$u(z) = \sum_{-\nu}^{\infty} c_{\mu}(z - \zeta_j)^{\mu/s} \qquad \nu > 0 \quad . \tag{8}$$

We will indeed have in this case a point of ramification of order s.

Similarly, we consider that which passes around the point at infinity, if for z infinite, the equation in u has mutiple roots: we set $z = 1/Z$ to reduce it to the case where $Z = 0$. *We are going to see that the solutions of (6) are the branches of a single analytic function.* Suppose this is not the case, i.e., by continuation of one of the elements $u(z)$, holomorphic about z_o, we would obtain only a part of the solutions. At every point z distinct from the points ζ_j, ζ'_k we would obtain just $p < m$ values for $u(z)$, moreover these are all distinct. For if at a point z distinct from the ζ_j, ζ'_k m values were obtained, it would be the same at every other such point, since two branches could only coincide at the points ζ_j.

Let $u_1(z), u_2(z), \ldots, u_p(z)$ be the values of these p branches at an arbitrary point z distinct from ζ_j, ζ'_k. They are holomorphic about every point z so considered. The sum

$$S_{\nu}(z) = [u_1(z)]^{\nu} + \ldots + [u_p(z)]^{\nu}$$

where ν is a positive integer, is holomorphic under the same conditions and besides which, is uniform in the plane minus the points ζ_j, ζ'_k. The points ζ'_k being certain poles of $u_j(z)$, could only be poles for $S(z)$.

When z tends towards a point ζ_j, certain functions $u_j(z)$ could not remain bounded, but then they form groups of s functions of the form (8). The sum of the powers of these s functions, is meromorphic at the point ζ_j, and hence $S_{\nu}(z)$ is as well. This is still true for z infinite. Thus each function $S_{\nu}(z)$ is meromorphic at a finite distance and at infinity, it is hence a rational fraction (I, 194). Following the known properties of the symmetric functions of the roots of an equation, the equation

$$[u - u_1(z)][u - u_2(z)] \ \ldots \ [u - u_p(z)] = 0$$

has for its coefficients, polynomials in $S_\nu(z)$ $\nu = 1,2,\ldots,p$. These coefficients are therefore rational fractions in z and consequently the function $u(z)$ in question, whose branches are $u_p(z),\ldots,u_1(z)$, would be defined by an equation $Q(z,u) = 0$, where Q is a polynomial in u and z, of degree $p < m$ in u. The polynomial $P(z,u)$ would be reducible.

We have thus reached the following proposition:

If the polynomial $P(z,u)$ *is irreducible, the equation* $P(z,u) = 0$ *defines* u *as an analytic function of* z. *It is an algebraic function with* m *branches if* m *is the degree of* $P(z,u)$ *with respect to* u. *The singularities are a finite number of poles and algebraic critical points provided by the zeros of the discriminant* $\Phi(z)$.

RIEMANN SURFACES

The Riemann surface on which the function $u(z)$ is uniform could be defined as follows. Let z_o be a point such that $P(z_o,u) = 0$ has all roots finite and simple. Let us join the points ζ_j to the point at infinity by simple curves C_j not pairwise crossing and not passing through z_o. The C_j form the boundary of a simply connected domain Δ which contains no other singularity of $u(z)$ other than poles. Each of the m branches of $u(z)$ that could be obtained via analytic continuation in Δ of the m elements with centre z_o, is meromorphic in Δ following the monodromy theorem. Let us consider m superposed planes endowed with lines C_j and correspond to each of them, one of the branches; let $u_q(z)$ be that which corresponds to the plane of rank q. If the branch $u_q(z)$ is holomorphic or meromorphic at point ζ_j, we suppress the line C_j in the plane of rank q. The lines remaining are the lines through which the sheets pass, in going from one to the other. If ζ_q is an algebraic critical point of order s for $u_q(z)$, i.e., if $u_q(z)$ is given by one of the formulae (7) or (8) for z close to ζ_q, the line C_j appears in s planes and the plane of rank q passes to that of rank q', when it is crossed in a certain sense, whilst in crossing it in another sense, one will pass into a plane of rank $q'' \neq q'$ if $s > 2$. We will obtain the disposition of the surface about the point at infinity and *at this point*, by considering the parts of the sheets outside of a great circle containing the points ζ_j in its interior. If a transformation $z = 1/Z$ is made, we will obtain one or more discs with one or more leaves ramified at the origin.

In particular we could take z_o in such a way that the lines joining z_o to the points ζ_j are all distinct and we could take as the C_j the half-lines extending beyond the points ζ_j, the segments z_o, ζ_j. In each of the planes, the lines of passage will be the boundaries of the of holomorphy with centre z_o, where the boundary half-lines abutting the poles, are suppressed. We could equally join the ζ_j to a point z' taken at a finite distance and such that the roots of $P(z',u) = 0$ are simple; it will therefore be necessary, should the question arise, to take account of the point ζ_j situated at infinity. The disposition of the surface at the point z' will be obtained as in the case where z' was the point at infinity.

Spherical representation

It is well known that instead of representing a complex number z in a plane, we could represent it on a sphere (I, 195). We could therefore attach to an algebraic function a system of m infinitely close superposed spheres, joined along the lines corresponding to the lines of passage of the superposed planes, and on the set of which the function will be uniform.

The above proof that showed that the branches of an algebraic function are the branches of one analytic function only, in turn shows that:

Theorem. *An algebraic function with* m *branches which is holomorphic about each point, except at a finite number of points which are either poles or algebraic critical points, is an algebraic function.*

To say that a point is algebraic critical, is to say that about this point, the branches group themselves into a circular system of the form (7) or (8), $(s = 1)$. By applying Liouville's theorem, we are confined to saying that at the singular points ζ at a finite distance, the product $(z - \zeta)^{\lambda}\, u(z)$ remains bounded for all branches $u(z)$, where λ is finite, and to giving a similar condition for the point at infinity if it is singular.

Remark. Every planar (or spherical) sheet of the Riemann surface evidently contains at least one line of passage, since the surface is all of one piece. *There is, in practice, at least two lines of passage on each sheet when these lines have been described by joining the points* ζ_j *to a point* z_i *distinct from the points* ζ_j.

Assume that there is only one of them on a sheet; call it C_j. The branch $u_n(z)$ being holomorphic at the z_i would indeed be meromorphic along C_j, $u_q(z)$ would be uniform about ζ_j, which is not so.

3 - Determination of ramification points. Puiseaux's method

The points of ramification corresponding to pairs of values z,u satisfy both equations of (4). The values of z are the roots of equation (5) obtained by equating the discriminant of $P(z,u)$ to zero, where u is the variable. Let us suppose that we had determined a pair ζ,u_1 of values satisfying equations (4). By means of the eventual change of u_1 to $\frac{1}{u_1}$ and ζ_1 to $\frac{1}{\zeta_1}$, we can assume these values to be finite and on putting $u = u_1 + U$, $\zeta = \zeta_1 + Z$, we could assume that they are zero. Hence let us assume that:

$$P(0,0) = 0 \ , \quad \frac{\partial P}{\partial u}(0,0) = 0 \quad .$$

If $\frac{\partial P}{\partial z}(0,0) \neq 0$, we could solve for z, the equation $P(z,u) = 0$ for u small enough; the solution which tends towards 0 when $u \to 0$ is holomorphic for such u, and we have

$$z = a_p u^p + a_{p\neq 1} u^{p+1} + \dots \ , \qquad p = 2 \quad , \tag{9}$$

from which we deduce

$$\sqrt[p]{\frac{z}{a_p}} = u \left(1 + \frac{a_{p+1}}{a_p} u + \dots\right)^{1/p} = u\left(1 + \psi(u)\right)$$

where $\psi(u)$ is holomorphic for $|u|$ sufficiently small. We could solve for u since the derivative of the second number is equal to 1 for $u = 0$, and we obtain

$$u = \left(\frac{z}{a_p}\right)^{1/p} + \alpha\left(\frac{z}{a_p}\right)^{2/p} + \dots \qquad p \geq 2 \quad , \tag{10}$$

where the series of the second member is convergent for $|z|$ sufficiently small. The point $z = 0$ is a point of ramification of order p, and (10) defines a circular system of order p.

Geometrically, we interpret this as saying that the *tangent* to the algebraic curve defined by $P(z,u) = 0$ is at the point in question, parallel to the u-axis. If $p = 2$, the branch of the curve defined by (9) has an ordinary point at the origin; if $p \geq 3$ we have a point of inflection. In the first case, the ramification point is of order 2.

The calculation of $a_p, a_{p+1}, \dots,$ hence a_1, α_2, α_3 could be made step by step, by bringing the expression (9) into $P(z,u)$, then ordering with respect to u and equating successive powers of u to zero.

THE CASE OF MULTIPLE POINTS

If we have simultaneously,

$$P(0,0) = 0 \ , \ \frac{\partial P}{\partial u}(0,0) = 0 \ , \ \frac{\partial P}{\partial z}(0,0) = 0 \quad ,$$

the point $(0,0)$ is said to be a multiple point of the curve $P(z,u) = 0$. (The numbers u_1 and ζ_1 in question, having the transformation $u = u_1 + U$, $\zeta = \zeta_1 + Z$ could be imaginary). If we assume that, for $z = 0$, n roots of $P(z,u) = 0$ are zero, then this equation is of the form

$$Au^n + \sum A_{\alpha,\beta} u^\alpha z^\beta \tag{11}$$

and, in the summation α is grater than n when $\beta = 0$. We know that the n solutions which tend towards zero when z tends towards zero, are holomorphic at the origin or admit this point as an algebraic critical point. They group themselves into a circular system and it would be convenient to consider a holomorphic branch as forming a system of order 1. Puiseaux has given a method which allows separation of these circular systems, a means of determining their order and a step-by-step calculation of the coefficients of their expansion, showing at the same time the existence of these expansions. Let us seek to determine, a priori, a solution of the equation under the guise of

$$u = \lambda z^{\mu}\left(1 + \varepsilon(z)\right) \quad , \quad \lim_{z=0} \varepsilon(z) = 0$$

where λ and μ are constants and μ is positive. By bringing this value into (11) and on ordering matters by commencing from terms of the smallest infinitesimal order, we should obtain at least two terms of the lowest degree which cancel each other out. Let

$$A_{\alpha',\beta'}u^{\alpha'}z^{\beta'}, \qquad A_{\alpha'',\beta''}u^{\alpha''}z^{\beta''}$$

be two of these terms, of which one could be Au^n (then $\alpha' = n$, $\beta' = 0$ for example). For all the other terms obtained by replacing $\varepsilon(z)$ by 0, the order with respect to z will be at least equal to the order of these. We must then have

$$\alpha'\mu + \beta' = \alpha''\mu + \beta'' \quad , \quad \alpha\mu + \beta \geq \alpha'\mu + \beta' \quad . \tag{12}$$

To obtain the possible values of μ, Puiseaux introduced a geometric construction which had been applied by Newton in a particular situation. In the plane with co-ordinates $O\alpha$, $O\beta$ (rectangular), we indicate the points $B(\alpha,\beta)$ whose co-ordinates are the pairs of exponents of the terms of equation (11). The first condition (12) expresses μ as the negative of the gradient of line Δ joining the points $B(\alpha',\beta')$ and $B(\alpha'',\beta'')$, the second expresses the point $B(\alpha,\beta)$ as being on or above this line (fig. 2).

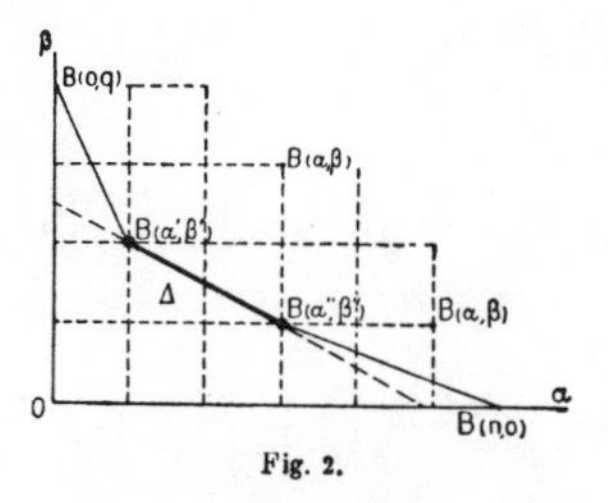

Fig. 2.

We therefore construct a polygon whose vertices are certain $B(\alpha,\beta)$ *and such that all other* $B(\alpha,\beta)$ *are on or above its sides (above signifying the side of the positive* β*). The gradients of the sides changed in sign, give the only possible values of* μ. This polygon is called *Newton's polygon* corresponding to equation (11).

To prove the existence of this polygon, and to construct it, we consider the lines joining the point $B(n,o)$ to the points $B(\alpha,\beta)$ corresponding to $\alpha < n$. The points exist since equation (11) contains a term independent of u, hence providing a point $B(o,q')$. The gradients of these lines have a maximum $-\mu_1$. Let $B(\alpha_1,\beta_1)$ be the point with the least abscissa situated on the line with gradient $-\mu_1$ passing through $B(n,o)$. If this point $B(\alpha_1,\beta_1)$ is the point $B(o,q')$, the polygon is completed, otherwise, the gradients of the lines joining this point $B(\alpha_1,\beta_1)$ to the points $B(\alpha,\beta)$ such that $\alpha < \alpha_1$, have a maximum $-\mu_2$, with $\mu_2 > \mu_1$. We consider the point $B(\alpha_2,\beta_2)$ of least abscissa situated on the line with gradient $-\mu_2$ passing through $B(\alpha_1,\beta_1)$. If this point is $B(o,q')$ the polygon is completed and is formed by the segments $B(n,o)$, $B(\alpha_1,\beta_1)$ and $B(\alpha_1,\beta_1)$, $B(o,q')$. Otherwise the construction is continued.

If $B(\alpha',\beta')$ and $B(\alpha'',\beta'')$ are two consecutive vertices of Newton's polygon, the value of μ obtained by means of this side, is a rational number equal to an irreducible fraction q/p

$$\mu = -\frac{\alpha'' - \beta'}{\alpha'' - \alpha'} = \frac{q}{p} \quad ;$$

for all other points $B(\alpha,\beta)$, we have

$$\alpha q + \beta p \geq \alpha' q + \beta' p \quad ,$$

The equality holding when the point is on the side in question. In this case,

$$(\alpha - \alpha')q = (\beta' - \beta)p$$

and these numbers are positive if $\alpha'' > \alpha'$. As q and p are primes, we have, for an integer k,

$$\alpha - \alpha' = kp \quad , \qquad \beta' - \beta = kq \quad .$$

If we make the transformation

$$z = Z^p \quad , \qquad u = UZ^q \quad ,$$

then the transformed equation (11) contains a factor Z to the power $\alpha' q + \beta' p$, and after suppressing this factor, takes the form

$$A_{\alpha',\beta'}U^{\alpha'} + \sum A_{\alpha,\beta}U^{\alpha} + A_{\alpha'',\beta''}U^{\alpha''} + ZQ(Z,U) = 0 \quad , \tag{13}$$

where $Q(Z,U)$ is a polynomial and the sigma corresponds to the values α,β for which the point $B(\alpha,\beta)$ is on the side in question. For $Z = 0$, equation (13) has $\alpha' - \alpha''$ non-zero roots, therefore, for Z tending towards zero, it has $\alpha'' - \alpha'$ roots $U(z)$ which tend towards the non-zero values of these roots. Hence for z tending towards zero, equation (11) has $\alpha'' - \alpha'$ roots of infinitesimal order $\mu = q/p$. As $\sum(\alpha'' - \alpha')$ is equal to n, the roots of (11) which tend towards zero correspond to the roots of the equations such as (13) which does not tend towards zero with Z.

We reduce matters to studying these roots. In (13), the differences $\alpha - \alpha'$ and $\alpha'' - \alpha'$ are multiples of p. If these equations are written as

$$U^{\alpha''}T(U) + ZQ(Z,U) = 0 \quad ,$$

the polynomial $T(U)$ is a polynomial in U^p, call it $\Psi(U^p)$. If ν is a simple root of $\Psi(\nu) = 0$, $T(U)$ admits for simple zero $\nu^{1/p}, \omega^p$, where ω is a primitive root of unity of order p and $\rho = 0,1,2,\ldots,p-1$, and we obtain for a principal part of u

$$\nu^{1/p}\ Z^{\ /p}\ \omega^{\rho} \quad ,$$

which provides a single circular system of order p. Consequently, *if the equation* $\Psi(\nu) = 0$ *only has simple roots, we obtain* $(\alpha'' - \alpha')/p$ *circular systems of order* p.

Let us assume that the equation $\Psi(\nu) = 0$ admits a root ν' of order $n' > 1$, $T(U)$ will have n' roots which will tend towards each number $(\nu')^{1/p}\omega^{\rho}$ when Z tends towards zero. Let U' be one of these numbers. We will put

$$U = U' + V$$

and reduce it to a new equation

$$W(V,Z) = 0 \quad ,$$

in which, when Z tends towards zero, n' roots tend towards zero. We can apply the above method. If the analogous equations to $\Psi(\nu)$ that will have been formed, only have simple roots, we will obtain for V, circular systems of orders at most equal to n'. If

$$V = V_1\ Z^{q'} + \ldots$$

is one of these circular systems, we see that by passing to U, then to u and z, that circular systems of order pp' are obtained. We note that the same result would be obtained, if U' is replaced by $U'\omega^{\rho}$. If one of the analogous equations to $\Psi(\nu) = 0$ has a simple root, a similar transformation to the one above must be made and then one starts again. These operations end after a finite number of transformations. For, if this were not so, as $n' \leq n$, then $n'' \leq n', \ldots$ and since the roots of $\Psi(\nu) = 0$ are necessarily simple if $n = 1$, then the numbers n, n', n'' will all equally commence from a certain operation. We could assume this to be so from the beginning. Since $n' \leq \alpha'' - \alpha' \leq n$, we would have $\alpha'' - \alpha' = n$ and $p = 1$, the Newton polygon would only have one side, and $\beta'' - \beta' = qn$. Equation (13) would have the form $A(U - U') + zQ(z,U) = 0$ and setting $U = U' + V$ we would recover an equation

$$AV^n + zQ_1(z,V) = 0$$

for which the same circumstances prevail, etc. The n solutions u will be given by

$$u = (U' + V)z^q \quad ,$$

then we would have, with q', q'' integers

$$V = (V' + \nu)z^{q'}, \ \nu = (\nu' + w)z^{q''}, \ \ldots \quad .$$

The differences of two of these solutions would be, with respect to z, of an infinitesimal order as great as one might wish, when it is finite as known on account of no. 2, or as can be seen, by forming the equation with the squares of the differences of the roots.

Puiseaux's method therefore admits, once the solutions of the equations are known, a means of determining the ramification points and the first coefficients of the corresponding expansions.

Remark. If the origin $(0,0)$ is a multiple point with distinct tangents, of order n, to each of these tangents except for Ou, the terms of lowest degree in $P(z,u) = 0$ are of the form

$$Au^n + \sum A_{\alpha,n-\alpha} u^{\alpha} z^{n-\alpha}$$

and on putting $u = zt$, the equation

$$At^n + \sum A_{\alpha,n-\alpha} t^{\alpha} = 0 \qquad (14)$$

has, by hypothesis, its roots distinct; one of the coefficients A_{oq^n} or $A_{1,n-1}$ is not zero, the Newton polygon has a single side if $A_{o,n} \neq 0$, two sides otherwise. The side of gradient -1 gives $p = 1$, $\alpha' = 0$ or 1 and the corresponding polynomial $T(U)$ is the first member of (14) where t is replaced by U and suppressing U if it is a factor; it has all its roots distinct. If $A_{o,n} = 0$, the Newton polygon has a second side, a gradient lower than -1, for this side $\alpha'' - \alpha'$ being equal to 1, $p = 1$, $T(U)$ is of the first degree. In both cases, the n branches of the algebraic function are holomorphic at the origin.

Moreover, we can verify directly, by setting $u = z(t' + t)$, that to each finite simple root of equation (14), there corresponds a holomorphic branch for $z = 0$.

4 - Examples

I - If we have $u^2P(z) = Q(z)$, where the polynomials P and Q have simple and distinct zeros, we have a function with two branches, namely a two-sheeted Riemann surface. The zeros of P and Q are ramification points of order 2, at a finite distance. If the degrees of P and Q have the same parity, the point at infinity is a simple point on the two leaves (a pole if the degree of Q exceeds that of P). If the degrees of P and Q are of different parities, the point at infinity is a ramification point of order 2.

II - Let $u^3 - 3u + 2z = 0$ (Briot and Bouquet). The point at infinity is manifestly of order 3. The ramification points at a finite distance correspond to $z^2 - 1 = 0$, they permute two solutions: these are the points $z = 1$, $u = 1$; $z = -1$, $u = -1$. If the curve $z = \frac{3u - u^3}{2}$ is sketched, where u is real, we see that for $z = 1$ or $z = -1$, it is the root that is real for z real, greater than 1 or less than -1, which does not permute with the others. Let us sketch the lines of passage on the three plane sheets by joining the three critical points to the origin by means of the segments 0,1; 0,-1; and the half-line joining the origin to the point at infinity, along the positive imaginary axis. Let $u_1(z)$ be the branch which is real for a large z real and positive, $u_2(z)$ that which under these conditions has argument $\frac{5\pi}{3}$ and $u_3(z)$ that which then has argument $\frac{\pi}{3}$. Following what was said above, $u_1(z)$ stays holomorphic for $z = 1$, and $u_3(z)$ which is real for z real and negative (seen by rotating z through the angle $-\pi$ about 0 on a great circle), stays holomorphic for $z = -1$. The lines remaining in the three planes of u_1, u_2, u_3 are represented in figure 3. The numbers appearing alongside each line of passage, on both sides of this line, indicate the index of the plane in which one passes when this line is crossed in the sense of the neighbouring arrow of this number. They indicate the side for which these planes must be joined along these lines in order to constitute a surface which is all of one piece.

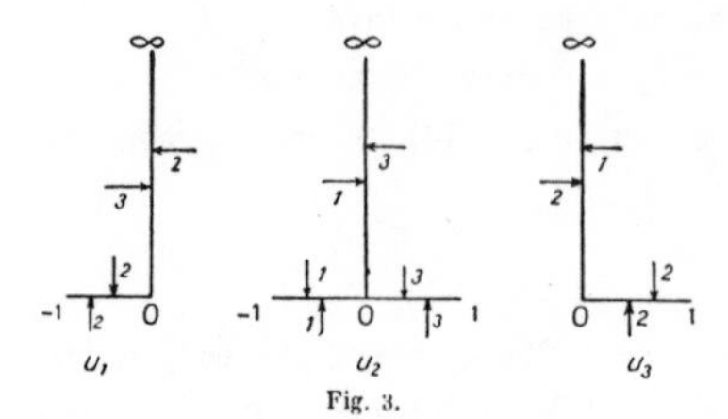

Fig. 3.

III - Let $u^3 - 3u^2 + z^6 = 0$ (Briot and Bouquet). The three branches are evidently uniform at infinity and admit this point for a double pole. The points of ramification correspond to $z^6 - 4 = 0$. For each of them, we have a double root and one simple, two branches permute about each of them. (For $z = 0$, two roots in u are zero, we have a double point $z = 0$, $u = 0$ with tangents identified, but on applying Puiseaux's method, we have $\mu = 3$, and on setting $u = Uz^3$, we obtain $V^3z^3 - 3V^2 + 1 = 0$, for which the roots are simple for $z = 0$. We have two holomorphic branches and zeros at the origin.) Let us sketch the lines of passage by joining the ramification points to the point at infinity along the extensions of the segments joining these points to the origin. We will obtain in each of the three sheets, the star set of holomorphy. If we consider the curve $z^6 = 3u^2 - u^3$, $u < 3$, we see that the root in u which is real for z real and positive, does not permute with the others about $z = 6\sqrt{4}$. Let $u_1(z)$ be the branch which is real for z real and positive. The two other branches permute about $6\sqrt{4}$. Let $u_2(z) = u_1(z)$, $\omega = \varepsilon^{2\omega i/3}$, for z real and large, with z taken to be below the real axis and let $u_3(z) = u_1(z)\omega^2$ under the same conditions. The equation in u and z does not change when z is replaced by $z\omega$; it follows that $u_1(z)$, which is real for z real and positive, near to zero, is again on the half-lines with arguments $2\pi/3$ and $-2\pi/3$ and consequently the points $6\sqrt{4}\,\omega$ and $6\sqrt{4}\,\omega^2$ are again points of holomorphy for $u(z)$, whereas the other three points $z^6 = 4$ are ramification points for $u_1(z)$. Similarly, $u_2(z)$ is real on the half-lines of arguments $\pm\pi/3$ and π as it is likewise at the point at infinity of these direction, as is seen by rotating z through $-\pi/3$ about the point at infinity. The points $z^6 = 4$ with arguments $\pm 2\pi/3$ and 0 are therefore ramification points for $u_3(z)$ and the corresponding lines of passage pass to $u_3(z)$. The lines of passage relative to $u_1(z)$ are also made to pass to $u_3(z)$. The stars relative to the three sheets are represented in figure 4; the numbers given along each line of passage indicate the index of the sheet through which is passed on crossing this line. Here, with ramification points of order 2, the passage in one sense or another, leads onto the same sheet.

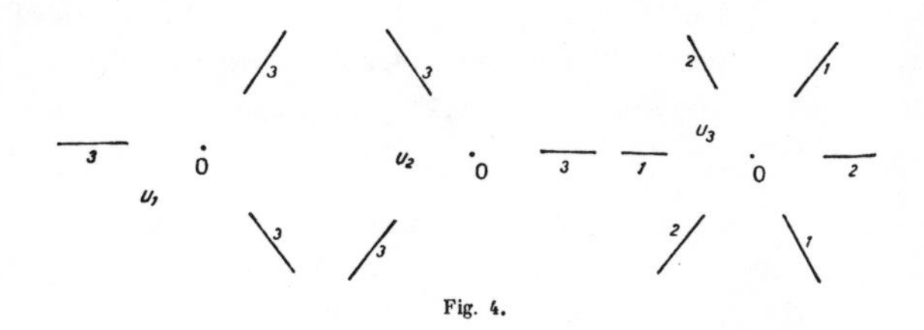

Fig. 4.

5 - Lüroth's theorem and surfaces

The Riemann surface does not change when the lines of passage are displaced. But it is possible to simplify the way of joining the sheets to each other by conveniently choosing these lines; we must then modify their relative order.

Lüroth's Theorem. *When the ramification points are all of order 2, i.e., the lines of passage only connect two sheets, it is possible to trace these lines in such a way that each sheet, except for two of them, only contains two lines of passage.*

As each line of passage only joins two sheets, we could represent these lines on a simple plane by indicating on each of them, the indices of the sheets of the surface that they connect. We can also assume that the lines of passage have been obtained by joining the ramification points to the point at infinity which is a point of holomorphy on all sheets. We will recover this case by changing, if needs be, z to $z_o + \frac{1}{Z}$. Let us assume then that the point at infinity is simple for all sheets and that the origin is also a point of holomorphy for all the branches. Let $u_1(z),\dots,u_n(z)$ be the m branches; *they are determined by their values in a circle with centre the origin not containing any ramification points,* and consequently by the given quantity of the lines of passage that had initially been traced in some manner. Let (i,k) and (j,ℓ) be two consecutive lines of passage (fig. 5) joining the points $\zeta_{i,k}$ and $\zeta_{j,\ell}$ to the point at infinity. Let us assume that, without changing the other lines, we replace one of them, e.g. (i,k), by another line, that is dotted in the diagram, and which goes off to infinity by passing between the line (j,ℓ) and the one preceding. This in no way changes the values of the branches on the sheets with indices other than i, j, k; but on the planes of indices i, j, k, ℓ, the lines of passage are changed. If on leaving the origin with the branch $u(z)$, we turn about ζ and return to 0, then we again return to 0 with the value $u_k(z)$; the dotted line is still then a line (i,k). But if a loop is described from 0, around $\zeta_{j,\ell}$, not intersecting the new lines of passage other than (j,ℓ), it will intersect the original line (i,k) (fig. 5).

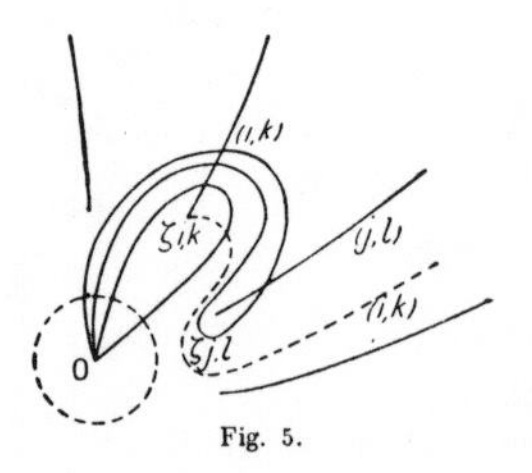

Fig. 5.

This line in the original system, is equivalent to a loop intersecting (i,k), then (j,ℓ) follows from a line intersecting (i,k). If j, ℓ, i, k are all different, $u_k(z)$ changes to $u_\ell(z)$; if $i = j$, $k \neq \ell$, $u_k(z)$ changes to $u_\ell(z)$; if $i = j$, $k = \ell$, $u_j(z)$ changes to $u_k(z)$. Hence: *when a line of passage* (i,k) *is displaced from a row towards the right, its indices do not change. If* (j,ℓ) *are the indices of the preceding line, these indices stay the same, except when* $j = i$, $k \neq \ell$, *being the case in which the line becomes a line* (k,i). We have a similar result when a line of passage is displaced from a row towards the left.

Let us consider then the lines of passage a priori indicated in the order where they have been met on turning about the point at infinity from the right to the left. One at least (and likewise two, following the above remark) takes the index 1. Let $(1,i)$ be such a line. If $(1,j)$ is the first line with index 1 that has been met when turning to the left from $(1,i)$, we could straight away send this line to the left of $(1,i)$ by successive displacements from a row towards the right. We repeat this operation for the first line $(1,k)$ that is met whilst turning from $(1,j)$, etc. All the lines with index 1 are then grouped one alongside the other, and to their left are found the lines not having index 1. If all these lines with index 1 do not have the same second index, the left-most one will have, for example, indices $(1,j)$, and ongoing from left to right from this line, we will have, for example, $(1,j)$, $(1,j)$, $(1,j)$,..., $(1,j)$, $(1,k)$...; the order of $(1,j)$, $(1,k)$ will be changed, which will give (j,k), $(1,j)$, then that of (j,k) and the line to its left $(1,j)$, which will give $(1,k)$, $(1,j)$, $(1,j)$. These operations are continued, either to lead to (k,j) $(1,j)$ ... $(1,j)$, or to $(1,k)$, $(1,j)$, $(1,j)$,... . In the second case, the number of lines of index 1 will have decreased by one unit. By continuing these operations, we will obtain a group of lines with like indices, $(1,k)$ say and to their left, lines all of whose indices differ from 1. By changing the numbering, we could assume that we have a group of λ consecutive lines of index (1,2), and to their left, lines whose indices all differ from 1; moreover, following no. 2, there are at least two lines with index (1,2). Amongst the lines to the left of those with index (1,2), one at least has index 2, otherwise sheets 1 and 2 will not be bound to the others. We can immediately take it to the left of the group of the λ lines with indices (1,2) and name (2,3) as the indices of this line. If $\lambda > 2$, we could diminish this number by 2 units by the following operations. We have the sequence of lines (1,2), (1,2),...,(1,2),(2,3) which could be written symbolically as $(1,2)^{\lambda}(2,3)$. On displacing these lines successively, we will obtain in turn, the following dispositions:

$$(1,2)^{\lambda-1}(2,3)(1,3) \quad ; \quad (1,2)^{\lambda-2}(1,3)(1,2)(1,3) \quad ;$$

$$(1,2)^{\lambda-2}(1,3)^2(2,3) \quad ; \quad (1,2)^{\lambda-6}(2,3)(1,2)(1,3)(2,3) \quad ;$$

$$(1,2)^{\lambda-3}(2,3)^2(2,3) \quad ; \quad (1,2)^{\lambda-3}(2,3)(1,3)(2,3)^2 \quad ;$$

$$(1,2)^{\lambda-2}(2,3)^3 \quad .$$

On repeating these operations for $\lambda > 4$, we see that we will reduce matters to the case where $\lambda = 2$. If $m > 3$, we operate in the same way, starting from (2,3) with lines of indices $(2,j)$, by staying to the left of the two lines (1,2). We will obtain one or two lines $(2,k)$, but there are effectively two, since, on circling the point at infinity starting from the plane of index 2, we return into this plane.

Continuing these operations if needs be, we arrive at the following disposition:

$$(1,2)^2(2,3)^2 \ldots (m-2, m-1)^2 \, (m-1, \,)^{\lambda} \quad .$$

Each plane with index less than m *is connected to the preceding plane along two lines of passage. The planes of indices* $m-1$ *and* m *are connected along* λ *lines, where* λ *is even.*

Lüroth's theorem is thus proved. It applies especially when the curve $P(z,u) = 0$, has only, to a finite or infinite distance, multiple points with distinct tangents, and when the points of contact of the tangents parallel to Ou (including the asymptotic parallel to Ou if there is one) are all ordinary. This is the general case. The number of ramification points is then equal to the number of tangents paralle to Ou, this is the *class* c of the curve. Hence we have, with the above hypotheses,

$$2(m - 2) + \lambda = c \quad .$$

EXAMPLE

Let us go back to the example of Briot and Bouquet, $u^3 - 3u^2 + z^6 = 0$; the lines of passage so determined, are depicted opposite (fig. 6,I) with their indices.

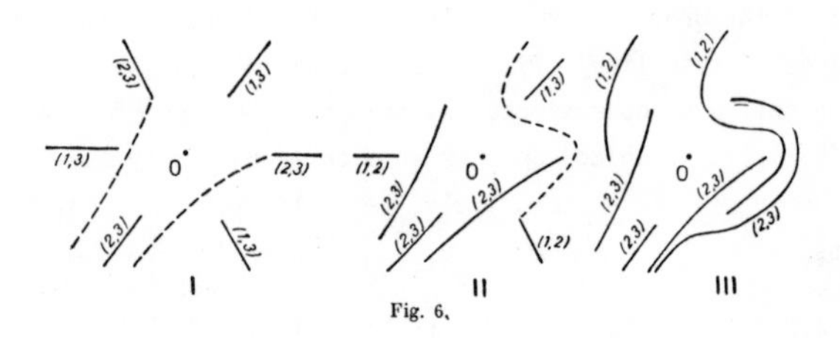

Fig. 6.

By displacing two of these lines (2,3) as indicated in the diagram, we change two lines (1,3) to lines (1,2) (fig. 6,II). By displacement along a line (1,2) (fig. 6,II), we obtain the final disposition (fig. 6, III). We have two lines (1,2) and four lines (2,3).

LÜROTH SURFACES

We could deform the lines of passage obtained by Lüroth's theorem, without them overlapping, by taking them to a finite distance. The two lines (1,2) could be replaced by a single line joining the two points $\zeta_{1,2}$ (fig. 7,I). The portion broken off from the original plan of index 1, will pass into the new plane 2. The portion broken off from the original plane 2 will becomes a portion of the new plane 1. Similarly, the two lines (2,3) will be modified by replacing them by a line joining the two points $\zeta_{2,3}$; and so on. Finally, the λ lines with indices $m-1$ and m will be replaced by $\lambda/2$ lines (fig. 7,II). For example, by starting from fig. 6,III, we will obtain the new lines represented in figure 8. We will say that the surface thus obtained in a Lüroth surface.

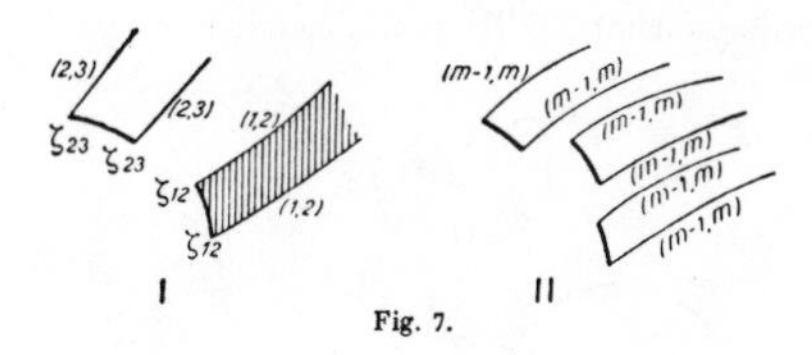

Fig. 7.

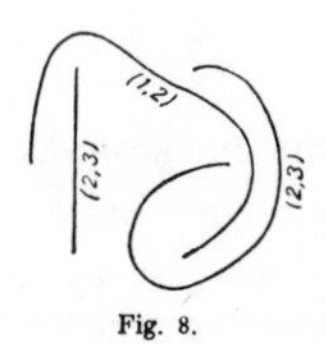

Fig. 8.

6 - <u>Topological transformation of the Lüroth surface</u>

The planar or spherical Riemann surface gives a real geometric realisation of the algebraic curve $P(z,u) = 0$. A pair of numbers (z,u) satisfying the equation corresponds to a single point of the surface, except for a finite number of pairs which correspond to multiple points with uniform or non-uniform branches passing through these points. Conversely, to a point of the surface, there corresponds a single pair (z,u). If we confine ourselves to finding such a geometric representation, we could replace the surface originally obtained by any other continuous surface corresponding to it pointwise in a bijective way. One such surface will be said to be *topologically equivalent to the first,* or equivalent to the point of view of *analysis situs*. The exact study of such transformations constitutes what we call the *topology*. Certain propositions of topology are quickly seen or proceed at once from theorems already known in function theory. Thus if two circular discs with the same radii are considered and a continuous bijective correspondence between the * points of the bounding circumferences, is established, then this correspondence could be extended to points of the surfaces of the discs: If P is a point of the first disc with centre O, with OP continued to intersect the circumference at M, to M there corresponds M' on the circumference of the second disc, with centre O', and we will associated to P the point P' of OM' such that O'P' = OP.

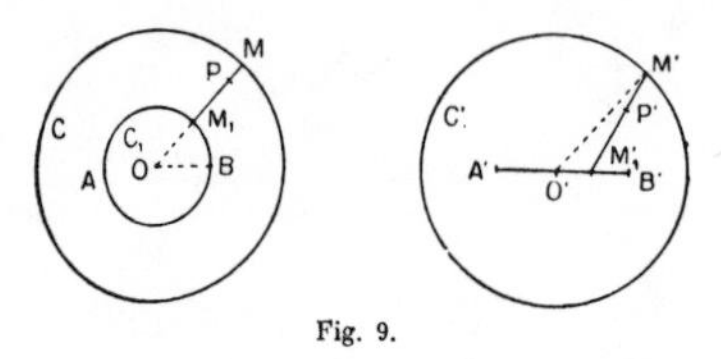

Fig. 9.

Applying the conformal representation, we deduce from it that if there is established a continuous bijective correspondence between the points of two simply connected curves composed of a finite number of analytic arcs, we can extend this correspondence to the interiors of these curves or to the exterior of one and to the exterior of the other. It follows that if a simple curve in the plane is taken, going from infinity to infinity and if a continuous bijective correspondence between points of this curve and those of a straight line taken in another plane, is established, then it is possible to extend this correspondence to the two planes (assuming that the curve is composed of a finite number of analytic arcs). We can, in fact, extend the correspondence to the two domains situated respectively on one side of the curve and on the side of the straight line, then to the others.(1) Now take a circular wreath and let the domain obtained by suppressing a circle C' of same radius as the exterior circle C, be a portion A'B' of a diameter, where A', B' are symmetric with respect to the centre O' (fig. 9). We can establish a bijective correspondence between the points of these domains and the points of their boundaries (each point of A', B', except A' and B', counting as two). It suffices to correspond to A' and B', the points A and B diametrically opposed on the inner circle C_1 of the wreath, to M in C, the point M' of C' such that the angle B'O'M' is equal to $\hat{BOM}$; to M_1 in C_1 such

(1) We could have suppressed the hypothesis on the analyticity of the curve arcs, but it would be necessary to appeal to the theorem of Caratheodory on the conformal representation or prove directly. The hypothesis stated will be satisfied in the application to Riemann surfaces.

that $0 < M_1\hat{O}B < \pi$, the point M_1' of A'B', above A'B' and such that the ratio of the lengths $A'M_1'$ and $M_1'B$ are equal to those of the arc AM_1 and M B of C_1 and to P taken along a radius M_1M of the wreath, the point P' of $M_1'M$ such that $\overline{M_1'P'} : \overline{P'M'} = \overline{M_1P} : \overline{PM}$, and finally to proceed in the same way from below the straight line AB and from A'B'.

Having done this, consider a Lüroth surface and the lines of passage indicated first of all, on a single plane, (1,2), (2,3), (3,4), (4,5), (4,5) for example. We can assume that these lines of passage are polygonal lines (fig. 10) and draw a polygonal L from a finite number of sides, going from infinity to infinity and containing these lines of passage.

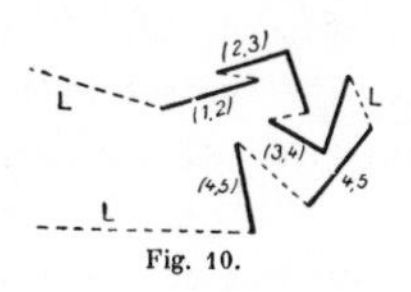

Fig. 10.

By that which preceded, we could replace the plane and L by a plane and a straight line ℓ, with the bijective correspondence. If the operation is no longer made for the simple plane, but simultaneously for other planes of the Lüroth surface, we will obtain an equivalent surface in which the lines of passage are aligned (fig. 11). The plane (1) joins the plane (2) along the segment (1,2), the upper bound of (1,2) in plane (1) is joined to the lower bound of (1,2) in plane (2) and the lower bound of (1,2) in (1), to the upper bound of (1,2) in (2). Let us take in plane (1) a circle C whose centre is in the middle of the segment (1,2), the diameter greater than the length of (1,2) and which the other segments of passage leave from its exterior, and the circle C_1 admits the segment (1,2) as one of its diameters.

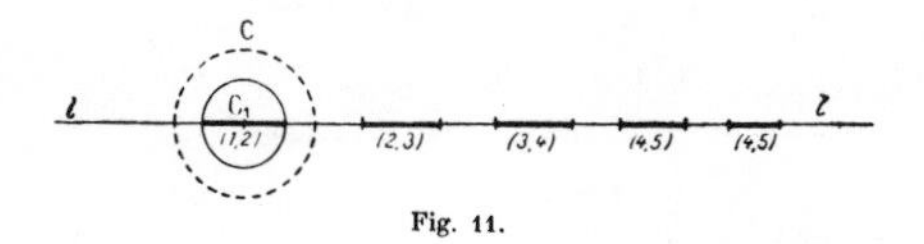

Fig. 11.

We could replace the domain formed by the portion of the plane (2) inside C, by the wreath formed between C and C_1, then the plane (1) restricted to the segment (1,2) by the interior of C_1, with the correspondence again extended on the boundaries. We obtain a surface equivalent to Lüroth's surface and which has at least one sheet. We could in the same way replace the new surface by another having at least one sheet, by operating in the same way from (2,3); and so on. We will recover a two-sheeted surface joined along the segmenets (4,5), (4,5) for example, taken along a similar line ℓ. We will then have a surface which can be considered as a single plane in which the two faces are distinguished; one passes from one face to the other along segments (4,5), but staying on the same side of the line ℓ. At the same time on the two surfaces, we could even successively replace by the superposed wreathes, the portions of these faces which are inside the circles centred in the middle of the segments of passage and radii a little more than the mean of the lengths of these segments. The surface will be replaced by a plane with two sides endowed with circular holes; one passes from one side to the other along these holes.

Now by making an inversion having as its invariant circumference, the circumference of one of its holes, we will finally arrive at a circular plane disc with two sides having $1/2\lambda - 1$ holes in the general Lüroth case. We pass from one face to another along the boundary of the disc and along the circumference of the holes. But the two faces of the disc could be separated by taking on both sides of the points P of the plane disc, on the normal to this plane, the segments PM, PM', where PM and PM' are continuous functions of the position of P, which are zero on the outer boundary and on the circumferences of the holes (fig. 12). We will obtain a veritable surface in the space, having the form of a cushion pierced with holes. In the case $\lambda = 4$, we have a single hole; we could take it as a surface to be a torus of revolution. It is possible to vary the method of these constructions of surfaces which are topologically equivalent to Lüroth surfaces. We could introduce spheres joined by clinders, a sphere with handles, etc. These are the Clifford surfaces.

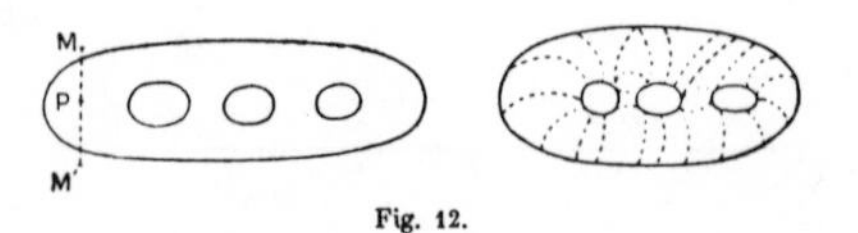

Fig. 12.

II. THE CONNECTIVITY OF RIEMANN SURFACES -- THE GENUS

7 - Theorems on connectivity

We shall consider a Riemann surface with m sheets planar or spherical. This surface is closed. We shall assume that the lines of passage are polygonal lines of a finite number of sides in the case of a plane surface, or lines composed of a finite number of circular arcs for a spherical surface. All of the lines that will be described on the surface will be of the same kind, such that two lines only intersect in a finite number of points. In order to follow the course of a line passing through a point of ramification (z_o, u_o), we can assume that the part of the surface inside a small circle of centre (z_o, u_o) had been transformed into a simple circle by the transformation $Z' = z - z_o$, where s is the order of the point of ramification.

The Riemann surface is closed. It will be replaced to an open surface by raising a small part of the surface inside a closed line about an ordinary point. We will say that we have in fact an opening. More generally, we shall consider the surface obtained by raising from the given surface, the interior of several small closed lines outside one to the others, where these lines could surround the ramification points. One such surface S will be bounded by several contours.

To make an incision on S is to describe a line Γ starting from a point of a controu C and returning to a contour or closing in on itself (fig. 13), and assume that it is no longer possible henceforth, to pass over this line. If the surface is realised on paper, we can suppose that it is effectively cut by a pair of scissors along Γ, and we stop as soon as a boundary or a cut already made, is reached. An incision might separate the surface, i.e., decompose it into two unconnected pieces. Having described an initial incision, we can describe a second, either on the surface if it has not been separated, or on one of the pieces; and so on. We retain the name of surface for every piece obtained in the sequence of these operations. Indeed it is understood that if the surface is planar, an incision could pass through the point at infinity.

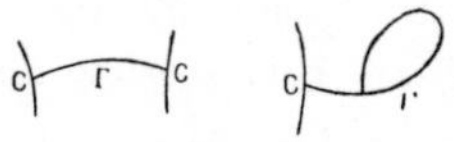

Fig. 13.

SIMPLY CONNECTED SURFACE

A surface is said to be simply connected when it is not possible to describe any incision on this surface without separating it.

Theorem I. *A simply connected surface is bounded by a single contour.*[1]

Let S be a surface bounded by several contours, and let C and C' be two of these contours. Let us cut S along a line Γ joining a point A of C to a point B of C' and let S' be the surface S thus cut. On S', A is replaced by two points A' and A" identified with A, but situated on both sides of the incision Γ; A' and A" could be joined on S' along C. If P is a point of S', hence of S, we could join it to A, on S, by a line L. If L does not cut Γ, P is joined to A' or to A" on S', hence to A'. If L cuts Γ at an initial point Q starting from P, Q could be joined to A' or A" along Γ, hence to A' by remaining on S'. In all cases, P could be joined to A', S' is all of one piece, Γ has not separated S, and S is not simply connected.

Theorem II. *A simply connected surface is divided by an incision into two simply connected pieces.*

The incision AB could join two points of the controu or close in one itself (fig. 14).

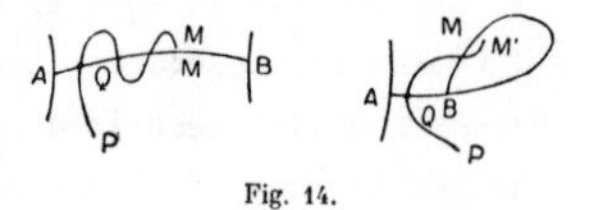

Fig. 14.

In both cases, we could take on both sides of the incision, two points M, M': every point of S can be joined to one of them by a line not intersecting the curve AB. For, if the line L joining P to M before the line of cut crosses this incision at an initial point Q, then Q could be joined to M or M' along AB. We hence obtain two connected surfaces: S' formed by the points that could be joined to M, and S" formed by points that could be joined to M'. S' for example is simply connected. Otherwise a certain

(1) In the case where the surface is a portion of a simple plane bounded by a single contour (composed of circular arcs or segments), it is indeed simply connected in the given sense of this word in volume I.

incision E, F would not separate it. Every point P of S' would again be joined to M (we could suppose that M is not an extremity of the incision), every point of S" could be joined to M' and consequently, if a part of the curve AB that contains M, M' is removed, S' and S" are joined and form nothing more than a similar surface. If EF are not on AB, we will eliminate AB and S will not be separated by the incision EF and will not be simple connected, contrary to hypothesis. If EF has an extremity E on AB, we will eliminate EA or EB (fig. 15) and the same contradicton will be obtained by considering on S the incision AEF or ABEF indicated by the bold lines. We proceed in the same way if E and F are on AB (fig. 15).

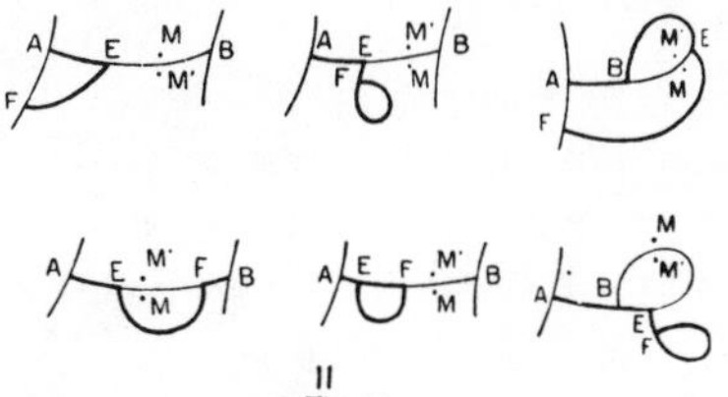

Fig. 15.

<u>Corollary</u>. *If* ν *successive incisions are made on* μ *simply connected surfaces, we obtain* $\nu + \mu$ *simply connected surfaces.*

(We could always cut the same surface or make cuts in some part of these surfaces, or in all of them if $\nu \geq \mu$.)

<u>Theorem III</u>. *Let us consider a finite number of surfaces. If by means of* ν *successive incisions, we could decompose this sytem into* α *simply connected pieces, then the number* $\nu - \alpha$ *is the same, in whatever way this is done.*

It has to be shown that, if another system of ν' incisions gives α' simply connected pieces, we have $\nu' - \alpha' = \nu - \alpha$. We could assume, by deforming the incisions infinitesimally, that the incisions of the two systems do not intersect themselves, neither on the boundaries of the surfaces, nor at their extremities. Let us consider on the system of surfaces, the complete set of the incisions of the first and second system. It provides a certain number of simply connected surfaces. We can count this number of surfaces in two different ways. Having made the ν incisions of the first system, the first incision of the second system is described; if p_1 is the number of its points of intersection with the incision of the first system, then this is equivalent to $p_1 + 1$ incisions in $p_1 + 1$ simply connected surfaces; this gives $p_1 + 1$ new simply connected surfaces. If the second division of the second system is then made, it likewise gives rise to $p_2 + 1$ new simply connected surfaces, where p_2 is the number of its intersections with an incision of the first system; and so on. The total number of the surfaces thus obtained is therefore

$$\alpha + \nu' + p_1 + p_2 + \dots + p_{\nu'} = \alpha + \nu' + P$$

where P is the total number of points of intersection of the incisions of the two systems. Similarly, we find $\alpha' + \nu + P$ pieces are obtained on the assumption that the count was made by taking the incision of the second system first. We gave then $\alpha + \nu' = \alpha' + \nu$.

8 - Order of connectivity of a surface

We say that a surface has a finite order of connectivity if, by means of a finite number ν of successive incisions, we can decompose it into simply connected surfaces. *The order of connectivity is the number*

$$N = \nu - \alpha + 2 \quad .$$

For a simply connected surface we have $\alpha = 1$ and $\nu = 0$, hence $N = 1$. When the order of connectivity is N, we also say that *the surface is* N-*times connected.*

Theorem 1. *If a surface is* N-*times connected and* $N > 1$, *an incision* Γ *which does not separate it, gives a* $(N-1)$-*times connection.*

For if S' denotes the surface obtained after the incision Γ, S' has finite order of connection, since if Γ is made in the set of pieces obtained by the incisions decomposing S into simply connected surfaces, we have pieces that are simply connected. Let N' be the order of connection of S'. We have $N' = \nu - \alpha + 2$ if ν incisions give in S', α simply connected pieces. But then in S, $\nu + 1$ incisions are made, giving α simply connected pieces, hence the order of S is

$$N = \nu + 1 - \alpha + 2 = N' + 1 \quad .$$

Consequently:

Corollary I. *A surface* N-*times connected could be made simply connected by* $N-1$ *successive incisions.*

The order of connection is therefore equal to the number of successive incisions that could be described without separating the surface, plus one. For example, a plane disc pierced with N - 1 holes is seen to be N-times connected. *When an incision is made on a surface without separating it, the number of contours of the surfaces increases or decreases by one.* It effectively decreased by one if the incision has its extremities on two different contours. It increases by one if its extremities are on the same contour or if there is only one contour (fig. 16).

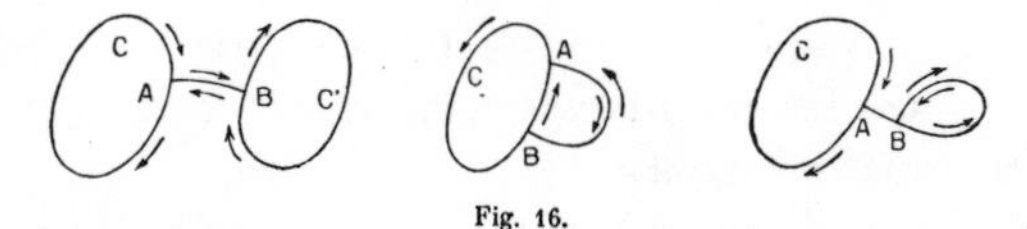

Fig. 16.

Let a Riemann surface be N-*times connected.* This means that if an opening is assigned to it, the surface S obtained is N-times connected. This number N could be such that it is independent of the position of the casement assumed to be quite small; this is seen by displacing the casement continuously. Then N - 1 successive incisions make S simply connected and the surface thus divided has a single contour (Theorem I of no. 7). It follows that N - 1 is even, since the parity of the number of controus changes with each operation. N is odd.

Every closed surface has an odd order of connection. When two surfaces are equivalent from a topological point of view (no. 6), every operation of division or every lifting of a piece on one of them, has its equivalent on the other; the connectivity could then be defined on surfaces that are equivalent to planar or spherical Riemann surfaces.

A torus has connectivity 3; it is made simply connected by making a small casement at an arbitrary point and cutting one part along a parallel circle passing through this opening, and the other part along a meridian circle. The surface so obtained is equivalent to a rectangle, this is evident when one takes the usual parametric representation in terms of two angles.

Theorem II. *If a small (simply connected) part of a surface* S *of connectivity* N, *not intersecting the controus of* S, *is removed, then the remaining surface* S' *has order of connectivity* N + 1.

The small portion raised is the interior of a simple closed curve C. If we consider this curve with respect to the set of simply connected surfaces of decomposition of S and join it to a contour, we obtain a finite number of pieces whose order of connectivity of S' exists. If incisions on S' are carried out to make this surface simply connected, the surface obtained will have no more than one contour; hence there exists an incision Γ which binds C to a point of a contour of S without separating S'. Let S" be the surface S' cut along Γ, and let N, N' and N" be the orders of connectivity of S, S', S". We have $N'' = \nu - \alpha + 2$, and since S" is deduced from S' by Γ which does not separate S', we also have $N' = \nu + 1 - \alpha + 2$. On S, the incision described by Γ taken along C, given two surfaces: the interior of C and S" and ν incisions on S" give α simply connected parts, hence

$$N = \nu + 1 - (\alpha + 1) + 2 = N' - 1 \quad .$$

Remark. The proof shows that it is sufficient for the order of connectivity of S' to exist in order that the corresponding order of S exists.

Corollary II. *If* n *small (simply connected) parts of a surface, one outside of the other and not intersecting the contours, are removed, then the order of connectivity increases by* n.

The preceding theorems (nos. 6, 7, 8), due in principle to Riemann, had been specified by the authors mentioned in the introduction to this chapter by Jordan (*Traite d'Analyse*, t. II) and by Appell and Goursat (*Theorie des fonctions algebraic*, t. I). They apply not only as mentioned, to surfaces equivalent to Riemann surfaces defined in no. 6, but directly to *ordinary* surfaces in geometry, notably to polyhedral surfaces.

Lhuiller's Theorem (1811). *Given a closed polyhedral surface with order of connection* N *having* S *vertices,* F *faces and* A *edges, we have*

$$A = F + S + N - 3 \quad .$$

For ordinary polyhedrons, $N = 1$, we recover Euler's formula. To prove this, we raise about each vertex a small part containing this vertex; the connectivity increases by $S - 1$, with one of these parts being the opening. We then slit the remaining surface along the edges; A incisions are made and F simply connected pieces are obtained; hence $N + S - 1$ is equal to $A - F + 2$.

9 - Genus--Riemann's formula

The genus of a Riemann surface (and more generally, a closed surface) of order of connectivity N, is the integer p defined by $2p + 1 = N$. We have seen that N is odd. Let us consider the Riemann surface of the algebraic function with m branches, defined by $P(z,u) = 0$. Let us regard it as spherical. The lines of passage have been obtained by joining the ramification points to a point z_o, a point of holomorphy on all the sheets. Let us raise on each of the small sheets, a small circle with centre z_o; with one of these circles being the opening, $m-1$ pieces have been raised. Similarly, let us raise the small circles centred at the points of ramification (we assume that all the circles raised were outside of each other). If we denote by r the order of any of the point of ramification, we have raised in all, $m-1+\sum(r+1-r)$ pieces. Let us cut the surface thus obtained along the lines of passage. We describe $\sum r$ lines and obtain m simply connected pieces; S had an order of connectivity, and hence the Riemann surface too, following the remark in no. 8; let N be this order and p the genus. Following the theorems of nos. 7 and 8, we have

$$N + m - 1 + \sum(r + 1 - r) = \sum r - m + 2$$

or

$$N = 2p + 1 = \sum(r - 1) - 2m + 3 \quad ,$$

$$p = \frac{\sum(r-1)}{2} - m + 1 \quad .$$

THIS IS THE FUNDAMENTAL FORMULA OF RIEMANN.

We see that the number $\sum(r-1)$ is even. In the case of a Lüroth surface (no. 5), $\sum(r-1)$ is equal to $2(m-2) + \lambda$, and we have

$$N = \lambda - 1 \, , \qquad p = \frac{\lambda}{2} - 1 \quad .$$

It follows that a "cushion" with p holes (no. 6) is of genus p, with order of connection $2p + 1$. We also say that the algebraic curve defined by $P(z,u) = 0$ is of genus p.

10 - Line of divisions. Examples

Generally, to describe the incision rendering the surface simply connected, we can use the following procedure due to Riemann. We work in such a way that the successive surfaces obtained have only one or two contours according to whether the connectivity is odd or even. When there is only one contour, one shows that the surface is connected by showing that the contour could be described continuously. For this surface S" arises from a surface S' which was connected and which had been cut along a line AB. Every point of S' could be joined to a point of the contour of S', hence every point of S" could be joined to a point of its contour which is continuous, hence also to a similar given point of the contour.

On making an opening in the given surface S and describing a cut not separating S, we obtain a surface of order of connection 2 , if p is the genus of S. By deforming the incision slightly, we could always assume that this incision has its extremities on the boundary of the opening, and the opening could always be reduced to a line. We have the curve C_1 which constitutes the two contours or boundaries of the new surface. If two points AB are taken respectively on the two boundaries then joining A to B, we cut along a curve D_1 and obtain a new surface with boundary (fig. 17). (When C_1 and D_1 are projected on the same plane, they appear to intersect each other in points other than A,B, but they are then on different sheets, which can be seen by not drawing the line of D_1 at such points.)

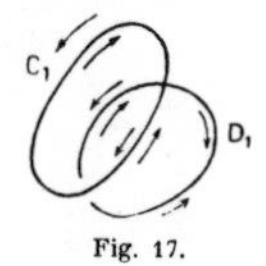

Fig. 17.

The surface thus cut, is connected, for it has only one continuous contour (fig. 17), its order of connection is $2p - 1$. If $p > 1$, we repeat by making an incision starting from a point of D_1 and returning there, or intersecting itself, and not separating. This is possible, for if the incision returns to another point of (C_1,D_1), it has been deformed. Let E_1C_2 be this cut; the connectivity is then $2p - 2$ and there are two boundaries. We join a point taken on the boundary of C_2 to a point facing the other boundary. We join them and cut it, so giving a single contour and a surface of order $2p - 3$ (fig. 18, I). If $p > 2$, we can continue for as long as the order of connection is not one. We start from a point of D_2 and make the cut E_2C_3, then D_3, etc. But after C_2,D_2 we could also make an incision starting from a point of E_1, and starting from a point opposite, then join two points opposite this cut, and consequently we can always start from E_1 (fig. 18, II). The set of the two incisions C_1,D_1 or C_2,C_2 etc., is often called

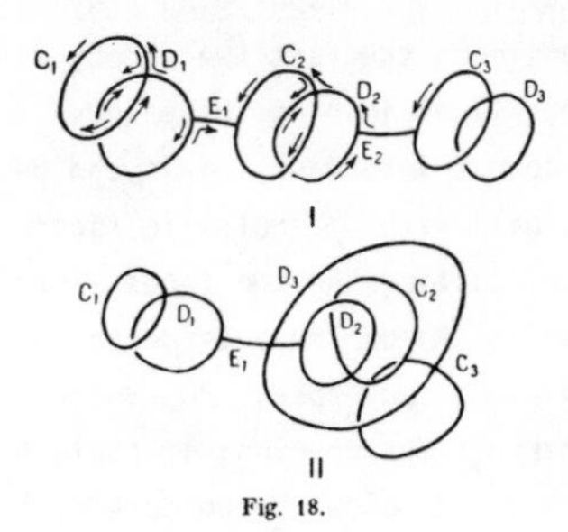

Fig. 18.

EXAMPLES

I - For a two-sheeted surface, assuming that the lines of passage emanate from z, we have for $\lambda = 4$ and $\lambda = 6$ the diagrams below: the lines of passage are indicated by the light lines and the incisions by the cold lines in one of the planes and by broken lines in the plane of the other sheet (fig. 19).

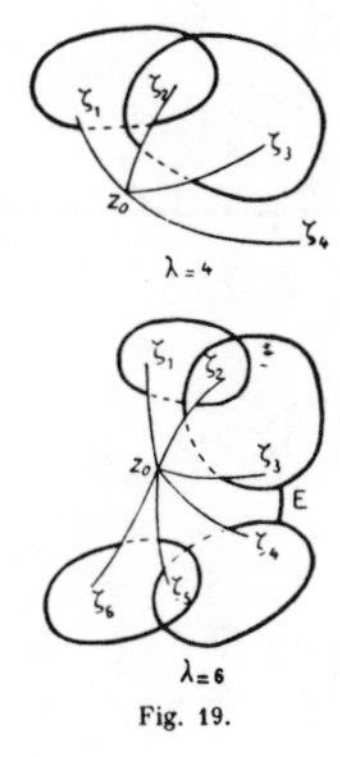

Fig. 19.

For $\lambda = 8$, the dispositions of the two kinds are given in figure 20.

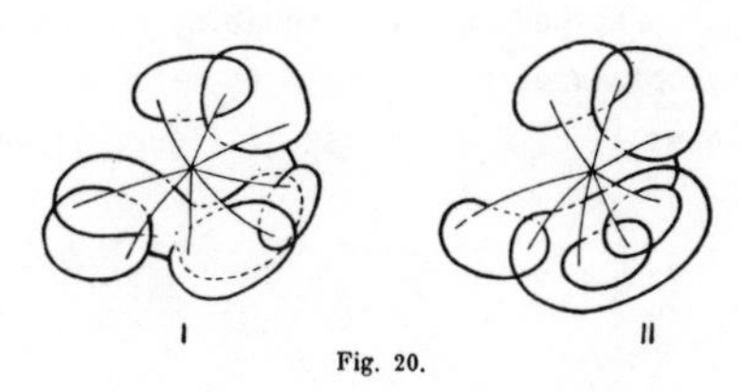

Fig. 20.

For the general Lüroth example, the disposition stays the same as in the two-sheeted case, one only considers the last two sheets containing the $\lambda/2$ lines of passage, the other lines joining the points ζ in pairs and passing to the other sheets that do not intervene. This can be seen on the surface equivalent to no. 6. The disc with p holes in figure 12, that has two faces, is made simply connected by cutting the two faces along the lines joining a point of each hole to a point A of the outside boundary, then by separating the two planes by cutting along each hole, $2p$ successive incisions are made. The point A plays the part of the opening; it could be replaced by a line E and the incisions joining the holes to the outside boundary are made to end there. We might imagine this operation to be made on the surface taking the form of a cushion (fig. 21, II). The incisions so made are like those of figure 20, II. They will be equivalent to those obtained if one applies the second way of operating (fig. 18, II) on the Lüroth surface defined by the lines of passage joining the points ζ pairwise (fig. 21, III).

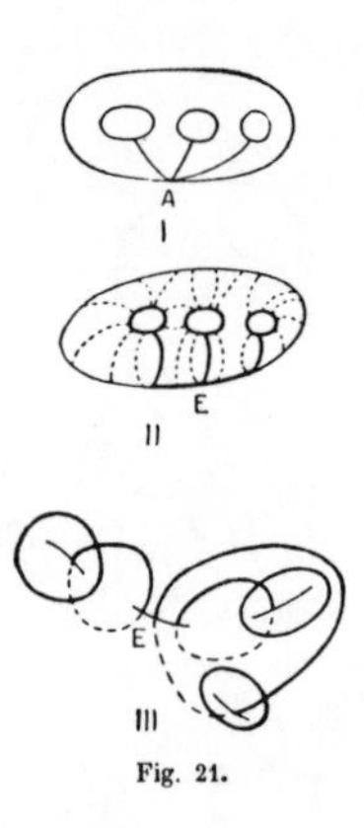

Fig. 21.

II - In the case studied in no. 4, where $u^3 - 3u + 2z = 0$, we have $\sum(r-1) = 4$ and $m = 3$, hence $p = 0$, and the surface is simply connected.

For $u^3 - 3u^2 + z^6 = 0$, the surface is reduced to the Lüroth form and we could apply that which preceded, but by operating on the surface with starred sheets initially obtained, we are able to make the description in figure 22, where the planes 1, 2, 3 are those of figure 4 (we have $p = 1$).

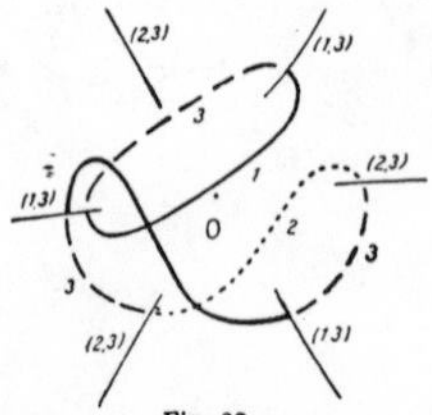

Fig. 22.

III. BIRATIONAL TRANSFORMATIONS - APPLICATIONS

11 - Cycles. Birational transformations of the plane to the plane, or Cremona transformations

We have seen in no. 2 that $P(u,z) = 0$ being an irreducible curve, where u is given by a series expansion about every point, that this point is an ordinary point or a ramification point. If z_o is finite, we have for u_o finite or not,

$$u = u_o + \sum_{1}^{\infty} a_n (z - z_o)^{n/s} \tag{15}$$

where

$$u = \sum_{-\rho}^{\infty} a_n (z - z_o)^{n/s} \;, \quad a_{-\rho} \neq 0 \;, \quad \rho > \tag{16}$$

and n, s, ρ are integers, s, ρ are positive, s is the order of the circular system if $s > 1$; for $s = 1$, we will say that we have a system of order 1.

If z_o is infinite, we place z by $\frac{1}{z}$; we hence obtain analogous expansions:

$$u = u_o + \sum_{1}^{\infty} a_n z^{-n/s} \tag{17}$$

$$u = \sum_{-\rho}^{\infty} a_n z^{-n/s} \;, \quad a_{-\rho} \neq 0, \quad \rho > 0 \quad . \tag{18}$$

If we introduce homogeneous co-ordinates Z, U, V defined to within a factor by $uV = U$, $zV = Z$, then we could represent the above systems parametrically. In (15) and (16), we first of all take $z = z_o + t^\rho$, then $V = 1$ in (15) and $V = t^\rho$ in (16), which will give

$$\begin{aligned} Z &= z_o + t^s \;, \quad U = u_o + \sum_{1}^{\infty} a_n t^n \;, \quad V = 1 \quad ; \\ Z &= z_o + t^{s+\rho}, \quad U = \sum_{-\rho}^{\infty} a_n t^{n+\rho} \;, \quad V = t^\rho \quad . \end{aligned} \tag{19}$$

In (17) and (18) we firstly take $z = 1/t^s$; then in (17), $V = t^s$, and in (18) $V = t^\delta$, where δ is the maximum of ρ and s. We will have

$$Z = 1 \ , \quad U = u_o t^s + \sum_1^{\infty} a_n t^{n+s} \ , \qquad V = t^s$$

$$Z = t^{\delta+s} \ , \quad U = \sum_{-\rho}^{\infty} a_n t^{n+\delta} \ , \quad V = t^s \ . \tag{20}$$

In homogeneous co-ordinates, the equation of the curve will take the form $Q(Z,U,V) = 0$, where Q is a homogeneous polynomial. This equation will be identically satisfied by the expansions (19) or (20) of various circular systems. About each point, the solutions are given by such systems.

CYCLES. DEGREE AND CLASS OF A CYCLE

The expansions (19) and (20) are of the form

$$Z = Z_o + \sum_1^{\infty} \alpha_n t^n \ , \quad U = U + \sum_1^{\infty} \beta_n t^n \ , \quad V = V + \sum_1^{\infty} \gamma_n t^n \tag{21}$$

where α_n, β_n and γ_n are constants such that the series converge in the same circle of positive radius. When t describes this circle, the point with homogeneous co-ordinates (21) describes a (of two dimensions in the 4-dimensional space). We will say that the expansions (21) define a *cycle* if the correspondence between the circle and this is bijective. This condition is satisfied, as is known, in the case of the systems (19) and (20). It is clear that eliminating t from the equalities (21), will not give in general, an algebraic relation. The cycles of (19) and (20) are algebraic cycles, under the condition that they arise from an algebraic function. If we assume, e.g., $V_o \neq 0$, we will obtain, by going anew to the ordinary co-ordinates,

$$z = z_o + \sum \alpha_n' t^n \qquad u = u_o + \sum \beta_n' t^n \ , \tag{22}$$

whence we deduce an expression $u - u_o$ as a function of $z - z_o$ similar to (15). But we could also express $z - z_o$ as a function of $u - u_o$.

The point Z_o, U_o, V_o is *the origin or vertex* of the cycle (21). A straight line with equation

$$\lambda Z + \mu U + \nu V = 0$$

passing through this point, and hence such that $\lambda Z_o + \mu U_o + \nu V_o = 0$, intersects the cycle in a certain number of points identified with its origin; the number of these points is the degree of the first non-zero term in the series

$$\sum_1^{\infty} (\alpha_n\lambda + \beta_n\mu + \gamma_n\nu)t^n \quad . \tag{23}$$

For λ, μ, ν arbitrary, but such that

$$\lambda Z_o + \mu U_o + \nu V_o = 0 \quad , \tag{24}$$

then this number has a determined value. *This is the degree of the cycle.* It is the smallest value of n for which α_n, β_n, γ_n are not proportional to Z_o, U_o, V_o. If r is the degree and if λ, μ, ν simultaneously satisfy the equation (24) and

$$\lambda\alpha_r + \mu\beta_r + \nu\gamma_r = 0 \quad , \tag{25}$$

then the straight line in question is called *the tangent to the cycle* to this cycle at the origin. The first non-zero term in the series in (23) has then a certain degree $r + \tau$, where τ *is the class of the cycle.* We see that by passing to the non-homogeneous form (22), the degree is the term of lowest degree that is non-zero in the two series appearing in (22); if r is this number, then the class is the smallest value n for which

$$(\alpha'\beta'_{s+r} - \alpha'_{s+r}\beta')$$

is non-zero. Now passing to the form in (15), we see that when written with the non-zero terms appearing:

$$u = u_o + a_n(z - z_o)^{q/s} + a_{q''}(z - z_o)^{q'/s} + \ldots$$

the degree is the smallest of the numbers q and s, and if $q > s$, the class is $q - s$, it is equal to $q' - s$ if $q = s$. The degree and class are therefore independent of the type of parametric representation.

HOMOGRAPHIC TRANSFORMATIONS

If a homographic transformation is made (Z, U, V replaced by linear homogeneous forms in Z', U', V' with constant coefficients and no zero determinant) with *a corresponding cycle evidently a cycle of the same degree and class,* then we will obtain

$$\lambda Z + \mu U + \nu V \equiv \lambda'Z' + \mu'U' + \nu'V'$$

where λ', μ', ν' are the new coefficients of the line. The transformation also preserves the contact of two cycles. In such a transformation, there corresponds to an algebraic curve, an algebraic curve which, as is known, has the same degree. The correspondence between the points (z,u) and (z',u')

of the two curves is bijective and continuous; the corresponding Riemann surfaces are hence also bijectively related. To every curve described to one, there corresponds a curve described on the other and consequently a system of incisions making one of the surfaces simply connected, does likewise to the other. *The two surfaces have the same connectivity and the same genus.* In particular if z is considered as a function of u instead of considering u as a function z, then the genus is invariant. For example, if z is a rational function of u, then the corresponding Riemann surface is the plane of u, and the genus is zero. Hence if we express u as a function of z, the genus is again zero. This is the situation occurring in the first example of Briot and Bouquet (no. 4).

CREMONA TRANSFORMATIONS

The birational transformations of the plane to the plane are, in homogeneous co-ordinates $Z, U, V,$ of the form

$$Z' = \Phi(Z,U,V) \ , \quad U' = \Psi(Z,U,V) \ , \quad V' = \Theta(Z,U,V)$$

where the second members are homogeneous polynomials of the same degree, such that it is possible to obtain inversely, Z, U, V (to within a factor) in the guise of polynomials in Z', U', V'. We show that these transformations are the product of quadratic transformations. These quadratic transformations are obtained by taking for $\Phi, \Psi, \Theta,$ polynomials of the second degree which, equated to zero, represent three curves of the second degree having three common points. In the general case, we could assume by means of a homographic transformation that these fixed points are the vertices of the triangle with co-ordinates

$$(Z = 0, U = 0; \ U = 0, V = 0; \ V = 0, Z = 0) \quad .$$

We will then have

$$Z' = zZU + bUV + cVZ \quad ,$$

where a, b, c are constants, and analogous expressions for U' and V'. This is the product of the particular transformation

$$Z' = VU \ , \quad U' = ZV \ , \quad V' = UZ \quad ,$$

and a homographic transformation. The inverse transformation to (26) is moreover

$$Z = U'V' \ , \quad U = V'Z' \ , \quad V = Z'U' \quad ;$$

the transformation (26) is reciprocal.

We note that the transformation $zz' = 1$, $u' = u$ which had been used in studying the points at infinity (no. 2), and which is evidently reciprocal, is a quadratic transformation of another kind to (26).

In a birational transformation, the points common to the base curve, $\Phi = 0$, $\Psi = 0$, for example, evidently play a special part: the point corresponding to one such point is, a priori, indeterminate. But when the point Z, U, V belongs to an algebraic curve, Z, U, V are given by cycles in the neighbourhood of such a point: Φ and Ψ become entire series in t, without constant terms, but not identically zero (under the conditions that one does not intend applying the transformation to the curve $\Phi = 0$, or to $\Psi = 0$), and the ratio Z'/U' has a limit when t tends towards zero. Hence there is yet again a continuous bijective correspondence between the Riemann surfaces corresponding to the given curve and its transformation. Consequently: *A Cremona transformation preserves the genus.*

TRANSFORMATION OF CYCLES IN QUADRATIC TRANSFORMATIONS

Let us take the transformation under the guise of (26). We consider a cycle whose vertex is not on one of the sides of the triangle of reference. We could take it in the form of (22) (non-homogeneous), or, by assuming as we say that $\alpha'_r \neq 0$ and by taking a new parameter T defined by

$$T = \alpha'_r\, t^r + \alpha'_{r+1}\, t^{r+1} + \dots \quad ,$$

it could be taken in the canonical form

$$Z = Z_o + T^r \,, \quad U = U_o + aT^r + bT^{r+\tau} + \dots \,, \quad V = 1, \quad bU_oZ_o \neq 0 \quad , \tag{27}$$

where r is the degree and τ the class.

The transformed cycle will be defined by

$$Z' = U_o + aT \;\; + bT^{r+\tau} + \dots \quad ,$$

$$U' = Z_o + T^r$$

$$V' = Z_oU_o + (U_o + aZ_o)T^r + bZ_oT^{r+\tau} + \dots + aT^{2r} + \dots \quad .$$

The independent terms and the terms in T^r not being proportional, the degree is r, and the tangent has as its equation

$$\begin{vmatrix} Z' & U_o & \alpha \\ U' & Z_o & 1 \\ V' & Z_oU_o & U_o + aZ_o \end{vmatrix} = 0 \quad .$$

Two cycles having the same vertex only are transformed to tangent cycles if they were tangents (the value of a the same). If $\tau < r$, the class is preserved.

Let us now consider a cycle whose vertex is on a side of the triangle of reference (between to vertices of this triangle). We can again take it in the form of (27), with Z_o or U_o zero. As the role of Z_o and U_o in (27) can be permuted, we see that we can limit matters by assuming $U_o = 0$. The tangent to the cycle will be the side $U = 0$ if $a = 0$. If $a \neq 0$, the cycle transformed has for its vertex $(0,1,0)$, its degree is again r, and its tangent, always given by the same equation, is $Z'Z_o - V' = 0$. Two cycles whose vertices are on $U = 0$, therefore transform to two cycles with the same vertex $(0,1,0)$, but the tangents to these cycles are distinct if the given cycles are with different vertices. If $a = 0$, the vertex of the transformed cycle is always $(0,1,0)$ and here we have

$$Z' = bT^{r+\tau} + \dots , \quad U' = Z_o + T^r , \quad V' = bZ_oT^{r+\tau} + \dots \quad ;$$

the degree of this cycle is $r + \tau$, and the tangent is again $Z'Z_o - V' = 0$.

Finally, let us look at the case where the given cycle is not tangent to a vertex of the triangle of reference. We shall assume that this cycle is not tangent to a side of the triangle. We could then take it under the guise of (27) by assuming $Z_o = U_o = 0$ and $a \neq 0$. The cycle transformed will be given by the above expressions, after dividing by the factor of T ,

$$Z' = a + bT^\tau + \dots , \quad U' = 1 , \quad V' = aT^r + bT^{r+\tau} + \dots \quad .$$

If $\tau < r$, the degree of the transformed cycle is τ. If $\tau \geq r$, the degree is again r. If $\tau = r$, the class of the new cycle depends on other coefficients and exponents; if $\tau > r$, the class becomes $\tau - r$ and so has been decreased.

12 - Nöther's Theorem. An application

We have seen in no. 3, that when a curve admits a multiple point at the origin, a branch of the curve is tangent at this point to a simple tangent not identified with the axis 0, is holomorphic at the origin. The corresponding cycle, deduced by $u = \sum a_p z^p$ is of the form (19) with $s = 1$, and is of the first degree. This stays the same if the branch corresponds to a simple tangent identified with $0u$, since it suffices to permute the role of u and z to recover the preceding case. The cycles with degree greater than 1 only arise from branches tangent to multiple tangents and which do not decompose into simple branches, in particular, turning points. A curve whose singularities are only multiple points with distinct tangents has only cycles of the first degree.

<u>Nöther's Theorem</u>: *By a birational transformation, it is possible to transform an algebraic curve to another curve having only multiple points with distinct tangents.*

To this extent, we first of all look for a curve having only cycles of the first degree. If the curve admits a cycle which is not of the first degree, we could make a homographic transformation such that the vertex of this cycle is the new origin, $z = 0, u = 0$, the axes $u = 0, z = 0$ and the line at infinity ($V = 0$) only intersect the curve at simple points distinct from the points at infinity of the axes, and that the axes and the line at infinity are not tangent to the curve. This does not entail any new multiple points and to each new contact with the cycles, the degree and the class of all the cycles are preserved.

Let us then make the quadratic transformation (26). Following the properties of this transformation, the degree of all the cycles other than those of the origin vertex ($z = 0, u = 0$), is preserved and each new contact does not enter for these cycles, but the cycles whose vertices were on the axes and on the line at infinity, yield cycles with vertices ($z = 0, u = 0$; $u = 0, V = 0$; $z = 0, V = 0$), but with distinct tangents and these cycles are of the first degree. Finally, the cycles of the origin vertex yield cycles, with different vertices. For each of these cycles, the degree is reduced if the class is less than the degree and is preserved in the other case, but if the class is greater than the degree, it is reduced. By applying the sequence of these transformations several times, we will then achieve a reduction of the degree of the cycles, of at least one of them or from those that will be obtained, not having its class equal to its degree. Hence it is convenient to acquire, in this case, the nature of the transformation.

Returning to the expression of u as a function of z, we have

$$u = a^{q}_{q'} z^{q/s} + a_{q''} z^{q''/s} + \dots \quad ,$$

where q' is equal to s, and the class is $q'' - s = s$. We will have a certain number of terms with integer exponents, then an initial fractional term:

$$u = Q(z) + a' z^{q+\rho/s} + \dots \tag{28}$$

where $Q(z)$ is a polynomial of degree at most q, without a constant term, and $0 < \rho < s$. Following a homographic transformation, we recover an analogous formula. We have, in effect, by suitably choosing the name of the new coordinates and taking the vertex of the cycle at the origin,

$$u' = \frac{c'u + d'z}{1 + cu + dz}, \quad z' = \frac{c''u + d''z}{1 + cu + dz} \quad ,$$

where c and d are constants. Replacing u by its value, we could extend as a function of z' and obtain u' as a function of z' under the form of an expansion in powers of $z'^{1/s}$ (since the degree s is indeed preserved). If the successive derivatives of u' as a function for z' for $z = 0$ are calculated, we have, by putting $u' = g(z,u)$, $z' = h(z,u)$ to simplify things:

$$\frac{du'}{dz'} = \frac{g'z + g'u \frac{du}{dz}}{h'z + h'u \frac{du}{dz}} , \ldots$$

and more generally, to the numberator of the derivative of order n, there will appear a polynomial in $\frac{du}{dz}, \ldots, \frac{d^n z}{du^n}$, whose coefficients are for $z = 0$, powers of the numbers c, d, c', d', c'', d'', and to the denominator

$$\left(h'z + h'u \frac{du}{dz}\right)^{2n-1}$$

which for $z = 0$ $(u = 0)$, reduces to

$$\left(d'' + c'' \frac{du}{dz}\right)^{2n-1} .$$

This number is not zero, if we assume, as is the case, that the tangent to the curve transformed by (28) is not one of the sides of the new triangle of references. In so far that the derivatives of u with respect to z are finite, it follows that the derivatives of u' with respect to z' are as well. The number q does not decrease; it is neither increased, since in the inverse transformation, we have an analogous property with $Q(z)$ having a term in z; q is therefore preserved.[1]

On the contrary, in the quadratic transformation (26), the value of q is reduced by 1. Let us take the transformation under the form

$$u' = \frac{z}{u} , \quad z' = z ,$$

which is possible. We then have

$$u' = \frac{1}{\phi} , \quad \phi = \frac{Q(z)}{z} + a'z^{q + \frac{\rho}{s} - 1} + \ldots$$

$$= a_q + \ldots + a'z^{q + \rho/s} + \ldots$$

and

(1) Another proof will be given in no. 14.

$$\frac{du'}{dz} = \frac{\phi'}{\phi^2}, \quad \frac{d^2u'}{dz^2} = -\frac{\phi''}{\phi^2} \quad \frac{2\phi'^2}{\phi^2}, \quad \ldots$$

$$\frac{d^q u'}{dz^q} = -\frac{\phi^{(q)}}{\phi^2} + \frac{\Pi(\phi,\phi',\ldots,\phi^{(q-1)})}{\phi^{q+1}}$$

where Π is a polynomial, which proves that, ϕ being equal to a_q for $z = 0$, the derivative of order q of u' will no longer be finite for $z = 0$, when those preceding are. Here, q is reduced by one. Thus, when the class is equal to the degree, with the reduction made by the polynomial in (28), q decreases. But when $q = 1$, the class is ρ; being lower than the degree is reduced.

The degrees of all the cycles will hence be taken to the end of a finite number of operations of the kind indicated. Moreover, if two cycles of the first degree with thesame vertex are tangents, these are of the form (28) with $s = 1$,

$$u = az + a'z^2 + \ldots + a''z^q + \ldots, \quad u = az + a'z^2 + \ldots + b''z^q + \ldots$$

Following the method of calculation of the derivatives of u' with respect to z', the homographic transformation preserves the contact: the degrees of the first terms, unequal in these two series, stay the same. They decrease by one with each quadratic transformation. In pursuing these transformations, the contact is removed. We will therefore arrive at a curve only having multiple points with distinct tangents. The theorem is thus established.

AN APPLICATION

A birational transformation does not change the genus; we can see that a curve of genus p could be transformed to another of the same genus, only having multiple points with distinct tangents. For this curve, we could take the axes (via a homographic transformation) in such a way that, for m the degree of the curve, the m points at infinity are all distinct and each of the asymptotes are not parallel to the axis of u, no tangent at a multiple point is parallel to Ou, and finally, the tangents parallel to Ou only have first order contact with the curve (i.e., the points of contact with these tangents are not points of inflexion). This last condition is realisable, as are the others, since the points of intersection of the given curve $Q(Z,U,V) = 0$ with the hessian

$$\frac{D\left(\frac{\partial Q}{\partial Z}, \frac{\partial Q}{\partial U}, \frac{\partial Q}{\partial V}\right)}{D(Z,U,V)} = 0$$

(where the notation is that of the functional determinant). These points are finite in number.

Under these conditions, the points of ramification of the algebraic function $P(z,u) = 0$ will be uniquely determined by the points where the tangent is parallel to the axis of u, their number is the class c of the curve and they are all of order 2, as in the Lüroth example; the genus p is given by

$$p = \frac{c}{2} - m + 1 \quad .$$

13 - Birational transformations of curves to curves. Bertini's theorem

Given an algebraic function defined by

$$P(z,u) = 0 \tag{29}$$

of degree m in u and of degree n in z, and two rational fractions

$$Z = \Phi(z,u) \ , \quad U = \Psi(z,u) \ ; \tag{30}$$

eliminating z and u from equations (29) and (30) give a condition

$$\Theta(Z,U) = 0 \tag{31}$$

which expresses the necessary and sufficient condition for the three equations to have a common solution in z and u. To study the correspondence between the points of the two curves $P = 0$, $\Theta = 0$, we appeal to the following property which will be established in no. 19: the number of points of the Riemann surface S of $P = 0$ for which the function Φ takes an arbitrary given value Z, is a number μ independent of Z. This is what is known as the order of the function Φ on the surface S. In the same way Ψ has an order ν. Hence, given z, we have μ points on S, hence μ values of Ψ and therefore of U. The polynomial Θ is of degree μ in U; similarly, it is of degree ν in Z.

Let Z_o be any value for which $\Theta(Z_o, U) = 0$ with two distinct solutions U_o, U'_o; (z_o, u_o) a pair of values for which Φ and Ψ take the values Z_o, U_o and (z'_o, u'_o) a pair of values for which Φ and Ψ take the values Z_o, U'_o. We could describe at the point (z,u) of the curve $P = 0$, a continuous path joining these two points (z_o, u_o), (z'_o, u'_o) and avoiding certain points or certain small regions prescribed in advance, for the surface S deprived of small regions, remains connected. To it there corresponds a path joining the corresponding points Z_o, U_o and Z_o, U'_o. Let us assume that Θ is reducible to a product of polynomials and that two of these polynomials of the decomposition are distinct: let $\Theta_1(Z,U)$ and $\Theta_2(Z,U)$ be these two polynomials. Let Z_o be a value of Z distinct from those providing the common points of the two curves $\Theta_1 = 0$, $\Theta_2 = 0$, and let U_o and U'_o be defined by $\Theta_1(Z_o, U_o)$ and $\Theta_2(Z_o, U'_o) = 0$. Following the above, we can describe a continuous path joining the points Z_o, U_o and Z_o, U'_o and avoiding the points common to the two curves $\Theta_1 = 0$ and $\Theta_2 = 0$, or likewise, the small regions about these common points, which is absurd.[1] It follows that:

If the polynomial $\Theta(Z,U)$ *is not irreducible, it is a power of an irreducible polynomial.*

If Θ is irreducible, then to an arbitrary value Z, there corresponds, in general, μ distinct values of U, as well as μ systems of values (z,u) or, if we wish, μ points of the curve $P = 0$; to each of these pairs there corresponds a single value of U, these μ points of the curve $P = 0$ are all distinct. To a pair (Z,U) of the curve $\Theta = 0$, there corresponds, in general, only a pair (z,u). The equations (30) hence only have a system of solutions (z,u) when the point (Z,U) is taken on $\Theta = 0$, and it results from the theory of elimination that z and u are rational fractions of Z and U. We thus have

$$z = \phi(Z,U) \ , \quad u = \psi(Z,U) \quad , \tag{32}$$

where ϕ and ψ are rational fractions. Thus, *via condition (29), the transformation is birational if* Θ *irreducible. We say that it is a birational transformation of a curve to a curve.* We also say, under these conditions, that equations (29) and (31), or the corresponding algebraic functions, belong to the *same class*. The birational correspondence of curves implies the bijective correspondence of Riemann surfaces, and consequently as we have already seen for the Cremona transformations, that these surfaces have the same connectivity and the same genus. *The genus of two curves in a birational correspondence is therefore the same.*

We have already seen (I, 99) some examples of birational transformations of a curve to a curve. We note that given the functions Φ and Ψ, if we take $P(z,u)$ arbitrary, then $\Theta(Z,U)$ will not, *in general, be an exact power,* and the transformation will be birational. For example, if we take $Z = z$, $U = u^2$, the transformation taken along a curve which is not symmetric with respect to Oz is birational, whilst for a curve which is symmetric with respect to Oz, with equation $P(z,u^2) = 0$, the result of elimination is $[P(Z,U)]^2 = 0$.

Bertini's Theorem: *Be means of a birational transformation from a curve to a curve, we may transform a curve having only multiple points with distinct*

---

(1) We could also apply the analytic continuation: it must be possible to pass by analytic continuation, from any element of the function defined by $\Theta_1 = 0$, to any element of the function defined by $\Theta_2 = 0$, now this is impossible since the points where the values of these functions are the same, are isolated.

tangents, to another whose singularities are only double points with distinct tangents.

Via a homographic transformation, we can assume that the given curve C with equation $P(z,u) = 0$, has distinct asymptotes which are not parallel to the axes, that the axes are not parallel to double tangents, to tangents passing through multiples, and, finally, that the points common to the tangents parallel to Oz and to tangents parallel to Ou, are all outside C.

The derivative u' of u with respect to z is given by

$$\frac{\partial P}{\partial z} + u' \frac{\partial P}{\partial u} = 0 \quad .$$

The abscissae ξ_j, $j = 1,2,\ldots,c$ of the tangents parallel to Ou, correspond to ramification points of the Riemann surface, where c is the class. Let us set

$$Q(z) = (z - \xi_1) \ldots (z - \xi_c)$$

and let $R(z)$ be a polynomial of degree c whose arbitrary coefficients will be constrained to certain conditions. Let us associate to the curve C, the curve Γ of the space defined by

$$X = u \ , \quad Y = u'Q(z) \ , \quad Z = R(z) \ ; \quad P(z,u) = 0 \quad .$$

We could choose $R(z)$ *such that* Γ *is effectively a skew curve, without multiple points.* In order that two distinct pairs of values (z,u), (z_1,u_2) give the same point of Γ, it is necessary to have $u_1 = u$, $z_1 \neq z$ and to have $u_1'Q(z) = u'Q(z)$, $u_1 = u$. *This is only possible for a finite number of values of* z. Otherwise there would correspond to every point (z,u), a point (z_1,u) for which $u_1'Q(z_1) = u'Q(z)$. When the point (z,u) coincides with a point of contact (α,β) of a tangent to C parallel to Oz, we would need to have $Q(z_1) = 0$ or $u_1' = 0$. Under the first hypothesis, we would have $z_1 = \xi_j$, hence $P(\xi_j,B) = 0$. A tangent to C parallel to Ou would intersect a tangent parallel to Oz on C, in a remote case. It would therefore be necessary to have $u_1' = 0$; there would exist a double tangent parallel to Oz, again a remote case, unless when (z,u) tends towards (α,β), the point (z_1,u) tends towards this same point α,β. For u near to β, we would have

$$z = \alpha + \gamma(u - \beta)^{1/2} + \ldots \ , \qquad z_1 = \alpha - \gamma(u - \beta)^{1/2} + \ldots \ ,$$

$$u' = \frac{1}{z'u} = \frac{2}{\gamma}(u - \beta)^{1/2} + \ldots \ , \quad u_1' = -\frac{2}{\gamma}(u - \beta)^{1/2} + \ldots \quad .$$

As $Q(\alpha)$ is non-zero, since α is distinct from the ξ_j, the limit equation $u_1'Q(\alpha) = u'Q(\alpha)$ is impossible.

Thus, the equations $u_1'Q(z) = u'Q(z)$, $u_1 = u$ only occur for a finite number of values of z and z_1. We could choose the coefficients of $R(z)$ such that, for these values, $R(z_1) \neq R(z)$. Two distinct points of C therefore yield two distinct points of Γ.

At a multiple point of C, u' has non-zero finite values that are all distinct, and $Q(z)$ is non-zero; the points of the Riemann surface corresponding to this point, provide distinct points of Γ.

The m points at infinity of C also yield m distinct points of Γ, for X, Y, Z are infinite, but Y/Z has a value Bu', where B is the ratio of the coefficients of z^c in $Q(z)$ and $R(z)$, and the m values of u' are finite and distinct. Thus we have proved that Γ *has no multiple points*. At a point X, Y, Z of Γ there corresponds a single pair (z,u) of C; *z, as is u, is therefore expressed rationally as a function of* X, Y, Z.

The curve Γ *is not planar*. For if we had

$$Du + Eu'Q(z) + KR(z) + L = 0 \quad ,$$

where D,...,L are constants, E would not be zero since u is not a polynomial. For z infinite, z/u is finite. The degree of Q is greater than one, since it is the class; we would find that u' has the same value as the m points at infinity.

Taking Γ to be a skew algebraic curve, let us consider the chords ST of Γ such that the tangents at S and T are concurrent. There is a finite number of chords which pass through a point S of Γ. They constitute an algebraic surface Σ which could be reducible. The triple secants to Γ passing through a point U of Γ, could be a finite or infinite number. There is indeed an infinite number; every chord US intersects Γ in at least one point T. These chords UST form part of Σ (which therefore contains a cone), there are therefore, only a finite number of points U of this kind. For the other points U of Γ only a finite number of triple secants pass through; when U varies, they constitute an algebraic surface Σ'. Finally, the tangents to Γ constitute an algebraic surface Σ''. If we take a reference point Ω outside of the plane OXY and outside of Σ, Σ' and Σ'', then the perspectives of Γ from Ω is a curve C' which corresponds birationally to Γ, and hence to C. Two distinct points of Γ could yield the same point M' of C' if the chord joining these points passes through Ω, but then the tangents at M' are distinct since Ω is not on Σ. The tangent at a point M' of C' is always determined by the tangent at the corresponding point of Γ since Ω is not on Σ''. All the cycles of C' therefore have distinct tangents when they have the same vertex and there are at most two with the same vertex. To establish that the singularities of C are uniquely double points with distinct tangents, it only remains to show that all the cycles are of the first degree. Now, if X_o, Y_o, Z_o are the co-ordinates of Ω, we have for the co-ordinates of a point of C'

$$\zeta = X_o - Z_o \frac{X - X_o}{Z - Z_o} , \quad \omega = Y_o - Z_o \frac{Y - Y_o}{Z - Z_o}$$

For a point at infinity at C, we have

$$u = kz + k_1 + \frac{k_2}{z} + \dots , \quad u' = k - \frac{k_2}{k^2} + \dots , \quad k \neq 0 ,$$

consequently ζ and ω are holomorphic about z infinite. The expansion of ζ with respect to $1/z$ does not include a first degree of $c > 2$; the expansion of ω includes one of $Q(z)/R(z)$ includes one. We can assume that the coefficients of $R(z)$ has been chosen such that this is the case. (The conditions imposed on $R(z)$ are realised if the coefficients of z^{c-1}, z^{c-2}, ..., are taken arbitrarily, where the first is non-zero, and then if the coefficient of z^c is taken to be sufficiently large). The cycles corresponding to the points at infinity of C are then the cycles of the first degree.

We can assume that Z_o had been taken in such a way that the roots $Z = R(z) = Z_o$ corresponding to the points (α,β) of C in which z is a holomorphic function of $u - \beta$, $z'_u \neq 0$, $R'(\alpha) \neq 0$ and that X_o and Y_o are such that $X - X_o$ and $Y - Y_o$ are non-zero at those points. We will then have, about the points at infinity at C', expansions of the form

$$\zeta = \frac{a}{u - \beta} + \dots , \quad \omega = \frac{b}{u - \beta} + \dots , \quad ab \neq 0 ,$$

which define cycles of the first degree.

At a point (α,β) of C where z is a holomorphic function of $u - \beta$, X and Z are holomorphic at $u - \beta$, ζ is holomorphic. If z'_c is non-zero, Y is holomorphic, hence ω is too about $u = \beta$ (we assume henceforth $Z - Z \neq 0$). If $z'_u = 0$, we have

$$z = \alpha + (u - \beta)^2 + \dots , \quad z'_u = 2\delta(u - \beta) + \dots , \quad \delta \neq 0 ,$$

but $Q(z)$ contains $z - \alpha)$ as a factor and Y consequently ω are holomorphic. Moreover, ζ'_u and ω'_u are both non-zero, which would imply that $X'(Z - Z_o) - Z'(X - X_o) = 0$, $Y'(Z - Z_o) - Z'(Y - Y_o) = 0$, which does not occur since $X' = 1$ and Ω is not on $\sum''$. The cycle is of the first degree. At a point of C where the tangent is parallel to Oz, we have

$$u = \beta + \delta'(z-\alpha)^2 + \dots, \quad u' = 2\delta'(z-\alpha) + \dots \quad , \quad \delta' \neq 0$$

$$Q(z) = Q(\alpha) + (z-\alpha)Q'(\alpha) + \dots, \quad Q(\alpha) \neq 0 \quad ,$$

X, U, Z are holomorphic at $(z-\alpha)$ and Y' is non-zero, ζ and ω and we see, as above, that the cycle is of the first degree. Bertini's theorem is thus proved.

14 - Intersection of two cycles. Bezout's theorem

Let us consider in ordinary co-ordinates z, u, two cycles having z_o, u_o as their common vertex and at a finite distance apart; we could take them under the guise of

$$u = u_o + \sum_{\mu}^{\infty} \alpha(z-z_o)^{n/r} \quad , \quad u' = u_o + \sum_{\mu'}^{\infty} \alpha'(z-z_o)^{n/r} \quad .$$

The first function u has r branches which are deduced from one of them by multiplying $(z-z_o)^{1/r}$ by the power $1,2,\dots,r-1$ of the root of $\omega^r = 1$ whose argument is $2\pi/r$. We shall call these branches $u_1, u_2, \dots, u_r$. Similarly, u' has r' branches u'_k, $k = 1,2,\dots,r'$.

The number of common points to the two cycles, which are identified with the vertex, is the infinitesimal order with respect to $z-z_o$ *of the product.*

$$\prod_{j,k} (u_j - u'_k) \tag{33}$$

for $j = 1,2,\dots,r$ *and* $k = 1,2,\dots,r'$. This infinitesimal order is actually an integer, since

$$(u_1 - u'_k)(u_2 - u'_k) \dots (u_r - u'_k)$$

is a polynomial in u'_k whose coefficients are the elementary symmetric functions of the u_j, hence of the entire series in $z-z_o$. The product of these polynomials in u'_k is also a symmetric function of the u'_k, it is an entire series in $z-z_o$.

Remark. One of the cycles could be a straight line $u-u_o = \alpha(z-z_o)$, where α is finite, that is, this line is not parallel to Ou. In this case, we see that the number of points of intersection of the second cycle with this line is equal to the degree of the cycle, if the line itself is not tangential, and to the sum of the degree and the class, if there is contact.

To study this intersection of two cycles, we shall assume that the common vertex is the origin; we reduce matters to this case by replacing $u-u_o$ by u and $z-z_o$ by z.

THE CASE WHERE THE TWO CYCLES ARE TANGENTS

We shall assume that when the exponent β of z is fractional, z^β denotes the function which is real and positive for z real and positive. We could write the two equations as

$$\bar{u} = \alpha_\mu z^{\mu/r} + \ldots + \varepsilon z^{\eta/r} + \ldots \quad ,$$

$$u' = \alpha'_u z^{\mu'/r'} + \ldots + \varepsilon' z^{\eta'/r'} + \ldots , \quad \alpha_\mu = \alpha'_\mu , \quad \frac{\mu}{r} = \frac{\mu'}{r'} , \ldots$$

where the terms preceding those in ε and ε' are the same in the two expressions and we have $\varepsilon' \neq \varepsilon$ if $\frac{\eta}{r} = \frac{\eta'}{r}$. The infinitesimal order of the terms in (35) is going to depend on the branches chosen; these are known as the *characteristic exponents*. Let us take the first cycle under the normal parametric form:

$$u = \alpha_\mu t^\mu + \ldots + \alpha_\nu t^\nu + \ldots + \varepsilon t^\eta + \ldots \quad , \qquad z = t^r \quad .$$

If d is the greatest common divisor of r and μ, then the function $z^{\mu/r}$ is a function with only r/d branches. This is the same for the following functions arising from $t^{\mu+1}$, $\alpha_{\mu+1} \neq 0$, as long as the exponents are divisible by d. If ν is the first exponent that is not a multiple of d, we say that ν is a characteristic exponent. The function $z^{\nu/r}$ does not take its initial value when the argument of z is increased by $2\pi r/d$. If we reduce the fractions μ/r and ν/r to have the same lowest common denominator r_1, it is necessary to take r_1 about $z = 0$ in order that the functions $z^{\mu/r}$ and $z^{\nu/r}$ take all their initial values. If d' is the greatest common divisor of r, μ, ν, r_1 is equal to r/d'. All the functions $z^{\lambda/r}$ occurring in u and for which λ is a multiple of d' (where it is understood that $\lambda = d'$ should this occur), take their initial value when the argument of z increases by $2\ r/d'$. If it exists, a smaller exponent τ of t which is not a multiple of d', is *the second characteristic exponent*. We take the greatest common divisor or r, μ, ν, τ, call it d''; all the functions $z^{\lambda/r}$ occurring in u and for which λ is a multiple of d'', will take the same value when z will have turned through $2\pi r/d''$, etc. As $d > d' > d'' > \ldots$, the number of the characteristic exponents is finite. Only those which are less than η if $\eta \leq \eta'$ will moreover occur.

Let us return then to the expressions of u and u' as functions of z and assume that $r'\eta \leq \eta' r$. *Let us suppose* $d < r$. If η is less than or equal to ν, all the terms $u_j - u'_k$ have order μ/r, save those points for which $j - k$ is equal to r/d or one of its multiples, terms for which the order becomes η/r. The infinitesimal order of (33) with respect to z is therefore

$$\frac{\mu}{r} r'(r-d) + \frac{\eta}{r} r'd = r'\mu + d\frac{r'}{r}(\eta-\mu) \quad .$$

If $\nu < \eta \le \tau$ where τ is the second characteristic exponent, the $r'd$ terms for which $j-k$ is equal to r/d or one of its multiples, are divided into two groups: $r'(d-d')$ terms for which $j-k$ is not equal to r/d' or one of its multiples which are of order ν/r, and the $r'd'$ other terms which are of order η/r. We obtain for the total order

$$r'\left(\frac{\mu}{r}(r-d) + \frac{\nu}{r}(d-d') + \frac{\eta}{r}d'\right) = r'\mu + d\frac{r'}{r}(\nu-\mu) + d'\frac{r'}{r}(\eta-\nu) \ .$$

If η is greater than τ and less than or equal to the following characteristic exponent, we will likewise obtain for the same total order

$$r'\mu + d\frac{r'}{r}(\nu-\mu) + d'\frac{r'}{r}(\tau-\nu) + d''\frac{r'}{r}(\eta-r) \quad ,$$

and so on.

If $d = r$, the terms of u and u' up to the term of degree ν/r are holomorphic and disappear in $u_j - u_k$; we must replace μ by ν, d by d', ν by τ and d' by d'', etc. in the above calculation. We obtain the above formula on setting $d = r$. Thus: *The infinitesimal order of intersection depends uniquely on* r, r', μ, η *and the differences* $\nu-\mu$, $\tau-\nu$, *etc., and the successive characteristic exponents.*

Theorem: *The order of intersection does not change if the role of* u *and* z *are interchanged.*

We have

$$u = \alpha_\mu t^\mu + \dots + \alpha_\nu t^\nu + \dots + \varepsilon t^\eta + \dots \quad , \qquad z = t^r \quad .$$

We set $u = T^\mu$, whence we deduce t as a function of T and then place it in $z = t^r$. We first or all have

$$T = at\left[1 + \dots + \frac{\alpha_\nu}{\alpha_\mu} t^{\nu-\mu} + \dots + \frac{\varepsilon}{\alpha_\mu} t^{\eta-\mu} + \dots\right]^{1/\mu} \qquad a = \sqrt[\mu]{\alpha_\mu}$$

and we apply the binomial formula. If $1 + Q(t)$ denotes the group of terms inside in the bracket, that precede the term in $t^{\nu-\mu}$, we first of all obtain, when ordered in t,

$$T = at\left[1 + \frac{1}{\mu}Q(t) + \dots + \frac{\alpha_\nu}{\mu\alpha_\nu} t^{\nu-\mu} + \dots\right] \quad ;$$

the terms of degree less than or equal to $\nu-\mu$ other than those written, can only arise from power of $Q(t)$; the exponents of these terms are all divisible by d, the greatest common divisor of μ and r, but $\nu-\mu$ is not it, hence the term written is the only term in $t^{\nu-\mu}$. The same reasoning applies to the term corresponding to the second characteristic exponent, and so on. Hence,

$$T = at\left[1 + \ldots + a't^{\nu-\mu} + \ldots + a''t^{\tau-\mu} + \ldots\right], \quad a' \neq 0, \quad a'' \neq 0 \quad ,$$

the term in $t^{\eta-\mu}$ could have disappeared, but then it will exist if the value of ε is changed. In order to pass from T to the inverse function, we employ the method of indeterminate coefficients. We have, for example,

$$t = \frac{1}{a}T(1 + b'T^d + \ldots + b''T^3 + c'T^{\nu-\mu} + \ldots)$$

and on identifying, we see that $c \neq 0$, etc.

We will then have

$$t = \frac{1}{a}T\left[1 + \ldots + c'T^{\nu-\mu} + \ldots + c'T^{\tau-\mu} + \ldots\right] \quad ;$$

the terms of degree less than $\nu -- \eta$ are terms of a multiple degree of d; those for which the degree is less than $\tau-\mu$ have a multiple degree of d', etc. The same reasoning as before shows that we will then have

$$z = t^r = a^{-r}T^r + \ldots + c'ra^{-r}T^{\nu-\mu\ \tau} + \ldots + c''ra^{-r}T^{\tau-\mu+r} \quad ;$$

the terms of degree less than $\nu-\mu+r$ have their degree divisible by d, those of degree less than $\tau-\mu+r$ have their degree divisible by d', but $\nu-\mu+r$, it cannot exist, to have a zero coefficient, but in each operation, the corresponding term has a coefficient, which is a linear function of ε, with the coefficient of ε non-zero. This term could only disappear for a single value of ε. Thus we will have

$$z = \alpha'\mu T^r + \ldots + \alpha_\nu' T^{\eta-\mu+r} + \ldots + \ldots + \varepsilon' T^{\eta-\mu+r} + \ldots \qquad u = T^\mu \quad ;$$

where ε' could be zero, but if this is so, the coefficient of the term with the same degree in the corresponding expression in the second cycle will not be zero. The order of intersection is

$$\mu'r + d\,\frac{\mu'}{\mu}(\nu-\mu) + d'\,\frac{\mu'}{\mu}(\tau-\nu) + d''\,\frac{\mu'}{\mu}(\eta-\tau) \quad ;$$

it has not changed since $\mu'/\mu = r'/r$ and $\mu'/r = \mu r'$. If this hypothesis, $\mu'r = \mu r'$, has not be realised, the first two terms of the expansions being unequal, the proposition would be obvious.

Remark: If we write the cycle as

$$u = \alpha_\mu t^\mu \left[1 + \ldots + \alpha_\nu t^{\nu-\mu} + \ldots + \alpha_\tau t^{\tau-\mu} + \ldots\right] , \quad z = t^r$$

it will become after inversion

$$z = \alpha'_r t^r \left[1 + \ldots + \alpha'_\nu t^{\nu-\mu} + \ldots + \alpha'_\tau t^{\tau-\mu} + \ldots\right] , \quad u = t^\mu$$

thus the reciprocity is complete.

THE INVARIANCE OF THE ORDER OF INTERSECTION IN THE HOMOGRAPHIC TRANSFORMATIONS

This invariance has been already proved when the cycles are not tangents, since the order is then the product of the degrees which are invariant. Thus let two cycles be tangential. If they are to one of the axes, we could, as before, assume that it is so to Oz. But moreover, by adding z to the expressions in u, u', the product in (33) is not changed; we could assume that the cycles are tangent to $z = u$. Similarly, when $\mu = r$, nothing changes in (33) by adding kz to u and u'. Hence, we can assume that we start from two cycles tangent to $u = z$; this gives a transformation

$$Z = \frac{c''u + d''z}{1 + cu + dz} , \quad U = \frac{c'u + d'z}{1 + cu + dz}$$

on the cycles

$$u = z + \ldots + \alpha z^\beta + \ldots + \gamma z^\delta + \ldots + \varepsilon z^\eta + \ldots ,$$

$$u' = z + \ldots. + \alpha z^\beta + \ldots + \gamma z^\delta + \ldots + \varepsilon' z^\eta , \quad \varepsilon' \neq \varepsilon ,$$

where β, δ, ..., are quotients by r and r' of the characteristic exponents; and we can assume that the transformed cycles are tangent to $Z = U$. As the terms of the lowest degree in the expansions

$$Z = (c''u + d''z)[1 - (cu + dz) + \ldots] ,$$

$$U = (c' + d'z)[1 - (cu + dz) + \ldots]$$

are $(c'' + d'')z$ and $(c' + d')z$, they are not simultaneously zero, since $c''d' - c'd'' \neq 0$, hence each is non-zero. We will assume, moreover, that $c'' \neq 0$ (otherwise, we would interchange Z and U), and we have $c'' + d'' = c' + d' = k$. By changing Z to kZ, U to kU, which will not change the order of intersection, we could then assume $k = 1$. If we replace u, in Z and U, by its expansion up to the term of degree less than β, we obtain in all the terms of degree less than β occurring in the expansions of U and Z, then the terms with integer exponents. If, in addition, we take in u the term of degree β, we obtain the term of degree β in Z and U; its coefficient in Z is $c''\alpha$ and in U it is $c'\alpha$. Likewise, with regards to the property of the characteristic exponents, the coefficients of the term of degree δ will be $c''\gamma$ and $c'\gamma$, etc.; those of z_η will be different depending on whether u and u' are taken. In the inversion of

$$Z = z + \ldots + c''\alpha z^{\beta} + \ldots + c''\gamma z^{\delta} + \ldots + \varepsilon_1 z^{\eta} + \ldots$$

we will obtain

$$z = Z + \ldots - c''\alpha Z^{\beta} + \ldots - c''\gamma Z^{\delta} + \ldots + \varepsilon_2 Z^{\eta} + \ldots$$

and on placing in U,

$$U = Z + \ldots + \alpha(c' - c'')Z + \ldots + \gamma(c' - c'')Z^{\delta} + \ldots + \varepsilon_3 Z^{\eta} + \ldots$$

and $c' \neq c''$, otherwise $d' = d''$ and $c''d' - c'd'' = 0$. Moreover, starting from u', we will have a different value for ε_3. The characteristic exponents and the position of η have not changed. There is indeed invariance for the homographic transformations as long as the vertices are at a finite distance.

For two cycles whose common vertex is at infinity, we make a homographic transformation taking it to a finite distance, and after this transformation, take the number of common points. Following that of above, whatever the means, we obtain the same result.

THE NUMBER OF POINTS OF INTERSECTION OF TWO ALGEBRAIC CURVES AT A COMMON POINT

The number of common points will be the sum of the numbers of common points to each cycle of the first curve with each cycle of the second. If the common point is a point of order q on the first curve, with distinct tangents of $q > 1$, and is of order q' on the second, with distinct tangents, and if all these tangents are distinct, all the cycles are simple and non-tangential, and the number of common points is qq'. In the general case, if 0 is the common point and if we consider each pair of cycles one belonging to the first curve and the other to the second curve, then the number of common points of each pair will be the infinitesimal order of (33); the total number will be the order of the product of these products. Each branch of a cycle of the first curve will be associated to each branch of the second, it will be necessary to consider the product

$$\prod_{j,k} (u_j - u'_k) \quad ,$$

where u_j will denote any one of the branches of the first curve and u'_k any one of the branches of the second.

Bezout's Theorem: *The number of points of intersection of two algebraic curves is equal to the product of their degrees.*

This number is the sum of the number of points of intersection of the two curves identified at each of their common points. To obtain it, we will assume that the axes have been taken such that the two curves do not have common points at infinity (achieved via a homographic transformation), that the axis Ou is not an asymptote and not parallel to the lines joining the common points pairwise. This is possible since the number of common points (each counted once) is (resulting from what will be siad) at most mm', where m and m' are the degrees, The equations of the two forms are of the form

$$P(z,u) \equiv a_o u^m + a_1(z)\, u^{m-1} + \ldots + a_m(z) = 0 \tag{34}$$

$$Q(z,u) \equiv b_o u^{m'} + b_1(z)\, u^{m'-1} + \ldots + b_{m'}(z) = 0 \quad , \tag{35}$$

where the $a(z)$ and $b(z)$ are polynomials of degree less than or equal to their index, and $a_o b_o \neq 0$. The abscissae of the common points are the roots of the equation

$$D(z) = 0 \tag{36}$$

obtained by equating to zero the reusltant of (34) and (35). We know, following elimination theory, that this resultant is also written

$$D(z) \equiv a_o^{m'} b_o^{m} \, \Pi\, (u_j - u'_j) \tag{37}$$

where the u_j are the solutions of (34) and the u'_k those of (35). If z_o is a solution of (36), necessarily finite by hypothesis, then the equations (34), (35) have for $z = z_o$, a common finite solution u_o and solutions that are all pairwise different and finite. We therefore have cycles with the same vertex z_o, u_o and others with vertices z z_o, u_1, ... for (34), and z_o, $u'_1, \ldots,$ for (35) where the differences of these u_j and u'_k are all different from zero for $z = z_o$. In (37), the order of the product with respect to $z - z_o$ is thus furnished by the terms corresponding to the cycles with common vertex z_o, u_o. This is the number of points of intersection of these cycles, but it is also the degree of multiplicity of the root z_o of equation (36). The number of points of intersection of the two curves is therefore equal to the degree of $D(z)$ which is mm'.

Let us assume, moreover, which we can, that the points at infinity of $P(z,u) = 0$ are all simple. For $|z|$ very large, we have $u_j = \alpha_j z + \ldots,$ where α_j are distinct from the angular coefficients of the asymptotes of $Q(z,u) = 0$; hence $Q(z,u_j)$ is of the order m' with respect to z and the product of these quantitites is of order mm'. Bezout's theorem is thus established.

It follows that $D(z)$ is of degree mm' as soon as the two curves do not have any common points at infinity.

15 - The class. Genus and multiple points

As is known, *the class* of an algebraic curve is the number of tangents that could be taken from this curve through an arbitrary point Q. This entails taking true tangents and not lines which join the point in question to a multiple point and which are not tangent to one of the branches of the curve at this point; the tangents at a point of inflexion are multiple tangents. In order to calculate this number, we can assume that the arbitrary point Q is neither placed on the tangents to the multiple points nor on the tangents of inflection, and via a homographic transformation, we could reduce matters to the case where Q is the point at infinity on the axis Ou, and where the line at infinity does not pass through any multiple point; Q is not, moreover, on the curve. Under these conditions, the tangents from Q are tangents parallel to Ou, these are the simple tangents. If $P(z,u) = 0$ is the equation of the curve, then the points of contact are the points, other than the multiple points, in which $\frac{\partial P}{\partial u}(z,u) = 0$. If we take the origin to one of these points, $P(z,u) = 0$ takes the form $z - ku^2 + \ldots = 0$ and we obtain a simple cycle; it intersects the curve $\frac{\partial P}{\partial u} \equiv -2ku + \ldots \equiv 0$, which also provides a simple cycle that is not tangential to $z = 0$, at a simple point identified with the origin. The class c is hence equal to the total number of the points common to the curve, which is $m(m - 1)$ if m is the degree of $P = 0$, minus a number of points of intersection which are identified with the multiple points of $P = 0$.

Let us take the origin to one of these points. For z in a neighbourhood of 0, the equation $P = 0$ can be written

$$P(z,u) \equiv (u - u_1)(u - u_2) \ldots (u - u_j) \ldots (u - u_m) = 0 \quad ,$$

where the u_j are various branches which are branches of various cycles. The first s of these functions are zero for $z = 0$, the remainder are non-zero. We have therefore

$$\frac{\partial P}{\partial}(z,u) \equiv \sum(u - u_1) \ldots (u - u_j - 1)(u - u_j + 1) \ldots (u - u_m) = 0 \tag{38}$$

if we show the branches of the cycles of which the first ρ are zero at the origin. The number of points of intersection identified with the origin is the infinitesimal order of the product

$$\Pi\,(u_j - u'_k) \quad ,$$

where j takes the values $1,\ldots,s$ and k, the values $1,\ldots,\rho$. Since in (39) the factors of rank greater than ρ are not zero for $z = 0$, the order of this product is that of

$$\prod_1^s \frac{\partial P}{\partial u}(z,u_j) \quad .$$

Now since the factors in (38) with index greater than s are non-zero at the origin, we can replace

$$\frac{\partial P}{\partial u}(z,u_j) = (u_j - u_1) \ldots (u_j - u_{j-1})(u_j - u_{j+1}) \ldots (u_j - u_m)$$

by the product of the first $(s - 1)$ terms. Finally, we need to consider the infinitesimal order of

$$\Pi\,(u_j - u_\ell)$$

for $j = 1,2,\ldots,s$ and $\ell \neq j = 1,2,\ldots,s$.

If there are q-cycles with origin the vertex, the u_j are the branches of these cycles. Each branch of one of these cycles is twice associated with each branch of the others and also with its proper branches.

Let us denote by $O(\alpha,\beta)$ the number of common points at the origin of the two cycles α,β and by $O(\alpha,\alpha)$ the infinitesimal order of the product

$$\Pi\,(u_j - u_\ell)\ , \quad j \neq \ell\ , \quad j = 1,2,\ldots \quad \ell = 1,2,\ldots$$

extended to the branches of the cycle α. This number $O(\alpha,\alpha)$ which following no. 14, depends on the characteristic exponents of the cycle, could be called the internal intersection of the cycle. We will have

$$\sum_q O(\alpha,\alpha) + 2.\sum O(\alpha,\beta) \quad ,$$

points of intersection at the origin. Operating in the same way with the other multiple points, we see that the class is given by the formula

$$c = m(m - 1) - \sum O(\alpha,\alpha) - 2\sum O(\alpha,\beta) \quad ,$$

where the summations are taken over all the multiple points.[1]

PARTICULAR CASES

I - For a multiple point of order q with distinct tangents, we have q cycles of the first degree, $O(\alpha,\alpha)$ is zero and

---

[1] The reasonings in no. 14 show that the internal intersection is invariant under a homographic transformation (the cycles are not tangent to Ou and Oz).

$$\sum 0(\alpha,\beta) = \sum 1 = \frac{1}{2}\, q(q - 1) \quad .$$

A multiple point of order q *with distinct tangents therefore decreases the class by* $q(q - 1)$. *As an ordinary double point lowers the class by two, the point of order* q *is equal, from this point of view, to* $\frac{1}{2}\, q(q - 1)$ *double points with distinct tangents.*

II - A turning point of the first type corresponds to a cycle of the form

$$u^2 - az^3 + \ldots = 0 \quad \text{or} \quad u = a'z^{s/1} + \ldots \quad ,$$

the degree is 2 and the class (of the cycle) is 1. We have two branches and the internal intersection is $2 \cdot \frac{3}{2} = 3$.

A turning point of the first kind lowers the class by 3.

III - For a turning point of the second kind corresponding to a cycle of the form $u = za^2 + bz^{s/2} + \ldots$ of degree 2 and class 2, the internal intersection is 5 and such a point will lower the class by 5.

THE GENUS AND MULTIPLE POINTS

Let us consider a curve of order m whose singularities are multiple points with distinct tangents and turning points of the first kind. We take Ou to be neither parallel to the tangents at the multiple points, the inflection tangents, nor the asymptotes, and we assume that the points at infinity are simple. The ramification points arise uniquely from the simple points in which the tangent is parallel to Ou; these are points of order 2, their number is the class c, and turning points which yield points or order 2. If R is the number of turning points, we therefore have $\sum(r - 1) = c + R$ and the genus given by Riemann's formula (no. 9) is

$$p = \frac{c + R}{2} - m + 1 \quad .$$

Now, following what we have just seen,

$$c = m(m - 1) - 3R - \sum q(q - 1) \quad ,$$

where the summation is taken over multiple points and q denotes the order of these points. It follows that

$$p = \frac{(m - 1)(m - 2)}{2} - R - \sum \frac{q(q - 1)}{2} \quad . \tag{40}$$

As p is positive or zero, we see that

$$R = \sum \frac{q(q - 1)}{2} \quad \frac{(m - 1)(m - 2)}{2} \quad .$$

The total number of turning points of the first kind and double points equal to the multiple points with distinct tangents, is at most equal to $\frac{1}{2}(m - 1)(m - 2)$. *When this maximum is attained, the genus is zero.*

16 - An application of Bertini's theorem to curves of genus 0 and 1

Given a curve Γ transformed by a (curve to curve) birational transformation to a curve Γ' with the same genus, and only having double points with distinct tangents, then the number d of double points of Γ' will be given by the formula in (40):

$$d = \frac{(m - 1)(m - 2)}{2} - p \ .$$

This number hence obtains the maximum value

$$\frac{(m - 1)(m - 2)}{2}$$

when $p = 0$ and this value decreases by one when $p = 1$. We have seen that in the first case, the curve Γ' is then (I, 98) and that in the second case, the co-ordinates of a point of Γ' could be expressed rationally by means of elliptic functions pt and $p't$. As the co-ordinates of a point of Γ are rational functions of those of Γ', we see that, whatever the nature of the singularities *every point of genus* 0 *is* . *The co-ordinates of a point of any curve of genus 1, could be expressed as rational functions of* pt *and* $p't$.

Going back to the proof in volume I (no. 99), we see that in the second case, the curve corresponds birationally to the cubic

$$U^2 = 4Z^3 - gZ_2 - g_3 \quad , \tag{41}$$

such that when we pass to the elliptic functions pt and $p't$, we can establish a bijective correspondence between the points of the curve and the values of t taken in a parallelogram of the periods. Conversely, if the co-ordinates of a point of a curve could be expressed in terms of elliptic functions: $z = \phi(t)$, $u = \psi(t)$, where ϕ and ψ are elliptic with the same periods, and if there exists a bijective correspondence between the points of the curve and the points of a periodic parallelogram, then ϕ and ψ are rational functions of pt and $p't$[1] and we conclude that there exists a

---

[1] This results from the Hermite decomposition formula (I, 251), from the additive formulae of the functions ζ and p (I, 233) and the equation

$$p'' = 6p^2 - \frac{g_2}{2}$$

which gives $p''' = 12\, pp'$, etc.

birational correspondence between Γ and the cubic (41), where g_2 and g_3 have suitable values *s*. As this cubic is of genus 1, the curve is of genus 1.

A curve corresponds bijectively to a line, and consequently it is of genus 0 (a fact exploited (I, 99) in the study of curves of genus 1); it has the maximum number of double points or the equivalent.

Every curve of genus zero and every curve of genus 1, admits birational transformations of its itself, depending on a parameter. For curves of genus zero, they are obtained as is known, by a homographic transformation on the parameter t occurring in the representation (hence they depend on three arbitrary parameters). For curves of genus 1, it suffices to operate on the corresponding cubic (41). Such transformations will be obtained by taking an arbitrary fixed point A on the cubic and corresponding to the variable point M of the curve, the point M' of the curve situated on the line AM. This correspondence is algebraic and bijective. When the representation in terms of the elliptic functions pt and $p't$ is introduced, we see that the geometric property of the points M, M', A, are translated due to the fact that the corresponding values t, t', t_o satisfy a relation $t + t' + t_o = 2m\omega + 2m'\omega'$. We deduce that $t + t' =$ constant. It is proved that all the desired birational transformations are thus obtained.

The curves of genus greater than 1 do not admit birational transformations of themselves that depend on an arbitrary parameter.

Chapter II

ON ABELIAN INTEGRALS, PROBLEMS OF INVERSION AND UNIFORMISATION

It is now possible to give here all of the major developments concerning the abelian integrals and the related problems. We will confine ourselves to matters concerning the integrals of the third kind, by dwelling in particular on those of the first kind. Having stated Abel's theorem together with some ideas on applications to geometry, we shall say something about the various inversion problems as they have been posed up to the time when Poincaré proved that the co-ordinates of the points of any algebraic curve can be expressed by fuschian functions of an auxiliary variable (*Comptes rendus de l'Academie des Sciences* t. XCIII, 1881). There is no question here of taking a glance at the theory of Poincaré and its developments; we refer the reader to the exposition of P. Fatou in volume II of the book by Appell and Goursat on the *Theory of Algebraic Functions*. But we shall give the proof, based on a theorem of Bloch, of Picard's theorem on the impossibility of uniformising algebraic functions in terms of functions having only isolated essential singularities.

I. ANALYTIC FUNCTIONS OF A RIEMANN SURFACE

17 - Uniform functions on a Riemann surface

Given an algebraic function defiend by $P(z,u) = 0$, we shall denote by S, the planar Riemann surface that corresponds to it. This surface is defined by the ramification points and the permutations of the solutions about these points. The consideration of the resulting lines of passage has served to describe, in a more or less convenient way following the choice of these lines, a contour that makes the surface simply connected. To a system of values (z,u), or if we wish, to a point of the curve Γ defined by the equations $P(z,u) = 0$, there corresponds a resulting point of S (except, as we have seen, for a finite number of multiple points providing points of holomorphy or meromorphy of the algebraic function, and then the correspondence holds for the neighbouring points). Conversely, to a point of S, there always corresponds a single pair (z,u) or a single point of Γ. *We shall call this pair* (z,u) *a point of* S *or* Γ, *and will also represent it by the letter* M. When z and u are real, M will be a point of the real part of Γ, the part that is depicted in analytic geometry.

A function is uniform on the Riemann surface or on part of this surface when it has a well determined value at each point; we will denote it by

$f(M)$ or $f(z,u)$. *We shall say that the function is holomorphic in a part of this sruface that does not contain a ramification point, when it is finite and derivable.* The definition of the derivative proceeds from the usual definition in terms of the variable z. About the point z_o, u_o, which is not a ramification point, we can consider a small enough part of the surface for which it is on the same sheet; in this circle the function will be a function of z, $F(z)$ say. The derivative will always be the limit of $[F(z+h) - F(z)]/h$ as $h \to 0$. It follows that $F(z)$ is holomorphic in this region. (A holomorphic function on a region of the Riemann surface is therefore an analytic function of z, which is finite and consequently uniform by virtue of the fact that the point z varies on the Riemann surface.) *We extend the definition to the ramification points.* If the function $f(z,u)$ is defined and holomorphic about a ramification point $M_o(z_o,u_o)$, then to study it at this point, we set $z = z_o + Z^r$, where r is the order of the point M_o in question, and $|z - z_o| < \rho$, where ρ is sufficiently small such that the region described on S by this circle in the sheets containing M_o, does not contain any other singular points. This transformation gives a correspondence between the circle with r sheets, ramified at M_o, and the circle $|Z| < \rho^{1/r}$ of the simple plane of the Z; the function $f(z,u)$ is transformed to $F(Z)$ which is holomorphic about the point $Z = 0$. (For the transformation is an ordinary conformal transformation if it is applied to a region of S situated on a single sheet.) *If* $F(Z)$ *is holomorphic at the point* $Z = 0$, *we say that* $f(M)$ *is holomorphic at the point* M_o. Under these conditions, we will have, about $Z = 0$,

$$F(Z) = A_o + A_1 Z + \dots + A_n Z^n + \dots$$

where the series is convergent for $|Z| < R^{1/r}$ and consequently we also obtain

$$f(z,u) = A_o + A_1 (z - z_o)^{1/r} + \dots + A_n (z - z_o)^{n/r} + \dots \qquad (1)$$

for $|z - z_o| < R$. *The point* $M_o(z_o,u_o)$ *will be said to be a zero of order* q *if* $A_o = A_1 = \dots = A_{i-1} = 0$ *and* $A_o \neq 0$. *The definition of a zero at an ordinary point is the usual definition.*

Remark: Following the equation (1), the derivative of $f(M)$, written $f'(M)$ or $f'(z,u)$, is no longer finite when $z \to z_o$, unless the ramification point z_o is not a zero of at least r. A holomorphic function on S at a ramification M_o, must not be regarded as having a derivative *with respect to* z at such a point.

POLES AND ESSENTIAL POINTS

A point M_o about which the function $f(M)$ is holomorphic will be said to be a pole if $1/f(M)$ is holomorphic at this point, which will consequently be a zero of this function. If it is a zero of order q, $f(M)$ will be a pole of order q at this point M_o. The expansion about this point will have the usual form if M_o is an ordinary point of S. If $M_o(z_o,u_o)$ is a ramification point of order r, we will have following (1),

$$f(z,u) = \sum_{-\infty}^{\infty} B_n(z - z_n)^{n/r} \quad . \tag{2}$$

If the point M_o is an isolated singularity of $f(M)$, i.e., if $f(M)$ is holomorphic about M_o, but is not holomorphic at this point, and if it is not a pole, it is an essential singularity. The expansion about this point is the ordinary Laurent expansion if M_o is an ordinary point of S; it is deduced from the Laurent expansion of $F(Z)$ if (z_o,u_o) is a ramification point. Therefore we have in this case

$$f(z,u) = \sum_{-\infty}^{\infty} A_n(z - z_o)^{n/r} \quad .$$

RESIDUES

We call the residue of a pole (z_o,u_o) which is not a ramification point, the ordinary residue, hence the coefficient of $1/(z - z_o)$ in the expansion about this pole. If z_o, u_o is a ramification point, the residue is not the coefficient of $(z - z_o)^{-1}$ in the expansion (2) about this point, but the product of this coefficient B_{-r} and the order r of the ramification point.

$$\int_{C+}^{o} f(z,u)dz$$

calculated is a circle C of centre z_o, in the direct sense (where the circle is considered on the Riemann surface, i.e., encircling the point z_o r times, once on each sheet), which is equal to $2i\pi$. B_{-r} is $2i\pi$ times the residue (to calculate the integral, we can set $z = z + \rho e^{i\phi}$, the coefficient of B_n gives

$$\int_0^{2\pi r} \rho\rho^{n/r} e^{in\phi/r} e^{i\phi} \, id\phi$$

and consequently of 0 if $n \neq r$, and $2i\pi r$ if $n = -r$; we could also make the transformation $z = z_o + Z^r$).

POINTS AT INFINITY

The definitions relative to the points at infinity situated on the various sheets or groups of sheets, arise by setting $z = 1/Z$. If the function is holomorphic about the point at infinity, a point of ramification of order r, we shall have about this point,

$$f(z,u) = f\left(\frac{1}{Z}, u\right) = \sum_{s}^{\infty} A_n Z^{n/r} = \sum_{s}^{\infty} A_n z^{-z/r} ,$$

and we shall have an essential point if $q = -\infty$, a pole of order $-q$ if q is negative, a point of holomorphy if $q = 0$ and a zero of order q if $q > 0$. For the same reasons as in the case of the functions considered on the simple plane (I, 194), we take as the value of the residue

$$- rA_r ,$$

such that the integral of $f(z,u)$ when a rotation is made on a great circle situated on the r sheets, in the inverse sense, is equal to $2i\pi$ times the residue at infinity.

18 - The Cauchy integral

The meaning of the integral

$$\int_C^o f(z,u)dz$$

where C is a rectifiable curve on the Riemann surface S and $f(z,u)$ continuous on this curve is quite evident. C is also a curve of the simple plane and at each point of C, the function $f(z,u)$ has a resulting value; these values form a continuous set. We will, moreover, only consider analytic curves C; they will intersect only at a finite number of points, the lines of passage passing from one sheet to another; these lines we will always assume to be polygonal.

If C is a closed curve described in a simply connected region of S and if $f(z,u)$ is holomorphic in this portion, we have

$$\int_C^o f(z,u)dz = 0 .$$

To prove this, we can assume that the curve C is simple, unless it is reduced to a finite number of curves of this kind. We could then decompose the interior of C in terms of the lines of passage. By adding and subtracting the integrals of $f(z,u)$ taken along the parts of these lines inside C, we recover the same integral taken on closed contours, where each of these contours is described on a single sheet. As $f(z,u)$ is then an ordinary holomorphic function of z, the integral on each of these closed contours is zero by Cauchy's theorem.

Similarly we see that (3) holds if C consists of several contours, a contour outside and a finite number of contours inside, where the integral is taken in such a way that the domain of holomorphy of $f(z,u)$ lies to the left. From the extension of Cauchy's theorem, we deduce

$$f(z_o,u_o) = \frac{1}{2i\pi r} \int_{C^+}^{o} \frac{f\ z,u}{z - z_o}\ dz \tag{4}$$

where the integral is taken in the direct sense on the total curve, as described, and where r denotes the number of sheets containing the point (z_o,u_o); r is therefore equal to 1 if the point (z_o,u_o) is an ordinary point on S and equal to r if it is a ramification point of order r. The proof is the same as in the case of the simple plane (I, 183).

If the function $f(z,u)$ possesses poles or essential points inside the contour C and is holomorphic everywhere else in C and on its boundary, we can remove from the interior of C some small circles having these poles or essential points as their centres and apply (3) to the contour formed by C and the circumferences of the circles. On the circumference of one of the small circles, the integral of $f(z,u)$ taken in the direct sense, is equal to the residue times $2i\pi$. We then have (see I, 192)

$$\int_{C^+}^{o} f(z,u)dz = 2i\pi \sum R \quad ,$$

where $\sum R$ denotes the sum of the residues of the residues of the poles and the essential points in question.

The formula in (4) extends to the case where $f(z,u)$ is no longer holomorphic in C, but has these, poles or essential points distinct from the point z_o, with the condition of adding to the first member the residues of $f(z,u)/(z - z_o)$ to these points. In all this we have considered a bounded domain. We could also consider, in a group of sheets forming a connected region of S containing the point at infinity, the exterior of a curve, i.e., a domain containing the point at infinity, and apply the above discussion by putting $z = 1/Z$, which reduces matters to a finite domain.

Theorem I: *If a function* $f(M)$ *is holomorphic in the domain obtained by removing a finite number of points from* S, *then the sum of the residues of this function at tis singular points is zero.*

If $u_1,u_2,\ldots,u_n$ are the m branches of the algebraic function $u(z)$, the function $f(z,u)$ has at the point z, the values, $f(z,u_j)$, $j = 1$ $j = 1,2,\ldots,m$. Every symmetric function of these values $f(z,u_j)$ is a uniform function of z; it is holomorphic about each point z that does not correspond to a singularity of $f(z,u)$ or to a ramification point and is therefore a holomorphic function throughout the plane, except at a finite number of points. It has, therefore, a finite number of poles or essential points. Let us consider in particular $V(z) = \sum^{\infty} f(z,u_j)$. The residue of $V(z)$ at one of the singularities of $f(z,u_j)$ is the sum of the residues of these functions: this can be seen if this point z_o is not a ramification point, where z_o could moreover be a point of holomorphy for $V(z)$. If z_o is a point of ramification, the branches of $f(z,u)$ group themselves in circular systems for each of which

$$f(z,u) = \sum_{qq}^{\infty} A_n(z - z_o)^{n/r} \quad ,$$

where r is the order of the point. This shows that the sum of the branches corresponding to this group will have the point z_o as a pole or essential point if $q < 0$ or $q = -\infty$, where the residue is the coefficient of $(z - z)^{-1}$ in this sum, hence rA_{-r} and is the residue of $f(z,u)$ at the point z_o. We also see that if $q > 0$, the sum in question is holomorphic. The same reasoning applies to each group and $V(z)$ will indeed have a residue equal to the sum of the residues at the point z_o. This also goes for the point at infinity. The sum of the residues of $f(z,u)$ is hence equal to the sum of the residues of $V(z)$, a sum which is zero following a well-known theorem (I, 194).

A CONSEQUENCE

We see that as a consequence of the above proof, all the sums

$$V_q(z) = \sum_{1}^{m} (u_j)^q f(z,u_j) \qquad q = 0,1,\ldots,m-1 \tag{5}$$

are uniform functions of z having only a finite number of poles or essential singularities. We could express $f(z,u)$ as a function of the $V_q(z)$ and of u. On solving the system in (5) with respect to $f(z,u_1)$, we obtain $f(z,u_1)$ as the quotient of two determinants, which we shall write by showing the lines of rank $q + 1$

$$\|V_q(z)\ u_2^q \ldots u_m^q\| \quad , \qquad \|u_1^q\ u_2^q \ldots u_2^q\| \quad .$$

If we multiply the two terms of this ratio by the denominator, we obtain for the denominator, the discriminant $D(z)$ of the polynomial in u, $P(z,u)$. Consequently,

$$D(z)\ f(z,u\ ) = \|V_q(z)\ u^{\ q} \ldots u_m^q\| \cdot \|u^{\ q}\ u^{\ q} \ldots u_m^q\| \quad .$$

The second member is a polynomial in u_1, of degree $m-1$, whose coefficients are functions of z and $u_2,\ldots,u_m$, that are symmetric in $u_2,\ldots,u_m$. Now $u_2,\ldots,u_m$ are the zeros of the polynomial in u, $P(z,u)/(u-u_1)$, whose coefficients are polynomials in z and u_1. Thus $D(z)\, f(z,u_1)$ is a polynomial in u_1 with uniform coefficients in z. On dividing this polynomial by $P(z,u_1)$ and bearing in mind that u_1 is a zero of this polynomial, we finally obtain the result:

Theorem II: *The uniform functions on* S *having only a finite number of singularities are of the form*

$$F_n(z) + uF_1(z) + \ldots + u^{m-1}\, F_{n-1}(z) \tag{6}$$

where the $F_j(z)$ *are uniform functions of* z *having only a finite number of singularities (poles or essential points) and* m *is the number of the sheets.*

19 - Rational fractions

A rational fraction $R(z,u)$ a quotient of two polynomials in z and u, taken under in an irreducible form and whose denominator is not divisble by $P(z,u)$ is a uniform function on S having only poles as singularities. For, about each point (z_o,u_o) of S, the numerator and denominator are expansions in series in $(z-z_o)^{1/r}$, where r is equal to one or the order of the point of ramification (z_o,u_o) and these series only have a finite number of terms with negative exponents. The quotient is therefore of this form.

Conversely, a uniform function on all of S, having only poles, has a finite number. Theorem II above applies but in expression (6) the $F_j(z)$ have only poles and the function is a rational fraction. That the $F_j(z)$ has only poles, results from the fact that the $V_p(z)$ then only have poles. Thus:

Theorem: *Every uniform function on* S, *only having poles as singularities, is a rational fraction. Every rational fraction on* S *could be expressed as a polynomial in* u, *of degree* $m-1$, *whose coefficients are rational fractions of* z.

If we assume that the rational fraction $R(z,u)$ is everywhere bounded on S, then the symmetric functions $\sum R(z,u_j)$, $\sum R(z,u_k)$ ... are everywhere bounded; $R(z,u)$ is therefore a root of an equation of degree m whose coefficients are rational fractions of z that are everywhere bounded, hence constants. $R(z,u)$ is constant. In a like way, but on applying Liouville's theorem (I, 184), we see that generally, *A uniform function on* S, *which only has a finite number of singularities, is a constant if it is everywhere bounded.*

<u>Theorem</u>: *The number of zeros of a rational fraction is equal to the number of its poles where each pole or zero is counted a number of times that is equal to its order of multiplicity.*

The derivative of $u(z)$ is given by

$$\frac{\partial P}{\partial z}\frac{du}{dz}+\frac{\partial P}{\partial z}=0 \quad ,$$

and it is a rational fraction; consequently, if $R(z,u)$ is rationalle on S,

$$\frac{dR}{dz}=\frac{\partial R}{\partial z}+\frac{\partial R}{\partial u}\frac{du}{dz}$$

is a rational fraction and hence also the logarithmic derivative.

$$R_1=\frac{1}{R}\frac{dR}{dz} \quad .$$

The sum of the residues of this fraction is therefore zero (theorem I of no. 18). Now a zero of order q of R yields in R_1 a simple pole of residue q, since about such point z_o, we have

$$R=A_q(z-z_o)^{q/r}+\ldots, \qquad \frac{dR}{dz}=\frac{q}{r}A_q(z-z_o)^{(q-r)/r}+\ldots$$

hence

$$R_1=\frac{q}{r}\frac{1}{z-z_o}+\ldots \quad .$$

Similarly, a pole of order q' of $R(z,u)$ yields in R_1, a simple pole with residue $-q'$. Finally, at a point of ramification z_o which is neither a zero nor a pole, we have

$$R=A_o+A_1(z-z_o)^{1/r}+\ldots, \qquad \frac{dR}{dz}=A_1\frac{1}{r}(z-z_o)^{(1-r)/r}+\ldots$$

this point is therefore a pole of R_1, or order $r-1$ and its residue is zero. The sum of the rsidues of R_1 is therefore $\sum q-\sum q'$; this is zero and we have $\sum q=\sum q'$.

<u>Corollary</u>: *Let* $R(z,u)$ *be a rational fraction, the number of points of* S *for which* $R(z,u)=a=$ *const. is independent of* a.

Since, whatever the value of z, this number is equal to the number of poles of $R(z,u)$. It is this proposition which had been applied in no. 13.

<u>Remark</u>: On account of this, we see at once that a rational fraction is determined up to the nearest additive constant when the principal parts of its poles are given (since the remainder are bounded, hence constant). Similarly, if the

poles and zeros are given, the function is determined up to the nearest constant factor. But we can no longer choose arbitrariliy those given.

AN APPLICATION OF BEZOUT'S THEOREM

In the proof of this theorem, we have used the fact that the degree of the resultant $D(z)$ of the polynomials $P(z,u)$ and $Q(z,u)$

$$P(z,u) \equiv a_o u^n + \dots + a_n(z) , \quad Q(z,u) \equiv b_o u^{n'} + \dots + b_{n'}(z)$$

is equal to mm'. Now

$$D(z) \equiv a_o^{n'} Q(z,u_1) \dots Q(z,u_n) \quad .$$

If (z_o,u_o) is a common point, we have for r of the solutions u_j,

$$Q(z,u_j) = B_q(z-z_o)^{q/r} + \dots \quad .$$

The order of multiplicity of the zero z_o of $D(z)$ is therefore q, it is the order of the zero of $Q(z,u)$ situated at the point (z_o,u_o). The degree of $D(z)$ is therefore the number of zeros of $Q(z,u)$, hence also the number of its poles. Let us suppose as we may, that the points at infinity of $P(z,u) = 0$ are all distinct, $Q(z,u)$ has on each of its sheets a pole at infinity of order of multiplicity m', and these are its only poles. The number of the poles of $Q(z,u)$ is therefore mm'; $D(z)$ is of degree mm'.

II. ABELIAN INTEGRALS

20 - Periods of Abelian Integrals

Let $R(z,u)$ be a rational fraction of z and u. An abelian integral is an integral

$$I(M) = \int_{M_o}^{M} R(z,u)\, dz \tag{7}$$

taken from M_o to M along a certain path γ on the Riemann surface S corresponding to an algebraic function $P(z,u) = 0$. *It is an abelian* associated to the curve Γ, $P(z,u) = 0$. When (z_o,u_o) and (z,u) are real and a real path is taken from M_o to M, i.e., when a real arc of Γ is described, we recover the abelian integrals that were previously studied in (I, 98). We will assume that the point M_o is an ordinary point of the surface and that it is not a pole of $R(z,u)$.

The integral in (7) is a function of γ and of the position of M in S. It is an analytic function of z when M varies. Its nature, in a neighbourhood of a point $A(a,b)$ in which it has a finite or infinite value, is obtained by expanding $R(z,u)$ in a neighbourhood of this point.

Firstly, let us assume that A *is not a ramification point. If* R(z,u) *is holomorphic at this point,* we have for $|z-a|$ sufficiently small,

$$R(z,u) = B_o + B_1(z-a) + \dots \quad .$$

When $M \to A$, I(M) has a finite limit I(A) and for M in neighbourhood of A,

$$I(M) = I(A) + B_o(z-a) + \frac{1}{2} B_1(z-a)^2 + \dots \quad .$$

I(M) *is holomorphic at the point* A. *If* R(z,u) *admits the point* A *as a pole,* we have

$$R(z,u) = \frac{B_{-q}}{(z-a)^q} + \dots + B_o + B_1(z-a) + \dots \quad .$$

I(A) does not have a meaning; we integrate up to D, a neighbouring point of A, then from D to M and we then have

$$I(M) - I(D) + \left[\frac{B_{-q}}{(1-q)(z-q)^{-1}} + \dots + B_{-1}\log(z-a) + B\,(z-a) + \dots\right] - k$$

where k denotes the value of the terms in the brackets at the point D. I(M) *has a pole of order* $q-1$ *at the point* A *if the residue* B_{-1} *of* R(z,u) *at* A *is zero, and has a critical transcendent point if* $B_{-1} \neq 0$. To simplify things, we say that in the second case, A *is a logarithmic point of* I(M).

Let us now assume that A *is a ramification point of order* r. *If* R(z,u) *is holomorphic at the point* A, *we have*

$$R(z,u) = B_o + B_1(z-a)^{1/r} + \dots$$

and the integral has a meaning at the point A, I(A). We have

$$I(M) = I(A) + B_o(z-a) + \frac{B_1 r}{1+r}(z-a)^{(1+r)/r} + \dots \quad .$$

The integral is holomorphic at the point A.

If A *is a pole of order less than* r, I(A) again has a meaning. If,

$$R(z,u) = \frac{B_q}{(z-a)^{q/r}} + \dots \;, \qquad q < r$$

we obtain

$$\mathrm{I(M)} = \mathrm{I(A)} + \frac{\mathrm{B}_q^r}{r-q}(z-a)^{(r-q)/r} + \dots , \qquad r-q > 0$$

and I(M) *is holomorphic at the point* A.

If A *is a pole of order greater than* r *and if the residue is zero,* I(A) no longer has a meaning. We introduce a neighbouring point D as above, and obtain

$$\mathrm{I(M)} = k' + \frac{\mathrm{B}_q^r}{r-q}\,\frac{1}{(z-a)^{(q-r)/r}} + \dots$$

without a logarithmic term. I(M) *has a pole at the point* A. *If* A *is a pole of order greater than or equal to* r*,* we have the same result as in the preceding case, but *with a logarithmic term.* We then say that A is *a singular logarithmic point.*

The same discussion applies to the integral about the point at infinity. This point could be a point of holomorphy, a pole, or a logarithmic point in the indicated sense. We thus see that *the integral* I(M) *is uniform about any point on* S, i.e., in a small circle with this point as its centre, when it is extended in this circle, *if all the residues of* $\mathrm{R}(z,u)$ *are zero.* For the contrary, it has logarithmic points, hence an infinity of braches, and likewise when we restrict to a neighbourhood of one of these points.

PERIODS

Now let us assume that I(M) has been calculated on two different curves joining M_o to M. The difference of the values so obtained remains the same when M is then displaced along a certain path. If the two curves of integration γ,γ' do not intersect at points other than M_o and M, the difference of the two integrals is equal to the integral of $\mathrm{R}(z,u)$ calculated on the closed contour $\mathrm{M}_o\gamma\mathrm{M}\gamma'\mathrm{M}_o$. If a region of S bounded by this contour is simply connected, then this integral is equal to $2i\pi$ times the sum of the residues of the poles of $\mathrm{R}(z,u)$ contained in this region. We are then led to consider the simply connected surface obtained by suitably intersecting S.

We shall call K the contour of the second type introduced in no. 10 and make S simply connected (figs. 18, II; 20, II; 21, III). We shall assume that it does not pass through the poles of $\mathrm{R}(z,u)$. We shall have, for example, the picture below which reproduces the third configuration (fig. 23). *The integral* I(M) *will have a well determined value at each point of* S *if the path* $\mathrm{M}_o\gamma\mathrm{M}$ *does not intersect* K *and if* $\mathrm{R}(z,u)$ *has all its residues zero.*

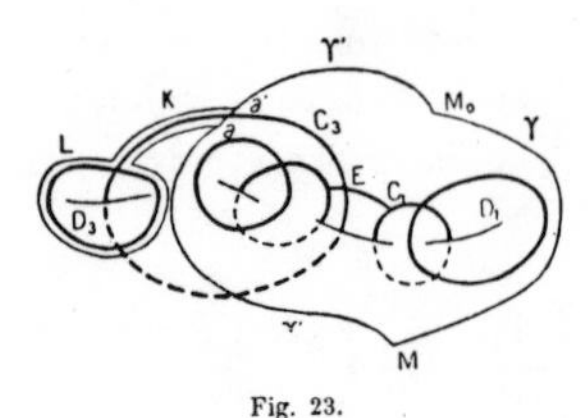

Fig. 23.

But let us take another path $M_o\gamma'M$ crossing K once at a point a, a', where a' is taken immediately before crossing K and a immediately after. We could join a to a' by a path passing along K. Let L be such a path (fig. 23). The integral calculated along the path $M_o\gamma'a'La\gamma'M$ will be I(M). It differs from the integral calculated along $M_o\gamma'a'a\gamma'M$ by the value of the integral relative to the path L. This value depends on the position of the point a, a'. Let us call C_j the cuts, being the parts of K whose extremities are on E and c_j the value

$$c_j = \int_{C_j} R(z,u)\,dz\ , \qquad j = 1,2,\ldots,p\ ,$$

calculated in a certain sense. Let D_j be the cut resting on C_j, and d_j the corresponding integral

$$d_j = \int_{D_j} R(z,u)\,dz\ , \qquad j = 1,2,\ldots,p\ ,$$

equally calculated in a certain sense.

When the point a, a' is on a curve C_j, we will take L alongside the curve D_j with the same index and twice a same part of C_j. The integral on L will be $\pm d_j$. If a, a' is a curve D_j, we can take L twice alongside the same part of D_j (on both sides of the cut) and once for C_j, then twice in the opposite sense for a part of E and certain curves C_j, D_j; the integral on L will be $\pm c_j$. Finally, if a, a' is on E, L will twice be alongside a certain number of curves C_j, D_j and a part of E, and these curves are taken in the opposite sense. We obtain 0. Hence for a curve γ cutting K once, we obtain

$$I(M) \pm c_j \quad \text{or} \quad I(M) \pm d_j\,, \quad \text{or} \quad I(M) \quad ,$$

where $I(M)$ is the integral taken on a direct curve, not intersecting K.

If γ cuts K at several points $a, a'; b, b'; c, c'$ say, we will consider the same difference between the integral taken on γ and on a direct curve $M_o\gamma a L a'\gamma b L' b'\gamma c L'' c'\gamma M$, where L is a curve joining a to a' along K, L' joins b to b' along K and L" joins c to c' along K. This difference will be equal to the value of the integrals along L, L', L" which are $\pm c_j$ or $\pm d_j$, j having three distinct values or otherwise depending on the position of the points a, b, c.

Finally, *all the values of the integral are of the form*

$$I(M) + \sum_1^{n^o} (m_j c_j + n_j d_j) \tag{8}$$

where p *is the genus and* m_j *and* n_j *are integers that can be positive, negative, or zero. This is the case if all the residues are zero.* If the m_j and n_j are given, we could effectively choose a path such that the integral is equal to (8). It is clear that the numbers c_j, d_j which do not depend on M_o, depend on the controu K. But if K is replaced by another contour K' to which there corresponds the numbers c_j' and d_j', the values of $I(M)$ will then be of a form similar to (8). It follows that the c_j', $j = 1,2,\ldots,$ and the c_j', $j = 1,2,\ldots,p$ will be linear homogeneous functions of the c_j, d_j with integer coefficients. Conversely, the c_j and d_j will be expressed in the same way according to the c_j' and d_j'. We will obtain two equivalent systems.

The c_j, d_j are called *the periods*, or *the cyclic periods of the abelian integral.*

When the residues of $R(z,u)$ are not all zero, we make $I(M)$ uniform by joining the poles of $R(z,u)$ with nonzero residues to a point of E, by means of simple curves K' that do not intersect between themselves and do not cut K outside their extremity on E. In the simply connected domain bounded by K and these curves, $I(M)$ is uniform when one integrates along a curve joining M_o to M keeping within the domain. For a curve of integration intersecting K or the curves K', we proceed as above. At a point of intersection d, d' with a curve K', we introduce a path along K' such that it goes around the pole situated at the extremity of K', the value of the integral on such a path is $\pm 2i\pi\rho$, if ρ is the residue of this pole. We also see that *the integral is given here, by the formula*

$$I(M) + \sum_1^{p} (m_j c_j + n_j d_j) + \sum 2i\pi\rho_k\, q_k$$

where the q_k, *like the* m_j *and* n_j, *are integers that can be positive, negative or zero, and the* ρ_k *are residues of the poles of* $R(z,u)$. *We have cyclic periods* c_j, d_j *and polar periods* $2i\pi\rho_k$.

21 - General Properties. The classification for those of the third kind

If a birational transformation is made from a curve to a curve

$$Z = \Phi(z,u) \quad , \quad U = \Psi(z,u) \quad ,$$

which transforms $P(z,u) = 0$ to $Q(Z,U) = 0$, the inverse transformation is

$$z = \phi(Z,U) \quad , \quad u = \psi(Z,U)$$

where ϕ and ψ are rational fractions. The abelian integral

$$I(M) = \int_{M_o}^{M} R(z,u)\, dz$$

taken on S, becomes

$$\int_{M_o}^{M'} R(\phi,\psi) \left[\frac{\partial\phi}{\partial Z} + \frac{\partial\phi}{\partial U}\frac{dU}{dZ}\right] dZ$$

taken along the transformed path from $M_o\gamma M$ on S' which is the Riemann surface of $Q(Z,U) = 0$. These integrals are equal. Indeed the second produces, as seen in no. 19, a rational fraction; it is also an abelian integral. These abelian integrals being equal to homologous points have the same cyclic and polar periods. Certain properties of the abelian integrals could be obtained by replacing the surface S by S'. In particular, by applying Noethen's theorem, we could assume that S' is a Lüroth surface. We will also use Bertini's theorem.

Integrals of the First Kind

The integrals of the first kind are those which are everywhere finite. They only admit cyclic periods. One birational transformation changes one integral of the first kind to another of the first kind.

Integrals of the Second Kind

These are those which are not finite everywhere, but which do not admit polar periods. If a birational transformation is made, one such integral is transformed to an integral of the second kind.

Integrals of the Third Kind

These are those which admit polar periods. A birational transformation changes one integral of the third kind to another of the third kind.

Theorem: *The sum of the polar periods of an integral of the third kind is zero.*

This sum is effectively $2i\pi$ times the sum of the residues of $R(z,u)$, a sum that is zero following theorem I of no. 18.

22 - Integrals of the first kind

We first of all seek the forms of the fraction $R(z,u)$ giving such integrals, by assuming that the points at infinity of the curve Γ, $P(z,u) = 0$, are distinct, and that 0_u is not an asymptotic direction:

$$P(z,u) \equiv a_o u^n + \dots + a_n(z) = 0 , \qquad a_o \neq 0 \quad .$$

In order that

$$I(M) = \int_{M_o}^{M} R(z,u)\, dz$$

always remains finite, it is necessary, first of all, that this is the case for z infinite. About the points at infinity of S, u is expandable in the form $u = \alpha z + \beta + \frac{\gamma}{z} + \dots$, hence $R(z,u)$ is of integer order in $1/z$; *it is necessary that* $R(z,u)$ *is of the order of* $z^{-2-\lambda}$ *where* λ *is an integer that is positive or zero.*

When z is finite, u is finite. At a pole at a finite distance, $I(M)$ can only be finite if the point of S is a ramification point and *the order of the pole is less than the order* r *of the ramification.* Under these conditions, the sums

$$V_q(z) = \sum_1^n (u_j)^q\, R(z,u_j) , \qquad q = 0,1,\dots,m-1 \quad , \tag{9}$$

which are rational fractions (no. 18), are in fact polynomials. For about a ramification point ζ, we have

$$u = u_o + A_\mu(z-\zeta)^{\mu/r} + \dots, \qquad u^q = u_o{}^q + qu_o{}^{q-1}A_\mu(z-\zeta)^{\mu/} + \dots,$$

$$R(z,u) = B_\nu(z-\zeta)^{-\nu/r} + \dots \quad ,$$

where the product $u^q R(z,u)$ has r braches whose sum is rational. In this sum, the terms with fractional exponents disappear; now the terms with negative exponents have an exponent greater than -1, hence fractional and they disappear. The sum remains bounded at the point ζ.

The $V_q(z)$ *are therefore polynomials.* At infinity, u^q is of order q, $R(z,u)$ is of order $-2-\lambda$, hence $V_q(z)$ is of order $q-2-\lambda$; consequently $V_q(z)$, $V_1(z)$ are identically zero, $V_2(z)$ is a constant, $V_q(z)$, $q \geq 3$ is of degree $q-2$ at most. Let us solve the system in (9) as in no. 18; we will have

$$\|V_q(z)\ u_n^{\ q}\ \ldots\ u_n^{\ q}\| = \|u_1^{\ q} u_2^{\ q}\ \ldots\ u_n^{\ q}\|\ R(z,u_1) \quad . \tag{10}$$

On both sides, the determinants contain the factor obtained by taking the product of the differences $u_j - u_k$, $j > 1$, $k > 1$, $j \neq k$. Let us cancel this common factor. On the second member, there will remain the product $(u_1 - u_2) \ldots (u_1 - u_n)$ which is the quotient of $\frac{\partial P}{\partial z}(z,u_1)$ by a_o. On the first member, the coefficient of $V_q(z)$ will be a symmetric polynomial in $u_2,\ldots,u_n$ of degree m-q-1. Now $u_2,\ldots,u_n$ are the roots of the equation

$$\frac{P(z,u)}{u-u_1} = a_o u^{n-1} + \left(a_1(z) + a_o u\right)\ u^{n-2} + \ldots$$

$$\ldots + \left(a_{n-1}(z) + \ldots + a_o u^{\ n-1}\right) = 0$$

whose coefficients are polynomials in z and u_1, where those of u^{n-q} are of degree $q-1$. The coefficient of $V_q(z)$ is therefore a polynomial in u_1 and z, of degree m-q-1, the product of $V_q(z)$ is of degree at most equal to m-q-1 + q-2 = m-3. Thus $R(z,u)$ *must be of the form*

$$\frac{Q(z,u)}{P_u{}'(z,u)}$$

where $Q(z,u)$ *is a polynomial of degree at most* $m-3$.

If this condition is satisfied, then the abelian integral stays finite at infinity. At this point, $Q(z,u)$ is of the order of z^{n-3} it suffices to show that $P_u{}'(z,u)$ is of the order of z^{n-1} to obtain the desired result. Now about the point at infinity of each sheet, we have

$$u_j = \alpha_j z + \beta_j + \ldots\ , \quad \text{and, since}$$

$$P(z,u) = a_o(u-u_1)\ \ldots\ (u-u_n), \quad \text{we shall have}$$

$$P_u{}'(z,u_1) = a_o(u_1 - u_2)\ \ldots\ (u_1 - u_n) \quad . \tag{11}$$

Each factor of this product is of order one with respect to z, since the α_{jj} are all different.

But $R(z,u)$ must satisfy conditions implying that for ramification points $I(M)$ remains bounded. We shall only deal with the case where the only singularities of Γ (all at a finite distance by hypothesis) are multiple points with distinct tangents and turning points of the first kind. As $R(z,u)$ only becomes infinite when $P_u{}' = 0$, but

$$P_u'du + P_z'dz = 0 \quad ,$$

then we can write

$$I(M) = \int_{M_o}^{M} - \frac{Q(z,u)}{P_z'(z,u)} \frac{du}{dz}\, dz = - \int_{M_o}^{M} \frac{Q'}{P_z'}\, du$$

which shows that $I(M)$ remains finite at the point in question if $P_z'(z,u)$ is not zero. It therefore suffices to examine the multiple points. *We shall assume that the point in question had been taken to the origin.* If the origin is a multiple point of order q, with distinct tangents, with one of the tangents as 0_u, then q branches $u_1, u_2, \ldots, u_q$ are zero and holomorphic for $z = 0$ and (11) is true; P_u' is of the order of $(u_1 - u_2) \ldots (u_1 - u_q)$, i.e., of order $q-1$ since the tangents are distinct. $Q(z,u)$ will then be of order $q - 1$ in z. It suffices that the origin is a multiple point of order $q - 1$ of $Q(z,u) = 0$ for this to be the case. This is also necessary. For if $Q_k(z,u)$ is the set of terms of the lowest degree in $Q(z,u)$, where these terms are of degree k, and if $u = m_j z$ is the tangent at 0 to the branch u_j, then we should have $Q(1,m_j) = 0$, for q distinct values of m_j, if $k < q - 1$; k is at most equal to $q - 1$. Thus *it is necessary and sufficient that the origin is a multiple point of order at least* $q - 1$ *of* $Q(z,u) = 0$.

If one of the tangents at the point of order q, situated at the origin, is Ou, there are q cycles of degree one, of which one is a tangent to Ou, having the origin as vertex. Here P_u' is of the order of $(u_1 - u_2) \ldots (u_1 - u_{q-1+r})$ where r is the number of branches of the cycle tangent to Ou, for which we have

$$u = \alpha z^{1/r} + \ldots \quad .$$

It follows that P_u' is of order $(q - 2 + r)/r$ or $\nu_o(1/r) + q - 2$ according to whether u_1 is or is not a branch tangent to Ou. This number is at most $q - 1$. It suffices then that $Q(z,u) = 0$ admits the origin as a multiple point of order $q - 1$. This is in fact necessary, since following what was said for the general case, it will first of all necessitate that the number k is at least $q - 2$. If it were only equal to $q - 2$, then $Q_{q-2}(z,u)$ would contain a term effectively in u^{q-2} and, for a branch u tangent to Ou, $Q(z,u)$ would be of the order of $(q-2)/r$, when P_u' is of this order plus 1. Then in this case, *it is necessary and sufficient that the origin is a multiple point of order at least* $q - 1$ *of* $Q(z,u) = 0$.

Now let us assume that we have a turning point of the first kind, transported to the origin. We have two branches u_1 and u_2 zero at the origin if the tangent is not Ou, and P_u' is of order of $u_1 - u_2$ hence of order

3/2. $Q(z,u)$ will be of order greater than 1/2; for this to be so, it is necessary and sufficient that $Q(0,0) = 0$, for the order will then be at least that of z or u. If the tangent at the turning point is Ou, we have

$$u = \alpha z^{2/3} + \ldots,$$

there are three branches u_1, u_2, u_3 which are zero at the origin, the order of P_u' is that of $(u_1 - u_2)(u_1 - u_3)$, i.e., 4/3. It is then necessary and sufficient that $Q(0,0) = 0$. By grouping these results, we see that:

Theorem: *The necessary and sufficient condition for an abelian integral assigned to a curve* $P(z,u) = 0$, *of degree* m, *which only has simple points at infinity, which does not admit* Ou *as an asymptote and whose only singularities are multiple points with distinct tangents and turning points of the first kind, to be an integral of the first kind, is that it should be of the form*

$$\int_{M_o}^{M} \frac{Q(z,u)}{P_u'(z,u)}\, dz \quad ,$$

where the curve $Q(z,u) = 0$ *is of degree at most* $m-3$ *and admits every multiple point of order* q *of* $P(z,u) = 0$ *for a multiple point of order at least* $q-1$ *(for a turning point* $q = 2$*).*

Such curves $Q(z,u) = 0$ do exist. We know (I, 98) that a curve of degree m' depends linearly on $1/2\, m'(m'+3)$ non-homogeneous parameters. To say that a point is a multiple point of order q', we need to show that the equation is satisfied and that the partial derivatives up to order $q'-1$ are also all non-zero at the point in question. This gives $1 + 2 + \ldots + q' = 1/2\, q'(q'+1)$ linear conditions. The number of conditions to be written in order that $Q(z,u) = 0$ is a curve responding to the need, is hence $\sum \frac{1}{2}(q-1)q + R$, where the sum is extended to the multiple points with distinct tangents and R is the number of turning points. Following formula (40) of no. 15, this sum is equal to

$$\frac{(m-1)(m-2)}{2} - p = \frac{m(m-3)}{2} - p + 1 \quad ,$$

where p is the genus of $\Gamma\left(P(z,u) = 0\right)$. Consequently, by taking $Q(z,u) = 0$ of degree $m-3$, there will remain at least $p-1$ arbitrary non-homogeneous parameters, or if one wishes, at least p arbitrary homogeneous parameters.

The abelian integrals of the first kind therefore exist as long as the genus is at least one. The curves $Q(z,u) = 0$ *which provide them are called adjoint curves of order* $m-3$.

23 - Integrals of the first kind that are linearly independent. Riemann's Formula

We say that integrals of the first kind $I_1,\dots,I_q$ are *linearly independent* if there do not exist constants $\lambda_1,\dots,\lambda_q,\lambda_{q+1}$ that are all nonzero such that

$$\lambda_1 I_1 + {}_2I_2 + \dots + \lambda_q I_q + \lambda_{q+1} \equiv 0 \quad .$$

If $Q_1,\dots,Q_q$ are the corresponding adjoints to these integrals, then we must not have

$$\lambda_1 Q_1 + \lambda_2 Q_2 + \dots + \lambda_q Q_q \equiv 0 \quad .$$

As the polynomials Q corresponding to the adjoints depend linearly on at least p homogeneous parameters, we see that under the conditions of the theorem in no. 22: *there exists at least* p *linearly independent integrals of the first kind. We are going to show that there are only* p *of them.*

RIEMANN'S FORMULA

In the ordinary plane of z, let $f(z)$ be a holomorphic function in a domain D that is bounded by simple rectifiable closed curves C, and holomorphic on C. If we present the real part and the coefficient of i, by setting

$$f(z) = X(x,y) + iY(x,y) \ , \qquad x = \mathfrak{R}z \ , \quad y = \mathfrak{I}z \ ,$$

then we can apply Riemann's formula (I, 142)

$$\int_{C^+} P\,dx + Q\,dy = \iint_D \left(\frac{\partial Q}{\partial x} - \frac{\partial P}{\partial y}\right) dx\,dy \quad ,$$

by taking $P = X\dfrac{\partial Y}{\partial x}$, $Q = X\dfrac{\partial Y}{\partial y}$, which gives

$$\int_{C^+} X\,dY = \iint_D \left[\frac{\partial X}{\partial x}\frac{\partial Y}{\partial y} - \frac{\partial X}{\partial x}\frac{\partial Y}{\partial y}\right] dx\,dy \quad . \tag{12}$$

If we take account of the conditions of monogeneity (I, 161), we see that the bracket is equal to

$$\left(\frac{\partial X}{\partial x}\right)^2 + \left(\frac{\partial X}{\partial y}\right)^2 = |f'(z)|^2$$

(We see that the formula represents the area of the transformed domain of D by the conformal transformation $Z = f(z)$, a transformation which in general is not *simple*.) It follows that the expression (12) is positive:

$$\int_{C^+} X dY > 0 \quad ,$$

provided $f(z)$ is not a constant; one passes along the curves constituting C by leaving the domain to the left. If the function $f(z)$ is holomorphic at infinity, we could apply formula (12) to a domain outside a simple closed curve C (outside of which and on which $f(z)$ is holomorphic) and inside a circle C' of large radius. We shall have

$$\int_{C^-} X dY + \int_{C^+} X dY > k > 0 \quad , \tag{13}$$

where k is the value of the integral of the second number of (12) taken in a region of D. If the radius of C' increases indefinitely, then equation (13) exists and the second integral of the left hand side tends towards zero. Because X is bounded and, if we put $x = r \cos\theta$, $y = r \sin\theta$, dY is equal to $(-Y' \sin\theta + Y' \cos\theta) r d\theta$ and the coefficient of $d\theta$ tends towards zero as $1/r$ when $r \to \infty$ since $f'(z)z^2$ has a finite limit when $r \to \infty$. (This result is indeed intuitive when seen from the above point of view: the conformal representation and area.) We have a similar result if the point at infinity is a boundary point of D. *It follows that, in all cases,*

$$\int_{C^+} X dY > 0$$

if C *is described with the area to its left.*

This formula remains true on the Riemann surface of no. 20: *if the function* $f(z,u)$ *is holomorphic at each point of the simply connected domain bounded by the contour* K, *including infinity, it is uniform, and we have*

$$\int_{K^+} X dY > 0 \quad , \qquad f(z,u) = X + iY \quad .$$

This can be seen by decomposing the domain into a finite number of pieces, each situated on a single sheet, by using the lines of passage and applying the above formula to each of them.

Let us take for $f(z,u)$, an integral I(M) of the first kind. When K is described, each part of K is covered twice, in the opposite sense. On both sides of E, the value of I(M) is the same, the regions of the integral relative to these two sides of E yield opposing values which disappear. If we take the Lüroth example, the circuit taken on K is made in the sense of the arrow (fig. 24).

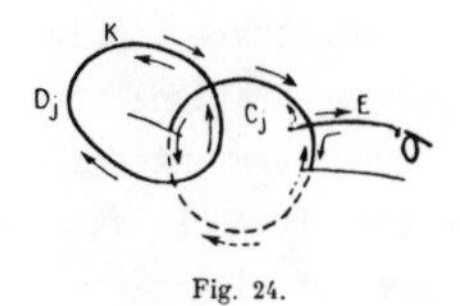

Fig. 24.

Let us call c_j and d_j the integrals of Q/P_u' taken on C in the sense of the first circuit on C_j starting from an original point taken on E, and on D_j in the sense of the first circuit made on D_j. Let us set

$$d_j = d_j' + id_j'' , \quad c_j = c_j' + ic_j'' , \quad d'_j = \Re d_j , \quad c_j' = \Re c_j \quad .$$

At two points on both sides of C_j, the values of the integral are I(M) and I(M) + d_j and at two points of D_j we have I(M) and I(M) - c_j. The corresponding regions of (14) are therefore

$$- d_j' \int_{C_j} d(\Im I(M)) = - c_j''d_j' , \qquad c_j' \int_{D_j} d(\Im I(M)) = c_j'd_j'' \quad ,$$

i.e. that these two contours provide the contribution

$$c_j'd_j'' - c_j''d_j' \quad .$$

We thus obtain RIEMANN's FORMULA

$$\sum_1^p (c_j'd_j'' - c_j''d_j') > 0 \quad .$$

CONSEQUENCES

We see that it is impossible for the real parts c_j', d_j' of the $2p$ periods to be all zero, as is the case for the coefficients c_j'' as is the case in the $2p$ periods, all the periods relative to the D_j and likewise all the d_j.

From this it follows that there could no longer exist p integrals of the first kind which are independent. For if there were $p+1$ of them, I_k say, and if $c(k,j)$ were the periods of I_k relative to the D_j, then the integral $\sum I\lambda_k$ where the λ_k, $k = 1,\ldots,p+1$ are constants, would have the numbers $\sum c(k,j)\lambda_k$ as the periods relative to the D_j. The p equations obtained by equating to zero these p periods would admit a system of solutions that are all non-zero: $\lambda_1,\ldots,\lambda_{p+1}$; there would exist an integral of the first kind all of whose periods c would be zero.

It is clear that if some integrals are linearly independent, then the integrals corresponding to them in a birational transformation are also independent (for, if they were not, the given integrals which are equal to them, would no longer be so). We see then that by taking account of the two results obtained and by the possibility of reducing matters to the case of multiple points with distinct tangents by Nöther's theorem, that, in all cases:

Theorem: *The number of abelian integrals of the first kind associated to a curve of genus p, is equal to p.*

NORMAL INTEGRALS

By taking homogeneous linear combinations of p independent integrals, we could try to form them from others whose periods relative to the C_j or to the D_j have given values. In particular, there exists integrals all of whose periods d_j are zero, save one which is equal to 1. We thus obtain p integrals, which are independent and which are called *normal integrals*.

24 - Integrals of the second kind and integrals of the third kind

By stating the same hypotheses as in theorem no. 22, we see that given a polynomial of any degree $Q(z,u)$, for which the abelian integral taken on $Q(z,u)/P_u'(z,u)$ remains finite at the singular points of the curve Γ, $P(z,u) = 0$, it is necessary and sufficient that the curve $Q(z,u) = 0$ admit each multiple point of order q of Γ as a multiple point of order $q-1$. We then say that the curve $Q(z,u) = 0$ is *an adjoint curve*, and the polynomial $Q(z,u)$ is an *adjoint polynomial.*

The adjoints of order $m-2$ depend linearly on $p+m-1$ homogeneous parameters. If a point $M_1(z_1,u_1)$ is taken on Γ, then the tangent to Γ at M cuts Γ at $m-2$ points there pass adjoints of order $m-2$ depending on at least $p+1$ parameters. Amongst these curves are found the curves $Q_1(z,u) = 0$, where Q_1 does not reduce to the product of $Q(z,u)$, an adjoint of order $m-3$, and the first member $au+bz+c$ of the tangent equation at M_1, since this product only depends on p parameters. The integral

$$J(M) = \int_{M_o}^{M} \frac{Q_1(z,u)}{(au+bz+c)P_z'(z,u)}\,dz$$

is not of the first kind. But it can only have a singularity at the point M_1. At this point M_1, distinct from the singular points, $au+bz+c$ is equal to zero along with its derivative with respect to z, P_u' is non-zero. We have a pole of order 2 of the rational fraction to integrate and as the sum of the residues of this fraction must be zero, in which z_1 is the only pole which could render one non-zero, then the residue at the point z_1 is zero and J(M) is *of the second kind*. This reasoning assumes that z_1 is not a ramification point. The result stands, in this case, as can be proved. J(M) has a simple pole at the point M_1.

By multiplying Q_1 by a constant, we could arrange it such that the residue of J(M) at the point M_1 is equal to 1. Let us assume this is the case. If $Q_2(z,u)$ is another polynomial of the same kind, we could choose the constant λ such that $Q_2 - \lambda Q_1$ is zero at the point M_1, hence contains a factor $au+bz+c$; the integral taken on $Q_2 - Q_1$ is then of the first kind. It follows that the integral taken on $Q_2(z,u)$ is equal to $\lambda J(M) + \sum \lambda_k I_k(M)$, where the $I_k(M)$ are p distinct integrals of the first kind. Every integral $J(M) + \sum \lambda_k I_k(M)$ is of the second kind; we can determine that λ_k such that the periods relative to the C_j are zero. (If the $I_k(M)$ are normal integrals, then it suffices to take for the λ_k the product, -1 times the period of J(M) relative to C_k). The periods of this integral, known as *the normal integral of the second kind relative to the point* M_1, are then $d_j = 0$, $c_j = e_j$, $j = 1,2,\ldots,p$. If the integral of $J(M)dI_k(M)$ is calculated on K, where J(M) is the normal integral in question, then we obtain

$$e_k = 2i\pi \frac{Q_1(z_1,u_1)}{P_u'(z_1,u_1)} \quad ,$$

where $Q_k(z,u)$ is the adjoint polynomial of order $m - 3$ which provides the normal integral $I_k(M)$.

LINEARLY INDEPENDENT INTEGRALS OF THE FIRST AND SECOND KIND

If p points of Γ are taken, where p is the genus, and if $J_1, J_2, \ldots, J_p$ are normal integrals of the second kind relative to these points, $I_1, \ldots, I_p$ the p normal integrals of the first kind, then the sum

$$\lambda_1 J_1 + \ldots + \lambda_p J_p + \mu_1 I_1 + \ldots + \mu_p I_p \quad ,$$

where the λ and μ are constants that are not all zero, is not a rational fraction of z and u. For if this were the case, then the periods would be zero. Crossing C_j, we would have $\mu_j = 0$; all the μ_j would be zero. The D_j would then give

$$\sum_1^p \lambda_k \frac{Q_j(z_k,u_k)}{P_u'(z_k,u_k)} = 0 \ , \qquad j = 1,2,\ldots,$$

where the (z_k,u_k) are given points. This system of homogeneous equations in $\lambda_k/P_u'(z_k,u_k)$ having a non-zero solution, the determinant

$$\| Q_j(z_1,u_1)\ Q_j(z_2,u_2)\ \dots\ Q_j(z_p,u_p) \| \tag{16}$$

would be zero. This would imply a relation between the columns, which would indicate that an adjoint of order $m - 3$ would pass the p points given, which in general is not the case, since the p normal integrals of the first kind are linearly independent.

Thus, in general, there do not exist constants λ and μ such that the sum in (15) is a rational fraction in z and u. We say that the integrals $I_1,\dots,I_p$, $J_1,\dots,J_p$ *are linearly independent*. The determinant of the $2p$ periods of these integrals is equal, up to a factor, to the determinant (16); it is non-zero. If $J(M)$ is an arbitrary integral of the second kind, then it can be added to expression (15) and the λ and μ are determined in order that the periods of the integral thus obtained are zero; it is then a rational fraction. *Every integral* $J(M)$ *of the second kind is hence the sum of an expression* (15), *where* J *and* I *are linearly independent,* λ, μ *are constants, and a rational fraction in* z *and* u.

Conversely, it can be shown that a rational fraction in z and u could be expressed in terms of a sum of integrals of the second kind.

INTEGRALS OF THE THIRD KIND

For such an integral, the sum of the polar periods is zero; there are at least two logarithmic points. If we take two points $M'(z',u')$ and $M''(z'',u'')$ on Γ $(P(z,u) = 0)$, the line $au + ba + c = 0$ which joins these points cuts Γ at $m - 2$ other points which determine an adjoint of order $m - 2$, $Q(z,u) = 0$. The integral

$$\int_{M_o}^{M} \frac{Q(z,u)}{(au + bz + c)P_u'(z,u)}\, dz$$

is an integral of the third kind for which M' and M'' are the only singular points provided $Q(z,u) = 0$ does not pass through M' and M''. Proceeding as before, we will define *a normal integral relative to* M' *and* M'': it will admit these points as logarithmic points with periods $2i\pi$ and $-2i\pi$ respectively and its periods d_j, $j = 1,2,\dots,p$ will be zero. The periods c_j are then given by

$$c_j = 2i\pi \int_{M'}^{M''} dI_j(M) \quad ,$$

where I_j is the corresponding normal integral of the first kind. Let $\Lambda(M',M'')$ be such a normal integral of the third kind. An integral of the third kind given a priori by critical logarithmic points M_k, $k = 1,2,\dots,q$, where M_k gives the polar period $2i\pi a_k$, $(a_1 + a_2 + \dots + a_k) = 0$, can be written as

$$a_1 \Lambda(M_1,M_2) + (a_2 + a_1) \Lambda(M_2,M_3) + \dots$$
$$+ (a_{k-1} + \dots + a_1) \Lambda(M_{k-1},M) + J \quad ,$$

where J is an integral of the second kind.

It follows that under these conditions, *every abelian integral is the sum of normal integrals of the third, second or first kind, and a rational fraction in* z *and* u. In terms of birational transformation, we could deal with the case left aside.

III. ABEL'S THEOREM - APPLICATIONS

25 - Differential form

Let us consider the curve Γ, $P(z,u) = 0$, a rational fraction of z and u, $R(z,u)$ and a family of algebraic curves depending on parameters, for example, two parameters a and b; let

$$\Theta(z,u,a,b) = 0 \tag{17}$$

where $\Theta(z,u,a,b)$ is a polynomial in z, u, a, b. Everything here extends to the case of q parameters. The curve Γ and the curve Λ with equation (17) have μ common points $M_k(z_k,u_k)$: the z_k are the roots of the resultant of P and Θ equated to zero, $\Phi(z,a,b) = 0$. Generally, to each z_k there will correspond a single value of u,

$$u_k = \Psi(z_k,a,b) \tag{18}$$

where Ψ is a rational fraction of a_k, a, b. Let us consider the sum

$$S(a,b) = \sum_1^{\mu} \int_{M_o}^{M_k} R(z,u)\, dz$$

where M_o is a fixed point Γ ; this sum is defined to within a sum of multiples of the periods. When a varies in the neighbourhood of one of its values, z_k is a function of a, holomorphic in general, as is u_k, hence also $S(a,b)$ which therefore admits a continuous derivative with respect to a. We have

$$\frac{\partial S}{\partial a} = \sum_1^{\mu} R(z_k,u_k) \frac{\partial z_k}{\partial a}$$

and following the relation $\Phi(z,a,b) = 0$,

$$\frac{\partial S}{\partial a} = - \sum_1^{\mu} \left[R(a_k, u_k) \frac{\partial \Phi}{\partial a} / \frac{\partial \Phi}{\partial z} (a_k, a, b) \right] \quad .$$

If we replace u_k by its value in (18), we obtain a rational fraction of z_k, a, b, which is a symmetric function of the z_k. This is therefore a rational fraction of a and b since $\Phi(z,a,b)$ is a polynomial in a and b. We would have the same result for the partial derivative with respect to b. The result could be written as a differential form. We have

$$dS = \Omega_1(a,b)da + \Omega_2(a,b)db \quad , \tag{19}$$

where Ω_1 and Ω_2 are rational functions. The calculation was made via certain hypotheses on the values of a and b, but the simplicity of the result obtained ascertained that is general, which we shall admit.

The equation in (19) constitutes *Abel's theorem in terms of differential forms. If there is only one parameter* a, we have $dS = \Omega_1(a)da$, where $\Omega_1(a)$ is a rational fraction, integrating yields

$$S(a) = R(a) + \sum A \log (a - \zeta) \quad ,$$

where $R(a)$ is a rational fraction and the A and ζ are numerical constants. When there are several parameters, e.g., two a and b say, direct integration of the total differential will introduce a function of a similar to that of above, but whose coefficients would be algebraic functions of b, plus a function of b such that the derivative of this sum, with respect to b, is $\Omega_2(a,b)$. We shall not undertake here to show that $S(a,b)$ is indeed composed of a similar type as in the case of one variable; this will be a consequence of the method to be outlined in no. 26.

THE CASE OF INTEGRALS OF THE FIRST KIND

If $S(a,b)$ *is an integral of the first kind*, then the functions $\Omega_1(a,b)$ and $\Omega_2(a,b)$ could not become infinite, since their integration would then provide an infinite integral; they will always be bounded, hence constants for a constant. These are therefore functions of a alone that are always bounded, hence constants and these constants are zero, otherwise $S(a,b)$ would become infinite. *In this case*, $S(a,b)$ *is a constant.*

26 - <u>The integral form of the theorem</u>

Firstly, let us consider an integral of the first kind $I(M)$ and a rational fraction $R(M)$. We assume that M does not pass over the contour K. At two opposite points on the contour K, the values of $I(A)$ differ by a period, hence the integral

$$\int_K I(M)\, d \log R(M)$$

is equal to the sum of the quantities of the form $\pm c_j \, [\log R(M)]_{D_j}$ and of the form $\pm d_j \, [\log R(M)]_{C_j}$. The brackets which represent the variation of the logarithm of a uniform function, are equal to a multiple of $2i\pi$ and we have

$$\frac{1}{2i\pi}\int_K I(M)\, d\log R(M) = \sum(m_j c_j + n_j d_j) \tag{20}$$

where m_j and n_j are integers. We have assumed that K had been deformed such that it does not contain poles of R(M). Following the generalized Cauchy theorem, the integral of the first member of (20) (taken in the direct sense), that carries a uniform function, a product of I(M) and the rational fraction

$$R_1' = \frac{\dfrac{\partial R}{\partial z} + \dfrac{\partial R}{\partial u}u'}{R}$$

is equal to the sum of the residues. We have seen in no. 19 that R_1 only has simple poles corresponding to the zeros and poles of R, where the residue is the order of the zero or the order of the pole with sign changed, and poles with zero residues. If the zeros are denoted by M_k and the poles by M_k', we have therefore

$$\sum I(M_k) - \sum I(M_k') = \sum(m_j c_h + n_j d_j) \quad .$$

If we take R(M) under the guise of

$$\frac{T(z,u)}{V(z,u)} \tag{21}$$

where T and V are polynomials of degrees equal to n and if m is of order of the given curve, we have, to within a sum of multiples of the periods,

$$\sum_1^{mn} I(M_k) = \sum_1^{mn} I(M_k') \quad .$$

Hence:

To within a sum of multiples of the periods, the sum of the values of an integral of the first kind at the points common to Γ *and to a curve* γ*, is equal to the sum of the values of this integral at the points common to* Γ *and to another curve* γ' *with the same degree* γ*. If* γ' *is fixed, and if* γ *varies continuously, and if* I(M) *is calculated on the curves described by the common points, then the sum of the integrals stays constant.*

Effectively, this sum varies continuously when $\sum(m_j c_j + n_j d_j)$ has only a countable set of values.

THE CASE OF INTEGRALS OF THE SECOND KIND

If the integral I(M) is of the second kind, admitting a single simple pole (α,β) and this point is distinct from the ramification points, it is at finite distance and is distinct from the poles and zeros of $R(z,u)$, then the above reasoning always applies, but it will be necessary to add the residue arising from the point (α,β).

If A is the residue of the integral I(M) at the point (α,β), we have, with the same notation as above,

$$\sum I(M_k) - \sum I(M_k') = - A\, R_1(\alpha,\beta) + \sum (m_j c_j + n_j d_j) \quad .$$

If we take $R(z,u)$ in the form (21), we see that the sum

$$\sum_{1}^{mn} I(M) + A\, T_1(\alpha,\beta)$$

will remain constant for all curves $T(z,u) = 0$ of the same degree, where T_1 is deduced from T as R_1 is from R always to within the periods. When the point (α,β) no longer satisfies the conditions imposed. T_1 is replaced by a more complicated function. If there are several poles for I(M), we will have a sum of terms similar to $A\, T_1(\alpha,\beta)$; the A's only depend on I(M) along with the (α,β).

If we take the curves

$$T(z,u,a,b) = 0$$

where T is a polynomial in z, u, a, b, then the values of the quantities $T_1(\alpha,\beta)$ will be rational at a, b. If we take a general integral of the second kind, it will be the sum of normal integrals of the two first kinds and a rational fraction. At the point M_k, this rational fraction takes values whose sum is a symmetric function of the z_k relative to the common points. These z_k are given by the resultant of $P(z,u)$ and $T(z,u,a,b)$ which is a polynomial in z, a, b. The symmetric function so envisaged is therefore a rational fraction of a and b. Consequently, we see that:

The sum of values of an integral of the second kind at the points common to Γ *and a curve depending rationally on parameters* $a,b,\ldots$ *is equal to a rational fraction of these parameters, where it is understood that it is calculated along curves described along the common points when the parameters vary continuously.*

We have thus recovered the statement that ascertained the differential form of the theorem.

THE CASE OF INTEGRALS OF THE THIRD KIND

In this case, I(M) is no longer uniform. Following the result of no. 24, it will suffice to study an integral having only two (logarithmic) singularities at the two points $M'(\alpha,\beta)$ and $M''(\gamma,\delta)$. We are allowed to assume that the contour K does not pass through these points. We may join one to the other by a curve K' not cutting K and join K' to E by a curve K''. In the simply connected domain having K, K', K'' as boundary, $I(M)$ is uniform (fig. 25; in this figure read K'' for K' along the vertical line).

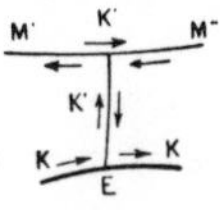

Fig. 25.

The polar periods at M' and M'' being opposite, $I(M)$ has the same value at two points, one opposite to the other on K'', whilst for two points which are face to face on K', the difference of the values is $2i\pi$. We will have to add

$$- \log \frac{R(\alpha,\beta)}{R(\gamma,\delta)}$$

to the expression obtained in the case of the integral of the first kind. Now, by taking $R(M)$ under the form (21), we see that

$$\sum_1^{mn} I(M_k) - \log \frac{T(\alpha,\beta)}{T(\gamma,\delta)}$$

will be preserved to within a sum of multiples of the periods, when the curve $T = 0$ of degree n will be replaced by another curve of degree n. If we assume that $T(z,u,a,b)$ is a polynomial in z, u, a, b, then $T(\alpha,\beta,a,b)/T(\gamma,\delta,a,b)$ will be a rational fraction in a and b. Hence, *the sum of the values of an integral of the third kind has only two singularities at the points common to* Γ *and to a curve depending rationally on parameters* $a,b,\ldots,$ *is the logarithm of a rational fraction of these parameters, where it is understood that when the parameters vary continuously, the integral is calculated along the curves described by the common points.*

Every abelian integral is a sum of normal integrals of the three kinds and a rational fraction. By proceeding as in the case for the integrals of the second kind, we can therefore state a general theorem:

<u>Theorem</u>: *Given an abelian integral* $I(M)$ *associated to* Γ *and a curve depending rationally on parameters* $a,b,\ldots,$ *then the sum* $\sum_1^{mn} I(M_k)$ *extended to the* mn *common points to* Γ *and* Λ *is equal to*

$$R_o(a,b,\ldots) + \sum_1 A_q \log R_q(a,b,\ldots) \quad ,$$

where the R_q *are rational fractions of* $a,b,\ldots$ *and the* A_q *are constants. The equality occurs to within a sum of multiples of the periods, a sum which remains constant when the parameters vary, if the integrals* $I(M_k)$ *are calculated along the curves described by the points* M_k.

A PRACTICAL FORMULA

To apply the theorem, it is convenient to actually calculate the residue of the integral in (20) corresponding to the poles of $I(M)$ and in the case of an integral of the third kind, the logarithmic terms. To do this, we can assume that on the Riemann surface bounded by K, it is taken along a simple contour C containing in its interior all the singularities of $I(M)$ and leaving outside, the poles and zeros of $R(M)$. Following Cauchy's theorem, the first member of (20) is equal to

$$\frac{1}{2i\pi} \int_C I(M)\, d \log R(M) + \sum I(M_k) - \sum I(M_k')$$

since $I(M)$ had been uniformised between C and K; likewise if it is of the third kind, considering the property of the sum of the polar periods. We have, on the other hand, on integrating by parts

$$\int_C I(M)\, d \log R(M) = [I(M) \log R(M)]_C - \int_C \log R(M)\, d\, I(M) \quad .$$

The part wholly integrated is zero, since $R(M)$ does not have any zeros or poles in C and consequently, $\log R(M)$ is holomorphic. To within $2i\pi$, the integral remaining is equal to the sum of the residues of the function that was to be integrated.

Finally, we shall have

$$\sum I(M_k) - \sum I(M_k') = \sum (m_j c_j + n_j d_j) + \sum 2\pi i\, \rho_k q_k + \sum \text{residues of } \frac{d\, I(M)}{dz} \log R(M) \quad , \tag{22}$$

where the residues are relative to the poles of the rational fraction, a factor of the logarithm.

27 - Geometric applications. Angular properties

Abel's theorem was applied to the study of the groups of common points to a given algebraic curve and to other algebraic curves subject to certain conditions. As a reference we suggest the book of Appell and Goursat already cited. We confine ourselves here to some given theorems obtained with the help of elementary abelian integrals having a geometric significance, and due mainly to G. Humbert and Laguerre. Here we shall take x and y the cartesian co-ordinates of a point on the curve Γ in question. If $M(x,y)$ is a point of the plane, then the angle Θ of OM with Ox is defined in each case by Laguerre's formula:

$$\Theta = \text{arc } tg \frac{y}{x} = \frac{1}{2i} \log \frac{x+iy}{x-iy}$$

(where x and y are real or complex). If M describes an algebraic curve Γ, $P(x,y) = 0$, we have on Γ,

$$\frac{d\Theta}{d\Theta} = - \frac{xP_z' + yP_y'}{(x^2 + y^2)P_y'}$$

where Θ is an integral of the third kind associated with Γ; its singularities are logarithmic points, being the intersection of Γ and isotropic lines emanating from the origin. The only period is π. Let us apply formula (22) by taking for R(M) a quotient $T(x,y)/V(x,y)$, where T and V are polynomials. If Θ denotes the values of Θ at the points common to Γ and C, $T(x,y) = 0$, and Θ' the values of Θ at the points common to Γ and C', $V(x,y) = 0$, we shall have

$$\sum\Theta - \sum\Theta' = kv - \sum \text{residues of} \left[\frac{xP_x' + yP_y'}{(x^2+y^2)P_y'} \log \frac{T(x,y)}{V(x,y)}\right] \quad . \quad (23)$$

Where k is an integer and the residues are relative to the poles of the factor of the logarithms. Let us assume that amongst the curves of the pencil $T - \lambda V = 0$, where λ is an arbitrary constant, one of them passes through all the points of intersection of Γ with the isotropic lines from the point 0. If λ_o is the value of λ corresponding to this curve, we have $T/V = \lambda_o$ at these points and, as the sum of the residues of a rational fraction is zero, the $\sum$ of the second member of (23) is zero. Following Laguerre, if we call the sum of the angles of these lines with a given direction, e.g., Ox, *the orientation of a pencil of lines,* a sum which is defined to within a multiple of π, we obtain the following theorem of Humbert:

Theorem: *If an algebraic curve Γ is cut by the curves C of a linear pencil containing a curve C_o which passes through the common points of Γ and the isotropic lines emanating from 0, then the orientation of the lines joining the point 0 to the common points of Γ and C, is constant.*

In particular, we see that the orientation of the system of the rays joining the given centre A of a circle to a point where an algebraic curve cuts this circle is independent of the rays of this circle. We also see that: the orientation of the system of the lines joining a given point A to the points where an algebraic curve Γ cuts a circle C is the same as that of the pencil of lines joining A to the common points of Γ and the radical axis of C and A, and the points at infinity of Γ.

TANGENTIAL PROPERTIES

If $\Pi(u,v) = 0$ is the tangential equation of Γ and Θ the angle of the tangent $ux + vy + 1 = 0$ with $0x$, we have

$$\Theta = -\text{arc } tg - \qquad , \qquad d\Theta = -\frac{v\,du + u\,dv}{u^2 + v^2}$$

$$\Pi_u'\,du + \Pi_v'\,dv = 0\,, \qquad \frac{d\Theta}{du} = -\frac{u\,\Pi_u' + v\,\Pi_v''}{(u^2 + v^2)\Pi_v'} \quad .$$

Θ is indeed an abelian integral of the third kind associated to the curve $\Pi(u,v) = 0$. The singularities correspond here to the tangents taken from the cyclic points to the curve and we obtain this result correlative to that of above:

The two systems of tangents common to C *and* Γ *at* C′ *at* Γ*, have the same orientation if, amongst the curves of the linear tangential pencil defined by* C *and* C′*, there is one that is tangent to the tangents taken to* Γ *by the cyclic points.*

This amounts to saying that the curve of the pencil in question admits as foci, all the foci of Γ. In particular, we obtain this theorem of Laguerre: if two curves C and C' are homofocal, then the system of tangents common to C to another curve C" and the system of tangents common to C' and to C" have the same orientation. We see this by letting C and then C' play the role of Γ. In the limiting case, it contains a theorem of Siebeck: If tangents are taken from a point M to a curve and if M is joined to the real foci as well, then the two pencils have the same orientation. Similarly we obtain this theorem of Laguerre: the system of the common tangents to two algebraic curves have the same orientation as the system of tangents taken to one of these curves by the real foci of the other.

28 - Properties relative to distance and areas

Let us take again the curve Γ, $P(x,y) = 0$, and let us consider the integral $x = \int dx$. For the points of intersection of Γ with C, $T(x,y) = 0$ and C', $V(x,y) = 0$, we shall obtain, since the periods are zero,

$$\sum x - \sum x' = \sum \text{residues of } \left[\log \frac{T(x,y)}{V(x,y)}\right]$$

where the sum of the second member is taken to the points of infinity Γ.

The second member is zero if C and C' have the same asymptotes distinct from those of Γ, or if, amongst the curves $T + \lambda V = 0$, one of them has the same asymptotes as Γ. If amongst the curves $T + \lambda V = 0$, one has the asymptotic directions of Γ, then the second member of (24) becomes linear in $1/\lambda - \gamma_o$ if T is replaced by $T + \lambda V$. By noting that we can reverse the roles of the axes, we obtain the following propositions, tenable with the above conditions (distinct asymptotes).

The centres of the means of the distances of the points common to two algebraic curves remains fixed, if one of these curves is displaced whilst remaining asymptotic to itself (Liouville). (We could take a curve and its asymptotes and cut through a line, a circle, etc.)

The centre of the mean distances of the points common to a curve Γ and to a curve of a linear pencil remains fixed if, amongst the curves of the pencil, one has the asymptotes of Γ. The centre of the mean distances of the points common to a curve Γ and to a variable curve of a linear pencil, describes a line if, amongst the curves of the linear pencil, one has the asymptotic directions of Γ.

PROPERTIES RELATIVE TO AREAS

We shall consider here the integral

$$S = \int x\,dy - y\,dx \quad ,$$

which gives twice the area of a sector when the elements are real. If the points at infinity of Γ are distinct, we have about one of these points

$$y = \alpha x + \beta + \frac{\gamma}{x} + \frac{\delta}{x^2} + \dots , \qquad xy' - y = -\beta - \frac{2\gamma}{x} - \frac{3\delta}{x^2} + \dots$$

and the integral is, in general, of the third kind $(\gamma \neq 0)$. Here we shall have:

$$\sum(S - S') = \sum \text{residues of } \left[\left(-\beta - \frac{2\gamma}{x} - \frac{3\delta}{x^2} - \dots\right) \log \frac{T(x,y)}{V(x,y)}\right].$$

For example, the residue will be zero if the term providing the logarithm is of the order of $1/x^2$, hence if C and C' have the same asymptotes distinct from those of Γ.

We thus obtain a theorem of G. Humbert:

Theorem: *If a curve* Γ *with distinct asymptotic directions is cut by a variable curve* C *whose asymptotes are fixed, then the radial vectors joining a point* 0 *to the common points of* Γ *and* C *span the areas whose sum is zero.*

29 - Arcs of the directions curves

The arc s of an algebraic curve Γ, $P(x,y) = 0$, is not in general given by an abelian integral since

$$\frac{ds}{dx} = \sqrt{1 + y'^2} = \pm \frac{\sqrt{P_z'^2 + P_y'^2}}{P_y'} .$$

Laguerre named the direction curves as those algebraic curves for which this expression is a rational function of x and y, $Q(x,y)/P_y'(x,y)$ say. Then the arc is given by an abelian integral associated to Γ and $Q(x,y)$ is an adjoint polynomial. These curves have been studied by G. Humbert.

We can easily verify that the inversion of a direction curve is a direction curve. *If an algebraic surface* $F(x,y,z) = 0$ *is cut by a plane at a constant angle that is neither zero nor right-angled, then the section is a direction curve.*

We can assume that the section plane is $z = 0$. We have

$$F(x,y,z) \equiv P(x,y)\Phi(x,y,z) + z\Psi(x,y,z) ,$$

where $P(x,y) = 0$ is the section for $z = 0$. By hypothesis, for $P(x,y) = 0$, we have

$$F_x'^2 + F_y'^2 + F_z'^2 = k^2 F_z'^2 ,$$

where k is a constant.

Now along $P(x,y) = 0$, we have

$$F_x' = P_x' \, \Phi(x,y,0) , \quad F_y' = P_y' \, \Phi(x,y,0) , \quad F_z' = \Psi(x,y,0)$$

hence

$$\left[P_z'^2 + P_y'^2\right] \Phi^2(x,y,0) \equiv (k^2 - 1) \; \Psi^2(x,y,0) .$$

CURVES OF DEGREE 3

Humbert has shown that the direction curves of degree three may be reduced to the form of the two curves

$$x^2 - 3xy^2 = 1 , \quad (x + r)^3 - 27(x^2 + y^2) = 0 .$$

The first is of genus 1 and has as its polar equation $r^3 \cos 3\Theta = 1$. The second is of genus 0 and its polar equation (with a pole at 0) is

$$r^{1/3} \cos \frac{C + \pi}{3} = 1 .$$

These curves are contained in the family of curves with polar equation

$$r^2 \cos n\Theta = 1 ,$$

for which

$$ds = r^{n+1}\, d\Theta = r^{n+1}\, d \text{ arc } tg \frac{y}{x} \quad ;$$

we have a direction curve if n is an odd integer. Halphen has shown that this is still the case when n is an irreducible fraction p/q, p and q odd.

If we apply Abel's theorem to the arc s of a direction curve, then the integral

$$s = \int \sqrt{1 + y'^2}\; dx$$

only has singularities at infinity. In order that a simple branch does not pass through a cyclic point, we have about the point at infinity

$$y = \alpha x + \beta + \frac{\gamma}{x} + \ldots \quad , \quad y' = \alpha - \frac{\gamma}{x^2} + \ldots$$

and the residue of ds/dx is zero. When this is the case for all infinite branches, s is of the first or second kind. We will apply Abel's theorem by calculating the residues of

$$\sqrt{1 + y'^2}\; \log \frac{T}{V} = \left(\sqrt{1 + \alpha^2} - \frac{\gamma'}{x^2} + \ldots\right) \log \frac{T}{V} \quad .$$

We see that if the curves C and C' have the same asymptotes, distinct from those of Γ, we see that the residue is zero, *the sum of the arcs of* Γ *taken over the points of intersection with a curve* C *will remain constant if* C *remains asymptotic to itself.* We may take concentric circles for the curve C.

When the integral is of the first kind, the theorem relating to these integrals applies, *the sum of the arcs of* Γ *taken over the points where* Γ *cuts an algebraic curve* C *remains constant when* C *is displaced arbitrarily. This will apply to the curve* $r^3 = \cos 3\Theta$.

IV. PROBLEMS OF INVERSION AND UNIFORMISATION

30 - On the Jacobi inversion problem and abelian functions

We know that the co-ordinates of a curve Γ of genus 1 could be expressed in terms of the functions pt and $p't$; these are rational functions of pt and $p't$, uniform functions of t. The function pt being the inverse function of the elliptic integral, we can see, via a birational transformation (no. 21), that the co-ordinates of a point of a curve Γ of genus 1 are expressed in terms of uniform functions of abelian integral associated to the curve Γ, where this integral is of the first kind. For a curve of genus 0, known to be unicursal (no. 16), the co-ordinates of a point

(z,u) of the curve are rational functions of a parameter t, and t is a rational function of z and u; $t = R(z,u)$. We could therefore consider t as being equal to the abelian integral $R(z,u)$ associated to Γ; here this integral is of the second kind and, moreover, could only have a simple pole on S.

Theorem: *The curves Γ of genus 0 or of genus 1 are the only ones for which the co-ordinates z and u can be expressed as uniform functions of a parameter which is an abelian integral associated to Γ.*

To prove this we could, if we wish, make a birational transformation. *A fortiori*, it is permissible to assume that the Riemann surface of Γ, S say, does not have any ramification points at infinity. If z and u are uniform functions of the integral

$$I = I(M) = \int_{M_o}^{M} R(z,u)\, dz \quad ,$$

then to a value of I, there must only correspond one point M; I must be univalent on S. We know (I, 193) that the derivative of a holomorphic function univalent in a domain is non-zero there. If $R(z,u)$ is holomorphic at a point M of S at a finite distance and *ordinary* on S, then I is holomorphic at z about this point, *hence at this point* $R(z,u) \neq 0$. *If* M *is a ramification point of order* r, with co-ordinates ζ, η, and if about this point $R(z) = B_\mu (z-\zeta)^{\mu/} + \ldots$, $\mu > -r$ we shall have

$$I = I(M) + \frac{B_\mu r}{\mu + r} (z-\zeta)^{(\mu+r)/r} + \ldots \quad .$$

By putting $z - \zeta = Z$, S is represented on a circle in the neighbourhood of M, and I must remain univalent, hence its derivative at Z cannot be zero for $Z = 0$, *we necessarily have* $\mu + r \leq 1$. If M tends to infinity, on one of the m sheets, *then the product* $z^2 R(z,u)$ *cannot tends towards zero,* otherwise we would have

$$R(z,u) = \frac{k}{z^3} + \ldots \, , \qquad I = I(\infty) - \frac{k}{2z^2} + \ldots$$

and by putting $z = 1/Z$, we could recover a function that is non-univalent.

Thus R(M) could only be zero at infinity, twice at most on each sheet, hence it has at most $2m$ zeros and every ramification point of order r is a pole of order $r - 1$ at least. The number of poles is at least $\sum(r-1)$, we therefore have $\sum(r-1) \leq 2$, and putting this into the Riemann formula giving the genus (no. 9), we have $p \leq 1$. The theorem is thus proved. Also we see that if $p = 1$, the only poles of $R(z,u)$ are ramification points of order $r - 1$, the zeros are the points at infinity, of order two; I is of the first kind and as there is only one for genus 1 to within a factor, then I is determined to within a factor. For $p = 0$, several cases are possible that might be studied.

Remark: We have also determined those conditions for which an abelian integral only takes the same value a finite number of times.

THE JACOBI PROBLEM

Let $I_1(M),\ldots,I_p(M)$ be p integrals distinct from the first kind corresponding to the same lower bound M_o. We consider the system of p equations

$$\sum_{k=1}^{p} I_j(M_j) = V_j \qquad j = 1,2,\ldots,p \quad ,$$

where the integrals which have the same upper bound are taken along the same path.

We propose to determine the upper bounds as functions of the V_j. We show that, in general, a system of values V_j only corresponds to a system of points M_k. Every rational symmetric function of the co-ordinates of the M_k is a uniform function of the V_j. They are these functions of p variables that are called *abelian functions*.

31 - On the Riemann inversion problem and uniformisation

To every algebraic function $P(z,u) = 0$, there corresponds a Riemann surface S composed of m overlapping sheets, points of ramification and lines of passage which could be described in various ways. The structure of the surface S is determined by the assignment of ramification points and say a system of lines of passage, a line K making S simply connected, or the same line K with the line E removed (no. 10), i.e., the lines C_j, D_j which leave S with a connectivity of order p. Let us consider this last case. The abelian integrals of the various kinds appearing as functions defined on S and having certain discontinuities or none at all (integrals of the first kind) and having certain periods relative to the curves $C_j D_j$. But we know that we might obtain integrals of the second kind without periods, hence rationals. In particular, u could be represented by such an integral.

Riemann made a direct study of the functions analogous to the integrals of the first two kinds, defined on a surface S given *a priori*, and showed that one could be found that provides a function $u(z)$ on the simple plane, with m branches, which is an algebraic function whose Riemann surface is S. Riemann's incomplete arguments were completed by Schwarz and Picard. The result is the following: *To a Riemann surface* S *given a priori, there corresponds a class of algebraic functions whose surface* S *is the Riemann surface.*

The problem of finding a parametric representation of a curve $P(z,u) = 0$, of genus greater than 1, in terms of uniform functions $z(t)$, $u(t)$, or *the problem of uniformisation of algebraic functions*, was solved by Poincaré in terms of the fuschian functions. The fuschian functions, as are the Schwarz functions (I, 222), are uniform and are defined in a fundamental domain and take the same value when the variable is subjected to the transformations of a group, which transforms the fundamental domain to other equivalent domains. These equivalent domains, adjacent to each other with overlapping, recover a circle for example. Two fuschian functions belonging to the same group of transformations are related algebraically. Conversely, to every algebraic curve $P(z,u) = 0$, we can associate a pair of fuschian functions satisfying this relation. If $z(t)$ and $u(t)$ are these functions, then the point $z(t)$, $u(t)$ will describe a Riemann surface S corresponding to $P(z,u) = 0$; as t varies in the fundamental domain, the function $z(t)$ will conformally represent the fundamental domain on S. From this last point of view, we see that the uniformisation problem is related to problems of conformal representation, but this conformal representation is not simple, the fundamental domain is represented by a surface which covers the plane several times.

Considered from this point of view, the Riemann inversion problem was generalised. Let us take a Riemann surface with an infinite or finite number of sheets, but of which a certain number could be bounded, which are related to each other about ramification points of finite or infinite order. If such a surface S is simply connected, we could make the conformal representation on a simple complete plane (including the point at infinity) or on the plane minus a point which can be assumed to be ignored at infinity, or on a circle. (In this general proposition, due to Poincaré, we assume that the correspondence occurs at the same points of finite order of ramification of S, where the surface is considered about such a point Z_o as extended on a simple circle by a transformation $Z = Z_o + \zeta^r$). According to whether one or another of these various cases are considered, we say that the surface S is of *elliptic* (applicable to the complete plane), *parabolid* (applicable to the punctured plane) or *hyperbolic type*. In the first case, the function $Z = f(z)$, which represents the plane S, is uniform everywhere and only has poles as singularities, it is a rational fraction; in the second case, it is a meromorphic (or holomorphic) function at every point at a finite distance. In the last case, it is a function that is meromorphic or holomorphic in a circle and non-extendable beyond.

32 - Bloch's theorem

We shall rely on this lemma which states the conditions for which a function holomorphic at a point and of non-zero derivative at this point, is uniform in a small circle about this point:

Lemma: *If* $f(z)$ *is holomorphic for* $|z| \leq 1$, $f(o) = 0$, $f'(o) = 1$ *and if* $|f'(z)| < M$ *for* $|z| < 1$, *then* $f(z)$ *takes once and only once in the circle* $|z| < 1/(1+M)$, *every value* Z *of modulus less than* $\frac{1}{2}(M+1)$.

As $|f'(o)| = 1$, we have $M \geq 1$. Following Rouche's theorem (I, 193), the number of roots of

$$Z - f(z) = Z - z + [Z - f(z)] = 0$$

for $|z| < \frac{1}{1+M}$ and $|Z| < \frac{1}{2(M+1)}$, will be the same as the number of roots of $Z - z = 0$, i.e., *one*, if it is shown that on the circumference $|z| = \frac{1}{1+M}$ we have $|z - f(z)| < |z - Z|$, hence *a fortiori*, if we have on this circumference

$$|z - f(z)| \leq \frac{1}{2(M+1)} \quad . \tag{25}$$

Now $|f'(z)| < M$, hence $|f'(z) - 1| < M + 1$ and since $f'(z) - 1$ is zero at the origin, we have following the Schwarz lemma (I, 184)

$$|f'(z) - 1| \leq (M + 1)\ |z|$$

hence on integrating from 0 to z along the segment $0, z$,

$$|f(z) - z| \leq (M + 1) \int_0^{|z|} r\, dr = (M+1)\,\frac{|z|^2}{2} \leq \frac{(M+1)}{2(M+1)^2}$$

which proves (25) and implies the conclusion.

Corollary: *If* $f(z)$ *is holomorphic for* $|z|R$, $f'(o) = 1$, *and* $|f'(z)| < M$ *for* $|z| < R$, *then the inverse function* $z = \phi(Z)$ *of* $Z = f(Z)$ *is holomorphic in a circle of radius at least* $\frac{R}{2(M+1)}$.

For if we set $z = \zeta(R - \varepsilon)$, then the function

$$F(\zeta) = \frac{f[(R-\varepsilon)\zeta] - f(o)}{R - \varepsilon}$$

satisfies the conditions of the lemma for $\varepsilon > 0$ small enough. The inverse function of $F(\zeta)$ is therefore holomorphic in this circle with centre the origin and radius $\frac{1}{2}(M+1)$. The function $\phi(Z)$, i.e., z is therefore holomorphic in the circle of centre $f(o)$ and of radius $(R-\varepsilon)/2(M+1)$, say in a circle of radius $R/2(M+1)$ since ε is arbitrary.

We could express the result by saying that *the Riemann surface described by* $Z = f(z)$ *when* $|z| < 1$ *contains on one of its sheets a disc of radius* $R/2(M+1)$ *at least* (this disc contains no ramification point in its interior. In short, we shall say that *on the Riemann surface described by* Z, *there is a disc with one sheet of radius* $R/2(M+1)$. This Riemann surface is the surface on which the inverse function of $Z = f(z)$ is uniform.

Bloch's Theorem: *If* $f(z)$ *is holomorphic for* $|z| < 1$ *and if* $f'(o) = 1$, *then on the Riemann surface described by* $f(z)$, *when* $|z| < 1$, *there exists a sheet of radius at least equal to a constant* B. *We have* $B > 0, 1$.

To prove this, we can assume that $f(z)$ is holomoprhic for $|z| \leq 1$, otherwise we could z by $z(1 - \varepsilon)$ and then let ε tend to zero. We know, on the other hand, that if a homographic transformation of the circle $|z| < 1$ to itself is made, then a function $f(z)$, that is holomorphic for $|z| < 1$, is transformed to another holomorphic function in the circle of radius 1 and the quantity $|f'(z)|(1 - |z|^2)$ is invariant (I, 167). With this in mind, let us denote by M, the maximum of $|f'(z)|(1 - |z|^2)$ for $|z| \leq 1$. As this is the maximum of a continuous function that is zero for $|z| = 1$ and equal to 1 for $z = 0$, then this maximum is at least equal to 1 and is attained inside the circle at one point ζ at least. Let us transform the circle to itself in such a way that the homologue of ζ is the origin $Z = 0$ say (we have

$$Z = \omega \frac{z - \zeta}{1 - z\zeta} \quad , \qquad |\omega| = 1 \ ) \quad .$$

The Riemann surface described by $F(z) = f(z)$ does not change. We have

$$F'(o) \ = |f'(\zeta)| \ (1 - |\zeta|^2) = M \ |F'(Z)| \ (1 - |Z|^2) \leq M \quad .$$

Let us apply the corollary to $F(Z)/M$ in the circle $|Z| < R$. The Riemann surface so described will contain a sheet with radius equal at least to

$$\frac{R}{2\left(1 + \dfrac{1}{1 - R^2}\right)} = \frac{R(1 - R^2)}{2(2 - R^3)} \tag{26}$$

and that described by $F(Z)$ will contain one of radius at least equal to this since $M > 1$. The theorem is thus proved and the maximum value of (26) corresponds to $R^2 = (5 - \sqrt{17})/2$ for $R = 1/2$, we obtain, 3/28.

The exact value of Bloch's constant B is still unknown. It has been shown that it lies somewhere between $\sqrt{3}/4$ and 0.472 (Ahlfors, 1938).

Corollary: *If a function* $Z = f(z)$ *is holomorphic about the point at infinity, which is an essential point, then the Riemann surface described by* Z *contains a sheet of arbitrarily large radius.*

Effectively, $f'(z)$ also admits the point at infinity as an essential point, hence, following the Weierstrass theorem (I, 191) or Picard's theorem on essential points (I, 225), there exists points that are as far removed as one wishes, for which $|f'(z)| > 1$. Let ζ be such a point, $f(z)$ is holomorphic for $|\zeta - z| < \frac{1}{2} |\zeta|$ provided $|\zeta|$ is sufficiently large.

Remark: If a point ζ is a pole of a function $f(z)$, then the surface S described by the values of $f(z)$ when $|\zeta - z| < \varepsilon$ contains circles with one sheet of radius as large as one likes, for small ε say, for S contains the exterior of a circle, or of a system of circles ramified at infinity. Therefore if $f(z)$ is meromorphic about the point at infinity, i.e., if the point at infinity is bounded by poles, then the Riemann surface described by $Z = f(z)$ contains circles with one sheet of arbitrairly large radius. This is still the case if $f(z)$ is holomorphic about the point at infinity which is a pole.

33 - Picard's theorem on uniformisation

It has been proved by E. Picard that it is impossible to uniformise algebraic curves of genus greater than one by uniform functions all of whose singularities are isolated. To establish this theorem it is permissible, via a birational transformation to reduce matters to the case where the curve in question has only multiple points, or likewise double points, with distinct tangents. We shall pursue Bloch's method by relying on the following:

Remark I: Let us outline on the Riemann surface S of $P(z,u) = 0$, a contour K making S simply connected and not passing through the zeros of the rational fraction $R(z,u)$ occurring in the abelian integral

$$I(M) = \int_{M_o}^{M} R(z,u)\, dz \quad .$$

If the genus is greater than 1 and if $I(M)$ is calculated from a point M_o not on K, to M, without crossing K, then $I(M)$ is not univalent, i.e., it takes the same value at distinct points.

The proof does not differ from that given in no. 30. We always assume that S does not have any ramification points at infinity and show that, if $I(M)$ is univalent, then $R(z,u)$ must be zero at most twice on each sheet, at infinity, when it has $\sum(r-1)$ poles at least, which implies that the genus is at most equal to 1.

Remark II: The Weierstrass theorem on the indetermination of a holomorphic function about an essential isolated point extends immediately to meromorphic functions which is a limit of its poles.

Let $g(z)$ be such a function that is meromorphic about the point at infinity which is such a limit. Outside of the circle $|z| > R$, of arbitrarily large radius, $g(z)$ takes the value ∞ infinitely often. Moreover, it approaches as near as one wishes to every finite value a, infinitely often. For if $g(z) - a$ only has a finite number of zeros for $|z| > R$, then the function $h(z) = 1/[g(z) - a]$ is holomorphic for $|z| > R' > R$, its maximum modulus increases indefinitely, and $g(z) - a$ tends to zero along a line to infinity. In the contrary case, $g(z) - a$ is zero an infinity of times for $|z| > R$.

Definition: *We shall say that, generally, a meromorphic function about a point* z_o *which is a limit point of its poles, admits this point as an essential singularity.*

Picard's theorem relating to the essential points considered up to the present, extends to these functions.

If we assume that the limit point of the poles is situated at infinity, which is possible, and if a is a finite arbitrary value, then two cases are possible: the function $f(z)$ takes the value a in a sequence of points having the point at infinity as the limit point, or $f(z)$ only takes the value a a finite number of times. In the second case $f(z) - a$ is no longer zero for $|z| > R$ if R is sufficiently great, hence $g(z) = 1/(f(z) - a)$ is holomorphic about the point at infinity and admits this point as an essential point in the order sense of the word, otherwise $g(z)$ would be at most a pole at infinity, hence also $g(z)$. Consequently, $g(z)$ takes every finite value c an infinity of times about this point at infinity, except one value b at most (I, 225). It follows that $f(z)$ takes every value other than a and $a + \frac{1}{b}$, an infinite number of time [b is non-zero since $g(z)$ is zero at the poles of $f(z)$]. Thus $f(z)$ takes every value except at most two, an infinite number of times about the point at infinity.

A uniform function about an essential point takes every finite or infinite value, an infinite number of times, except at most two exceptional values.

Picard's Theorem on Uniformisation: *If the genus of the curve* $P(z,u) = 0$ *is greater than one, then there do not exist any pairs of functions* $z = g(t)$, $u = h(t)$, *holomorphic or meromorphic about the point at infinity, where this point is an essential point for one of them, and such that* $P(g,y) \equiv 0$.

Let us assume that we have $z = g(t)$, $u = h(t)$, where $g(t)$ for example has an essential point at infinity, and that $P(g,h) \equiv 0$. An abelian integral of the first kind, associated with P, will become a function of t,

$$V = \int_{z_o,u_o}^{g\ t\ ,h\ t} R(z,u)\,dz = \int_{t_o}^{t} R(f,g)g'(t)\,dt = k(t) \quad ,$$

and $k(t)$ will be finite for t finite, $|t) > T$, and is holomorphic in the neighbourhood of each point. This function $k(t)$ cannot be uniform about the point at infinity, but then, when t takes the same value, the point (z,u) takes the same position, v has increased by one period τ say, hence v/τ increases by one when t passes around the point of infinity.

Let us take another integral of the first kind distinct from v, v_1 say: if it is not uniform at t, then there exists one number τ_1 such that v_1/τ_1 increases by 1 when one passes around the origin. Then $v/\tau - v_1/\tau_1$ is an abelian integral of the first kind which is uniform after the change of variable. We can then assume that v is chosen in such a way that $k(t)$ is uniform about the point at infinity. *Let us show that the point at infinity could be neither an essential point nor a pole for* $k(t)$. Let us assume that it is.

The corollary of Bloch's theorem will apply. There exists a circle $|v - v'| < D$, where D is arbitrarily large, in which the inverse function of $k(t)$ is holomorphic. There will be a bijective correspondence between this circle and a domain Δ of the t plane, containing t' such that $v' = k(t')$; the correspondence is defined by $v = k(t)$. Let M' be the point of the Riemann surface S of $P(z,u) = 0$ which corresponds to t' and K a cutting of S making it simply connected and not passing through the zeros of $R(z,u)$. Let us calculate v, starting from M' on a path $M'M$ not intersecting K. When M so describes S cut by K, then $|v - v'|$ stay less than a number H which only depends on $P(z,u) = 0$, K and $R(z,u)$. We shall assume that we have taken $D > H$, and we are allowed to take v' such that M' is not on K. When M describes S cut by K, v takes values belonging to $|v - v'| < D$, t remains in Δ and at a point of Δ, there corresponds a point M of S.

It follows that $v - v'$ *is univalent in* S *cut by* K. For if v, calculated along the paths $M'M_1$, $M'M_2$ has the same value at M_1 and M_2, then to these paths there corresponds lines $v', v; v', v$ of $|v - v'| < D$, hence lines t', t and t', t of Δ. When a certain closed path t, t', t is described in Δ, the point $z = g(t)$, $u = h(t)$, would pass from M_1 to M_2 and the functions g and h would not be uniform. *Thus* $v - v'$ would be univalent in S cut by K. This is impossible following remark I above since the genus is greater than 1.

Thus the point at infinity is a point of holomorphy of $k(t)$. When $t \to \infty$, $v = k(t) \to v_o$. We see that this is also impossible. This results from the fact that $g(t)$ admits the point of infinity as an essential point approaching every finite value;[1] if it was $h(t)$ alone that admits the point at infinity as an essential point, one would reason starting from $h(t)$. Let us take a particular value z' such that all points of S for which $z = z'$ are ordinary points in which $R(z,u)$ is holomorphic and with non-zero derivative. If (z',u') is one of these m points, v' an arbitrary value of v at this point, then $v - v'$ remains holomorphic in a circle of centre z', its derivative is non-zero at the centre z', then the function

(1) We could also use Picard's theorem instead of the Weierstrass theorem that we use here.

$v - v'$ of z represents conformally and bijectively a small circle of centre z' on a small domain of centre v'; we could therefore describe a circumference $|z - z'| = \rho$ on which $|v - v'| > \alpha > 0$. We would carry out this operation for each of the m points in question, then take on S, another m points (z'',u'') corresponding to the same z'' outside of the above m small circles.

Following Weierstrass's theorem applied to $g(t)$, there exist values t' of t sufficiently large, in which z is arbitrarily near to z', and values t'' also sufficiently large for which z is also as near as desired to z''. We can join a point t' to a point t'' by an arc of a curve on which $|t|$ is at least equal to the minimum of the numbers $|t'|$ and $|t''|$, hence for which v tends to v_o when $|t'|$ and $|t''|$ tend to infinity. To this arc of the curve, there corresponds an arc of the curve Γ on S that joins a point M' to a point M", an arc on which v will be as near as one wishes to v_o and, the point M' will remain in one of the m circles with centres (z',u') and radii ε, and M" in one of the m circles with centres (z'',u'') and radii ε described on S, when ε is as small as is desired. Therefore Γ will be joined to a point M' arbitrarily near to one of the centres (z',u') with a point outside the small circles $|z - z'| = \rho$, Γ will intersect one of the circumferences, and the value of v at this point of intersection will differ from the value at M' by a quantity whose modulus will be at least half of α. This is absurd, since these two values are also as near as so desired to v_o.

The theorem is thus proved completely.

OTHER PICARD THEOREMS

The first part of the proof provides at once the theorem of Picard of 1912 and a similar theorem of Landau (I, 223). Take the equation $P(z,u) = 0$, with genus greater than 1; let us assume that we can set $z = a_o + a_1 t + \dots$, $u = b_o + b_1 t + \dots$ (b_o and b_1 may be seen to be determined by a_o and a_1), then the functions given by these Taylor expansions about the origin remain holomorphic for $|t| < T$. Every abelian integral of the first kind will be holomorphic in this circle. Bloch's theorem (and no longer its corollary) will apply and will limit T. If H has the same significance as above, we shall have

$$BT|R(a_o,b_o)a_1| < H \quad ,$$

where B is Bloch's constant.

Bloch's theorem provides a proof of Schottky's theorem (I, 224) independent of the modular function; we are going to give an outline of it. If $f(z)$ does not take the values 0 and 1 for $|z| < 1$, then the function

$$F(z) = \frac{\log f(z)}{2i\pi} \quad ,$$

where the logarithm has its value reduced at the origin, is holomorphic for $|z| < 1$ and does not take positive, negative or zero integer values. The function

$$G(z) = \sqrt{F(z)} - \sqrt{F(z) - 1} \quad ,$$

again holomorphic, does not take the values 0 and $\sqrt{m} \pm \sqrt{m-1}$, where m is an integer -1, as can be easily seen; finally $H(z) = \log G(z)$ is holomorphic for $|z| < 1$ and does not take the values

$$\pm \log (\sqrt{m} + \sqrt{m-1}) + 2i\pi n \quad ,$$

where n is an integer, positive, negative or zero. The maximum distance between two of these excluded points is a number less than 8, the largest circle has a sheet of the Riemann surface described by the point $Z = H(z)$ is a radius less than 4. By applying Bloch's theorem to $H(z)$ in the circle of centre z tangent inside to $|z| = 1$, we obtain, where B is Bloch's constant:

$$B|H'(z)| \, (1 - |z|) < 4$$

or

$$H'(z) = B' \frac{\Theta(z)}{1 - |z|} \quad , \qquad |\Theta(z)| < 1 \quad ,$$

where B is a constant. On integrating, we find:

$$|H(z)| < |H(o)| + B' \log \frac{1}{1 - |z|} \, , \qquad |z| < 1 \quad .$$

$H(o)$ only depends on $|f(o)|$. By returning to $f(z)$:

$$\log f(z) = \frac{i\pi}{2} \left[e^{2H} + e^{-2H} + 2 \right]$$

we obtain the theorem in question.

Chapter III

ANALYTIC FUNCTIONS OF SEVERAL VARIABLES

THE METHOD OF THE MAJORANT FUNCTIONS

The basic operations on power series easily extend to power series in several independent variables. This has led to the importance of functions of several variables which can be expanded as power series within the neighbourhood of those points where they exist. These are known as *analytic functions.* When such a real function of real variables is given, we can extend its definition to the complex elements. In finding solutions to differential equations, Cauchy exploited power series, and to show that those series converged in a certain domain, implicitly defined beforehand, he formulated a method known as *the calculus of limits.* This consisted of replacing the known and unknown quantities, written in series, by *bounded* series converging under the weakest conditions and for which the calculations were the easiest. We usually refer to these new series as the majorants, and Cauchy's method in its entirety together with the applications, especailly those of Briot-Bouquet, Meray, Weierstrass, Sophie Kowaleska, and Poincaré, is now known as the method of series or majorant functions. This method will be the main subject of this chapter.

The general theory of analytic functions of one complex variable has only quite recently been extended to those of several complex variables. There are considerable difficulties, even in the case of just two complex variables. Poincaré extended Cauchy's integral theorem to these functions. In the case of two variables alone, the geometric representation of the variables must appear in two planes; the simplest singularities, the zeros, are no longer isolated and the analogous problem to conformal representation becomes complicated. We shall only look at some aspects of this theory. Let us note that the theory of multiform functions the most simple in two variable, which are *the algebraic functions of two variables,* has given rise to the famous works of E. Picard and the Italian geometers, in particular Castelnuovo and Enriques.

I. POWER SERIES IN SEVERAL VARIABLES

34 - Series with coefficients and positive variables

Let us consider first of all a double series of the form

$$\sum A_{m,n} X^m Y^n \qquad (1)$$

where the $A_{m,n}$ and the variables X and Y are non-negative, m and n are non-negative integers. To simplify matters, we shall only insert a single summation sign. The point (0,0) is excluded. If the series converges for the pair X_o, Y_o, i.e., at the point $M_o(X_o, Y_o)$ in the usual geometric representation (fig. 26), then it converges for $X \leq X_o$, $Y \leq Y_o$, and hence in the rectangle $OP_oM_oQ_o$ with sides parallel to the axes, and on its sides. If it diverges at the point $M_1(X_1, Y_1)$ then it diverges for $X \geq X_1$, $Y \geq Y_1$ and hence in the quarter of the plane $X'M_1Y'$ (fig. 26).

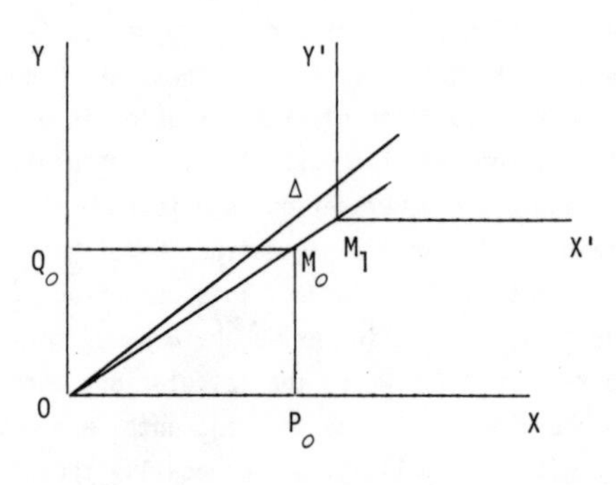

Fig. 26

If we set $Y = kX$ where k is a finite positive number, then the resulting series could be assigned to X without altering its convergence or divergence, since everything is positive (I, 21). We obtain a power series in one variable that will have either a radius of convergence zero, one finitely positive, or one infinitely positive:

$$\sum_1^{\infty} B_p X^p \quad , \qquad B_p = \sum_{m+n=p} A_{m,n} \, k^n \quad . \tag{2}$$

If the radius of convergence of (2) is infinite, (1) converges on every half-line Δ, $Y = kX$ and hence also for arbitrary X and Y. If the radius of convergence of (2) is zero, (2) diverges for $X > 0$, hence therefore (1) diverges on Δ, hence for all $X > 0$ and $Y > 0$.

Finally, if the radius of convergence is R, finite positive, then the series (2) converges for $X < R$ and diverges for $X > R$, therefore (1) converges in the rectangle $X < R$, $Y < kR$ and diverges in the quarter of the plane $X > R$, $Y > kR$. Another line Δ', $Y = k'R$, $k' \neq k$, will then give the same since there will be a segment of Δ' in the rectangle ORM, kR (fig. 27) and a half line of Δ' in the quarter of the plane $X > R$, $Y > kR$.

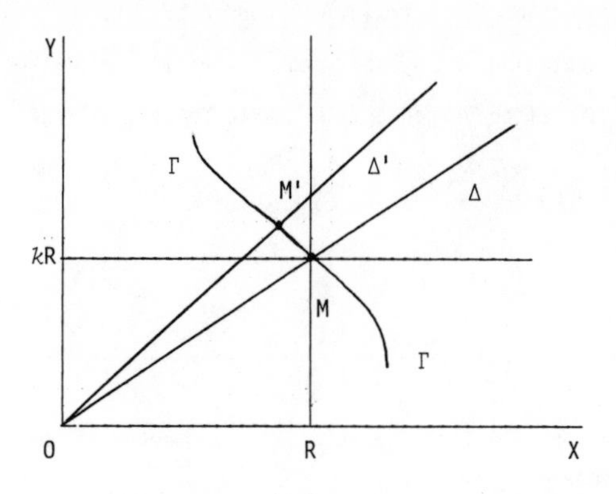

Fig. 27

We will obtain a point M' on Δ' bounding the interval of convergence from the half-line of divergence and it is clear that k' tends to k as M' tends towards M. Hence there exists a continuous curve Γ bounding a domain D in the quarter of the plane in which the series (1) converges, whilst outside of Γ, the series diverges.

At each point of Γ we could have convergence or divergence. The curve Γ could end at points at a finite distance on OX and OY, or even could have OX or OX as asymptotic directions. When Γ meets OX at a point A, the series (1) could converge for $Y = 0$ beyond this point. The same remark applies for OY.

Examples

The series

$$\sum_{o}^{\infty} \sum_{o}^{\infty} \frac{1}{m!n!} X^m Y^m = e^x e^y \quad ,$$

converges for any X, Y; the series

$$\sum_{o}^{\infty} \sum_{o}^{\infty} M \left(\frac{X}{a}\right)^m \left(\frac{Y}{b}\right)^n = \frac{M}{\left(1 - \frac{X}{a}\right)\left(1 - \frac{Y}{b}\right)} \quad , \quad M > 0, \quad a > 0, \ b > 0 \tag{3}$$

converges for $X < a$, $Y < b$ and diverges otherwise. The domain of convergence is a rectangle. For

$$\sum_{o}^{\infty} \sum_{o}^{\infty} M \frac{(m+n)!}{m!n!} \left(\frac{X}{a}\right)^m \left(\frac{Y}{b}\right)^n = \frac{M}{1 - \left(\frac{X}{a} + \frac{Y}{b}\right)} , \quad M > 0, \quad a > 0, \ b > 0 \tag{4}$$

the domain of convergence is a traingle, bounded by OX, OY and the line $\frac{X}{a} + \frac{Y}{b} - 1 = 0$. Replacing X by X^2 and Y by Y^2 in (4), we obtain a series whose domain of convergence is a quarter of the ellipse:

$$X \geq 0, \quad Y \geq 0, \quad \frac{X^2}{a} + \frac{Y^2}{b} - 1 < 0 \quad .$$

For the series

$$\sum_{o}^{\infty} \frac{X^n Y^n}{a^n} = \frac{1}{1 - \frac{XY}{a}} \quad , \quad a > 0 \quad , \tag{5}$$

the domain of convergence is the portion of the quarter of the plane situated below the semihyperbola $XY = a$.

THE DETERMINATION AND PROPERTIES OF Γ

Following Cauchy's theorem (I, 17), the radius of convergence of the series (2) is the inverse of

$$\overline{\lim_{p=\infty}} \; \sqrt[p]{B_p} \quad . \tag{6}$$

Now, if b_p is the maximum of the numbers $A_{m,n}k^n$, for which $m+n = p$, we have $b_p \leq B_p \leq (p+1)b_p$ and consequently, since the pth root of $p+1$ tneds to 1, we could replace (6) by

$$\overline{\lim_{p=\infty}} \; \sqrt[p]{b_p} \quad . \tag{7}$$

In particular, in order that the convergence holds for any X, Y, it is necessary and sufficient that it holds for any X if Y = X, therefore

$$\lim_{(m+n)=\infty} \sqrt[m+n]{A_{m,n}} = 0 \quad . \tag{8}$$

This notation means that we take $m+n = p$ successively, then the roots of order p of the numbers $A_{m,n}$ for which $m+n = p$, and then the upper bound of the set of these numbers for p infinite. It is clear that we could replace the $p+1$ numbers corresponding to $m+n = p$ by their maximum, in a way that the first member of (8) is (7) for $k = 1$, and the result is still clear.

Following what was said before, the curve Γ crosses the parallels to the parallels to the axes, at most once. If M is a point of Γ, then Γ is found, in the broad sense of angles with vertex M, between sides parallel to the axes, which do not contain OM. This curve has another general property, discovered by Fábry, that arises from the known property of the arithmetic and geometric means. We know, or as we can easily verify, that if

x, y, α, β are positive then we have $x^{\alpha}y^{\beta} \leq \alpha x + \beta y$ when $\alpha + \beta = 1$. Assume we have taken three points $M(X,Y)$, $M_1(X_1,Y_1)$ and $M_2(X_2,Y_2)$ with positive co-ordinates such that

$$X = X_1^{\alpha}X_2^{\beta} \quad , \quad Y = Y_1^{\alpha}Y_2^{\beta} , \quad \alpha > 0, \quad \beta > 0. \quad \alpha + \beta = 1 \quad . \tag{9}$$

If m and n are non-negative, we will have

$$X^mY^n = (X_1^mY_1^n)^{\alpha}(X_2^mY_2^n)^{\beta} \leq \alpha X_1^mY_1^n + \beta X_2^mY_2^n \quad , \tag{10}$$

and consequently, if $F(X,Y)$ denotes the sum of the series (1) which is assumed to be convergent at the points M_1 and M_2, we see that by multiplying both sides of (10) by $A_{m,n}$ and summing any number of terms, (1) converges at the point M and that

$$F(X,Y) \leq \alpha F(X_1,Y_1) + \beta F(X_2,Y_2) \quad .$$

If, on leaving e.g. the abscissae of M_1 and M_2 fixed, we allow these points to tend towards the points of Γ with co-ordinates X_1,Y_1 and X_2,Y_2, then we find that every point M defined by (9) is found within the domain of convergence or on Γ. This property of Γ takes a simple form if we set

$$x = \log X \ , \quad y = \log Y \quad ,$$

and consider the transformation γ of Γ. Again this will be a curve crossing the parallels to the axes, only once, and going from infinity to infinity (since if e.g. X in decreasing stays greater than $X_o > 0$, then Y increases indefinitely).

If m_1 and m_2 are on γ (fig. 28), then every point m of the segment m_1m_2, i.e., a point for which $x = \alpha x_1 + \beta x_2$, $y = \alpha y_1 + \beta y_2$, $\alpha > 0$, $\beta > 0$, $\alpha + \beta = 1$ will be on γ, or in the corresponding domain to the domain of convergence. γ *will be convex towards* $y > 0$, and hence will enjoy the properties of convex curves (I, 39); it will have a tangent except at a countable number of points, hence Γ will also have the same. The property of γ easily proves that if Γ has an asymptotic direction parallel to OY and is not an asymptote to OY, then Γ reduces to a semi-parallel to Oy, hence Γ will too. Similarly if Γ admits an asymptotic direction parallel to OX and is not an asymptote to OX, then Γ is a semi-parallel to OX. Here is an example of a case encountered when setting $X = \text{const.}$; this is also a method of determination of Γ. We take

$$\sum_1^{\infty} \left(\frac{X^nY}{n}\right)^n \quad .$$

Cauchy's rule shows that for $X \leq 1$, we have an entire function at Y that is also convergent. For $X > 1$, there is divergence for any Y.

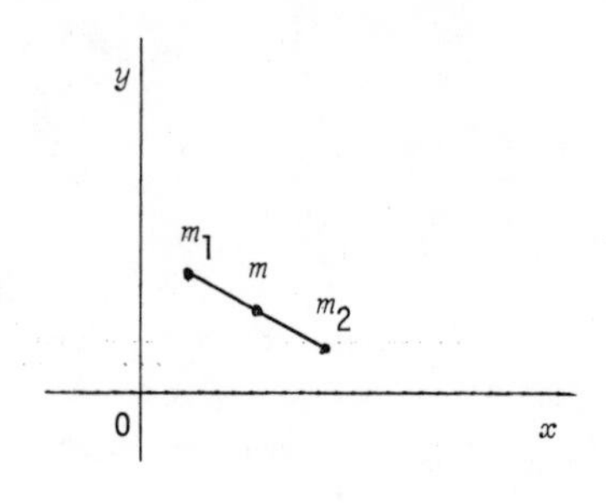

Fig. 28

POWER SERIES IN THREE VARIABLES

We can repeat matters in the case of a power series in three positive variables and with positive coefficients:

$$\sum A_{m,n,p} X^m Y^n Z^p \quad .$$

Representing the point X,Y,Z in three dimensional space, we obtain a domain of convergence bounded by a surface. If we put $x = \log X$, $y = \log Y$ and $z = \log Z$, we can show by a similar to that of above, that the above, that the transformed surface is convex; if we take three points on the surface, forming a triangle, then all the points of this triangle are always on the same side of the surface, broadly speaking.

POWER SERIES IN q VARIABLES

We will obtain similar properties when the geometric representation is made in q-dimensional space, the definition of convexity may then be extended in purely analytical terms.

35 - Series with arbitrary terms. The associated circles of convergence

Let now

$$F(x,y) = \sum a_{m,n} x^m y^n \tag{11}$$

by a power series in two variables, for which the variables x and y are complex (they could be real if specified). When we wish to specify the values of the integers m and n, we will take a double summation with initial values m and n. We can add to this series, the series of modules

$$\sum A_{m,n}X^mY^n, \quad A_{m,n} = |a_{m,n}| \, , \quad X = |x| \, , \quad Y = |y| \, , \tag{12}$$

for which the above applies. The series in (11) will converge absolutely when the point $M(X,Y)$ is in the domain of convergence of the series (12).

If the point M is outside the domain of convergence of the series (12) and its frontier Γ, then this series diverges, hence the series (11) does not converge absolutely in the sense of the theory of double series (I, 21). This supposes $X \neq 0$ and $Y \neq 0$. We note that *the series (11) assigned to a simple series in an arbitrary way could not converge at a point outside the curve* Γ *and in the domain of convergence.* For if the series (11) converges for $x = x_o$, $y = y_o$ and a certain ordering of the terms, then the general term is bounded, $A_{m,n}X_o^mY_o^n < K$, whence we have, for $X < X_o$ and $Y < Y_o$,

$$\sum A_{m,n}X^mY^m < K\left(\frac{X}{X_o}\right)^m\left(\frac{Y}{Y_o}\right)^n \leq \frac{K}{\left(1 - \frac{X}{X_o}\right)\left(1 - \frac{Y}{Y_o}\right)} \, .$$

There is absolute convergence of (12) for $X < X_o$, $Y < Y_o$, hence the point X_o, Y_o is on Γ or in the domain of convergence (we have assumed $X_oY_o \neq 0$).

But the series (11) could converge by means of certain groups of terms. For example, the double series (4) as given above, converges throughout the domain

$$\left|\frac{x}{a} + \frac{y}{b}\right| < 1$$

when X is replaced by x, Y by y and it is assumed that all the terms for which $m + n = p$, $p = 0,1,\ldots,$ are grouped at a single term.

THE CIRCLES ASSOCIATED TO THE CONVERGENCE

To each point M of the curve Γ bounding the domain of convergence of the series of modules (12), there corresponds two numbers R and R' which are the abscissa and the ordinate of M, such that the series (11) converges absolutely when the point P representing the complex number x, is found in the circle $|x| < R$, and the point Q representing the number y is found in the circle $|y| < R'$. We shall call these circles C and C' in the two distinct planes, the plane of $x = x' + ix''$ ($x' = \mathrm{R}x$, $i = \sqrt{-1}$), and the plane of $y = y' + iy''$ ($y' = \mathrm{R}y$) respectively, we call these the circles associated to the convergence (fig. 29). If P and Q are both outside the circles of convergence of their respective planes, then the series (11) no longer converges absolutely. But when M is displaced along Γ, the radii C and C' vary; generally one increases and the other decreases.

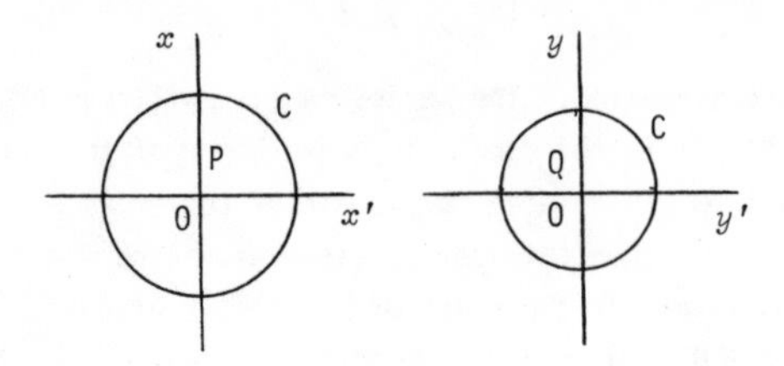

Fig. 29

If the series (12) converges everywhere, there is no longer any need to consider the associated circles of convergence.

THE CASE OF MORE THAN TWO VARIABLES

We have a similar situation. For example, for three variables, we will have a series

$$\sum a_{m,n,p} x^m y^n z^p \quad ,$$

where the series of moduli will provide the domain of absolute convergence: the point $|x|$, $|y|$, $|z|$ will be a certain surface outside of which the general term of the series will no longer tend to zero. To each point of the bounded surface, there corresponds three associated radii of convergence.

THE CASE OF TWO REAL VARIABLES

If we assume that x and y are real in (11), then the locus of points of the plane Oxy whose co-ordinates x and y render (11) to be absolutely convergent, is clearly obtained by indicating in this plane the curve Γ (x = X, y = Y) and then reflecting about Ox to give Γ', then a reflection of $\Gamma\Gamma'$ about Ox. The shaded region is the domain of absolute convergence (fig. 30). The series could then converge at certain points of Γ, from 0 and from 0 . A similar result, which is susceptible to a simple geometric interpretation, will be obtained in the case of three real variables.

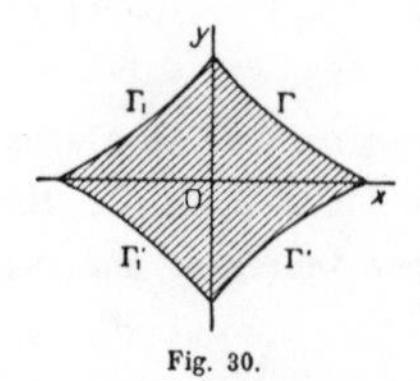

Fig. 30.

36 - Properties of the summation. Analytic functions

An absolutely convergent power series could be assigned once and for all to a simple series. The notion of uniform convergence will apply and the usual results will be entailed. If R and R' are the associated radii of convergence and $\rho < R$, $\rho' < R'$, then the series (12) converges for $X = \rho$, $Y = \rho'$; we could then take enough terms such that it would remain less than ε say. In (11), each term is of modulus less than the corresponding term of (12) if $|x| \leq \rho$, $|y| \leq \rho$ (when in (12) we take $X = \rho$, $Y = \rho'$). There is then uniform convergence following the Weierstrasses rule. We deduce that the sum $F(x,y)$ of the series (11) is a continuous function of x,y when x and y belong to two circles respectively inside the two associated circles of convergence.

In the case of real variables, this indicates that $F(x,y)$ is contained in every rectangle with sides parallel to the axes (fig. 30) and centre the origin, and it is completely within the region bounded by Γ, Γ', Γ_1, Γ_1'. The function $F(x,y)$ is therefore contained in this domain.

PARTIAL DERIVATIVES

Given a fixed values inside a circle C, let us take y to be a variable in the associated circle to C, C' say. The series (11) is a function of y. We could assign it to y since there is absolute convergence; we obtain then a power series at y. It is differentiable with its derivative obtained by term-by-term differentiation. We could also carry out this differentiation on (12), knowing that its derived series will have the same circle of convergence. Hence for all x in C and y in C', we have

$$\frac{\partial F}{\partial x} = \sum a_{m,n} n x^m y^{n-1} \quad .$$

For this series the curve Γ is the same as for (11), as the result of what was said. We can continue these operations, by taking derivatives with respect to x,y etc. The function $F(x,y)$ is infinitely differentiable with respect to x and y in the associated circles of convergence, with the partial derivatives obtained by differentiating term by term. We have

$$\frac{\partial^{p+q} F}{\partial x^p \, y^2} = \sum \frac{m!n! \; a_{m,n}}{(m-p)!(n-q)!} \; x^{m-p} \; y^{n-q} \quad , \tag{13}$$

with the summation commencing from $m \geq p$ and $n \geq p$. This always generalises in the case of 3, 4 or more variables. For the case of two real variables, the formulae giving the partial derivatives are true in the domain of absolute convergence in figure 30.

Taylor's formula. Let us recall the case of two complex variables. The following would generalize for any number of variables. If x_o and y_o are two points respectively inside two associated circles of convergence C and C', x and y are two points inside these circles respectively, with centres x_o and y_o and radii $R - |x_o|$ and $R' - |y_o|$ (fig. 31).

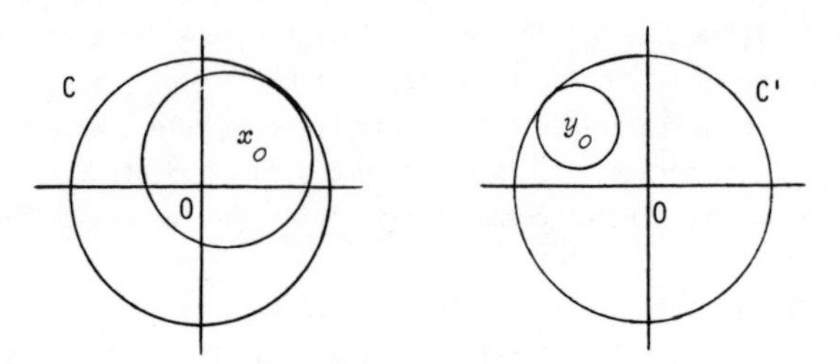

Fig. 31

(R is the radius of C and R' the radius of C'), then seeing that these circles are tangent to the associated circles, we could write

$$F(x,y) = \sum\sum a_{m,n} [x_o + (x - x_o)]^m [y_o + (y - y_o)]^n \quad .$$

As the series

$$\sum\sum A_{m,n}[X_o + h]^m [Y_o + k]^n, \quad X_o = |x_o|, \; Y_o = |y_o|,$$

$$h = |x - x_o|, \quad k = |y - y_o|$$

is convergent, we can see that the series occuring in $F(x,y)$ is a quadruple series

$$\sum\sum\sum\sum a_{m,n} C_m^p x_o^{m-p} (x - x_o)^p C_n^q y_o^{n-q} (y - y_o)^q \tag{14}$$

converging absolutely. We could assign (14) to a power series in $x - x_o$ and $y - y_o$; we will obtain

$$F(x,y) = \sum\sum b_{p,q} (x - x_o)^p (y - y_o)^q$$

with

$$b_{p,q} = \frac{1}{p!q!} \sum\sum \frac{m!n!}{(m-p)!(n-q)} a_{m,n} x_o^{m-p} y_o^{npq} \quad .$$

We recover expression (13) from the partial derivative. *Thus, with the above hypotheses:*

$$|x - x_o| < R - |x_o|, \quad |y - y_o| < R' - |y_o| \ ,$$

we have

$$F(x,y) = \sum_{o}^{\infty} \sum_{o}^{\infty} \frac{1}{p!q!} \frac{\partial^{p+q}F}{\partial x^p \partial y^p} (x_o,y_o)(x - x_o)^p \ (y - y_o)^q \qquad (15)$$

The sum of a power series may be expanded as a Taylor series about every point x_o, y_o *belonging to two associated circles of convergence. This power series in* $x - x_o$ *and* $y - y_o$ *is absolutely convergent in the circles with centres* x_o, y_o *that are internally tangential to the two associated circles.*

In particular, every power series is a Taylor series. The series (15) could in a certain case, be absolutely convergent in circles of radii greater than $R - |x_o|$ and $R' - |y_o|$ respectively. They will then admit the analytic continuation of the function $F(x,y)$.

ANALYTIC FUNCTIONS

Given a function $F(x,y)$ of two complex variables: $x = x' + ix''$, $y = y' + iy''$, defined in a certain four dimensional region, we shall say that this function is analytic in this domain when for x_o, y_o being any system of values of x and y corresponding to a point in the region, $F(x,y)$ is equal to the sum of a power series in $x - x_o$, $y - y_o$ for $|x - x_o| < R$, $|y - y_o| < R'$ where R and R' depend on x_o, y_o, but are positive. From before, $F(x,y)$ will admit derivatives of all orders in x and y, these being partial derivatives which will also be analytic in the domain in question and the expansion about x_o, y_o, will be a Taylor expansion of the form (15).

Operations. The sum of two or more analytic functions in the same domain is analytic since the sum of two absolutely convergent power series is clearly an absolutely convergent power series when the variables are constrained to satisfy simultaneously the conditions assuming the convergence of the given series. (For example, if Γ and Γ' are the curves corresponding to the two given series, then the curve Γ'' corresponding to the sum will bound a region which will contain at least the common portion of the regions bounded by Γ and Γ'.) Since the product of two absolutely convergent series is a series which could be ordered in any way, then the product of two power series in x and y could be ordered as a power series once x and y satisfy the conditions of absolute convergence. It follows that the product of two analytic functions is again an analytic function. *All the combinations of addition, subtraction, multiplication and partial differentiation applied to analytic functions of two variables yield analytic functions.*

The division of an analytic function by another will provide an analytic function under certain conditions. We see this as for the case of one variable (I, 22) by appealing to the theorem on the substitution of one or more power series into another. *If the series*

$$F(x,y) = \sum\sum a_{m,n} x^m y^n \tag{16}$$

converges (absolutely) for $|x| < R$, $|y| < R'$, *and if we set*

$$x = G(u,v) = \sum_0^\infty \sum_0^\infty b_{p,q} u^p v^q , \quad y = H(u,v) \quad \sum_0^\infty \sum_0^\infty c_{p,q} u^p v^q , \tag{17}$$

then these series G *and* H *both converge for* $|u| < \rho$, $|v| < \rho'$ *and if* $|b_{0,0}| < R$, $|c_{0,0}| < R'$, *then* $F(x,y)$ *will become a function of* u *and* v *which will be expandable as a power series if* $|u|$ *and* $|v|$ *are sufficiently small.*

In fact, if we replace the variables and the coefficients by their moduli, then we obtain

$$A_{m,n} X^m Y^n , \qquad \sum_0^\infty \sum_0^\infty B_{p,q} U^p V^q , \qquad \sum_0^\infty \sum_0^\infty C_{p,q} U^p V^q$$

and the second series will have a sum less than R, if U and V are sufficiently small, since this sum is continuous and less than R for U and V zero. We have an analogous result for the third series. For U and V sufficiently small, we could replace X and Y in the first series by sums of the second and third series, and since everything is positive, we could assign the series so obtained to U and V. It follows that the operation which amounts to replacing x and y in (16) by the series (17) and then assigning to u and v, will be justified under the same conditions. It follows that: *An analytic function of* x *and* y *in a region* D *will be an analytic function of* u *and* v *in a region* Δ *if* x *and* y *are replaced by the analytic functions* u *and* v *in* Δ *such that the point* x,y *remains in* D.

This still is true whatever the number of variables in the change of variables. *If we take an analytic function of* x *and* y $F(x,y)$ *say which is non-zero in a domain* D, *then the function* $1/F(x,y)$ *will be analytic in* D. Since for the point x_0, y_0 in D, we will have about this point

$$F(x,y) = F(x_0,y_0) + u , \qquad u = \sum\sum a_{m,n} (x - x_0)^m (y - y_0)^n$$

where the double summation in u does not contain a term corresponding to $m = n = 0$; u is zero for $x = x_0$, $y = y_0$ but for $|u| < |F(x_0,y_0)|$, $1/F(x,y)$ is a power series in u. Consequently,

If $F(x,y)$ *and* $G(x,y)$ *are analytic in* D *and if* $F(x,y)$ *is non-zero there, then the quotient* $G(x,y)/F(x,y)$ *is analytic in* D.

All of this holds for any number of variables.

Determination of an analytic function. An analytic function in a domain D is completely determined by its value and the values of all its partial derivatives at a single point in the domain. This means that two analytic functions in D which at a point are equal and have their partial derivatives of all orders equal at this point, are equal throughout D; or else an analytic function in a domain D is identically zero if it is zero along with all its derivatives at a point M_o of D.

Now such a function is given by its Taylor series about every point M of D, the series converging in a system of circles $|x - x_j| < \rho(M)$, $j = 1,2,\ldots,q$, if there are q variables. *A fortiori* it converges and gives the value of the function and by differentiation its partial derivatives in a sphere $\sigma(M)$ of the domain D having M as its centre. If Δ is a part of D containing M_o and M, then it can be covered by a finite number of spheres $\sigma(M_k)$, $k = 1,2,\ldots,p$, and we can assume, by increasing their number if needs be, that M_o can be joined to M by a chain of these spheres such that the first is $\sigma(M_o)$, the last $\sigma(M)$ and that each sphere of the chain, other than $\sigma(M_o)$, has its centre in the preceding sphere. In $\sigma(M_o)$, the function is given by a Taylor series with zero coefficients, it is therefore identically zero along with its partial derivatives; the Taylor series at the centre M_1 of the second sphere is again with zero coefficients, therefore the function is identically zero in $\sigma(M_1)$, etc.

37 - The case of two real variables. An example relating to harmonic functions

If we take the particular case of an analytic function of two real variables x,y, $F(x,y)$ say, then it will be defined in a domain D of the plane Oxy. If D contains the domain, then about the origin, the Taylor series of $F(x,y)$ will converge in the domain in question in no. 35 (fig. 30).

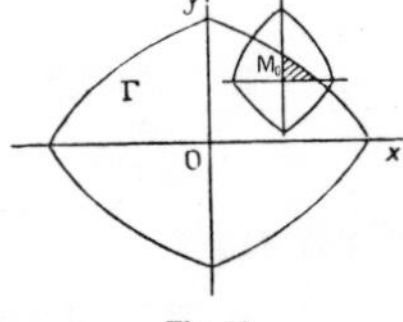

Fig. 32.

In this domain, the series will define an analytic function which will coincide with $F(x,y)$ in every part of D inside the domain of convergence and forming a domain containing the origin 0. It is possible to extend the definition of $F(x,y)$.

We would also define a function $F(x,y)$ by giving a power series in x and y for example (or more generally, in $x - x_o$, $y - y_o$) and analytically continuing as in the case of functions of one complex variable (I, 199). It will again be seen that if we take a point $M_o(x_o,y_o)$ in the domain of convergence Δ of the series (fig. 32), then the Taylor series relative to this point M_o will be (absolutely) convergent in a domain with centre M . This domain Δ_o contains the part of Δ contained in the quarter of the plane bounded by the half-lines parallel to Ox, Oy passing through M_o and which does not intersect Ox and Oy. Since if, for example, $x_o > 0$ and $y_o > 0$, then the Taylor series at the point M_o has its terms less in moduli to those of

$$\sum |a_{m,n}| \ (x_o + |x - x_o|)^m \ (y_o + |y - y_o|)^n$$

and this series converges in the shaded part (fig. 32). It follows that Δ_o together with its boundary would not be inside Δ.

If Δ_o has points outside Δ, then the function will have been continued. If this first continuation is possible, then it is possible to continue, in order to make a radial extension in the various directions, to extend along a broken line. If, in the course of continuation, a domain is reached which encroaches on Δ without the values obtained coinciding with those of $F(x,y)$ in the common portion, then we will have defined a multiforma function.

It can be shown that the given condition of analyticity for the functions of one variable (I, 78), extends to the functions of two variables. This will be of the form

$$\left|\frac{\partial^{p+q}}{\partial x^p \partial y^q}\right| < M \ p!q!\rho^p\rho^q$$

where M, ρ and ρ' are fixed, $p \geq 0$, $q \geq 0$. This is a necessary and sufficient condition.

AN EXAMPLE RELATING TO HARMONIC FUNCTIONS

We know that if $f(z)$ is a holomorphic function in a domain D, then its real part is a harmonic function being a solution of Laplace's equation (I, 161). This harmonic function is analytic. To see this, it suffices to consider a Taylor series

$$\sum_{o}^{\infty} a_n z^n , \qquad z = x + iy \qquad i = \sqrt{-1} \qquad x = \mathrm{R}z \tag{18}$$

and show that the real part $P(x,y)$ is analytic in a circle with centre the origin. Now if $a_n = \alpha_n + i\beta_n$, we have

$$P(x,y) = \sum_0 [\alpha_n(x^n - C_n^2 x^{n-2}y^2 + \ldots) - \beta(C_n^1 x^{n-1}y - \ldots)] \quad (19)$$

If on replacing each term by its modulus, a convergent series is obtained, then we shall have an absolutely convergent power series. Now the general term so obtained is

$$\begin{aligned} &|\alpha_n|(|x|^n + C_n^2|x|^{n-2}|y|^2 + \ldots) + |\beta_n|(C_n^1|x|^{n-1}|y| + \ldots) \\ &\qquad \leq |a_n|(|x| + |y|)^n \quad . \end{aligned} \quad (20)$$

It will suffice to have $|x| + |y|$ less than the radius of convergence of the series (18) in order that (19) converges absolutely. The domain of absolute convergence of (19) is at least the square $|x| + |y| < R$, where R is the radius of convergence of (18). *A harmonic function is analytic.* It is as long as $f(z)$ is. Its analytic continuation could be made for as long as that of $f(z)$ is possible.

If, in the expression (20), we set $|y| = k|x|$, where k is taken to be finitely positive, then we obtain

$$|x|^n \quad (|\alpha_n|u_n + |\beta_n|v_n)$$

with

$$u_n = 1 + C_n^2 k^2 + \ldots, \qquad v_n = C_n^1 k + \ldots, \qquad u_n + v_n = (1+k)^n$$

u_n and v_n each comprise half (to within a unit) of the $n+1$ terms of $v_n + u_n$. In this sum $u_n + v_n$, the ratio of one term to the preceding one is $(n-p+1)k/p$, where the maximum term corresponds, as soon as n is sufficiently large, to a value of p whose ratio to n is as near as is desired to $k/(1+k)$, and the ratio of the terms near to this maximum term tend towards one if $n \to \infty$. We deduce from this the somewhat crude, albeit sufficient, inequalities

$$u_n > H\frac{(1+k)^n}{n} \quad , \qquad v_n > H\frac{(1+k)^n}{n} \text{ [1]}$$

where H is a positive number independent of n, $n \geq 1$.

---

[1] In practice, the ratios of u_n and v_n to $u_n + v_n$ tend to 1/2, as is easily seen.

Thus

$$|x|^n(|\alpha_n|u_n + |\beta_n|v_n) > |x|^n(|\alpha_n| + |\beta_n|)\frac{H}{n}(1+k)$$

$$\geq |a_n|\frac{H}{n}(|x| + |y|)^n \quad .$$

The last member of this inequality is the general term of a power series in $|x| + |y|$ whose radius of convergence is that of the series (18) (I, 17). The series with the general term (20) diverges for $|x| + |y| > R$. *The domain of absolute convergence of the series* $P(x,y)$ *is the square* $|x| + |y| < R$.

38 - Majorant functions

In the calculations of no. 36, we replaced power series such as

$$F(x,y) = \sum\sum a_{m,n}x^m y^n \tag{16}$$

by the series of the moduli, which has positive terms and the same domain of absolute convergence. This series of moduli is a *majorant series of the series* (16). In a more general way, we will say that the series with positive terms

$$C_{m,n}X^m Y^n \;, \qquad X = |x| \;, \qquad Y = |y| \;, \tag{21}$$

majorises the series (16) or is *a majorant series of* (16), if for all pairs m,n, $m \geq 0$, $n \geq 0$, we have $|a_{m,n}| \leq C_{m,n}$. *The sum* $G(X,Y)$ *of* (21) *will be a majorant function of* $F(x,y)$. It is clear that the series (16) will converge absolutely in the domain of convergence of (21). The series of moduli is the strictest majorant; its value is less than that of all the others. But, in general, it does not lend itself to a calculation of its sum by means of a simple formula.

In the calculations, we utilize simple majorants whose region of convergence is smaller than that of (16).

Let R and R' be two radii of convergence associated to the series (16). If $\rho < R$, $\rho' < R'$, then the series (16) converges absolutely for $|x| = X = \rho$ and $|y| = Y = \rho'$; the sum of the moduli of its terms is a number M. We have therefore

$$|a_{m,n}|\rho^m\rho'^n \leq M \;, \qquad m \geq 0 \;, \quad n \geq 0 \quad .$$

We take

$$C_{m,n} = \frac{M}{\rho^m\rho'^n}$$

which yields the majorant

$$G(X,Y) = \sum_{o}^{\infty} \sum_{o}^{\infty} M \left(\frac{X}{\rho}\right)^m \left(\frac{Y}{\rho'}\right)^n = \frac{M}{\left(1 - \frac{X}{\rho}\right)\left(1 - \frac{Y}{\rho'}\right)}$$

tenable for $X < \rho$, $Y < \rho'$. If $a_{oo} = 0$, we will again obtain a majorant by curtailing M from $G(X,Y)$, etc. *A fortiori*, the series

$$\frac{M}{1 - \left(\frac{X}{\rho} + \frac{Y}{\rho'}\right)} = \sum M \left(\frac{X}{\rho} + \frac{Y}{\rho'}\right)^p$$

is a majorant since in the expansion to the power p, the term corresponding to the general term of $G(X,Y)$ will be this same term multiplied by a number greater than or equal to 1. In this new majorant, the domain of convergence will now only be the triangle bounded by $\frac{X}{\rho} + \frac{Y}{\rho'} = 1$, when for $G(X,Y)$ we have the rectangle $X < \rho$, $Y < \rho'$.

All of this extends to series in q variables. The partial derivations of a majorant clearly majorise the corresponding partial derivatives. Likewise, a power of G majorises the corresponding power of F.

II. IMPLICIT ANALYTIC FUNCTIONS--ANALYTIC CURVES AND SURFACES

39 - A function of a single variable

We have already seen that if the equation $F(x,y) = 0$ is satisfied at a point $M_o(x_o,y_o)$, if $\frac{\partial F}{\partial y}(x_o,y_o)$ exists and is non-zero and if certain continuity conditions are realised (I, 120), then there exists a unique implicit function $y = \phi(x)$ taking the value y_o for x_o; it is continuous at this point and is such that $F(x,\phi(x)) \equiv 0$.[1] We are going to establish that when $F(x,y)$ is analytic in a domain containing the point M_o, then $\phi(x)$ is analytic in a domain containing α_o. We may suppose that M_o is the origin. Hence $F(x,y)$ is a power series zero at the origin and $\frac{\partial F}{\partial y}(0,0)$ is non-zero. We can divide $F(x,y)$ by the coefficient of y and write the equation on solving for y

$$y = a_{10}x + a_{20}x^2 + a_{1,1}xy + a_{0,2}y^2 + \dots \tag{22}$$

where the series in the second member is convergent for $|x| \le \rho < R$ and $|y| \le \rho' < R'$. We seek a solution of the form of the power series

$$y = c_1x + c_2x^2 + \dots + c_nx^n + \dots \quad . \tag{23}$$

---

[1] Moreover, the proof previously given (I, 120) assumed the variables and functions to be real, but it can be extended.

Let us assume that the series (23) converges for $|x| < \alpha$; it defines a holomorphic function of x. Let us replace y by this series in the two members of (22). We could order the second member with respect to x, if $|x|$ is sufficiently small, on account of the theorem on the substitution of a series into a series (No. 36). If the function y of x defined by (23) is a solution, then both members (sides) of (22) are equal, and hence have the same coefficients. Conversely, if the two series so obtained in the two members of (22) have the same coefficients, then the function y defined by (23) will be a solution, provided that $|x| < \alpha$ and that $|x|$ is as small as possible, for the second member of (22) ordered at x. Now the coefficients c_1, c_2 are obtained iteratively by the equations

$$c_1 = a_{1,0} \qquad c = a_{2,0} + a_{1,1}c_1 + a_{0,2}c_1^2 \quad ,$$

$$c_3 = a_{1,1}c_2 + 2a_{0,2}c_1c_2 + a_{3,0} + a_{2,1\ 1} + a_{1,2}c_1^2 + a_{0,3}c_1^3 \quad ,$$

Generally, the coefficient of x^n is c_n in the first member of (22). In the second member, there only arises terms whose sum of degrees at and y is at most n, hence the terms

$$a_{p,q}x^p y^q = a_{p,q}x^{p+q}\ (c_1 + \dots + c_m x^{m-1} + \dots)^q \quad .$$

In the power of order q of the series, only the terms up to $c_{n-p-q+1}$ will yield terms of degree $n-p-q$. It follows thatwe will have

$$c_n = \mathrm{P}_n\ (c_1 c_2, \dots, c_{n-1}, a_{i,j}) \tag{24}$$

where the second member is a polynomial in $c_1, c_2, \dots, c_{n-1}$ and at $a_{i,j}$ with $i + j \leq n$, *this polynomial has integer coefficients that are positive or zero.*

Consequently, if we replace the $a_{i,j}$ in the second member of (22) by positive numbers greater than or equal to their moduli, and make the same calculation, then c_1 will be replaced by a positive number greater than or equal to $|c_1|$, hence c_2 will be replaced by a positive number greater than its modulus. Similarly, if the $c_1, c_2, \dots, c_{n-1}$ in the second member of (24) are replaced, just as for the $a_{i,j}$, by positive numbers greater than or equal to their moduli, then the value c_n will be positive and greater than or equal to the modulus of its original value. Hence

If the series in the second member of (22) is replaced by a majorant series, then the series (23) satisfying the equations (24) will be replaced by a majorant. If it is seen that the majorant thus obtained for (23) converges in a certain interval, then a fortiori, the first series (23) will converge and will provide a solution of (22). It is necessary to observe that, from the moment the series (22) and (23) are assumed to have positive elements, then the possibility of ordering at x following the substitution of y by the series (23) in the second member of (22), arises uniquely from the fact that x and y belong to the domain of convergence of this second member of (22); this will be the case since the second member will be equal to the first. Finally, *if (22) is replaced by an equation* $Y - G(X,Y)$, *where* $G(X,Y)$ *is a majorant of the second member of (22), if it is known to be solved at* Y, *and of the solution obtained is analytic for* $X < \alpha$, *then this function* $Y(X)$ *will necessarily have positive coefficients* [since it is possible to obtain them by the equations (24)], *and for* $|x| < \alpha$, *the series (23) obtained by the equations (24), will be a solution of (22).*

Such is the mechanism of the calculus of limits or the method of the majorant functions. It is necessary to choose a good majorant permitting the calculations. Taking M to be the sum of the series of moduli of the second member of (22) for $X = |x| = \rho$ and $Y = |y| = \rho'$, we take as the majorant

$$\frac{M}{\left(1 - \frac{X}{\rho}\right)\left(1 - \frac{Y}{\rho'}\right)} - M - M\frac{Y}{\rho'}$$

[the second member of (22) having neither an independent term, nor a term in y]. The limiting equation is

$$Y = \frac{M}{\left(1 - \frac{X}{\rho}\right)\left(1 - \frac{Y}{\rho'}\right)} - M\left(1 + \frac{Y}{\rho'}\right) \tag{25}$$

It is of the second degree in Y:

$$\left(\frac{Y}{\rho'}\right)^2 (M + \rho') - \rho' \frac{Y}{\rho'} + \frac{MX}{\rho - X} = 0 \quad .$$

On taking the solution which is zero for $X = 0$,

$$Y = \frac{\rho'^2}{2(M + \rho')}\left[1 - \left(1 - \frac{4M(M + \rho')X}{\rho'^2(\rho - X)}\right)^{1/2}\right] \quad .$$

We know that $\sqrt{1 - u}$ is expandable for $u < 1$ (u is positive), and has all its coefficients negative, except the first, 1, on taking

$$u = \frac{4M(M + \rho')X}{\rho'^2(\rho - X)}$$

we see that u is a power series in X if $X < \rho$, hence Y be will expandable as a power series if $X < \rho$ and $u < 1$. It will, moreover, be less than $\rho'^2/2\,(M + \rho')$, hence less than ρ'. The condition $u < 1$ yields

$$X < \rho \left(\frac{\rho'}{\rho + 2M}\right)^2 \tag{26}$$

which implies $X < \rho$. The limiting equation (25) hence has an analytic solution by means of condition (26).

On calculating ρ *and* ρ' *as described, we see that equation (22) admits the function (23) as a solution, where the* c_n *are calculated by identification [formula (24)] providing* $X = |x|$ *satisfies condition (26).*

We will observe that a majorant is known for (23), hence a limit of error which will be committed in taking as the nearest value of the solution, the polynomial obtained by neglecting in (23), the terms of degree greater than n. This will be the same for all applications of this method.

We could put equation (22) under a guise which will alternatively show that the solution obtained is the only continuous solution equal to 0 for $x = 0$. Let us put, in effect, $y = z + y$, where y_1 is the determined solution. If $F(x,y) = 0$ is the given equation, i.e., the equation deduced from (22) by putting all the terms in the first member, then we obtain

$$F(x,y_1 + z) = 0 \quad ,$$

and with y_1 replaced by the series, we have a power series in x and z that could be ordered at z. There is no term independent of z since the equation is satisfied for $z = 0$. We have therefore

$$F(x,y) = F(x,y_1 + z) \equiv zG_1(x,z) \equiv (y - y_1)G(x,y) \quad , \tag{27}$$

where $G(x,y)$ is an absolutely convergent power series for x and y sufficiently small. The term independent of x and y in $G(x,y)$ is one, since the term in y is one. *This form of* $F(x,y)$ *due to Weierstrass, shows that* $G(x,y)$ *cannot be zero for* $|x|$ *and* $|y|$ *sufficiently small; the only solution for* $|x|$ *and* $|y|$ *sufficiently small, is* $y = y_1(x)$.

Returning to the general case of $F(x,y) = 0$, we see that if $F(x,y)$ is analytic in a region D and if $F(x_0,y_0) = 0$, $\frac{\partial F}{\partial y}(x_0,y_0) \neq 0$, we will obtain a holomorphic solution for $|x - x_0|$ sufficiently small. If x is a point of this circle, y_1 the value of y at this point, then we again have $F(x_1,y_1) = 0$. In certain cases, we could extend the solution by starting from this new point x_1,y_1 which belongs to D, if $\frac{\partial F}{\partial y}(x_1,y_1)$ is non-zero. Let us restrict our attention to the case where $F(x,y)$ is *given by a convergent power series for any* x,y. *The solution* $y = f(x,x_0,y_0)$ *obtained by starting from the point* (x_0,y_0), *will satisfy* $F(x,y) = 0$ *throughout the circle of convergence,* since $F(x,f(x,x_0,y_0))$ is a holomorphic function of x in this circle, a function which is identically zero in a part of this circle, hence in all of this circle. We see, moreover, that the principle of permanence applies (I, 203). *The analytic continuation of* $f(x,x_0,y_0)$ *will be satisfied throughout the equation.* But it will be necessary that $F(x,y)$ decomposes and that there are other solutions. We see that this generalises what had been said for the algebraic functions.

40 - The case of p equations in p unknowns

The above material generalises to the case of a system of equations in p unknowns and n free variables: $f_j(y_1,y_2,\dots,y_p,x_1,x_2,\dots,x_n) = 0$, $j = 1,2,\dots,p$; which we assume to be satisifed at a point which will be brought back to the origin, and whose functional determinant is non-zero at this point. *If the functions* f_j *are analytic in a domain containing the point in question, then we will have a system of analytic solutions.* To prove this, we will confine ourselves, for the sake of simplicity, to a system of two analytic equations in u,v,x,y,z about the origin. These equations are written as

$$au + bv + cs + dy + ez + \dots = 0 \quad ,$$

$$a'u + b'v + c'x + d'y + e'z + \dots = 0 \quad ,$$

with the condition relative to the functional determinant which assumes the form $ab' - ba' \neq 0$. As in the general, non-analytic case, we first of all solve this system at u and v by eliminating v and u in the two terms of the first order (the first equation is multiplied by b', the second by $-b$, and on adding, this yields an equation for which v does not occur in the first order terms). We will take then the system in the reduced form

$$\begin{cases} u = \sum a_{m,n,p,q,r} x^m y^n z^p u^q v^r \\ v = \sum b_{m,n,p,q,r} x^m y^n z^p u^q v^r \end{cases} \quad . \tag{28}$$

The series in the second member has neither constant terms nor first order terms in u and v. They converge absolutely in the associated circles of convergence, we could therefore assume that they converge absolutely for $|x| = |y| = |z| = \rho$ and $|u| = |v| = R$. We shall let M be the tangent of the two sums of the moduli of terms of these series, for these values of the variables. If we seek the zero solutions at the origin under the guise of power series

$$u = \sum c_{s,t,w} x^s y^t z^w \quad , \qquad v = \sum d_{s,t,w} x^s y^t z^w \quad , \tag{29}$$

then the formal calculation of the coefficients could again be made iteratively as in no. 39, on noting that having replaced u and v by these values in the two members of (28), this being done in the second member for the products of the series $u^q v^r$ and ordered at x,y and z, the coefficients of the various powers of x,y,z are the same in the two members. We have first of all, the coefficients of x,y,z in (29), e.g.

$$c_{1,0,0} = a_{1,0,0,0,0} \quad , \qquad d_{1,0,0,0} = b_{1,0,0,0} \quad .$$

We will then obtain the c and d whose sum of indices is z, by means of the a and b, and the c and d whose sum of indices is 1. Generally, when the c and d, whose sum of indices is $n'-1$, has been calculated, we will obtain those whose sum of indices is n' by equating the coefficients of the powers of x,y,z whose sum is n'. As in the second member the terms containing u or v, are of degree greater than one with respect to the five variables x,y,z,u,v, the terms of degree n' in x and y in the second member only contain the c and d whose sum of indices is at most $n'-1$. We will have

$$c_{s,t,w} = P_{s,t,w}\,(c_{i,j,k}, d_{i,j,k}, a_{m,n,p,q,r})$$

and a similar expression for $d_{s,t,w}$, where the second member is a polynomial with positive coefficients with respect to a, c and d, and whose sum of indices is at most $s+t+w-1$. It again follows that by majoring the second members of the equations (28) and making the same formal calculation, the series obtained for U and V will major the second members of (29) and that the convergence of these majorants of the series (29) will guarantee that of these series and the legitimacy of the operations.

We will obtain some solutions for the equations (28), that are valid in the region of convergence of the solutions of the bounding equations. If we take as majorants for the two series (28)

$$\frac{M}{\left(1-\frac{U+V}{R}\right)\left(1-\frac{X+Y+Z}{\rho}\right)} - M\left(1+\frac{U+V}{R}\right)$$

then the two equations obtained are

$$U = V = \frac{M}{\left(1-\frac{U+V}{R}\right)\left(1-\frac{X+Y+Z}{\rho}\right)} - M\left(1+\frac{U+V}{R}\right)$$

or

$$U = V \quad , \qquad x = X+Y+Z \quad ,$$

$$U = \frac{M}{\left(1-\frac{U}{R'}\right)\left(1-\frac{x}{\rho}\right)} - M\left(1+\frac{U}{R'}\right) \quad , \qquad R' = \frac{R}{2} \quad .$$

We recover the same equation as in the case of one unknown (equation (25)). We will obtain the same conclusion: *the series (29), calculated as above, are absolutely convergent for* x,y,z *less in modulus to*

$$\frac{\rho}{3}\left(\frac{R}{R+4M}\right)^2 \quad ;$$

they yield under these conditions, a solution of the system (28).

If we denote the analytic solutions so obtained by $u_1(x,y,z)$ and $v_1(x,y,z)$, then we can put $u = u_1 + t$, $v = v_1 + \tau$ and, after substituting into the equations (28) where all the terms in the first member are admitted, we will obtain for x,y,z and τ sufficient small, the absolutely convergent power series whose terms will contain t or z in factor. By replacing t and z by their value, $t = u - u_1$, $\tau = v - v_1$, we have on combining equations (28),

$$[u - u_1(x,y,z)]\, g(x,y,z,u,v) = 0 \quad ,$$

$$[v - v_1(x,y,z)]\, h(x,y,z,u,v) = 0 \quad ,$$

where g and h are power series in five variables whose constant term is 1 and which converge absolutely when the moduli of the variables are sufficiently small. *This form puts into evidence that for* x,y,z,u,v *sufficiently small in modulus, the only solutions are* u_1 *and* v_1.

We can extend the solution, in the particular case where the first members of the equations $f_j(y_1,\ldots,x_n) = 0$, $j = 1,2,\ldots,p$ are expandable in absolutely convergent power series, whatever the variables. This occurs especially if these first members are polynomials.

41 - Analytic curves

We say that an arc of a curve is analytic when the cartesian co-ordinates of this arc can be expressed in terms of analytic functions of one parameter. In the case of a curve in three-dimensional space, we will have

$$x = f(t) \quad , \qquad y = g(t) \quad , \qquad z = h(t) \quad , \tag{30}$$

where the functions f,g,h are analytic functions of t in a certain region D.

The real branches of the curve will most often correspond to the real values of the parameter. The representation (30) will be a proper representation of the curve if to a point of thecurve there corresponds, in general, only one value of t in D. We will consider this case. We will pass from one proper representation to another by a simple conformal transformation of the domain D onto another domain Δ.

In the neighbourhood of a point t_o, f, g and h will be expandable as a power series

$$\begin{cases} x = x_o + \alpha_1(t-t_o) + \alpha_2(t-t_o)^2 + \ldots \\ y = y_o + \beta_1(t-t_o) + \beta_2(t-t_o)^2 + \ldots \\ z = z_o + \gamma_1(t-t_o) + \gamma_2(t-t_o)^2 + \ldots \end{cases} \tag{31}$$

converging for $|t-t_o|$ sufficiently small. If $\alpha, \beta_1, \gamma_1$ are not zero simultaneously, then the corresponding point M_o will be called an *ordinary* point of the curve, the line

$$\frac{x-x_o}{\alpha_1} = \frac{y-y_o}{\beta_1} = \frac{z-z_o}{\gamma_1}$$

Will be the tangent to the curve at the point M_o. This definition of ordinary points is independent of the parametrix representation. This property results from the passage of one representation to another: if τ is the new parameter then we will have around t_o, τ_o,

$$t - t_o = S_1(\tau - \tau_o) + \ldots \qquad S_1 \neq 0 \quad .$$

The values of the coefficients of $\tau - \tau_o$ in the new expansions will be $\alpha_1 S_1, \beta_1 S_1, \gamma_1 S_1$.

We can easily see that the definition of the ordinary points is independent of the position of the co-ordinate axes. It is equally invariant under a homographic transformation. The tangent also remains the tangent.

A point which is not ordinary is said to be a singular point. For such a point, the expansions (31) do not have terms of the first degree in $t-t_o$, but there will exist an integer p which will be the least for which the coefficients $\alpha_p, \beta_p, \gamma_p$ will not be zero simultaneously. The tangent will then be the line

$$\frac{x - x_o}{\alpha_p} = \frac{y - y_o}{\beta_p} = \frac{z - z_o}{\gamma_p} \quad .$$

In the case of a plane curve of the plane Oxy, the equations (31) reduce to the first two. Those are the equations of a cycle (no. 11), but this cycle will not be algebraic in general. What was said concerning the cycles and their intersection, did not entail the fact that they were algebraic and so this follows on. We will return to the case of skew curves in the theory of contact (Chapter IV). But let us remark now that it so happens that for two distinct values of the parameter t, t_o and t_1 say, the functions $f(t)$ and $g(t)$, $h(t)$ take the same value, we will obtain *a multiple point* of the curve, but this point will be ordinary for the values t_o and t_1. We will have two ordinary branches passing through the same point. This is the case for algebraic curves.

As we know, a plane curve can be defined by an equation $F(x,y) = 0$. If $F(x,y)$ is analytic about the point x_o,y_o and if $F(x_o,y_o) = 0$ and $\frac{\partial F}{\partial y}(x_o,y_o) \neq 0$, then the arc of the curve passing through $M_o(x_o,y_o)$ will be analytic following the result of no. 39. It will be the same if $\frac{\partial F}{\partial x}(x_o,y_o) \neq 0$, when $\frac{\partial F}{\partial y}(x_o,y_o) = 0$, for x will then be an analytic function of y. We could then analytically continue the solution so obtained so long as $F(x,y)$ remains analytic at the attained points (x,y) and so long as the two first order partial derivatives are not zero simultaneously. The study made (I, 128) in the case when F and the two derivatives are simultaneously zero at the point x_o,y_o, continues to apply: by a series of transformations, a finite number in general, we will discover a new implicit equation for which one first partial derivative is non-zero. These transformations preserve the analytic functions. The axes of the curves obtained will therefore be analytic (instead of using Taylor's formula, we take here the most convenient expansion, since no question arises): in particular, through a double point with distinct tangents, there will pass two ordinary arcs. In the case of a multiple point of greater order, we could separate the cycles by Puiseaux's method (no. 3), but we are no longer certain of arriving at a result at the end of a finite number of operations, unless knowing in advance that there is no decomposition.

Similarly an arc of a skew curve defined by two equations $F(x,y,z) = 0$, $G(x,y,z) = 0$ satisfied for $M_o(x_o,y_o,z_o)$, analytic about this point and such that the functional determinant

$$\frac{D(F,G)}{D(x,y)}(x_o,y_o,z_o) \quad ,$$

for example, is non-zero; it will be analytic about this point, since under these conditions, x and y will not be analytic functions of z for z near to z_o. We will recover the parametric form (31) by setting $z = t$; the point M will be ordinary. It will hence suffice that one of the three functional determinants of F and G with respect to x,y or y,z or z,x, is non-zero, in order to obtain an analytic arc passing through M_o, and M_o will be an ordinary point.

42 - Analytic surfaces

A region of a surface is analytic when the cartesian co-ordinates of a point $M(x,y,z)$ of the surface are analytic functions of two paramters u,v: we have

$$x = f(u,v) \quad , \quad y = y(u,v) \quad , \quad z = h(u,v) \quad ,$$

where f,g,h are analytic in a domain D. If we restrict to a real surface and the real points of the surface, it will be most often assumed that these are the real values of u and v which provide these real points, but it may be interesting to consider matters otherwise. In all cases, we will assume that the parametric representation is proper: at a point M there will correspond in general only a pair of numbers (u,v). The point $M_o(u_o,v_o)$ of the surface will be an ordinary point of one of the functional determinants

$$A = \frac{D(g,h)}{D(u,v)} \quad , \quad B = \frac{D(h,f)}{D(u,v)} \quad , \quad C = \frac{D(f,g)}{D(u,v)}$$

is non-zero at this point (u_o,v_o). We could then solve two of the equations (32) in u and v for $|u-u_o|$ and $|v-v_o|$ sufficiently small and we obtain analytic functions of x and y. One of the co-ordinates x,y,z is then an analytic function of the other two. For example

$$z = F(x,y) \quad . \tag{33}$$

Conversely, given a surface of the form (33), where $F(x,y)$ is analytic for $|x-x_o|$ and $|y-y_o|$ sufficiently small, is a portion of the parametric surface all of whose points are ordinary. We could then put under the form of (32) by taking $x=u$ and $y=v$. Under the form (33), we see that the definition of the ordinary points is independent of the parametric representation, under the form (32); we see at once that it is independent of the axes and is preserved under a homographic transformation.

The points M_o which are not ordinary, are called *singularities*. They can be isolated as the vertex of a cone or form lines (a turning line, for example). But it may happen that for two different curves $v=\phi_1(u)$, $v=\phi_2(u)$, we obtain the same values of x,y,z in the equations (33) and that the points obtained are ordinary. We will have a crossing of two ordinary sheets of the surface.

An equation

$$\Phi(x,y,z) = 0$$

which is satisfied at a point $M_o(x_o,y_o,z_o)$ belonging to a region where the function $\Phi(x,y,z)$ is analytic, will define z, about M_o, as an analytic function of x and y if $\Phi'_z(x_o,y_o,z_o) \neq 0$. We will hence obtain an analytic region. Consequently it suffices that one of the three first order partial derivations of Φ, with respect to x,y,z respectively, is non-zero at M in order that the surface is analytic about M_o and that M_o is an *ordinary* point. When the three partial derivatives in question are zero at M_o, we could have a crossing of ordinary sheets or a possibly isolated singular point.

In particular, when Φ is a polynomial in x,y,z, the surface is said to be algebraic or z is said to be an algebraic function of x and y. All of this applies.

We can extend the definition of the tangent plane at an ordinary point in the case of imaginary points. Under the guise (33), the tangent plane at M_o, was

$$z - z_o = \frac{\partial F}{\partial x}(x_o, y_o)(x - x_o) + \frac{\partial F}{\partial y}(x_o, y_o)(y - y_o) \quad ;$$

under the guise of (32) this will be

$$A(x - x_o) + B(y - y_o) + C(z - z_o) = 0 \quad ;$$

and for the surface defined by $\Phi = 0$, this will always be

$$(x - x_o)\Phi'_x + (y - y_o)\Phi'_y + (z - z_o)\Phi'_z = 0 \quad ,$$

where the partial derivatives arealways taken at M_o.

An arc of a curve of an analytic plane only cuts one line of its plane at a finite number of points (we assume that the analytic arc corresponding to the values of t is taken on a continuous path). A skew analytic arc will only cut a plane at a finite number of points; one part of the analytic surface will only cut a line at a finite number of points (always under the same conditions). This results from the fact that an analytic function of one variable, only has a finite number of zeros in a domain in the interior of which and on the boundary of which, it has no singularities.

43 - Lagrange's formula. The methods of Laplace and Hermite

Consider an equation of theform

$$y = a + xf(y) \tag{34}$$

where $f(y)$ is an analytic function of y in an interval containing the number a (we would take a domain contain a in the complex case). If we put $y - a = Y$, the equation is written as

$$Y = xf(a + Y) \quad . \tag{35}$$

The conditions of the theorm of no. 39 prevail and thus there exists a solution $Y = y - a = \phi(x)$, that is analytic for $|x|$ sufficiently small and is equal to zero for $x = 0$.

Let $F(y)$ be a function of y, analytic for y near to a. If we replace y by $a + \phi(x)$, $F(y)$ becomes a function $\Phi(x)$ which is analytic for $|x|$ sufficiently small, following the theorem on the substitution of series. *We propose to find the coefficients of the expansions of* $\Phi(x)$ *about the origin, i.e., the quantities*

$$\frac{1}{n!}\Phi^{(n)}(0) \quad .$$

To calculate the derivatives of Φ at the origin, we depend on the fact that Y defined by (35) is a function of x *and of* a. *It is an analytic function of* x *and* a *for* $|x|$ *sufficiently small and* a *near to its initial value.* For if we put $a = \alpha + z$, $f(a + Y)$ becomes $f(\alpha + z + Y)$ which is analytic at z and Y for $|z|$ and $|Y|$ sufficiently small and the equation (35) is an equation with one unknown Y and two variables z and x which recover those of no. 40. We would therefore consider $y = a + Y$ as a function of x and a, $\phi(x,a)$, where this function has partial derivatives of all orders. Laplace's method consists in utilising the partial derivatives with respect to a. Let us derive the identity of (34) assuming y replaced by $\phi(x,a)$. We will have

$$[1 - xf'(y)]\frac{\partial y}{\partial x} = f(y) , \qquad [1 - xf'(y)]\frac{\partial y}{\partial a} = 1 ;$$

$$\frac{\partial y}{\partial x} = f(y)\frac{\partial y}{\partial a} .$$

We also have

$$F(y) = F[\phi(x,a)] = \Phi(x,a) ,$$

hence

$$\frac{\partial \Phi}{\partial x} = F'(y)\frac{\partial y}{\partial x} , \qquad \frac{\partial \Phi}{\partial a} = F'(y)\frac{\partial y}{\partial a}$$

and consequently, following (36),

$$\frac{\partial \Phi}{\partial x} = f(y)\frac{\partial \Phi}{\partial a} .$$

We will then have

$$\frac{\partial^2 \Phi}{\partial x^2} - f'\frac{\partial y}{\partial x}\frac{\partial \Phi}{\partial a} + f\frac{\partial}{\partial a}\left(\frac{\partial \Phi}{\partial x}\right) = f'f\frac{\partial y}{\partial z}\frac{\partial \Phi}{\partial a}\; f\frac{\partial}{\partial a}\left(\frac{\partial \Phi}{\partial a}f\right)$$

$$= \frac{\partial f}{\partial a}\cdot f\frac{\partial \Phi}{\partial a} + f\frac{\partial}{\partial a}\left(\frac{\partial \Phi}{\partial a}f\right)$$

$$= \frac{\partial}{\partial a}\left[f\frac{\partial \Phi}{\partial a}\right] .$$

More generally we will have

$$\frac{\partial^n \Phi}{\partial x^n} = \frac{\partial^{n-1}}{\partial a^{n-1}}\left[f^n\frac{\partial \Phi}{\partial a}\right] . \qquad (37)$$

Effectively, the equality occurs for $n = 1$ and $n = 2$. If we differentiate with respect to x, we will obtain in the second member the derivative of order $n-1$ with respect to a of

$$\frac{\partial}{\partial x}\left[f^n \frac{\partial\Phi}{\partial z}\right] = nf^{n-1}\frac{\partial f}{\partial x}\frac{\partial\Phi}{\partial a} + f^n\frac{\partial}{\partial a}\left(\frac{\partial\Phi}{\partial x}\right)$$

$$= nf^{n-1}\frac{\partial f}{\partial n}\cdot f\frac{\partial\Phi}{\partial a} + f^n\frac{\partial}{\partial a}\left(f\frac{\partial\Phi}{\partial a}\right)$$

$$= \frac{\partial}{\partial a}\left(f^n f\frac{\partial\Phi}{\partial u}\right) \quad .$$

The formula (37) is therefore general. For $x = 0$, y reduces to a, hence $f(y)$ is $f(a)$ and $\frac{\partial\Phi}{\partial a}$ is equal to $F'(a)$. Consequently,

$$\Phi^{(n)}(0) = \frac{d^{n-1}}{da^{n-1}}\left[F'(a)(f(a))^n\right] \quad .$$

The expansion of $F(y)$ *is*

$$F(y) = F(a) + \sum_1^\infty \frac{x^n}{n!}\frac{d^{n-1}}{da^{n-1}}\left[F'(a)(f(a))^n\right] \quad .$$

This is Lagrange's formula.

EXAMPLES

I. If we consider the equation

$$y = a + \frac{1}{2}(y^2 - 1),$$

we obtain

$$y = a + \sum_1^\infty \frac{x^n}{2^n n!}\frac{d^{n-1}}{da^{n-1}}(a^2-1)^n \quad .$$

Now, we have in this case,

$$y = \frac{1}{x} - \frac{1}{x}\sqrt{1 - 2ax + x^2}$$

which means that, for x and a sufficiently small, we obtain a double series at a and x, that is absolutely convergent. We can then differentiate with respect to a, which gives

$$\frac{1}{\sqrt{1 - 2ax + x^2}} = 1 + \sum_1^\infty x^n P_n(a) \quad ,$$

where $P_n(x)$ is the Legendre polynomial of degree n (I, 100). The formula obtained is the Legendre expansion (I, 101).

II. If we consider Kepler's equation

$$u = a + e \sin u \quad ,$$

we obtain for e sufficiently small (where e is the eccentricity)

$$u = a + e \sin a + \ldots + \frac{e^n}{n!} \frac{d^{n-1}}{da^{n-1}} (\sin^n a) + \ldots \quad .$$

III. The solution with respect to x of

$$y = a_1 x + a_2 x^2 + \ldots \qquad a \neq 0 \quad ,$$

where the series converges for $|x| < R$, can only be made by writing the equation in the form

$$x = y\, f(x) \quad , \qquad f(x) = \frac{1}{a_1 + a_2 x + \ldots} \quad ;$$

we will obtain

$$x = \sum_1^{\infty} \frac{y^n}{n!} \left[\frac{d^{n-}}{dx^{n-}} \quad (f(x))^n \right]_{x=0} \quad .$$

HERMITE'S METHOD

We shall assume that $f(y)$ is a holomorphic function of y for $|y - a| \leq R$. Following Rouche's theorem, the function of y

$$G(y) = y - a - xf(y)$$

admits the same number of zeros as $y - a$, hence *a zero* in the circle $|y - a| < \rho \leq R$ providing that on the circumference of this circle, we have $|xf(y)| \leq \rho' < \rho$. If $H(y)$ is holomorphic for $|y - a| \leq \rho$, we will have, on applying the residue formula (I, 192) and denoting the zero of $\mathfrak{G}(y)$ by y:

$$\frac{H(y)}{G(y)} = \frac{1}{2i\pi} \int_{C^+} \frac{H(\xi)}{G(\xi)}\, d\xi = \frac{1}{2i\pi} \int_{C^+} \frac{H(\xi)\, d\xi}{\xi - a - xf(\xi)} \quad . \tag{38}$$

Now for ξ on C, we could expand a uniformly convergent series

$$\frac{1}{\xi - a - xf(\xi)} = \frac{1}{\xi - a} \sum_o^{\infty} \left(\frac{xf(\xi)}{\xi - a} \right)^n \quad ,$$

and the third member in (38) could be replaced by a series whose general term is, following the Cauchy formulae,

$$\frac{x^n}{2i\pi}\int_{C^+}\frac{H(\xi)[f(\xi)]^n}{(\xi-a)^{n-1}}d\xi = \frac{x^n}{n!}\frac{d^n}{da^n}\,[H(a)(f(a))^n] \quad .$$

We have then the relationship

$$\frac{H(y)}{G(y)} = H(a) + \sum_1^\infty \frac{x^n}{n!}\frac{d^n}{da^n}\,[H(a)(f(a))^n] \quad . \tag{39}$$

Let us apply this formula to the function

$$H(y) = (1 - xf'(y))F(y) = G'(y)F(y) \quad ,$$

where $F(y)$ is a given function that is holomorphic for $|y-a| \leq \rho$. We will obtain

$$F(y) = F(a)[1 - xf'(a)] + \sum_1^\infty \left\{\frac{x^n}{n!}\frac{d^n}{da^n}(Ff^n) - \frac{x^{n+1}}{n!}\frac{d^n}{da^n}(Ff^nf')\right\} \quad .$$

We can regroup the terms corresponding to the same power of x, since the two series obtained by taking one or the other of the two terms in the bracket are convergent (this would be obtained, to within a factor for the second, by applying (39) to the functions F and Ff'). On making this regroupment, we recover Lagrange's formula

$$F(y) = F(a) + \sum_1^\infty \frac{x^n}{n!}\frac{d^{n-1}}{da^{n-1}}\,[F'(a)(f(a))^n] \quad .$$

The proof shows that the formula is true if f and F are holomorphic for $|y-a| \leq \rho$ and if $|xf(y)| \leq \rho' < \rho$ for $|y-a| = \rho$. It will be necessary then to take $|x|$ at most equal to the quotient of ρ by the maximum of $|f(y)|$ for $|y-a| = \rho$, $M(\rho)$ say. We are led to find to the maximum of $\rho/M(\rho)$ when ρ varies from 0 to the smallest radii of holomorphy of f and F about a. In particular if $f(y)$ is an entire function, this maximum will be attained for a value r and will be finite (if we put aside the uninteresting case where $f(y)$ admits a as a double zero, in the case where the solution to the equation will be $y = a$). Then the formula will be true for $|x|$ less than this maximum if $F(y)$ is holomorphic for $|y-a| < r$.

Hermite has applied this method to Kepler's equation.

44 - A remark on the solutions of the implicit equations

In the proof of Lagrange's formula by Laplace's method, we have used the fact that the solution was analytic not only at x, but also with respect to the parameter a. This is a general fact. If the equations of a system contain a finite number of parameters, then we may consider these parameters as new variables (and if they are given a very weak field of variation, then the domain of validity of the solutions will not change by a great deal) and the solutions will be analytic not only with respect to the variables, but also with respect to the parameters.

III. AN APPLICATION OF THE METHOD OF MAJORANT FUNCTIONS TO DIFFERENTIAL EQUATIONS

45 - The case of one equation

Let

$$y' = f(x,y) \tag{40}$$

be a solvable differential equation whose second member is expandable as a power series about the point (x_o, y_o). *We are first of all going to establish that this equation admits a unique holomorphic solution which takes the value* y *for* $x = x_o$. We may assume here that $x_o = 0$, $y_o = 0$; we are reduced to this case by setting $x = x_o + X$, $y = y_o + Y$, which does not change y': we have $Y' = y'$. Let the equation then be

$$y' = a_{0,0} + a_{1,0}x + a_{0,1}y + \dots \quad , \tag{41}$$

where the series of the second member is absolutely convergent for $|x| \le r$ $|y| \le R$. We are going to seek a solution, zero for $x = 0$, in the form

$$y = c_1 x + c_2 x^2 + \dots + c_n x^n + \dots \quad . \tag{42}$$

If this series converges for $|x| < \alpha$, if the sum of the moduli of those terms is less than R, then we may replace y by this series in the second member of (41) and order at x. If the series obtained has its terms equal to those of y', then we will have found a solution for $|x| < \alpha$: the sum of (42). We are going to show that the calculation of the coefficients c_n, made by assuming the above hypotheses hold, is made in terms of such relation that if the second member of (41) is replaced by a majorant, then the series (42) is also replaced by a majorant. It will suffice, therefore, to show that we could define a majorant that effectively allows us to find a solution with the above properties.

The formal calculation of the coefficients, starting from the equation

$$c_1 + 2c_2 x + 3c_3 x^2 + \dots + nc_n xn^{-1} + \dots$$
$$= a_{0,0} + a_{1,0}x + a_{0,1}(c_1 x + \dots) + \dots$$

gives

$$c_1 = a_{0,0} \quad , \qquad 2c_2 = a_{1,0} + a_{0,1}c_1 \quad \ldots$$

$$nc_n = a_{0,1}c_{n-1} + a_{1,1}c_{n-2} + a_{0,2}\,(c_{n-2}c_1 + \ldots + c_1c_{n-2})$$

$$+ \ldots + a_{0,n-1}c_1^{n-1} \quad .$$

For the same reasons as in no. 39, we also see that c_n is a polynomial with respect to the c_j, $j \leq n-1$ and with respect to certain $a_{m,p}$ with positive coefficients. It is clear then, that if the $a_{m,p}$ are replaced by the positive numbers $A_{m,p}$ greater than their moduli, c_1 will be replaced by a number $C_1 \geq |c_1|$, and hence then c_2 by $C_2 \geq |c_2|$, and so on.

If we put $M = |a_{0,0}| + |a_{0,1}|R + |a_{1,0}|r + \ldots$, we will take as a majorant of the second member of (41),

$$\frac{M}{\left(1 - \frac{X}{r}\right)\left(1 - \frac{V}{R}\right)}$$

and the limiting equation will be

$$Y'\left(1 - \frac{Y}{R}\right) = \frac{M}{1 - \frac{X}{r}} \quad , \tag{43}$$

which can be integrated straight away. We obtain

$$\left(1 - \frac{Y}{R}\right)^2 = \frac{2Mr}{R}\log\left(1 - \frac{X}{r}\right) + \text{const.}$$

and the solution which is zero for $X = 0$, is

$$Y = R\left[1 - \left(1 + \frac{2mr}{R}\log\left(1 - \frac{X}{r}\right)\right)^{1/2}\right] \quad . \tag{44}$$

The second member is a holomorphic function of X as long as $X < r$ and that the root is non-zero, hence

$$X < r(1 - e^{-R/2\,Mr}) \quad . \tag{45}$$

Under these conditions, the function (45) satisfies equation (43) and its expansion necessarily coincides with that which would yield the direct calculation of the coefficients. The coefficients of the expansion of (44) are then positive and, when they are substituted into (43) solved with respect to Y', they produce in the second member, a series with positive coefficients which converges since the sum is Y'. We first of all show at once that the expansion of Y indeed has its coefficients positive and it is clear that $Y < R$. Consequently: *With the numbers* r, R *and* M *determined as above, then the differential equation (41) admits a holomorphic solution (42), where the* c_n *are calculated by identification; this series converges when* $|x| = X$ *satisfies condition (45). This is the only holomorphic solution in this circle and it is zero at the origin.*

46 - Uniqueness of the solution. The dependence of the initial conditions. The integrating factor

We are going to see that equation (40) has, under the indicated conditions, a unique analytic solution about the point (x_o, y_o). But the argument only shows that there does not exist any analytic solutions, singularities at the origin, satisfying the initial conditions, or non-analytic solutions. Briot and Bouquet have partly established this impossibility of the existence of other solutions. Their arguments were completed by E. Picard. *To prove the uniqueness of the solution, we shall follow the method of Weierstrass and Poincaré.* We shall thus seek the analytic solution obtained depending on the initial conditions x_o, y_o. We assume that in equation (40), the second member is analytic for $|x-\alpha| \le r$, $|y-\beta| \le R$ and we seek a solution which takes the value y_o for $x = x_o$, where x_o is near to α and y_o to β. Here again we shall simplify matters by taking $\alpha = \beta = 0$. The equation therefore becomes

$$\frac{dy}{dx} = \sum\sum a_{m,n} x^m y^n \quad , \tag{46}$$

where the double series is absolutely convergent by hypothesis, for $|x| \le r$, $|y| \le R$. Let us set $x = x_o + \xi$, $y = y_o + \eta$. If we have

$$|x_o| + |\xi| \le r \ , \qquad |y_o| + |\eta| \le R \quad ,$$

then on replacing x and y by $x_o + \xi$, $y_o + \eta$, then we can order the series of the second member of (46) at ξ, η. We will obtain

$$\eta' = \sum\sum b_{p,q} \xi^p \eta^p \quad , \tag{47}$$

and we can apply the above method to this equation. But we shall seek the limiting equation by taking as the majorant of the second member

$$\frac{R}{\left(1 - \frac{|x_o| + X}{}\right)\left(1 - \frac{|y_o| + Y}{R}\right)} \ , \qquad M = \sum\sum |a_{m,n}| r^m R^n \quad .$$

[We indeed have a majorant of the second member of (47) since the second member is none other than $\sum\sum a_{m,n}(x_o + \xi)^m (y_o + \eta)^n$.] The limiting equation will then be

$$\frac{dY}{dX} = \frac{M}{\left(1 - \frac{|x_o| + X}{r}\right)\left(1 - \frac{|y_o| + Y}{R}\right)}$$

the integral which is zero for $X = 0$ will be

$$Y = (R - |y_o|) \left\{1 - \left[1 + \frac{2MrR}{(R - |y_o|)^2} \log \frac{r - |x_o| - X}{r - |x_o|}\right]^{1/2}\right\} \tag{48}$$

The integral of (47), zero for $\xi = 0$, is holomorphic in the circle obtained by setting the root of the second member of (48), equal to zero. In this circle is an analytic function not only of ξ, but also of x_o and y_o, for the $b_{p,q}$ are power series in x_o and y_o and when the majorant has been taken, the terms of these series have been replaced by their moduli. We also see in (48) that the second member is indeed a power series in $|x_o|$, $|y_o|$ and X (the logarithm is $\log(1-u)$ with $u = X/(r - |x_o|)$, hence the logarithm is a power series in X and $|x_o|$ with all negative coefficients, etc.). We take

$$|x_o| < \frac{r}{2}, \qquad |y_o| < (1-k)R, \qquad 0 < k < 1 \quad . \tag{49}$$

X will then be subjected to the condition

$$X < \frac{r}{2}(1 - e^{-\nu}), \qquad \nu = \frac{k^2 R}{2Mr} \quad .$$

Thus:

Theorem 1. *If* x_o *and* y_o *satisfy conditions (49), then the solution of (46), which is equal to* y_o *for* $x = x_o$ *and which is holomorphic at this point, is holomorphic in the circle*

$$|x - x_o| < \frac{r}{2}(1 - e^{-\nu}) = r', \quad \nu = \frac{k^2 R}{2Mr} \quad . \tag{50}$$

It is expandable as a power series in x_o, y_o $x - x_o$ *and is absolutely convergent by the inequalities (49) and (50);* $|y - y_o|$ *remains less than* kR.[1]

This theorem of Poincaré and Weierstrass already shows that the equation cannot be verified by an analytif function which, being singular at the origin, would tend to zero when x tends to zero. For this solution would be holomorphic at a point x' as near as is wished to the origin and there y will take a value of y' correspondingly near to zero. Now for as long as

---

[1] Since Y given by (48) is a series in $1/(R - |y_o|)$ and is maximum for $|y_o| = (1-k)R$.

$|x'| < r'$, where r' is the number defined in (50), and $|y'| < (1-k)R$, the solution in question will coincide with the solution determined by the initial conditions $x_o = x'$, $y_o = y'$. This solution is holomorphic for $|x - x'| < r'$, therefore at the origin, whence we derive a contradiction.

AN EQUATION CONTAINING PARAMETERS

If the second member of (46) depends analytically on a finite number of parameters, then the solution will be an analytic function of these parameters; in particular, it will always be continuous in a neighbourhood of the origin. For we may assume that the parameters vary from certain initial values which can be assumed to be zero. The second member of (46) is then of the form

$$\sum\sum\sum\sum a_{m,n,p,q}x^m y^n \lambda^p \mu^q$$

if there are two parameters; and the series converges absolutely for $|x| \leq r$, $|y| \leq R$, $|\lambda| \leq b$, $|\mu| \leq b$. We will then obtain a majorant of the form

$$M/\left(1 - \frac{|x_o| + X}{r}\right)\left(1 - \frac{|y_o| + Y}{R}\right)\left(1 - \frac{|\lambda|}{b}\right)\left(1 - \frac{|\mu|}{b}\right) \quad ,$$

and nothing will be changed in the calculations other than the coefficients of η will depend on x_o, y_o, λ, μ. If we take $|\lambda| < b/2$ and $|\mu| < b/2$, then the radius of convergence obtained for the integral which takes the value y_o for $x = x_o$, is deduced from that of above by changing M to 4M. We will have here

$$y = y_o + \phi(x - x_o, x_o, y_o, \lambda, \mu)$$

where the function ϕ is analytic with respect to the five variables appearing. There will in particular be *a continuity of the solution* under the indicated conditions: it will vary little for weak variations of the λ, μ, x_o, y_o. This property of the solutions has been presented by Poincaré (*Les nouvelles méthodes de la mécanique celeste*, 1892); it is important for applications, since practically, we only know the values approaching numerical constants that enter into the equation, and equally the values approaching the initial conditions.

If $y' = F(x,y,\lambda,\mu)$ is the equation and $y = \phi(x,\lambda,\mu)$ a solution taking the value y_o for $x = x_o$, then this function will admit derivatives with respect to λ and μ. If we set, for example, $\Lambda = \frac{\partial\phi}{\partial\lambda}(x,\lambda,\mu)$, we see that Λ is a solution of the linear differential equation obtained by differentiating with respect to λ, the two members of the given equation;

$$\frac{d\Lambda}{dx} = \frac{\partial F}{\partial y}(x,y,\lambda,\mu) + \frac{\partial F}{\partial \lambda}(x,y,\lambda,\mu) \quad .$$

These are the equations of the form that Poincare called *equations aux variations*. The product of Λ and a small increment in λ, will yield the principal part of the variation corresponding to y.

A SHEAF OF SOLUTIONS

Following Theorem I, the solution taking the value y_o for x_o, is given by

$$y = y_o + (x - x_o)\phi(x - x_o, y_o, y_o) \tag{51}$$

where ϕ is a power series in $x - x_o, x_o, y_o$ that is absolutely convergent. We shall take $|x_o| < r'/2$ and the value of the number k is taken to be 1/4. We must then assume in (51), that $|x - x_o| < r'$, $|y_o| < 3R/4$ and $|y - y_o|$ will be less than $R/4$. *If we assume* $|y_o| < R/2$, then the solution y given by (51) will have its modulus less than $3R/4$ and at an arbitrary point x' such that $|x'| < r'/2$, this solution S will take a value y' of modulus less than $3R/4$. The solution S', which corresponds to the initial values x', y', a solution given by

$$y = y' + (x - x')\psi(x - x', x', y') \quad ,$$

will coincide with S about the point x', y' since it will only have one holomorphic solution at a point, it is holomorphic throughout the circle $|x - x'| < r'$, hence *a fortiori* for $|x| < r'/2$, just as for S. These two solutions coincide. In particular for $x = x_o$, S' takes the value y_o. We have

$$y_o = y' + (x_o - x')\phi(x_o - x', x', y') \quad .$$

But x was any point of the circle $|x| < r'/2$ and y' the value of y at this point. Hence:

Lemma. *If* $|x_o| < r'/2$, $|y_o| < R/2$, *then the holomorphic solution which takes the value* y_o *for* $x = x_o$, *satisfies the implicit equation*

$$y_o = y + (x_o - x)\phi(x_o - x, x, y) \tag{52}$$

in the circle $|x| < r'/2$. *In (51), we can invert the roles of* x,y *and* x_o, y_o.

If we set $x_o = 0$, then the second member of (52) becomes a function $\Theta(x,y) = y - x\phi(-x,x,y)$ which is zero at the origin, expandable as a power series for $|x| < r'/2$ and $|y| < 3R/4$ and whose partial derivative with respect to y is equal to the one at the origin. Just as every point x,y for which $|x| < r'/2$, $|y| < R/4$, appertains to a holomorphic solution whose modulus is less than R/2 at the origin, we have the result:

Theorem 11. *For any* y_o *of modulus less than* R/2, *there exists a solution of equation (46) which is holomorphic for* $|x| < r'/2$ *and which satisfies the relation*

$$y_o = \Theta(x,y) \quad , \tag{53}$$

where $\theta(x,y)$ *is the function defined above. If* x *and* y *are arbitrary, but* $|x| < r'/2$ *and* $|y| < R/4$, *there exists a solution defined by (53) which takes the value* y *for this value* x.

INTEGRATING FACTOR

If $|x| < r'/2$ and $|y| < R/4$, then the holomorphic solution taking this value y for this x, satisfies (53) where y_o has the value of Θ at this point x,y. We have then for this solution

$$\frac{\partial\Theta}{\partial x} + \frac{dy}{dx}\frac{\partial\Theta}{\partial y} \equiv 0 \ , \qquad \frac{dy}{dx} = f(x,y) \quad ,$$

and consequently

$$f(x,y) = -\frac{\dfrac{\partial\Theta}{\partial x}}{\dfrac{\partial\Theta}{\partial y}} \tag{54}$$

when the point x,y belongs to the domain in question.

If the two members of equation (46) taken in the form

$$dy - f(x,y)dx = 0 \quad ,$$

are multiplied by $\frac{\partial\Theta}{\partial y}$, then it becomes

$$\frac{\partial\Theta}{\partial y}\,dy + \frac{\partial\Theta}{\partial x}\,dx = 0$$

or

$$d\Theta(x,y) = 0 \quad .$$

We call $\frac{\partial\Theta}{\partial y}(x,y)$ *an integrating factor.*

UNIQUENESS OF THE SOLUTION

Let us suppose that the equation admits a solution $\Phi(x)$ that is zero for $x = 0$. If we replace y by $\Phi(x)$ in $\Theta(x,y)$, we obtain a function $\Theta(x,\Phi,x)$ whose derivative is

$$\frac{\partial\Theta}{\partial x} + \frac{\partial\Theta}{\partial y}\,\Phi'(x) \quad .$$

This derivative is zero following (54), since $\Phi'(x) = f(x,y)$. It follows that $\Theta(x, (x))$ is constant and this constant is zero since $\Theta(x,y)$ is zero at the origin along with $\Phi(x)$. The function $y = \Phi(x)$ is hence a solution of the implicit equation $\Theta(x,y) = 0$. This solution is unique since Θ satisfies the conditions of no. 39 (the uniqueness having been proved in no. 39 for all continuous functions of real or complex variables).

47 - Systems of equations of the first order

The results relating to a differential equation easily extend to a system of such equations. To simplify matters, let us take three equations of the form

$$\frac{dx}{dt} = f(x,y,z,t), \quad \frac{dy}{dt} = g(x,y,z,t), \quad \frac{dz}{dt} = h(x,y,z,t) \quad . \tag{55}$$

(If we have p equations, then they would be written as $x'_j = f_j(x_1, x_2, \ldots, x_p, t)$ $\quad j = 1,2,\ldots,p$). We assume that the functions f, y, h are expandable as a power series in $x - \alpha$, $y - \beta$, $z - \gamma$, $t - \delta$ that are absolutely convergent for $|x - \alpha| \leq a$, $|y - \beta| \leq b$, $|z - \gamma| \leq c$, $|t - \delta| \leq t$. Under these conditions, there exists a system of solutions

$$x - \alpha = \phi(t-\delta), \quad y - \beta = \psi(t-\delta), \quad z - \gamma = \chi(t-\delta)$$

where the functions ϕ, ψ, χ are holomorphic for $|t-\delta|$ sufficiently small, and are zero for $t = \delta$. To show this, we proceed as in no. 45. We can assume that $\alpha = \beta = \gamma = \delta = 0$. The second members of equations (55) are power series in four variables

$$\frac{dx}{dt} = \sum A_{m,n,p,q} x^m y^n z^p t^q \quad ,$$

$$\frac{dy}{dt} = \sum B_{m,n,p,q} x^m y^n z^p t^q$$

$$\frac{dz}{dt} = \sum C_{m,n,p,q} x^m y^n z^p t^q \quad .$$

We proceed to satisfy the equations by taking

$$x = \sum_1^\infty D_j t^i \ , \quad y = \sum_1^\infty E_j t^i \ , \quad z = \sum_1^\infty K_j t^j \quad . \tag{57}$$

The calculation of the coefficients D_j, E_j, K_j is made iteratively by means of the three equations. We equate the coefficients of t^j in the two members of equations (56) where x, y, z are replaced by their values. When the D, E, K have been calculated up to the index $j - 1$, we obtain jD_j, jE_j, and jK_j by means of the polynomials with positive coefficients, with respect to the D, E, K with indices less than j and with respect to the A, B, C. It follows that the method of the majorant functions applies. We take the same majorant

$$\frac{M}{\left(1-\frac{X}{\rho}\right)\left(1-\frac{Y}{\rho}\right)\left(1-\frac{Z}{\rho}\right)\left(1-\frac{T}{\tau}\right)}$$

for the three second members of the equations (56) where ρ denotes the smallest of the three numbers a, b, c.

$$\frac{dX}{dT} = \frac{dY}{dT} = \frac{dZ}{dT} = \frac{M}{\left(1-\frac{X}{\rho}\right)\left(1-\frac{Y}{\rho}\right)\left(1-\frac{Z}{\rho}\right)\left(1-\frac{T}{\tau}\right)} \quad .$$

As $X = Y = Z = 0$ for $T = 0$, we see that $X = Y = Z$ and X is given by the equation

$$\left(1-\frac{X}{\rho}\right)^3 dX = \frac{M\tau dT}{-T} \quad ,$$

from which we extract X

$$X = \rho\{1 - [1 + \frac{4Mt}{\rho} \log(1-\frac{T}{\tau})]^{1/4}\} \quad .$$

This function is holomorphic for T less than the zero of the root

$$T < t(1 - \rho^{-\rho/4M\tau}) \quad .$$

It follows that under the indicated conditions, the system (56) has a system of holomorphic solutions when $|t| = T$ *satisfies the inequality (58), these solutions are zero for* $t = 0$. On returning to the system (55), we will have a system of solutions on taking for $t = \delta$ the assigned values α, β, γ when f, g, h are analytic about the point $\alpha, \beta, \gamma, \delta$.

Being in possession of this method, we see how it can be modified to study the way in which the solutions depend on the initial conditions and deduce from this, the uniqueness of the solutions and the existence of the first integrals. Taking the hypothesis $\alpha = \beta = \gamma = \delta = 0$, we seek solutions taking the values x_o, y_o, z_o for t_o. We therefore set $x = x_o + \xi$, $y = y_o + \eta$, $z = z_o + \zeta$, $t = t_o + \lambda$ to revert to the above case, but we let x_o, y_o, z_o, t_o appear in the transformed equations of which the first is

$$\frac{d\xi}{dt} = \sum A_{m,n,p,q}(x_o+\xi)^m(y_o+\eta)^n(z_o+\zeta)^p(t_o+\lambda)^q \quad . \tag{59}$$

We order the second members as series in $\xi, \eta, \zeta, \lambda$ which is possible if $|x_o| + |\xi| \le \rho$, etc. The calculation by identification of the coefficients of ξ, η, ζ will give these coefficients under the guise of the series in x_o, y_o, z_o, t_o. Now these series will be majored if the second members of the equations are majored such that (59) is considered as a series in eight variables $x_o, \xi, \ldots, t_o, \lambda$. We will assume that here $|x_o| < U$, $|y_o| < U$, $|z_o| < U$ and $U + X \leq \rho$, $U + Z \leq \rho$, $|t_o| + |\lambda| \leq t$. We will take as the common majorant

$$M/\left(1 - \frac{U+X}{\rho}\right)\left(1 - \frac{U+Y}{\rho}\right)\left(1 - \frac{U+Z}{\rho}\right)\left(1 - \frac{T+|t_o|}{t}\right), \quad T = |\lambda|$$

such that the limiting equations yield $X = Y = Z$ and matters reduce to solving the same equations as before where X changes to $X + U$ and T to $T + |t_o|$. The zero solution at the origin is

$$X = (\rho - u)\ \{1 - \left[1 + \frac{4M\tau\rho^3}{(\rho-u)^4} \log\left(1 - \frac{T}{\tau = |t_o|}\right)\right]^{1/4}\} \quad .$$

The outcome is the same as in the one variable case. Taking $U = (1-k)\rho$, $0 < k < 1$, $|t_o| < \tau/2$, we see that: *If* $|t_o| < \tau/2$ *and* $|x_o|$, $|y_o|$, $|z_o|$ *are less than* $(1-k)\rho$, *then the equations (56) admit solutions* x, y, z *which for* $t = t_o$, *take the values* x_o, y_o, z_o *and which are holomorphic in the circle*

$$|t-t_o| < \tau' = \frac{\tau}{2}\,(1 - \rho^{-v}) \ , \qquad v = \frac{k^4\rho}{4M\tau} \quad .$$

These solutions are expandable as absolutely convergent power series in x_o, y_o, z_o, t_o, $t-t_o$ *under the given conditions and* $|x-x_o|$, $|y-y_o|$, $|z-z_o|$ *stay less than* k_ρ.

The uniqueness of the analytic solutions follows. *An analogous result would be obtained if there were parameters.*

If we write the solutions as

$$x = x_o + (t-t_o)\phi(t-t_o, x_o, y_o, z_o, t_o)$$

$$y = y_o + (t-t_o)\psi(t-t_o, x_o, y_o, t_o)$$

$$z = z_o + (t-t_o)\chi(t-t_o, x_o, y_o, t_o) \ ,$$

then we can see that if $|t_o| < \tau'/2$ and $|x_o|$, $|y_o|$, $|t_o|$ are less than $\rho/2$, then we could change the x, y, z with x_o, y_o, z_o. It follows again that for $t_o = 0$: *For any* x_o, y_o, z_o *of modulus less than* $\rho/2$, *there exists a system of holomorphic solutions for* $|t| < \tau'/2$ *which satisfies the relations*

$$x_o = F(x,y,z,t) , \quad y_o = G(x,y,z,t) , \quad z_o = H(x,y,z,t) , \tag{60}$$

where the functions F, G, H *are zero at the origin, are expandable as power series for* $|t| < \tau'/2$ *for* $|x|, |y|, |z|$ less than $\zeta\rho/4$, *and such that*

$$\begin{cases} F'_x(0,0,0,0) = 1 , & F'_y(0,0,0,0) = 0 , & F'_z(0,0,0,0) = 0 \\ G'_x(0,0,0,0) = 0 , & G'_y(0,0,0,0) = 1 , & G'_z(0,0,0,0) = 0 \\ H'_x(0,0,0,0) = 0 , & H'_y(0,0,0,0) = 0 , & H'_z(0,0,0,0) = 1 . \end{cases} \tag{61}$$

Every point x, y, z, t *with* $|t| < \tau'/2$ *and* $|x|, |y|, |z|$ *less than* $\rho/4$ *belongs to a system of solutions which forms part of the sheaf (60).*

It results that for f, g, h as the second members of equations (56), we have the relations

$$F'_x f + F'_y y + F'_z h + F'_t = 0$$
$$G'_x f + G'_y y + G'_z h + G'_t = 0$$
$$H'_x f + G'_y y + G'_z h + H'_t = 0 .$$

It follows that if x, y, z are replaced in F, G, H by a system of solutions that is zero at the origin $t = 0$, then F, G, H are constant in the neighbourhood of the origin, hence zero, and as the conditions (61) imply that the system $F = 0$, $G = 0$, $H = 0$ is solvable uniquely in x, y, z then this solution is none other than the holomorphic system. *The uniqueness is proved. The functions* F, G, H *are called the first integrals.*

48 - The equation of order n

A solvable differential equation

$$y^{(n)} = f(x,y,y',y^n,\ldots,y^{(n-1)}), \tag{62}$$

is equivalent to a system of equations:

$$\frac{dy}{dx} = y_1, \quad \frac{dy_1}{dx} = y_2, \ldots, \quad \frac{dy_{n-2}}{dx} = y_{n-1} \quad ,$$

$$\frac{dy_{n-1}}{dx} = f(x,y,y_1,y_2,\ldots,y_{n-1}) \quad .$$

By applying the result obtained we see that we could assign the values y_o, $y'_o,\ldots,y_o^{(n-1)}$ say, for $y,y',\ldots,y^{(n-1)}$ for $x = x_o$. If $f(x,y,\ldots,y^{(n-1)})$ is analytic in a domain containing this point $x_o,y_o,\ldots,y_o^{(n-1)}$, then there exists a unique solution $y = \phi(x,x_o,\ldots,y_o^{(n-1)})$ taking for $x = x_o$, the value y_o, whilst y' takes the value $y_o',\ldots,$ and the derivative of order n-1 of ϕ, takes the value $y_o^{(n-1)}$, and this solution is holomorphic about the point x_o. This function is equally analytic with respect to x_o, $y_o,\ldots,y_o^{(n-1)}$. *We say that* $x_o,y_o,\ldots,y_o^{(n-1)}$ *are the initial conditions.*

CONTINUATION OF THE SOLUTION

The solution obtained starting from the determined initial conditions, can be analytically extended. If we extend along a line Γ for x_o, then the extended function will satisfy the differential equation so long as the function $f(x,y,\ldots,y^{(n-1)})$ is analytic about the point $x,y,\ldots,y^{(n-1)}$. But generally this is not the case, as singularities of the solutions are anticipated. *The singularities depend in general on the initial conditions* $x_o,y_o,\ldots,y_o^{(n-1)}$; *they are then called moving singularities.* It is this mobility that makes the overall study of solutions of differential equations difficult.

Example. The equation $y' = y^2$ has a second member that is everywhere holomorphic. The solution which takes the value y_o for $x = x_o$, is, as can at once be seen,

$$y = \frac{y_o}{1 - y_o(x-x_o)} \quad .$$

It admits the moving pole $x = x_o + \frac{1}{y_o}$. The radius of holomorphy of the solution about the point x_o is $1/|y_o|$, it is as small as $|y_o|$ is large.

THE CASE OF NON-SOLVABLE EQUATIONS

Let us assume that the differential equation cannot be solved. For example, let us restrict matters to a second order equation for brevity; suppose we have

$$F(x,y,y',y'') = 0 \quad .$$

In order to apply the theorem relating to the existence of solutions, we need firstly to solve for y'', or at the very least, see if the conditions for applying the theorem are satisfied when having found a solution. If the function $F(x,y,y',y'')$ is analytic about the point x_o,y_o,y_o',y_o'', and if the partial derivative with respect to y'' will be an analytic function of x, y,y' for $|x-x_o|$, $|y-y_o|$ and $|y'-y_o'|$ sufficiently small. There will exist a unique holomorphic solution taking the value y_o for x_o, for which its derivative at this point is equal to y_o'.

The case where $F(x,y,y',y'')$ *is a polynomial.* If the equation is of the form

$$F(x,y,y',y'') = 0 \quad ,$$

where F is a polynomial in x,y,y',y'', then to a system of values x_o, y_o, y_o' of x, y, y', there will correspond values of y'' which are roots of the algebraic equation $F(x_o,y_o,y'_o,y'') = 0$. If y''_o is a *simple finite root* of this equation then $F'_{y''}$ will not be zero at this point, we can solve with respect to y'' and y'' will be analytic about the point x_o, y_o, y'_o. The differential equation will therefore limit a unique holomorphic solution for which y', y', y'' will take the values y_o, y'_o, y''_o for $x = x_o$. If we extend this solution, it will continue to satisfy the equation for as long as it exists. This holds for each finite simple root of the equation in y''. Every solution $y = \phi(x)$ of the equation will appear in the set of solutions thus obtained, unless we do not always have

$$F(x,\phi,\phi',\phi'') \equiv 0 \ , \quad \frac{\partial F}{\partial y''}(x,\phi,\phi',\phi'') = 0 \quad . \tag{63}$$

One such solution $y = \phi(x)$ is either a *singular solution* or a *multiple solution*. If ϕ'' is eliminated from the algebraic equations (63), we see that ϕ must satisfy the equation

$$D(x,\phi,\phi') = 0 \quad . \tag{64}$$

The singular solutions, if there are any, are therefore solutions of equation (64) which is again algebraic and of lower order. By fixing x_o, the solutions of the second order equation depend on y_o and y'_o which are arbitrary, hence with arbitrary constants. Those of equation (64) no longer depend on one constant and amongst them, it is again necessary to select those that satisfy the first equation (63). In all cases, the singular solutions from a set less rich than the integrals obtained directly by the theorem of existence.

In the case of an algebraic equation of the first order,

$$F(x,y,y') = 0 \quad ,$$

where F is a polynomial in x, y, y', a singular integral will be a solution of this equation and will satisfy $F'_{y'} = 0$. On eliminating y', we obtain

$$D(x,y) = 0 \quad ,$$

where $D(x,y)$ is a polynomial. *The only singular integral possible is the algebraic curve defined by* $D(x,y) = 0$, which can moreover be decomposed. If it is not an integral, which can be proved directly, there is no singular integral. We will return to the singular integrals in no. 70 and in Chapter VIII.

49 - The case of linear systems

A system of n differential equations solvable with respect to the derivatives of the unknown functions $y_1,\ldots,y_n$ is said to be linear if the second members are linear with respect to the y_j. It is of the form

$$\frac{dy_j}{dx} = A_{j,1}y_1 + A_{j,2}y_2 + \dots + A_{j,n}y_n + A_{j,n+1}, \quad j = 1,2,\dots,n \tag{65}$$

where the $A_{j,k}$ are functions of the variable x.

Theorem. *The solutions of a linear system can only admit the singularities of the coefficients* $A_{j,k}$ *as its singularities.*

Preliminary Remark. Clearly we assume here that the $A_{j,k}(x)$ are analytic functions of x and that there exists a domain where at least one of the branches of each of these functions is holomorphic.

Let then x_o be a point for which each function $A_{j,k}(x)$ has at least one holomorphic branch. We know that the system has then a system of holomorphic solutions in a circle of centre x_o. We are going to see that the radius of the circle of holomorphy with centre x_o is, for all solutions, at least equal to the shortest distance from x_o to the singular branches of the $A_{j,k}(x)$ in qeustion. We may assume $x_o = 0$ and that the values of y_i are zero at this point. We assume that all the branches of $A_{j,k}(x)$ are holomorphic for $|x| \leq R$. We could then take as majorants of the $A_{j,k}(x)$, the function $\frac{M}{1 - \frac{X}{R}}$ for $|x| \leq R$. The limiting equations of the system (65) could then be taken in the form

$$\frac{dY_1}{dX} = \frac{dY_2}{dX} = \dots = \frac{dY_n}{dX} = \frac{MR}{R-X}(Y_1 + Y_2 + \dots + Y_n + 1) \quad ,$$

with $X = |x|$ and $Y_j = |y_j|$. We will then have $Y_1 = Y_2 = \dots = Y_n$, since the Y_j are all zero for $X = 0$. Hence we will obtain

$$\frac{dY}{dX} = \frac{MR}{R-X}(nY + 1)$$

where Y is the common value of these functions. The integral, zero for $X = 0$, is given by

$$nY + 1 = \left(1 - \frac{X}{R}\right)^{-nRM} \quad ;$$

Y is holomorphic for $X < R$, hence also the $y_j(x)$.

If we take any polygonal line Γ from x_o and if, when extending the branches of the $A_{j,k}(x)$ along this line no singular points of the functions are encountered on a segment $x_o x_1$ of Γ, then each point x of Γ will be a centre of a circle not containing the singularities of the $A_{j,k}(x)$. We could cover the segment $x_o x_1$ of Γ with a finite number of these circles, then augment their number to obtain a chain of circles, the first with centre x_o, the last with centre x_1, such that each centre (with the exception of x_o) is in the preceding circle. The solutions $y_j(x)$ are holomorphic in the circle with centre x_o which contains the second centre x'. At this point x' they coincide with the holomorphic solutions calculated with the initial conditions x', $y_j(x')$, hence these new solutions extend into the second circle; and so on. The solutions are extendable along Γ as long as no singularities of $A_{j,k}(x)$ are encountered.

Particular cases. Remarks

I. It could happen that there exist several distinct domains in which the $A_{j,k}$ are holomorphic, whilst certain $A_{j,k}$ admit boundaries of these domains as singular lines. We will then obtain solutions corresponding to each of these domains.

II. If all the $A_{j,k}$ are uniform functions, then the solutions could be multiform. Thus the solutions of

$$y' = k\frac{y}{x}$$

are

$$y = Cx^k \quad ,$$

where C is an arbitrary constant. This solution has an infinity of branches once k is not real and rational, and a finite number for k real, rational but not integer.

III. In particular, if the $A_{j,k}$ are rational fractions, then the only singularities of the solutions will be poles of these fractions. But the solutions could be multiform. If the $A_{j,k}$ are entire functions, or especially polynomials, then the solutions are entire functions.

AN APPLICATION TO LINEAR EQUATIONS OF ORDER n

A linear equation of order n is an equation of the form

$$A_o(x)\frac{d^n y}{dx^n} + A_1(x)\frac{d^{n-}y}{dx^{n-1}} + \ldots + A_{n-1}(x)\frac{dy}{dx} + A_n(x)y + A_{n+1}(x) = 0 \quad .$$

If we divide both members by $A_o(x)$, then the equation reduces to a system of linear equations by proceeding as in no. 48. The coefficients of these equations are the ratios $A_1(x)/A_o(x),\ldots,A_{n+1}(x)/A_o(x)$. The only possible singular points of the solutions will be the singular points of these functions. For example, if all the $A_j(x)$ are polynomials, then the possible singular points of the solutions are the zeros of $A_o(x)$.

IV. ON THE THEORY OF HOLOMORPHIC FUNCTIONS OF TWO VARIABLES FOLLOWING POINCARÉ

50 - Monogeneous functions of two variables

Let $f(z,z')$ be a function of two variables defined in a domain D. D is a four dimensional domain: by showing the real part and the coefficient of i in z, z', f, we have $z = x + iy$, $z' = x' + iy'$ and $f(z,z') = P(x,y,x',y') + iQ(x,y,x',y')$, hence two functions of four variables.

The function $f(z,z')$ is said to be monogeneous at a point x,y of D when it admits at this point, the first partial derivatives with respect to z and z'. Following what was said for the functions of one variable, for that itself is similar, it is necessary and sufficient that $P(x,y,x',y')$ and $Q(x,y,x',y')$ are differentiable with respect to x,y on one hand, and with respect to x',y' on the other, and that the following relations are satisfied.

$$\frac{\partial P}{\partial x} = \frac{\partial Q}{\partial y}, \quad \frac{\partial P}{\partial y} = -\frac{\partial Q}{\partial x}, \quad \frac{\partial P}{\partial x'} = \frac{\partial Q}{\partial y'}, \quad \frac{\partial P}{\partial y'} = -\frac{\partial Q}{\partial x'} \quad .$$

By eliminating Q, we find that P must satisfy four second order equations.

HOMOMORPHIC FUNCTIONS OF TWO VARIABLES

These are the functions $f(z,z')$ defined *and continuous in* z,z' in a domain D, and at each point they are monogeneous. When z is assigned a certain value, we obtain a holomorphic function of z'; similarly for z' constant, $f(z,z')$ is a holomorphic function of z.

Following no. 36, an analytic function of z and z' is holomorphic.

DOUBLE INTEGRALS

Let us assume that $f(z,z')$ is holomorphic when z belongs to a domain A of the z-plane, and z' belongs to a domain A' of the z'-plane. Let Γ and Γ' be two rectifiable curves; one situated in A and the other in A'. Let us indicate on Γ, a set of dividing points $z_o, z_1, \ldots, z_p, z_{p+1}, \ldots, z_n$ and on Γ' a set of points $z'_o, z'_1, \ldots, z'_q, z'_{q+1}, \ldots, z'_m$, where $z, z_n = Z_o$ are the extremities B and C of Γ and $z'_o, z'_m = Z'_o$ are the extremities B' and C' of Γ'. *The sum*

$$\sum_{p=0}^{n-1} \sum_{q=0}^{m-1} (z_{p+1} - z_p)(z'_{q+1} - z'_q)\, f(\zeta_p, \zeta'_q) \quad , \tag{66}$$

where ζ_p *is an arbitrary point taken on* Γ *between* z_p *and* z_{p+1} *and* ζ'_q *the same on* Γ' *taken between* z'_q *and* z'_{q+1}, *tends to a limit when the largest chords* (z_{p+1}, z_p) *of* Γ *and* z'_q, z'_q *of* Γ', *tend to zero.* This limit is denoted by

$$\int_{\Gamma}\int_{\Gamma'} f(z,z')dzdz' \quad .$$

In order to establish the existence of this limit, we shall confine ourselves to the case where, along Γ, y is a monotonic function of x and conversely x is a monotonic function of y, and similarly along Γ, $y' = \psi(x')$ and $x' = \chi(y')$. We would then pass to the more general cases by dividing the arcs Γ and Γ'. By setting $z_p = x_p + iy_p$, $z'_q = x'_q + iy'_q$, $\zeta_p = \xi_p + i\eta_p$, $\zeta'_q = \xi'_q + i\eta_p$, $f = P + iQ$, we see that the expression (66) is the sum of eight expressions such that

$$\sum_{p=0}^{n-1}\sum_{q=0}^{m-1} (x_{p+1} - x_p)(x'_{q+1} - x'_q)P(\xi_p,\eta_p,\xi'_q,\eta'_q) \quad . \tag{67}$$

Along Γ, we have $\eta_p = \phi(\xi_p)$ and on $\Gamma', \eta'_q = \psi(\xi'_q)$, where the functions ϕ and ψ are continuous, $P(\xi_p,\phi(\xi_p),\xi'_q,\psi(\xi'_q))$ is a continuous function of ξ_q, ξ'_q and the expression (67) has as a limit the double integral

$$\iint P(x,\phi(x),x',\psi(x'))dxdx'$$

extended to the rectangle $x_o < x < X_o$, $x'_o < x' < X'_o$, $X_o = RZ_o$, $X'_o = RZ'_o$. By calculating this double integral in the usual way, we obtain

$$\int_{x_o}^{X_o} dx \int_{x'_o}^{X'_o} P(x,\phi(x),x',\psi(x'))dx' \quad .$$

Proceeding in the same way for the other seven, we find by regrouping the terms, that

$$\int_{x_o}^{X_o} dx \left[\int_{x'_o}^{X'_o} [P(x,\phi,x',\psi) + iQ(x,\phi,x',\psi)]dx' \right.$$
$$\left. + \int_{y_o}^{Y'_o} [P(x,\phi,x,y') + iQ(x,\phi,\chi,y')]idy' \right]$$

and a similar term. Now the bracket is the curvilinear integral

$$\Phi(z) = \int_{\Gamma'} f(z,z')dz', \qquad z = x + i\phi(x) \quad ;$$

the second term will be similarly

$$i\int_{y_o}^{Y_o} \Phi(z)dy, \qquad z = \theta(y) + iy \quad .$$

Finally, we have

$$\int_\Gamma \int_{\Gamma'} f(z,z')dzdz' = \int_\Gamma \Phi(z)dz = \int_\Gamma dz \int_{\Gamma'} f(z,z')dz' \quad .$$

We can, moreover, reverse the order of integration in the last member. It is clear that, if $|f(z,z')| < M$ on Γ, Γ', and of L is the length of Γ and L' that of Γ', then the modulus of the double integral is at most MLL'.

51 - Extension of the Cauchy formula. Applications. Weierstrass's Theorem

Let us suppose that $f(z,z')$ is holomorphic when z belongs to a domain D and z' to a domain D'. Let x be a point of the domain D, x' a point of D', Γ a simple closed curve belonging to D, bounding a domain contained in D and containing the point x; Γ' a simple closed curve belonging to D' and bounding a domain contained in D' and containing x'. We assume that Γ and Γ' have the properties invoked in proving the theorem on the double integral. We will obtain

$$\int_\Gamma \int_{\Gamma'} - \frac{f(z,z')dzdz'}{(z-x)(z'-x')} = \int_{\Gamma'} \frac{dz'}{z'-x'} \int_\Gamma \frac{f\ z,z'\ dz}{z-x}$$

We assume that the integrals are taken in the direct sense. In the second member, the function $f(z,z')$ where z' is constant is a holomorphic function of z, the interior integral is then equal to $2i\pi\, f(x,z')$ following the fundamental theorem of Cauchy (I, 183). The second member therefore reduces to

$$2i\pi \int_{\Gamma'} f(x,z') \frac{dz'}{z'-x'}$$

and a second application of Cauchy's theorem yields its value $-4\pi^2 f(x,x')$. Consequently,

$$f(x,x') = -\frac{1}{4\pi^2} \int_\Gamma \int_{\Gamma'} \frac{f(z,z')dzdz'}{(z-x)(z'-x')} \tag{68}$$

This formula, analogous to that of Cauchy, implies the same consequences. Letting x_o be a point of D and x'_o a point of D', let us take for Γ and Γ', two circles with these points as centres and contained, along with their interiors, in D and D' respectively. If x belongs to the interior of Γ and x' to that of Γ', then $|x-x_o| < |z-x_o|$ and $|x'-x'_o| < |z'-x'_o|$ uniformly if z is on Γ and z' on Γ', hence

$$\frac{1}{(z-x_o)(z'-x')} = \frac{1}{(z-x_o)(z'-x'_o)} \frac{1}{\left(1-\frac{x-x_o}{z-x_o}\right)\left(1-\frac{x'-x_o}{z'-x'_o}\right)}$$

$$= \sum_o^\infty \sum_o^\infty \frac{(x-x_o)^m(x'-x'_o)^n}{(z-x_o)^{m+1}(z'-x'_o)^{n+1}}$$

The convergence is uniform for z on Γ and z' on Γ'. We could carry this expression into (68) and integrate term by term. As the rest of the series is uniformly less than ε on Γ and Γ' if enough terms have been taken, then the rest of the integral is less than $\varepsilon MLL'$, where M is the bound of $|f(z,z')|$. We thus obtain the Taylor expansion of $f(x,x')$

$$f(x,x') = \sum_o^\infty \sum_o^\infty a_{m,n}(x-x_o)^m(x'-x'_o)^n$$

with

$$a_{m,n} = \frac{1}{4\pi^2} \int_\Gamma \int_{\Gamma'} \frac{f(z,z')dzdz'}{(z-x_o)^{m+1}(z'-x'_o)^{n+1}} \quad . \tag{69}$$

Thus, the holomorphic functions are analytic. The associated radii of convergence corresponding to the series expansion about a point x_o, x'_o are the radii of the largest circles with centres x_o, x'_o contained in D and D'. The partial derivatives of $f(z,z')$ of all orders, are again holomorphic in D and D'. There are given, as in the one variable case, by the integrals obtained by differentiating (68) within the sign of integration. This results from (69) and from the fact that the double integral taken along the boundaries of a domain where $f(z,z')$ is holomorphic (the boundaries included) is zero (for the interior integral is zero). The Cauchy inequalities (I, 184) extend; they provide a majorant of the series expansion. Poincaré has extended the theory of the residues.

THE THEOREM OF WEIERSTRASS

Following the theorem on the continuity of the roots established in no. 2, a holomorphic function $f(x,y)$ in a domain containing the point x_o, y_o and which is zero at this point, where y_o is a zero of order p of $f(x_o, y_o)$, has the following property. There exists a circle $|x-x_o| \leq r$ in which the equation in y, $f(x,y) = 0$ admits exactly p roots such that $|y-y_o| < \rho$. These p roots $y_1, y_2, \ldots, y_p$ are functions of x. *Let us first of all show that they are the roots of an algebraic equation*

$$y^p + A_1(x)y^{p-1} + \ldots + A_{p-1}(x)y + A_p(x) = 0 \tag{70}$$

whose coefficients $A_j(x)$ *are holomorphic for* $|x-x_o| < v$.

As we have already seen in the theory of algebraic functions, it suffices to show that the sums $y^k + y^k + \ldots + y_p^k$, k a positive integer are holomorphic functions of x. Let Γ be the circle $|\xi - x_o| = r$ and Γ' the circle $|\eta - y_o| = \rho$, and x a number such that $|x - x_o| < r$. Following Cauchy's theorem on residues, we have

$$\frac{1}{2i\pi}\int_{\Gamma'} \eta^k \frac{\frac{\partial f(x,\eta)}{\partial \eta}}{f(x,\eta)}\, d\eta = y_1^k + \ldots + y_p^k = \Phi(x) \quad .$$

Let us consider the double integral

$$I = \int_{\Gamma'} d\eta \int_{\Gamma} \eta^k \frac{\frac{\partial f(\xi,\eta)}{\partial \eta}}{f(\xi,\eta)} \frac{d\xi}{\xi - x} \quad .$$

The interior integral is equal to

$$2i\pi\eta^k \frac{\frac{\partial f(x,\eta)}{\partial \eta}}{f(x,\eta)}$$

following Cauchy's theorem, since $f(x,\eta)$ is non-zero in Γ and on Γ. It follows that $I = -4\pi^2\Phi(x)$. Now I is a holomorphic function of x power $|x - x_o| < r$, since $1/(\xi - x)$ could be expanded as a uniformly convergent power series in $x - x_o$ since $|\xi - x_o| = r$ on Γ. The roots $y_j(x)$ are indeed solutions of an equation (70) with holomorphic coefficients. The first member of (70) is a function $y(x,y)$ holomorphic for $|x - x_o| < r$ and arbitrary y. Let

$$h(x,y) \equiv \frac{f(x,y)}{g(x,y)} \quad ,$$

$h(x,y)$ is a holomorphic function of x,y for $|x - x_o| < r$ and $|y - y_o| < \rho$.

This can again be seen by considering the double integral

$$J = \int_{\Gamma'} d\eta \int_{\Gamma} \frac{h(\xi,\eta)}{(\xi - x)(\eta - y)}\, d\xi$$

which is holomorphic at x,y as can be seen as above [$y(x,y)$ is non-zero on the cricles Γ, Γ' following the properties of the roots of $f(x,y) = 0$]. Now when $|\eta - y_o| = \rho$, $h(x,\eta)$ is holomorphic at x for $|x - x_o| \leq r$, since $g(x,\eta)$ is non-zero under these conditions. The interior integral is therefore equal to

$$2i\pi \frac{h(x,\eta)}{\eta - y} \quad .$$

For a given x in $|x-x_o| < r$ and $g(x,y)$ as holomorphic functions of y have the same zeros (with the same order of multiplicity), hence $h(x,y)$ is holomorphic in y for $|y-y_o| \leq \rho$ and

$$J = -4\pi^2 h(x,y)$$

$h(x,y)$ is therefore holomorphic and we obtain the

Theorem of Weierstrass: *The function* $f(x,y)$ *can be decomposed into a product*

$$g(x,y)h(x,y)$$

of holomorphic functions; $g(x,y)$ *is the polynomial (70) in* y, *and* $h(x,y)$ *a function which is non-zero for* $|x-x_o| < r$ *and* $|y-y_o| < \rho$.

Chapter IV

THE THEORY OF CONTACT-ENVELOPES

The study of contact of curves and surfaces and the determination of envelopes was developed at the same time as the theory of algebraic or transcendental curves and surfaces, that is to say, as early as the foundations of analytic geometry. The introduction of such considerations in analysis enters into the theory of ordinary and partial differential equations. For the latter, the integration methods of Lagrange and Cauchy depend on the theory of contact and envelopes. Lagrange was the first to state the condition for analytic curves to have a given order of contact. Moreover, the theory of contact and envelopes is a direct application of the theory of implicit functions and analytic functions.

I. THE THEORY OF CONTACT

52 - Definition of the order of contact

We shall consider two curves or surfaces of three dimensional space (and eventually two plane curves), that have a common point. We assume that this point is *an ordinary point*. The sense of this word has been given when it applies to analytic curves or surfaces (nos. 41 and 42). But the definition can be extended to the case where the coordinates, instead of being analytic functions of one or two parameters (in the case of curves and surfaces respectively), are expandable by Taylor's formula, under the sufficient conditions for which the calculations and definitions remain tenable. For example, we know that given two plane curves defined by the expansions

$$y = \sum_{2}^{\infty} a_n x^n , \qquad y_1 = \sum_{2}^{\infty} b_n x^n ,$$

converging in a small interval $|\alpha| < \alpha$, providing

$$a_2 = b_2,\ a_3 = b_3, \ldots, a_p = b_p, \qquad a_{p+1} \neq b_{p+1} \tag{1}$$

the two curves have contact of order p at the origin. This definition holds if y and y_1, instead of being expandable as power series, have only continuous derivatives up to order $p + 1$. For we could then write

$$y = a_2 x^2 + \ldots + a_p x^p + (a_{p+1} + \varepsilon) x^{p+1} ,$$

$$y_1 = b_2x^2 + \ldots + b_px^p + (b_{p+1}+\varepsilon')x^{p+1} \quad ,$$

where ε and ε' tend to zero with $|x|$. The conditions (1) imply that the difference $y-y_1$ is of infinitesimal order with respect to x. We will obtain a contact of order p. But if we have $b_{p+1} = a_{p+1}$, it will be necessary to appeal to other terms of Taylor's formula and hence make alternative hypotheses on the derivatives.

We will always assume in advance that the matters in question, i.e., the curves and surfaces, are in contact, i.e., for curves, they have the same tangent at a common point, and for surfaces, they have the same tangent plane at this point. In the case of a curve and a surface, the tangent to the curve is to be located in the tangent plane to the surface.

In the case of analytic curves or surfaces, it is important to establish definitions tenable for the complex elements. *The difference* of two points is taken, instead of their distance; this was Jordan's approach. The difference of two points M, M' with coordinates x, y, z and x', y', z' respectively, will be the number $|x'-x| + |y'-y| + |z'-z|$. *This difference depends on the coordinate cases.* Let (M,M') denote this difference for a system of axes and let (M,M') denote the new difference after a coordinate transformation (not rectangular in general), then at once it can be seen that we will have $(M,M')' < K(M,M)$ where the finite number K only depends on the transformation. Conversely, we will have $(M,M') < K_1(M,M')$ for K_1 finite; hence the difference is defined to within a positive factor. We clearly have $(M,M'') \leq (M,M') + (M',M'')$.

Definition: *Let us consider two figures* Γ_o, Γ_1 *(curves or surfaces) having an ordinary common point* O *and which are in contact at this point. We say that* Γ *has a contact of order* n *with* Γ_1 *at* O *if when every point* M *of* Γ_1 *tends to zero, we can assign a point* M_1 *of* Γ_1 *such that the difference* (M,M') *is infinitesimally small with respect to* $(O,M)^n$ *and* n *is the largest integer with this property.*

NOTATION

To avoid taking ε constantly, we will denote by $o(u)$ each function of u which is infinitesimally small with respect to u (when u tends to zero), hence such that $\lim\limits_{u=o} \frac{o(u)}{u} = 0$ (Bachmann's notation).

In the above terms, we then assume

$$(M,M_1) = o[(O,M)^n] \quad .$$

A PROPERTY

It can be seen straight away that the contact is preserved under an invertible pointwise transformation.

53 - The contact of two curves

If we consider a curve Γ passing through the origin, which is an ordinary point, we can suppose that the origin corresponds to the value 0 of the parameter t. The equations will then be of the form

$$\begin{cases} x = \alpha_1 t + \alpha_2 t^2 + \ldots + \alpha_p t^p + o(t^p) \quad , \\ y = \beta_1 t + \beta_2 t^2 + \ldots + \beta_p t^p + o(t^p) \quad , \\ z = \gamma_1 t + \gamma_2 t^2 + \ldots + \gamma_p t^p + o(t^p) \quad ; \end{cases} \tag{2}$$

these expressions come about by Taylor's formula, where x, y, z *have continuous derivatives up to order* p *about* $t = 0$. The point is ordinary, $|\alpha_1| + |\beta_1| + |\gamma_1| \neq 0$. We can assume $\alpha_1 \neq 0$ and solve the first equation with respect to t. With t having continuous derivatives up to order p with respect to x (I, 121), we will have

$$t = \tau_1 x + \ldots + \tau_p x^p + o(x^p)$$

and putting this into y and z, which will have derivatives with respect to x, we will obtain the expansions

$$\begin{cases} y = a_1 x + a_2 x^2 + \ldots + a_p x^p + o(x^p) \quad , \\ z = b_1 x + b_2 x^2 + \ldots + b_p x^p + o(x^p) \quad , \end{cases} \tag{3}$$

with

$$a_j = \frac{1}{j!} \frac{d^j y}{dx^j}(o) \ , \qquad b_j = \frac{1}{j!} \frac{d^j z}{dx^j}(o) \quad .$$

In the analytic case, we will have convergent series in place of those Taylor expansions. If a second curve Γ_1 is tangent to Γ at the origin, then the directing parameters of the tangent at 0 are proportional to those of Γ. For 0 an ordinary point, we could then solve with respect to the same coordinate x, and Γ_1 could be placed under the form:

$$\begin{cases} y_1 = c_1 x_1 + c_2 x_1^2 + \ldots + c_p x_1^p + o(x_1^p) \\ z_1 = d_1 x_1 + d_2 x_1^2 + \ldots + d_p x_1^p + o(x^p) \quad , \end{cases} \tag{4}$$

where the c_j and d_j are again given as for Γ.

Let us assume that the contact of Γ with Γ_1 is of order $n \leq p - 1$. We must have $(M,M_1) = o[(0,M)^n]$, then *a fortiori* for the points M, M_1,

$$|x-x_1| = O[(0,M)^n], \quad |y-y_1| = o[(0,M)^n], \quad |z-z_1| = o[(0,M)^n] \quad .$$

Now

$$(0,M) = |x| + |y| + |z| = (1 + |a_1| + |b_1|)|x| + o(|x|) \quad ,$$

is such that the above conditions may be written as $|x-x_1| = O(x^n)$, $|y-y_1| = o(x^n)$, $|z-z_1| = o(x^n)$. We then have $x_1 = x + o(x^n)$, when it follows that

$$y_1 = c_1 x + c_2 x^2 + \dots + c_n x^n + o(x^n) \quad ,$$

$$z_1 = d_1 x + d_2 x^2 + \dots + d_n x^n + o(x^n) \quad ,$$

and the conditions relative to $y-y_1$ and $z-z_1$ imply that we must have

$$a_1 = c_1, \; a_2 = c_2, \dots, \; a_n = c_n, \quad b_1 = d_1, \; b_2 = d_2, \dots, \; b_n = d_n \quad .$$

Moreover, one of the differences $a_{n+1} - c_{n+1}$, $b_{n+1} - d_{n+1}$ must not be zero, otherwise on taking $x_1 = x$, we would have

$$(M,M_1) = o[(0,M)^{n+1}] \quad .$$

These necessary conditions are sufficient. For if they are realised, we have on taking $x_1 = x$,

$$(M,M_1) = [o(0,M)^n] \quad ,$$

and there is no correspondence such that n is replaced by $n+1$ in this equality, since this implies $a_{n+1} = c_{n+1}$ and $b_{n+1} = d_{n+1}$. Thus: When Γ and Γ are given as in (3) and (4), such that there is contact of order n, $(n \leq p-1)$, it is necessary and sufficient that we have

$$\begin{aligned} &c_j = a_j, \quad b_j = d_j \qquad j = 1,2,\dots,n \\ &|c_{n+1} - a_{n+1}| + |d_{n+1} - b_{n+1}| \neq 0 \quad . \end{aligned} \tag{5}$$

The conditions (5) could be written by setting the derivatives

$$\frac{d^j y}{dx^j} \quad \frac{d^j y_1}{dx_1^{\,j}}, \quad \frac{d^j z}{dx^j} \quad \frac{d^j z_1}{dx_1^{\,j}}, \quad j = 1,2,\dots,n \quad ,$$

$$\left|\frac{d^{n+1} y}{dx^{n+1}} \quad \frac{d^{n+1} y_1}{dx_1^{n+1}}\right| \quad \left|\frac{d^{n+1} z}{dx^{n+1}} \quad \frac{d^{n+1} z_1}{dx_1^{n+1}}\right| \neq 0 \quad ,$$

where these derivatives are taken at the origin.

A CONSEQUENCE

The order of contact of Γ *with* Γ_1 *is the same order of contact as that of* Γ_1 *with* Γ. We can speak of the order of contact of Γ and Γ_1 without anymore precision.

Remark I. For the real plane curves, we see that if the contact is of even order, then the curves cross each other; this is not the case if the order is odd.

Remark II. By assuming that only $x(t)$, $y(t)$ and $z(t)$ are differentiable up to order p at the point $t = 0$ (I, 41), we can write the equations in (2). We could in fact invert the first equation and write y and z under the form (3). We will then obtain the conditions (5). The same remark applies to everything that follows.

THE CASE OF TWO PLANE CURVES WHERE ONE IS IN AN IMPLICIT FORM

We shall always take the common point to be the origin, and the common tangent will assume to be arbitrary. We will have

$$x = \alpha_1 t + \alpha_2 t^2 + \dots \qquad |\alpha_1| + |\beta_1| \neq 0 \quad ; \qquad (\Gamma)$$
$$y = \beta_1 t + \beta_2 t^2 + \dots$$

extended such that in the non-analytic case, the limiting expansion will go sufficiently far. On the other hand,

$$F(x_1,y_1) = 0 \qquad (\Gamma_1)$$

with the conditions

$$\left|\frac{\partial F}{\partial x_1}(0,0)\right| + \left|\frac{\partial F}{\partial y_1}(0,0) \neq 0\right| \neq 0 \quad ,$$

$$\alpha_1 \frac{\partial F}{\partial x_1}(0,0) + \beta_1 \frac{\partial F}{\partial y_1}(0,0) = 0 \quad .$$

Let us take M arbitrary on Γ, in a neighbourhood of the origin; we see that $(0,M)/|t|$ remains taken between two fixed numbers. We can replace (0,M0 by t. If $M_1(x_1,y_1)$ is on Γ_1, we have

$$F(x,y) = F(x,y) - F(x_1,y_1) = (A+\varepsilon)(x-x_1) + (B+\varepsilon')(y-y_1) \quad ,$$

where A and B are the first partial derivatives of F at the origin, hence $|A| + |B| \neq 0$ and ε and ε' tend to be zero with $(0,M)$ as long as we assume $(M,M_1) = o(0,M)$. We have then, for K finite,

$$|F(x,y)| < K(M,M_1) \quad .$$

If the order of contact is n, then we could take M_1 such that $(M,M_1) = o((0,M)^n)$, hence $|F(x,y)| = o[(0,M)^n]$. Let us suppose that conversely,

$$|F(x,y)| = o[(0,M)^n] \tag{7}$$

(where M is on Γ). If $A \neq 0$, we could assign the point M_1 of Γ_1 to M such that $y_1 = y$; x_1 will indeed be determined by $F(x,y) = 0$ since $A \neq 0$ and (6) will yield

$$(M,M_1) = |x-x_1| < K'|F(x,y)| = o[(0,M)^n] \tag{8}$$

If n was replaced by $n+1$ in (7), and similarly in (8), then the contact would not be of order n, but at least $n+1$. As we can replaced $(0,M)$ by t, we see that:

The curves Γ *and* Γ_1 *having been defined in the above expressions, the necessary and sufficient condition for the contact to be of order* n *is that the function of* t, $F(\alpha_1 t + \alpha_2 t^2 + \ldots + \beta_1 t + \beta_2 t^2 + \ldots)$ *is infinitesimally small with respect to* t^n, *but not with respect to* t^{n+1}.

Consequently, the first n derivatives of the function of t above, must be zero for $t = 0$ and the derivative of order $n+1$ must not be zero.

Example. If we have

$$F(x,y) \equiv y - ax + bx^2 + cxy + dy^2 + \ldots$$

then the tangent $y = ax$ has a contact of one if $b + ca + da^2 \neq 0$. Let us consider a tangential circle (in rectangular axes)

$$y - ax + k(x^2 + y^2) = 0 \quad ,$$

which on taking x as a parameter, may be written as

$$y = ax - k(1 + a^2)x^2 + \ldots \quad .$$

We will have to consider the function

$$(-k(1 + a^2) + b + ca + da^2)x^2 + \ldots$$

where the contact will be of order 1, except when $k(1 + a^2) = b + ca + da^2$, where the contact will be of order at least 2 and the circle will be the osculating circle.

THE CASE OF A CURVE Γ GIVEN BY $F(x,y,z) = 0$, $G(x,y,z) = 0$, WHERE Γ_1 IS GIVEN IN PARAMETRIC FORM.

As in the preceding case, we shall see that if

$$x = f(t), \quad y = g(t), \quad z = h(t) \quad ,$$

are the equations of Γ_1, the (ordinary) common point corresponding to $t=0$ is the origin, and the tangent is the same for both curves, then we shall have contact of order n if the derivatives (from the second onwards, where the first is zero) of $F(f,g,h)$ and of $G(f,g,h)$ are zero up to order n, where one of the derivations of order $n+1$ is non-zero.

Example. We have the curve

$$x = t + \alpha_2 t^2 + \ldots, \quad y = \beta_2 t^2 + \ldots, \quad z = \gamma_3 t^3 + \ldots,$$

and the circle (rectangular axes) $z = 0$, $x^2 + y^2 - 2Ry = 0$. If we take the function

$$(1 - 2R\beta_2)t^2 + \ldots \quad ,$$

then the contact will be of order at least 2 if $2R\beta_2 = 1$.

Remarks

I. For two indeterminate (non-solvable) plane curves, we may suppose that they have been solved at y and y_1 (if they are not tangent to Oy). We will equate the derivatives of y and y_1 with respect to x and x_1, calculated by the usual method (the derivatives of the implicit functions).

II. Let us consider a real curve taken under the guise of (3). The arc of the curve, computed from the origin, in the sense of x increasing, for example, is expandable about the origin as a function of x. We have

$$s'^2 = 1 + (a_1 + 2a_2 x + \ldots)^2 + (b_1 + 2b_2 x + \ldots)^2 \quad ,$$

whence s', then s, is deduced. We see that in the expansion of s as a function of x, the coefficients up to the term in x^n, only arise from the a_p and b_p up to the index n. If we then express x as a function of s, then the coefficients of this expansion up to those of s^n, only arise from these same a_p, b_p. Finally, if we express x, y, z as a function of the arc, this gives the normal parametric representation, the coefficients of the expansions in x, y, z, up to those of s^n, only arise from the a_p and b_p with indices equal to at most n. It follows that if two curves have a contact of order n, then their normal parametric representations will have the same coefficients up to the term in s^{n+1}, since by taking as the corresponding points on the two curves, the points M and M_1 obtained for the same values of the arc s, we would at once see that the contact would be at least of order $n+1$.

So, when the representations are normal, the contact of order n occurs if the coefficients of the powers of the parameter s, are the same in the expansions of x, y and z up to the terms in s^n, but not beyond.

54 - Contact of two surfaces

Let us first of all take two surfaces under the solvable form with respect to z:

$$z = f(x,y) = ax + by + \dots , \quad z_1 = g(x_1,y_1) = a_1x_1 + b_1y_1 + \dots ;$$

we assume that the common point is the origin and that f and y are series expandable about the origin by Taylor's formula. Finally, we assume that the tangent planes to the origin are the same. M is a point of the first surface S, $M_1(x_1,y_1,z_1)$ is a point of the second surface. We have

$$(0,M) = |x| + |y| + |z| < (1 + |a|)\,|x| + (1 + |b|)\,|y| + o(|x| + |y|) ,$$

hence $(0,M)$ could be replaced by $|x| + |y|$. If the order of contact of S with S_1, is n, then we have for the corresponding points.

$$|x-x_1| + |y-y_1| + |z-z_1| = o[(|x| + |y|)^n] ,$$

therefore

$$x_1 = x + o[(|x| + |y|)^n] , \quad y_1 = y + o[(|x| + |y|)^n] .$$

It follows that we can replace x_1 by x and y_1 by y in z_1, with the supplementary condition $o[(|x| + |y|^n]$, and in order that $|z-z_1|$ is still of the same order, it necessitates that the coefficients of the terms of degree n and less, in the expansions of z and z_1, are then the same. But if they are the same up to degree $n+1$, the order of contact would be at least $n+1$. We again deduce, as in the case of curves, that: *These will contact of order* n *if we have, at the common point:*

$$\frac{\partial^{p+q} f(x,y)}{\partial x^p \partial y^q} = \frac{\partial^{p+q} g(x_1,y_1)}{\partial x_1^p\ \partial y_1^q}$$

for all p *and* q *such that* $p+q \le n$, *provided that at least one of these equations does not occur for* $p+q = n+1$. *The order of contact of* S_1 *with* S *is the same as that of* S *with* S_1.

The equal number is $\frac{(n+1)(n+2)}{2} - 1 = \frac{n(n+3)}{2}$, whereas for two curves it is only $2n$.

We can always return to the case that was studied. If the surfaces are in an indeterminate form, it is sufficient to calculate the partial derivatives of z with respect to x and y for each then by the known techniques (I, 121). If the two surfaces are in a parametric form, such that

$$x = \phi(u,v), \quad y = \psi(u,v) , \quad z = \chi(u,v) , \tag{9}$$

then one of the functional determinants of (z,y), (z,x) or (x,y) with respect to and v, is non-zero since the point is ordinary and about this point, we could calculate u and v as a function of x and y, for example, which reduces to the case studied.

THE CASE WHERE ONE OF THE SURFACES IS IN AN IMPLICIT FORM AND THE OTHER IS IN A PARAMETRIC FORM

Let us suppose that the surface S is given by the expressions (9) where the coordinate origin, corresponding to $u = v = 0$, is the common point of S and S_1, given by

$$F(x_1, y_1, z_1) = 0 \quad .$$

We again see that for M on S and M_1 on S_1, we have

$$F(x,y,z) = (A + \varepsilon)(x - x_1) + (B + \varepsilon')(y - y_1) + (C + \varepsilon'')(z - z_1) \tag{10}$$

where A, B, C are the partial derivatives of F at the origin, all three being non-zero. It follows that if the order of contact is n, then

$$F(x,y,z) = o[(0,M)^n] \quad . \tag{11}$$

Conversely, if this equation exists and if for example, $A \neq 0$, then we shall associated M_1 to M such that $y_1 = y$, $z_1 = z$. The number x_1 will be determined by $F(x_1, y, z) = 0$, which is possible since $A \neq 0$ is the derivative of F with respect to x at the origin. Then following (10), we will have $(M, M_1) < K|F(x,y,z)|$. The order of contact will be at least n. But it will not be $n+1$ if equation (11) does not exist when n is replaced by $n+1$. We thus reduce matters to writing (11).

As one of the functional determinants of the system (9) is non-zero at the origin as the system is solvable at u and v, then starting from the first two equations, for example, we extract

$$|x| + |y| + |z| < K(|u| + |v|) < K'(|x| + |y|) \quad .$$

We can replace (0,M) by $|u|$ and $|v|$, and (11) becomes

$$F[\phi(u,v), \psi(u,v), \chi(u,v)] = o[(|u| + |v|)^n]$$

this condition cannot occur when n is replaced by $n+1$. Hence:

The coefficients of the expression of the first member $F(\phi,\psi,\chi)$ *according to the powers of* u *and* v, *must be zero up to the terms of degree* n, *whilst at least one of the terms of degree* $n+1$ *must not vanish.*

Example. Let

$$z = ax^2 + 2bxy + cy^2 \ \ldots, \qquad |a| + |b| + |c| \neq 0 \quad , \tag{12}$$

be the equation of a surface S tangential to the plane Oxy at the origin. We see that the tangent plane $z=0$, has a contact of order 1 and that the surface has a contact of order 2 at least with the paraboloid

$$z = ax^2 + 2bxy + cy^2 \quad .$$

This contact is only of order 2 if in equation (12), there exists a term of at least degree 3 in x and y. If we consider any surface of the second degree tangential to the origin at the plane Oxy, then its equation will be of the form

$$z = Ax^2 + A'y^2 + A''z^2 + 2(By + B'x)z + 2B''xy \quad . \tag{13}$$

In order to study the contact of this quadric and the surface S, we replace z in the function derived from (13), by the expansion (2), by going through all the terms of the second member; this yields for the second degree terms

$$Ax^2 + A'y^2 + 2B''xy - ax^2 - 2bxy - cy^2 \quad .$$

The contact is then of order 1, in general. It will be of order 2 if

$$A = a, \quad A'' = b, \quad B'' = c,$$

but will only be of order 3 for certain surfaces. The equation of the quadrics having a second order contact with S is therefore

$$z = ax^2 + 2bxy + cy^2 + 2z(By + B'z) + A''z^2$$

and if S is such that there is contact of order 3, then B and B'' are determined, but there remains an arbitrary parameter A''.

55 - The contact of a curve and of a surface

Let Γ be a curve in the parametric form

$$x = f(t), \quad y = g(t), \quad z = h(t) ,$$

and S a surface

$$F(x_1,y_1,z_1) = 0 \quad .$$

We suppose that Γ is tangent to S at a point O, that could be thought of as the origin. We shall assume that the point O is ordinary on Γ and S and by means of changing t to t_o+t, we will assume that O is obtained on Γ for $t=0$. For $M(x,y,z)$ on Γ and $M_1(x,y,z)$ on S, we have

$$F(x,y,z) = (A+\varepsilon)(x-x_1) + (B+\varepsilon')(y-y_1) + (C+\varepsilon'')(z-z_1) \tag{14}$$

and one of the numbers A, B, C, A for example, is non-zero. We have

$$|F(x,y,z)| < K(M,M_1)$$

and since $x = \alpha_1 t + \ldots$, $y = \beta_1 t + \ldots$, $z = \gamma_1 t + \ldots$, with $|\alpha_1| + |\beta_1| + |\gamma_1| \neq 0$, the ratio $(0,M)/|t|$ remains between two fixed numbers. It follows that if the contact of Γ with S is of order ,

$$|F(f,g,h)| = o[(0,M)^n] = o(t^n) \quad ,$$

whereas if it were of order $n+1$, we need to replace n by $n+1$. Conversely, suppose

$$F(f,g,h) = o(t^n) \quad ; \tag{15}$$

on defining M_1 by the condition $z = z_1$, $y = y_1$, which gives $F(x,y,z) = 0$, an equation providing x_1 since $A \neq 0$, we will have following (14),

$$(M,M_1) < K'|F(f,g,h)| = o(t^n) = o[(0,M)^n] \quad .$$

The contact will be at least of order n. It follows that:

In order for the contact to be of order n, *it is necessary and sufficient that on replacing* x, y, z *in* $F(x,y,z)$ *by* $f(t)$, $y(t)$ *and* $h(t)$ *respectively, we will obtain equation (15), but not the analogous equation where* n *is replaced by* $n+1$. *The first* n *derivatives with respect to* t *of* $F(f,g,h)$, *must be zero for* $t=0$ $(t=t_o$ *in the case where* 0 *is obtained for* $t=t_o)$, *but the derivative of order* $n+1$ *must be non-zero.*

When the curve Γ is effectively given in the parametric form, we make the substitution $F(f,g,h)$ and use the expansion. If Γ is given by the intersection of two surfaces, then we can seek a parametric representation; it will most often suffice to obtain the first terms of the expansion of this representation.

Remark. If it happens that the curve and surface are real, then the sign of $F(f,g,h)$ changes when the surface is crossed: if the contact is of even order, the sign changes, the curve crosses the surface at the point 0, and it does not cross it if the contact is odd.

Example. Let S be a surface which is assumed to be tangent at 0 to the plane $0xy$, taken under the form

$$z = ax^2 + 2bxy + cy^2 + \ldots , \qquad |a| + |b| + |c| \neq 0 \quad .$$

A tangent to the surface at the point 0 is given by

$$x = \alpha t, \quad y = \beta t, \quad z = 0 \quad .$$

The order of contact is that of

$$(a\alpha^2 + 2b\alpha\beta + c\beta^2)t + \dots$$

[we have divided the expression $F(f,g,h)$ by t]. Hence the contact is of order 1, except for the tangents to the curve section of the surface by the plane Oxy; these tangents, *real or imaginary* if S is analytic, are called the *asymptotic tangents*. For these tangents, the order of contact will be at least equal to 2.

56 - Vectorial notations

We know that a curve defined parametrically by $x = f(t)$, $y = g(t)$, $z = h(t)$ could also be considered as the locus of the extremity M of a tied vector at the origin for example. This vector denoted by $\vec{M}$ or $\overrightarrow{M(t)}$ has as its components f, g, h. If these functions f, g, h are analytic, then the vector $\overrightarrow{M(t)}$ will be an analytic vector, it will be expandable about $t = t_o$ as a power series

$$\overrightarrow{M(t)} = \overrightarrow{M(t_o)} + (t - t_o)\frac{d\overrightarrow{M(t)}}{dt_o} + \dots + \frac{(t - t_o)}{n!}\frac{d^n\overrightarrow{M(t)}}{dt_o^n} + \dots$$

for which the notation $\dfrac{d^n\overrightarrow{M(t)}}{dt_o^n}$ signifies the value of the n-th order derivative of the vector for $t = t_o$. This formula condenses the three relevant formulae for the series expansion of x, y, z. Such an analytic vector is said to be *complex*.

If f, g, h are not analytic but can be expanded by Taylor's formula about t_o, up to the term of degree n in $(t - t_o)$, then the derivatives of order n exist for three functions, we will have expansions such that

$$x = f(t_o) + (t - t_o)f'(t_o) + \dots + \frac{(t - t_o)^n}{n!}f^{(n)}[t_o + \theta(t - t_o)] \qquad 0 < \theta < 1 \quad . \tag{16}$$

In the expansions of y and z, the Lagrange remainder will correspond to the values θ' and θ'' which are different from θ in general. *We cannot write*

$$\overrightarrow{M(t)} = \overrightarrow{M(t_o)} + \dots + \frac{(t - t_o)^n}{n!}\frac{d^n\overrightarrow{M}}{dt^n}(t_o + \theta_1(t - t_o)) \quad ;$$

the coefficient of $(t - t_o)^n/n!$ will only be a vector whose components will be the values of $f^{(n)}$, $g^{(n)}$, $h^{(n)}$ in the complementary terms. (This fact is similar to that encountered for functions of a complex variable.) But if the n-th order derivative is continuous, the remainder in (16) is of the form

$$\frac{(t-t_o)^n}{n!}\,[f^{(n)}(t_o) + \varepsilon(t-t_o)] \quad ,$$

where $\varepsilon(t-t_o)$ tends to zero with $t-t_o$. We will have in this

$$\overrightarrow{M(t)} = \overrightarrow{M(t_o)} + (t-t_o)\,\overrightarrow{\frac{dM(t)}{dt_o}} + \ldots + \frac{(t-t_o)^n}{n!}\left(\overrightarrow{\frac{d^nM(t)}{dt_o^n}} + \overrightarrow{\varepsilon(t-t_o)}\right)$$

where $\overrightarrow{\varepsilon(t-t_o)}$ is a vector tending to zero with $t-t_o$. We could in fact denote it by $\overrightarrow{o(1)}$ in conforming with Bachmann's notation (the ratio of the modulus with the unit (vector) tends to zero). It is generally in this form that Taylor's formula may be applied to vectors. For simplicity, we will denote the derivative of order p of the vector $\vec{M}$ by $\vec{M}^p$ ($\vec{M}'$, $\vec{M}''$ for the first two derivatives) and we will not in general use the index 0 to indicate that these derivatives are taken for $t = t_o$. We will simply write

$$\vec{M} = \vec{M}_o + (t-t_o)\vec{M}' + \ldots + \frac{(t-t_o)}{n!}(\vec{M}^n + \overrightarrow{o(1)}) \quad .$$

In the same way, a surface S is the locus of the extremity of a vector $\overrightarrow{M(u,v)}$ bound to a fixed point and depending on two parameters. If we take axes, the components of the vector will be $x = f(u,v)$, $y = g(u,v)$, $z = h(u,v)$. The partial derivatives of the vector have as components, the partial derivatives of the same order of x, y, z. We will denote them by $\overrightarrow{M'_u}$, $\overrightarrow{M'_v}$, $\overrightarrow{M''_{uu}}$, $\overrightarrow{M''_{uv}}$, $\overrightarrow{M''_{vv}}$, ... By limiting to two terms, the Taylor formula will be

$$\begin{aligned}\vec{M} = \vec{M}_o &+ (u-u_o)\overrightarrow{M'_u} + (v-v_o)\overrightarrow{M'_v} \\ &+ \frac{1}{2}\,[(u-u_o)^2\overrightarrow{M''_{uu}} + 2(u-u_o)(v-v_o)\overrightarrow{M''_{uv}} + (v-v_o)^2\overrightarrow{M''_{vv}}] \\ &+ \overrightarrow{o[(u-u_o)^2 + (v-v_o)^2]} \quad .\end{aligned}$$

If the functions f, g, h are analytic at u and v, then the vector is said to be analytic; it could then be complex and it is expandable in a double power series.

57 - Osculation

Let us consider a fixed figure Γ, which can be either a curve or surface, and a figure Γ_1 tangent to Γ at a point 0 and which again depends on parameters λ_j, $j = 1,2,\ldots,p$. We may seek a determination of these parameters in order that the order of contact of Γ_1 and Γ is maximum. *We then say that* Γ *osculates* Γ *at the point* 0. If for certain choices of 0, the contact is of a much higher order, there is *hyper-osculation and* Γ_1 *by perosculates* Γ.

In the case where Γ is a plane curve, the tangent at a point does not depend on any parameter and the contact is of order *one* in general. The points where it is of order at least two (and where there will be hyperosculation) are the inflexion points. The tangential circles to Γ at a point M, depend on one parameter. Amongst these circles, one has a contact of order at least two. We have seen it in no. 53, it is the osculating circle; the points of Γ in which the contact of the osculating circle will be of order greater than 2, are the points where the osculating circle is one of *hyperosculation*. This will be the case for the vertices of an ellipse on a hyperbola and for the vertex of a parabola where the contact is of order three. More generally, if Γ is a plane curve in the parametric form $x = f(t)$, $y = g(t)$, and is $F(x,y,\lambda_1,\ldots,\lambda_p) = 0$ is the equation of Γ_1, then the order of contact is provided by the order, less one, of the first derivative of $F(f,g,\lambda_1,\ldots,\lambda_p)$ which is non-zero for $t = t_o$, where t_o provides the point envisaged. By hypothesis, the first derivative is zero. As there are p parameters, we could annihilate the following p derivatives. Let us assume that we obtain a single system of numbers λ_j; we will have osculation with a contact of order $p+1$. The equation at t,

$$F(f,g,\lambda_1,\ldots,\lambda_p) = 0$$

gives the common points of Γ and Γ_1. If the $p+1$ first derivatives of F are zero with F, for $t = t_o$, then the equation admits t_o as a root of order $p+2$. If we are in the analytic case, which we will assume, then the theorem on the continuity of the roots (no. 2) shows that if $\lambda_1,\ldots,\lambda_p$ are close to the values providing the osculation, then the p roots, and only p, (other than the double root $t = t_o$ corresponding to the contact) are close to t_o. Thus:

The osculating curve is the limiting position of a curve of the family which intersects Γ *at* p *points close to the point of contact* M_o *when these points are identified with* M_o.

(We have assumed that F, f, y are analytic; the neighbouring points in question could be imaginary.)

Remark. We could also take a curve Γ_1 depending on $p+1$ parameters and passing through M_o, without assuming it to be tangent to Γ. The osculating curve will be the limit of the Γ_1 passing through $p+1$ points close to M_o when these points tend towards M_o.

THE CASE WHERE Γ IS A SKEW CURVE AND Γ_1 A SURFACE

If we again take Γ in parametric form and S_1 in an implicit form, then what we have still holds. If $x = f(t)$, $y = g(t)$, $z = h(t)$ are equations of the curve Γ and $F(x,y,z,\lambda_1,\ldots,\lambda_p) = 0$ that of the surface S_1 tangent to

Γ at the point M_o, then *the osculating surface* S_1 *will in general be determined and will have a contact of order* $p+1$ *with* Γ. *This will be the limiting position of a surface* S_1 *passing through* M_o *and through* p *neighbouring points when these points are identified with* M_o. *The same remark as above may also apply.*

In the applications, the parameters λ_j will enter *linearly* into F; the determination of the osculating surface will be made by the resolution of a linear system.

THE OSCULATING PLANE

If $Ax + By + Cz + D = 0$ is the equation of a tangent plane to the curve $x = f(t)$, $y = g(t)$, $z = h(t)$ at the point $M_o(t = t_o)$, hence satisfying the two conditions

$$Ax_o + By_o + Cz_o + D = 0 \ , \qquad Ax_o' + By_o' + Cz_o' = 0 \tag{17}$$

(x', x'' will be the derivatives of $f(t)$, etc.), then this plane no longer depends on a single parameter. By stating that the order of contact is two, the plane will be determined, in general; this will be the osculating plane. We have therefore to supplement the two previous equations by that obtained by equating to zero, the second derivative of $Ax + By + Cz + D$, with respect to t when x, y, z are replaced by $f(t)$, $g(t)$, and $h(t)$ respectively. We obtain

$$Ax_o'' + By_o'' + Cz_o'' = 0 \quad . \tag{18}$$

This equation and the second equation (17) determine quantities proportional to A, B, and C, if one of the determinants $y_o'z_o'' - z_o'y_o''$, $z_o'x_o'' - x_o'z_o''$ and $x_o'y_o'' - y_o'x_o''$, is non-zero. We can then take

$$A = y_o'z_o'' - z_o'y_o'', \quad B = z_o'x_o'' - x_o'z_o'', \quad C = x_o'y_o'' - y_o'x_o''$$

and the osculating plane has as its equation [D being provided by (17)]

$$A(x - x_o) + B(y - y_o) + C(z - z_o) = 0 \quad .$$

We can see straight away from (17) and (18) that in the determinantal form, we have

$$\begin{vmatrix} x-x_o & y-y_o & z-z_o \\ x_o' & y_o' & z_o' \\ x_o'' & y_o'' & z_o'' \end{vmatrix} = 0 \quad .$$

If the numbers (19) are zero, then all the tangent planes at M , have a contact of order two. Following (18), we need to consider the relation

$$Ax''_o{}' + By''_o{}' + Cz''_o{}' = 0$$

which will determine the plane of *hyperosculation*. Otherwise, we shall just continue. When A, B, C are zero, we say that *the tangent at* M_o *is stationary*; it has a contact of order at least two with the curve. We say that the osculating plane at a point M_o is stationary when the order of contact with the curve is greater than 2. At such a point M_o, we will need to have

$$Ax_o'' + By_o'' + Cz_o'' = 0 \quad .$$

Following what was said, the osculating plane is the limit of a tangent plane at M_o passing through M close to M_o when $M \to M_o$, or of a plane passing through M_o and through M and M' close to M_o when M and M_o tend towards M_o.

Following condition (18), the osculating plane is determined by the point M_o and the vectors $\vec{M}'_o$ and $\vec{M}''_o$, when the tangent is not stationary.

THE OSCULATING SPHERE

A sphere tangent at a point M_o to the curve Γ again depends on two parameters; the coordinates of the centre in the normal plane to the curve at the point M_o. If we take the arc s as a parameter on Γ, then $\vec{M}'$ is the unit vector $\vec{t}$ of the positive tangent, $\vec{M}''$ is taken along the principal normal and is equal to $\overrightarrow{n/R}$ where $\vec{n}$ is the unit vector of the principal normal and R the radius of curvature. We have

$$\vec{M}''' = \frac{d}{ds}\,\frac{\vec{n}}{R} = \vec{n}\,\frac{d}{ds}\left(\frac{1}{R}\right) + \frac{1}{R}\,\frac{\overrightarrow{dn}}{ds} \quad ,$$

and on applying the second Frenet formula,[1] we obtain

$$\vec{M}''' = -\,\frac{\vec{t}}{R^2} + n\,\frac{d}{ds}\left(\frac{1}{R}\right) + \frac{\vec{b}}{RT}$$

where 1/T is the torsion and $\vec{b}$ the unit binormal vector. Hence we obtain

$$\vec{M} = \vec{M}_o + s\vec{t} + \frac{s^2}{2}\,\frac{\vec{n}}{R} + \frac{s^3}{6}\left(-\frac{\vec{t}}{R^2} + \vec{n}\,\frac{d}{ds}\left(\frac{1}{R}\right) + \frac{\vec{b}}{RT}\right) + o(s^3) \quad , \tag{20}$$

by taking, as we may, the origin of the arcs at the point M_o. A sphere of

---

[1] These formulae will be recalled in Chapter XI. The sign chosen for the torsion is that used by Darboux.

centre A can be written in the vectorial form

$$(\vec{M} - \vec{A})^2 - \rho^2 = 0$$

if M is one of its points and ρ is its radius. On substituting $\vec{M}$ into the expansion (20), we obtain a scalar function of s. We need to state that the term independent of s and the term in s, are zero, to guarantee the contact, since the term in s^2 and s^3 are zero to guarantee the osculation. We have

$$(\vec{M}_o - \vec{A})^2 - \rho^2 + 2(\vec{M}_o - \vec{A})\vec{t}s + \left(1 + (\vec{M}_o - \vec{A})\frac{\vec{n}}{R}\right)s^2$$

$$+ \left[\frac{1}{3}(\vec{M}_o - \vec{A})\left(-\frac{\vec{t}}{R^2} + \vec{n}\frac{d}{ds}\left(\frac{1}{R}\right) + \frac{\vec{b}}{RT}\right) + \frac{\vec{n}\vec{t}}{R}\right]s^3 + o(s^3) \quad .$$

The annulment of the constant term indeed indicates that the sphere passes through M_o, and that of the term in s, indicates that the centre is in the normal plane. By annihilating the term in s^2, we have

$$(\vec{A} - \vec{M}_o)\vec{n} = R \quad , \tag{21}$$

which indicates that the projection of the centra A onto the principal normal, is of distance +R from M_o; it is the point known as the centre of curvature. Finally the term in s^3 equal to zero; when the zero terms along with the factor of 1/3 are suppressed and (21) is taken into account, this gives

$$\vec{b}(\vec{A} - \vec{M}_o) = -R^2T\frac{d}{ds}\frac{1}{R} = T\frac{dR}{ds}$$

The projection of A *onto the binormal has* $T\frac{dR}{ds}$ *as its abscissa on this line. The radius of the osculating sphere is*

$$\sqrt{R^2 + T^2\left(\frac{dR}{ds}\right)^2} \quad .$$

THE CASE WHERE Γ IS ASKEW CURVE AND Γ_1 IS ANOTHER CURVE

Let the curve Γ be taken in a parametric form and the curves Γ_1 be given by two equations $F(x,y,z,\lambda_1,\ldots,\lambda_p) = 0$ and $G(x,y,z,\lambda_1,\ldots,\lambda_p) = 0$ and let Γ and Γ_1 be tangent at a point M_o. Then the conditions which determine the λ_p will be those obtained by annihilating the derivatives of the defined functions on replacing in F and G, x,y,z, by their expressions as functions of the parameter t. We have to consider an even number of equations; the problem will, in general, only be solvable if p is even and equal to $2q$.

The most simple case is that of the circle. A tangential circle depends on two parameters, there will be an osculating circle. We could determine it directly. Buy by observing that the osculating plane at the point M_o has a contact of order 2 with the curve Γ and that the osculating sphere has a contact of order 3, we see that the intersection of these two figures gives a line having a contact of order 2. The osculating circle is therefore the tangential circle situated in the osculating plane and centred at the centre of curvature.

II. THE THEORY OF ENVELOPES

58 - Envelopes of plane curves

The means of determining envelopes in the plane are already known. We shall confine ourselves to specifying the principles on which this study lies.

We consider the family of curves

$$f(x,y,\lambda) = 0 \quad , \tag{22}$$

depending on the parameter λ. We consider matters in a domain D where the function $f(x,y,\lambda)$ is analytic in x,y,λ. Let $C(\lambda)$ be the curve corresponding to the value λ of the parameter. *We intend to determine a curve* Γ *tangent to the curves* $C(\lambda)$, *where each curve* $C(\lambda)$ *is tangent to* Γ *at a point* $M(\lambda)$ *which varies continuously with* λ. The coordinates x,y of $M(\lambda)$ are continuous functions of λ. I say that, *in order for* Γ *to exist, it is necessary that these coordinates satisfy equations (22) and*

$$f'_\lambda(x,y,\lambda) = 0 \quad . \tag{23}$$

Let us assume that at a point of Γ, (23) does not hold. The equation (22) considered as an equation in λ would be solvable, and the equation of the curves $C(\lambda)$ could be written in a neighbourhood of this point as

$$\lambda = \phi(x,y) \ , \tag{24}$$

where ϕ is analytic. We may suppose that, e.g., $\frac{\partial\phi}{\partial y}$ is non-zero at the point in question, for if $\frac{\partial\phi}{\partial y}$ and $\frac{\partial\phi}{\partial x}$ were zero about this point, ϕ would be constant. On a curve $C(\lambda)$, dy/dx would be finite in a neighbourhood of the point; the equation of Γ would be of the form $y = \psi(x)$ and for the contact condition, we would need to have

$$\frac{\partial\phi}{\partial x} + \frac{\partial\phi}{\partial y}\,\psi'(x) = 0 \quad . \tag{25}$$

The function $\phi(x,\psi(x))$ would be constant, Γ must coincide with a curve $C(\lambda)$ and could not be tangent to the other curves $C(\lambda)$. Thus:

In order for the envelope to exist, it is necessary that its points M *satisfy the equations in (22) and (23).* Let us assume that this system is normally solvable at x and y, and the functional determinant is non-zero at the point in question. We have in particular,

$$\left|\frac{\partial f}{\partial x}\right| + \left|\frac{\partial f}{\partial y}\right| \neq 0 \quad ,$$

which implies that the point is ordinary on $C(\lambda)$. x and y are then functions of λ, $x = y(\lambda)$, $y = h(\lambda)$ say; their derivatives exist and are given by

$$f'_x x' + f'_y y' = 0 \ , \quad f''_{\lambda x} x' + f''_{\lambda y} y' + f''_{\lambda^2} = 0 \quad ,$$

and hence both are non-zero if $f''_{\lambda^2} \neq 0$. Then the condition $x'f'_x + y'f'_y = 0$ expresses the contact. Thus:

If the resolution of the system (22) and (23) at x *and* y *is possible (with the functional determinant non-zero) and if* $f''_{\lambda^2} \neq 0$, *then there exists a curve* Γ *tangent to the* $C(\lambda)$ *at its ordinary points.*

Let us show that, *in this case, the characteristic points defined by equations (22) and (23) are the limiting positions of the common points to two neighbouring curves.* Let $C(\lambda)$ and $C(\lambda + \delta\lambda)$ be two neighbouring curves. The common points to these curves are also given by (22) and

$$\Phi(x,y,\delta\lambda) = \frac{1}{\delta\lambda}\ [f(x,y,\lambda + \delta\lambda) - f(x,y,\lambda)] = 0 \quad .$$

Now for λ fixed and $\delta\lambda$ varying in a neighbourhood of zero, $\Phi(x,y,\delta\lambda)$ is analytic in $x,y,\delta\lambda$ and reduces to $f'_\lambda(x,y,\lambda)$ for $\delta\lambda = 0$. The functional determinant of f and Φ with respect to x and y, for $\delta\lambda = 0$, is that of f and f'_λ, and is non-zero by hypothesis. The system $\Phi(x,y,\delta\lambda) = 0$, $f(x,y,\lambda) = 0$ (λ fixed) satisifed for the point of contact of $C(\lambda)$, Γ and $\delta\lambda = 0$, therefore admits a *unique solution* $x(\delta\lambda)$, $y(\delta\lambda)$ which tends toward the point of contact when $\delta\lambda \to 0$. Thus:

In the case in question where

$$\frac{D(f,f'_\lambda)}{D(x,y)} \neq 0 \ , \qquad f''_{\lambda^2} \neq 0 \quad ,$$

the point of contact of $C(\lambda)$ *with its envelope* Γ, *is the limiting position of a single point common to* $C(\lambda)$ *and to* $C(\lambda + \delta\lambda)$, *which tend towards this point of contact when* $\delta\lambda \to 0$.

In particular, if $C(\lambda)$ is algebraic of degree n, $C(\lambda)$ and $C(\lambda + \delta\lambda)$ have, in general, n^2 common points; there will be, *in general,*

n^2 points of contact of $C(\lambda)$ with its envelope. But if the $C(\lambda)$ have fixed points, this number is less. For example, if $C(\lambda)$ is a circle, there are only two common points at finite distance, instead of four.

But the functional determinant of (22) and (23) at x *and* y, *could be zero, where the two curves (22) and (23) are tangential; this implise* $f''_{\lambda^2} = 0$. The coordinates $x(\lambda)$, $y(\lambda)$ of the common point could again define a curve tangent to $C(\lambda)$. But then the contact will be of higher order; in general, of order two. The characteristic point will no longer be the limit of a neighbouring point of intersection of $C(\lambda)$ and $C(\lambda + \delta\lambda)$, but of two neighbouring points. This is what happens in the case of osculating circles to a curve Γ. If x and y, the coordinates of a point of the curve are expressed as functions of the arc s and if R is the radius of curvature and θ the angle of the tangent with 0_x, then the osculating circle has as its equation

$$(X - x + R\ \mathrm{Sin}\theta)^2 + (Y - y - R\cos\theta)^2 - R^2 = 0 \quad .$$

On differentiating with respect to the arc, we have, after reduction,

$$(X - x)R'\sin\theta - (Y - y)R'\cos\theta = 0 \quad .$$

The functional determinant at X, Y for $X = x$, $Y = y$ is zero. Here, two neighbouring osculating circles intersect in two imaginary points which have the point of osculation as their limit when one of the circles tends to the other.

We shall not make a detailed study of cases of this kind, where the envelope can be considered as a multiple curve.

Let us suppose that, in a general way, the locus of common points to the two curves (22), (23), or part of this locus, is a curve on which y is a differentiable function of x, with continuous derivative, and that, for the system of values $x, y(x)$ and λ which satisfy (22) and (23), the derivative $f''_{\lambda^2}(x,y,\lambda)$ is non-zero. We could extract λ as a function of x and y from (23), $\lambda(x,y)$ say, and on taking into (22), we will have

$$f(x,y(x),\ \lambda(x,y(x))) \equiv 0 \quad ,$$

hence

$$f'_x(x,y,\lambda) + y'(x)f'_y(x,y,\lambda) + f'_\lambda[\lambda'_x + \lambda'_y y'] = 0 \quad ,$$

an equation which reduces to $f'_x + y'f'_y = 0$, and expresses the contact of the curve $y = f(x)$ and of $C(\lambda)$, if f'_x and f'_y are not both zero. Thus we can briefly say that:

If λ *is eliminated from equations (22) and (23), then the curve* $R(x,y) = 0$ *will be formed by the envelope and the locus of the singular points.*

In the case where $f(x,y,\lambda)$ is a polynomial in λ, we could make the elimination by algebraic methods. This is the case encountered in practice.

More especially, if $f(x,y,\lambda) = 0$ *is an algebraic curve, when* f *is a polynomial in* x *and* y, *and is also algebraic in* λ, then the elimination of λ from f and $f'\lambda$ entails writing the discriminant, $R(x,y) = 0$, which defines an algebraic curve, an envelope and locus of singular points. We note that if we take account of the factors containing $f'_\lambda(x,y,\lambda)$, then the elimination could also provide curves $C(\lambda)$ which are said to be stationary; these are curves for which f'_λ is zero at every point. For example, the envelope of the tangents to a curve is that curve, but if $y - F(\lambda) - F'(\lambda)(x - \lambda) = 0$ is the equation of its tangents ($y = F(x)$ being the curve), then the derived equation $-F''(\lambda)(x - \lambda) = 0$ is satisfied not only for $\lambda = x$, which gives the curve, but also for the λ roots of $F''(\lambda) = 0$ which give the tangents of inflexion. The stationary curves do not form part of the envelope.

THE CONDITIONS FOR VALIDITY OF THE ABOVE RESULTS

We are lead to consider the case where $f(x,y,\lambda) = 0$ is analytic. But what was said remains true if f has first and second partial derivatives continuous.

Remarks

I. We know that in order to determine the envelope of a curve

$$f(x,y,\lambda,\mu) = 0$$

depending on two parameters bound by the relation $g(\lambda,\mu) = 0$, we have to determine the locus of points x,y satisfying the relations

$$f = 0, \quad g = 0, \quad f'_\lambda g'_\mu - f'_\mu g'_\lambda = 0 \quad .$$

For, if we imagine the equation $g(\lambda,\mu) = 0$ solved, with respect to μ for example, then there no longer remains one parameter λ and we have to consider $f = 0$ and $f'_\lambda + f'_\mu \mu'_\lambda = 0$ and μ'_λ is given by the derivation of $g(\lambda,\mu) = 0$; we have $g'_\lambda + g'_\mu \mu'_\lambda = 0$.

II. For a curve taken in the parametric form

$$C(\lambda) \qquad x = f(t,\lambda) \ , \qquad y = y(t,\lambda) \quad ,$$

where t is the parameter defining each curve $C(\lambda)$ matters are reduced to determining λ as a function of t, such that the curve, so defined, is tangent to the curves $C(\lambda)$. If $\lambda(t)$ is this function, we will need to have

$$\frac{\partial f}{\partial t}\left(\frac{\partial g}{\partial t} + \frac{\partial g}{\partial \lambda}\lambda'\right) - \left(\frac{\partial f}{\partial t} + \frac{\partial f}{\partial \lambda}\lambda'\right)\frac{\partial g}{\partial t} = 0 \quad ,$$

which leads to the condition

$$\frac{D(f,g)}{D(\lambda,t)} = 0 \quad ,$$

which is the sought-after relationship. We will note the symmetry in t at λ. The envelope of the curves $C(\lambda)$ obtained by assuming λ constant in the given equations, is the same as the envelope of the curves $\Gamma(t)$ obtained by supposing t is constant and λ variable.

There are well-known applications of the theory of envelopes to the expansions of plane curves, with the determination of the point equation of a given curve by its tangential equation, and to the envelopes of circles.

59 - Envelopes of surfaces depending on a parameter

We follow the same line of ideas as above. Let

$$f(x,y,z,\lambda) = 0 \quad , \tag{26}$$

be the equation of the surfaces of the family, where λ is the parameter. For each λ we have a surface $S(\lambda)$. Let us imagine a surface Σ which would be tangent to each surface $S(\lambda)$ along a curve $C(\lambda)$, with the condition of continuity analogous to that given in the case of the curves. If M is one of the points of $C(\lambda)$, then its coordinates must satisfy equation (26). Moreover, they must satisfy the equation

$$f'_{\lambda}(x,y,z,\lambda) = 0 \quad . \tag{27}$$

For if at a point M of $S(\lambda)$ the first member of (27) is non-zero, we could write the equation of the surfaces $S(\lambda)$, in the neighbourhood of the point x,y,z,λ as

$$\lambda = \phi(x,y,z) \quad . \tag{28}$$

Let us assume, for example that $\frac{\partial\phi}{\partial z}$ is non-zero; the surface Σ must have an equation of the form $z = \psi(x,y)$ and would have a tangent plant. The conditions of contact would be

$$\frac{\partial\phi}{\partial x}(x,y,\psi) + \frac{\partial\phi}{\partial z}(x,y,\psi)\psi'_x = 0 \ , \qquad \frac{\partial\phi}{\partial y} + \frac{\partial\phi}{\partial z}\psi'_y = 0 \qquad ;$$

the partial derivatives of the function $\phi(x,y,\psi)$ would be zero, this function would be constant, Σ would coincide with a surface $S(\lambda)$ in a neighbourhood of the point in question, and would not have any common points with the other surfaces. This is impossible. The surfaces (28) will not have envelopment. Thus: *if the envelope exists, then the coordinates of its points must satisfy equations (26) and (27).* Let us consider such a point.

Let us suppose that one of the functional determinants of f, f'_λ with respect to x,y or u,z or z,x, is non-zero at a point. We could extract x and y, for example, from these equations, as functions of z and λ. We will define a surface Σ and x and y will have partial derivatives, certainly in the analytic case, and provided that the second partial derivatives of $f(x,y,z,\lambda)$ exist in the general case. The partial derivatives of x and y with respect to z and λ will be given by

$$f'_x \frac{\partial x}{\partial z} + f'_y \frac{\partial y}{\partial z} + f'_z = 0 , \qquad f'_x \frac{\partial x}{\partial \lambda} + f'_y \frac{\partial y}{\partial \lambda} = 0 \quad ,$$

$$f''_{\lambda x} \frac{\partial x}{\partial z} + f''_{\lambda y} \frac{\partial y}{\partial z} + f''_{\lambda z} = 0 \quad , \quad f''_{\lambda x} \frac{\partial x}{\partial \lambda} + f''_{\lambda y} \frac{\partial y}{\partial \lambda} + f''_{\lambda^2} = 0 \quad .$$

As the functional determinant of z and x with respect to z and λ is $\frac{\partial y}{\partial \lambda}$ and that of z and, $\frac{\partial y}{\partial \lambda}$, then the last relation shows that if $f''_{\lambda^2} \neq 0$, these functional determinants are not both zero, *the point* x,y,z *is ordinary on* Σ, and the first two relations then indicate the contact of Σ and $S(\lambda)$.

As in the case of plane curves, we will see that the contact curve of Σ and $S(\lambda)$ is the limiting position of a curve common to $S(\lambda)$ and $S(\lambda + \lambda S)$ when $S\lambda$ tends towards zero.

In practice, one eliminiates, if possible, λ *from (26) and (27). In theory this is possible: if* f''_{λ^2} *is non-zero; we can in general extract* λ *from equation (27) and carry it into (26).* We can repeat what was said previously on this point, in the case of curves. By proceeding thus, we can obtain, in addition to the envelope, the locus of singular points and if the elimination is complete, the stationary surfaces $S(\lambda)$ on which $f'_\lambda \equiv 0$.

The curve defined by equations (26) and (27) is called the *characteristic curve*. The surface Σ is the locus of these characteristics $C(\lambda)$. These curves $C(\lambda)$ traced on the surface Σ could have an envelope. The points of this envelope satisfy the derived equation of $f'_\lambda(x,y,)$, hence $f''_{\lambda^3} = 0$ (see no. 61). We will see later that they are, in general, singularities of Σ forming what is known as a "amête de rebroussement".

In the case of algebraic surfaces, where $f(x,y,z,\lambda)$ is algebraic in x,y,z,λ, Σ will be algebraic, real or complex, the will be real or complex.

Examples

I. *A developable surface* - A plane $Ax + By + Cz + D = 0$ depending on one parameter (A,B,C,D are functions of λ) has as its envelope the locus of the straight line defined by the equation of the plane and the derived $A'x + B'y + C'z + D'' = 0$. This ruled surface is called a *developable surface*.

The characteristic line has as its envelope, the locus of the point defined by the two preceding equations to which we add the derived equation $A''x + B''y + C''z + D'' = 0$.

II. *An evenlope of spheres* - If $(\vec{M} - \vec{A})^2 - R^2 = 0$ is the equation of the sphere with centre $\vec{A}$ and radius R, then A and R are functions of λ. The derived equation $(\vec{M} - \vec{A})\vec{A}' + RR' = 0$ is that of a plane perpendicular to the vector $\vec{A}'$; the characteristic curve is then a circle situation in a plane perpendicular to the tangent at A instead of A. This circle is real or imaginary. On differentiating a second time, we obtain a second plane which meets the characteristic circle in two real or imaginary points; they are the points of contact of the circle with its envelope.

When the sphere has a constant radius, the plane of the circle passes through A. The circle has a constant radius equal to that of the sphere; it is real (we assume it is the locus of the real centre). The surface envelope is called a *canal surface*. For example a torus of revolution is the envelope of the spheres which are tangent to it along the meridian circles. It is a canal-surface; in this case, the meridian circles pass through two fixed points, they do not have a true envelope.

Remarks

I. If we are given surfaces depending on two parameters λ and μ bound by an equation, say

$$f(x,y,z,\lambda,\mu) = 0 \qquad g(\lambda,\mu) = 0 \quad ,$$

then we can assume that μ, for example, has been taken from the second equation to the equation of the surface, that no longer depends on a single parameter. This is permitted if $g'_\mu(\lambda,\mu)$ is non-zero at the point in question. We have then

$$f(x,y,z,\lambda,\phi(\lambda)) = 0 \quad , \quad g(\lambda,\phi(\lambda)) = 0 \quad ,$$

and the envelope is determined by adding to the first equation, the derived equation

$$f'_\lambda(x,y,z,\lambda,\phi(\lambda)) + \phi'(\lambda)f'_\mu(x,y,z,\lambda,\phi(\lambda)) = 0 \quad .$$

As $\phi'(\lambda)$ is given by

$$g'_\lambda(\lambda,\phi) + \phi'(\lambda)y'_\mu(\lambda,\phi) = 0 \quad ,$$

we see that the points of the envelope, if they exist, satisfy the three equations

$$f(x,y,z,\lambda,\mu) = 0 \quad , \qquad g(\lambda,\mu) = 0 \quad ,$$

$$f'_\lambda g'_\mu - f'_\mu g'_\lambda = 0 \quad .$$

Within all of this lies the existence of the necessary derivations. From these equations, we could extract x,y,z as functions of λ and μ, or even eliminate λ and μ. The surface obtained will be the envelope or a singular locus, or will contain a portion which will be the enveloper and another which will be a singular locus. In calculating, we will be lead to suppress, if needs be, the stationary surfaces.

III. If the surfaces are given in the parametric form

$$x = \phi(u,v,\lambda) \ , \quad y = \psi(u,v,\lambda) \ , \quad z = \chi(u,v,\lambda) \quad , \tag{29}$$

then the characteristic curve $C(\lambda)$ is a curve of the surface $S(\lambda)$. On this curve, v for example, will be a function of u and λ. We have, therefore, to find a function $v = \theta(u,\lambda)$ such that the surface Σ

$$x = \phi[u,\theta(u,\lambda),\lambda], \quad y = \psi[u,\theta(u,\lambda),\lambda], \quad z = [u,\theta(u,\lambda),\lambda]$$

is tangent to $S(\lambda)$ along the curve $\lambda = \text{const.}$ The tangent plane to $S(\lambda)$ is defined by the vectors

$\vec{M}'_u$ with components $\phi'_u, \psi'_u, \chi'_u$;

$\vec{M}'_v$ with components $\phi'_v, \psi'_v, \chi'_v$,

respectively tangent to the line $v = \text{const.}$ and $u = \text{const.}$ The tangent plane to Σ is defined by the vectors $\vec{M}'_u + \vec{M}'_v\theta'_u$ with components $\phi'_u + \phi'_v\theta'_u$, $\psi'_u + \psi'_v\theta'_v$, $\chi'_u + \chi'_v\theta'_u$; $\vec{M}'_v\theta'_\lambda + \vec{M}'_\lambda$ with components $\phi'_v\theta'_\lambda + \phi'_\lambda$, $\psi'_v\theta'_\lambda + \psi'_\lambda$, $\chi'_v\theta'_\lambda + \chi'_\lambda$. These four vectors must be coplanar. The third is the plane of the first two, and in order for the fourth to be there, it is necessary and sufficient that $\vec{M}'_\lambda$ is located there. Hence

$$\vec{M}'_u, \ \vec{M}'_v, \ \vec{M}'_\lambda \quad ,$$

must be coplanar, and the mixed product

$$(\vec{M}'_u, \ \vec{M}'_v, \ \vec{M}'_\lambda)$$

must be zero. The result is written in determinant form. The points of Σ will be defined by

$$\vec{M} = \vec{M}(u,v,\lambda) \ , \qquad (\vec{M}'_u, \ \vec{M}'_v, \ \vec{M}'_\lambda) = 0 \quad ,$$

i.e., by equations (29) and by

$$\begin{vmatrix} \phi'_u & \phi'_v & \phi'_\lambda \\ \psi'_u & \psi'_v & \psi'_\lambda \\ \chi'_u & \chi'_v & \chi'_\lambda \end{vmatrix} = 0 \quad .$$

We see again that u,v and λ act symmetrically.

60 - The envelopes of surfaces depending on two parameters

Let us consider a surface $S(\lambda,\mu)$ depending on two parameters; let

$$f(x,y,z,\lambda,\mu) = 0 \tag{30}$$

be its equation. We will assume that f is analytic with respect to the five variables, or even this function has a sufficient number of partial derivatives in order that the following considerations. *We seek to determine a surface* E *which is tangent at a point or a finite number of points to each surface* $S(\lambda,\mu)$ *and such that if* M *is an ordinary point of* E, *then through this point there passes, in general, a surface* $S(\lambda,\mu)$. *Moreover, when* λ *and* μ *are given any values close to those providing the contact at* M, *we require a surface* $S(\lambda,\mu)$ *tangent to* E *at a neighbouring point of* M. Under these conditions, it is necessary that M is located on the surfaces defined by

$$f'_\lambda(x,y,z,\lambda,\mu) = 0 \quad , \quad f'_\mu(x,y,z,\lambda,\mu) = 0 \quad . \tag{31}$$

Let us assume, in effect, that at a point M of $S(\lambda,\mu)$ are of the conditions, e.g., the first, is not satisfied. We could extract λ from equation (30) for x,y,z,λ,μ close to their values at M. We will obtain $S(\lambda,\mu)$ under the form

$$\lambda = \phi(x,y,z,\mu) \quad .$$

If we let μ keep its value on $S(\lambda,\mu)$ and if λ varies, then we ought to obtain surfaces tangent to E in the neighbourhood of M. Now the argument that has already been made twice shows that the only surfaces tangent to $S(\lambda,\mu)$ in this neighbourhood are these surfaces themselves, which correspond to different λ and are without common points; M could not belong to E.

The points of E *must therefore satisfy equations (30) and (31).* If we assume that these equations are satisfied at a point $M(x,y,z)$ of $S(\lambda,\mu)$ and *if the functional determinant of the first members of (30) and (31) with respect to* x,y,z *is non-zero at this point,* then we can express x,y,z as functions of λ and μ; we obtain part of the surface of E. For x,y,z have partial derivatives; the point M is ordinary on $S(\lambda,\mu)$ since the functional determinant is non-zero at M and about M implies $|f'_x| + |f'_y| + |f'_z| \neq 0$, and on differentiating (30) and taking into account (31)

$$f'_x x'_\lambda + f'_y y'_\lambda + f'_z z'_\lambda = 0 \ , \qquad f'_x x'_\mu + f'_y y'_\mu + f'_z z'_\mu = 0 \quad , \tag{32}$$

which is a condition of contact if M is ordinary on E. Now $x'_\lambda, \ldots,$ are given by

$$f''_{\lambda x} x'_\lambda + f''_{\lambda y} y'_\lambda + f''_{\lambda z} z'_\lambda + f''_{\lambda\lambda} = 0 \quad ,$$

$$f''_{\mu x} x'_\lambda + f''_{\mu y} y'_\lambda + f''_{\mu z} z'_\lambda + f''_{\lambda\mu} = 0 \quad ,$$

and by the first equation (32). The derivatives with respect to μ are given by similar equations. *These derivatives with respect to* λ *and* μ *of* x,y,z *will only be proportional if*

$$\frac{D(f'_\lambda, f'_\mu)}{D(\lambda,\mu)} = 0 \quad . \tag{33}$$

We can then say that, *in general,* we will obtain about M a part of the envelope by means of the hypothesis that the functional determinant of the first members of (30) and (31) is non-zero.

In the exceptional case where (33) occurs, it may happen that there is no surface envelope, but there is a curve envelope.

If we determine the envelope E *by eliminating* λ *and* μ *between (30) and (31),* which implies the new hypothesis that the first member of (33) *is non-zero, we will obtain an equation defining a surface which will be, in various parts, an envelope or singular locus.*

For, if $z = F(x,y)$ is an ordinary part of the surface E obtained by this elimination of the system (30), then (31) is satisfied when z is replaced by this function and λ and μ by their values drawn from (31) where z is again replaced by this function. We have then

$$f(x,y,F,\lambda,\mu) \equiv 0 \quad ,$$

and on differentiating, which is possible since λ and μ have derivatives, there remains

$$f'_x + f'_z F'_x = 0 \quad , \qquad f'_y + f'_z F'_y = 0 \quad ,$$

by taking account of (31). This expresses the contact if f'_x and f'_y and f'_z are not all simultaneously zero at the point in question, which is then an ordinary point.

If we return to the case where the envelope could be obtained by solving the system (30), (31) for which the functional determinant is non-zero at a point; hence in its neighbourhood, then we see as before in the case of curves, that *the point of contact* M *of* $S(\lambda,\mu)$ *and of* E *is the limiting position of the common point to the surfaces* $S(\lambda,\mu)$, $S(\lambda + S\lambda,\mu)$ *and* $S(\lambda,\mu + \mu S)$ *when the increments of* $S\lambda$ *and* $S\mu$ *tend to zero.*

This type of property is often used in geometry to find the point of contact.

<u>Example</u> - The Envelope of Spheres

If we take the spheres under the vectorial form

$$(\vec{M} - \vec{A})^2 - R^2 = 0 \quad ,$$

where the tied vector $\vec{A}$ whose extremity A is the centre and the radius R are functions of λ and μ, then the envelope is given by the intersection of the sphere and the two planes

$$(\vec{M} - \vec{A})\vec{A}'_\lambda + RR'_\lambda = 0 \ , \qquad (\vec{M} - \vec{A})\vec{A}'_\lambda + RR'_\mu = 0 \quad ,$$

whose line of intersection is normal to the vectors $\vec{A}'_\mu$ and $\vec{A}'_\lambda$, is therefore parallel to the normal at A to the surface locus of this point A. It is the limit of the radical axis of the sphere of centre A and of the spheres which are infinitesimally near. *When* R *is constant,* this line becomes the normal at A. The surface envelope is obtained in this case by taking along the normal to the locus surface σ of A, from one part to another of A, the vectors of modulus R; this is called a *surface parallel* to σ.

SOME REMARKS AND EXAMPLES

I. If a family of surfaces depends on three parameters bound by a relation, then there is in fact just two parameters. The determination of the envelope will be made by generalising what had been said in the case of curves and surfaces with one parameter. If

$$f(x,y,z,\lambda,\mu,\nu) = 0$$

is the equation of the surface and

$$g(\lambda,\mu,\nu) = 0$$

is a relation between the three parameters, then we will combine these equations with the two equations

$$\frac{f'\lambda}{g'\lambda} = \frac{f'\mu}{g'\mu} = \frac{f'\nu}{g'\nu} \quad ,$$

and we will eliminate λ,μ,ν between these four equations.

This is, in fact, the calculation one makes in going from the tangential equation of the non-developable surface to its pointwise equation. We seek the envelope of a plane

$$ux + vy + wz + h = 0 \tag{34}$$

whose coefficients are bound by a homogeneous equation

$$G(u,v,w,h) = 0 \quad . \tag{35}$$

On dividing by h and h^{α}, where α is the degree of homogeneity of G, we are in the preceding case. But by preserving the four parameters, we could combine with (34) and (35) the three equations

$$\frac{x}{G'_u} = \frac{y}{G'_v} = \frac{z}{G'_w} = \frac{1}{G'_h} \quad ,$$

which reduces to two if we take into account (34) and (35).

When equation (35) is the tangential equation of a curve and not that of a surface, then the envelope of the planes (34) is a curve. We find ourselves in the case indicated above where condition (33) is satisifed for all the common points to the surfaces (30) and (31).

II. If the surfaces are given in a parametric form, e.g., by using vectorial notation, and if

$$\vec{M} = \overrightarrow{M(u,v,\lambda,\mu)}$$

defines a surface for each pair (λ,μ), then we could directly determine u and v as functions of λ and μ to obtain an envelope. We will easily see that the four vectors

$$\vec{M}'_u, \vec{M}'_v, \vec{M}'_\lambda, \vec{M}'_\mu \quad ,$$

must be in the same plane, which gives two conditions by equating to zero the mixed product of the first two vectors with each of the last two. The condition corresponding to condition (33) is in this case: $\vec{M}'_\lambda$ and $\vec{M}'_\mu$ are parallel.

We again see the symmetric roles played by the four parameters u,v,λ,μ.

III. The surface envelope E of the surface $f(x,y,z,\lambda,\mu) = 0$ is tangent along a curve of the envelope of the surfaces with one parameter obtained by assuming that λ and μ are bound by $g(\lambda,\mu) = 0$, where g if any differentiable function. For this surface is defined by $f = 0$ and $f'_\mu g'_\lambda - f'_\lambda g'_\mu = 0$, and intersecting it with the surface $f'_\mu = 0$, for example, we obtain a line belonging to E.

61 - The envelope of space curves depending on one parameter

Let us consider these curves $C(\lambda)$ as deinfed by the intersection of surfaces

$$f(x,y,z,\lambda) = 0 , \qquad g(x,y,z,\lambda) = 0 \quad . \tag{36}$$

We look to see if there exists a curve Γ tangent at each of its points at one $C(\lambda)$ and such that if M is a point of Γ and λ the value of the parameter for which $C(\lambda)$ is tangent to Γ at M, then for neighbouring values of λ, the curves $C(\lambda)$ are tangent to Γ at a neighbouring point of M. The previous argument shows that the points M must satisfy, besides (36), the derived equations

$$f'_\lambda(x,y,z,\lambda) = 0 \quad , \quad g'_\lambda(x,y,z,\lambda) = 0 \quad . \tag{37}$$

(For if, for example, $f'_\lambda \neq 0$ at the point $M(x,y,z,\lambda)$, we can solve the first equation in λ in the neighbourhood of this point and take in $y(x,y,z,\lambda) = 0$. The system becomes $\lambda = F(x,y,z)$, $G(x,y,z) = 0$ and this is possible in the neighbourhood of M.)

The system of four equations (36), (37) in x,y,z,λ will not have variable solutions in general. *A family of space curves depending on one parameter does not have an envelope, in general.* But if one of the equations (36), (37) is a consequence of the others, *then the equations reduce to three.* If, in those three remaining equations which are satisfied at a point, the functional determinant with respect to x,y,z is non-zero, then we can solve for x,y,z about this point. We obtain functions x,y,z of λ, having derivatives if the first derivatives of f,g,f'_λ for example are continuous; we have defined a curve arc, on which we obtain, by differentiating at λ and taking into account (37)

$$x'f'_x + y'f'_y + z'f'_z = 0 , \qquad x'g'_x + y'g'_y + z'g'_z = 0 \quad .$$

As above, we see that M is necessarily ordinary on $C(\lambda)$. As for x',y',z' they could not be zero, following their calculation, if $f''_{\lambda^2} = 0$. The arc so obtained is then effectively tangent to $C(\lambda)$, as long as $f''_{\lambda^2} \neq 0$. We again remark that, when the functional determinant of the four functions appearing in (36) and (37) is simultaneously zero with these equations, then $f''_{\lambda^2} = 0$ implies $g''_{\lambda^2} = 0$. Thus:

With the possibility of calculating the solutions by the general methods, there is an arc of the envelope as long as the second derivatives with respect to f and g with respect to λ are non-zero.

<u>Remark</u>. It would be worth remarking on the subject of the surface described by the curves $C(\lambda)$ in the case where there is an envelope Γ. In the general case, if M and M' are two neighbouring points of Γ, then the curves $C(\lambda)$ and $C(\lambda')$ will not have any common point neighbouring M and M'. For example, Γ will be a line and the $C(\lambda)$ circles, with constant radius to fix matters, tangent to this line for which the plane turns about the line whilst the point of contact is displaced. This can be seen in the simple case when Γ is a cusp. Moreover, if we wish to eliminate λ from equations (36), then a singular case is found along C, since the equations (37) occur. But it could happen in certain cases, that the envelope Γ is an ordinary line on the surface S being a locus of the $C(\lambda)$. The neighbouring curves $C(\lambda)$ will intersect and *we will be in a similar situation to those of plane curves*. We can express this in brief by saying that the curves $C(\lambda)$ are on a surface S, *by requiring that the curves* C *pass through the ordinary points of* S. In this case, we will have equations of curves in the form

$$F(x,y,z) = 0 \quad , \qquad g(x,y,z,\lambda) = 0 \quad ,$$

and the envelope, if it exists, will be given by these equations to which we include

$$g'_\lambda(x,y,z,\lambda) = 0 \quad .$$

In the algebraic case, where F and g will be polynomials, common points (real or imaginary) will exist, and there will be an envelope or a singular locus.

THE CASE WHERE THE CURVE IS GIVEN IN PARAMETRIC FORM

If u is the parameter on the curve $C(\lambda)$, we have

$$\vec{M} = \overrightarrow{M(u,\lambda)} \quad ,$$

and we must determine u as a function of λ in order that the tangent vector to the curve Γ so obtained is colinear to that of $C(\lambda)$. We must then have

$$\vec{M}'_u \frac{du}{d\lambda} + \vec{M}'_\lambda = k(\lambda)\vec{M}'_u$$

or

$$\vec{M}'_u = h(\lambda)\vec{M}'_\lambda \quad .$$

It will be necessary then that the exterior product $\vec{M}'_u \wedge \vec{M}'_\lambda$ is zero. We will thus have two relations which will be compatible by introducing the coordinates

$$\frac{x'_u}{x'_\lambda} = \frac{y'_u}{y'_\lambda} = \frac{z'_u}{z'_\lambda} \quad .$$

Example. THE CASE OF LINES. Let us take them to be of the form

$$x = az + p \, , \qquad y = bz + q \quad ,$$

where a, p, b, q are functions of λ, and z is the parameter on the line $D(\lambda)$. We will have the condition

$$\frac{a'z + p'}{a} = \frac{b'z + q'}{b} = \frac{0}{1}$$

which implies that

$$a'z' + p' = 0 \, , \quad b'z + q' = 0 \quad ,$$

are compatible and that suffices. Hence, it is necessary and sufficient that

$$a'q' - b'p' \equiv 0 \quad ,$$

and we then have

$$z = \frac{-p'}{a'} = \frac{-q'}{b'} \quad .$$

THE CASE WHERE THE CHARACTERISTICS OF THE SURFACES DEPEND ON ONE PARAMETER

If $f(x,y,z,\lambda) = 0$,

$$f'_\lambda(x,y,z,\lambda) = 0$$

are the equations of the characteristic curve, then the conditions that must be satisfied by the numbers x,y,z,λ in order that there should be an envelope, reduce to these equations to which we include

$$f''_{\lambda^2}(x,y,z,\lambda) = 0 \quad .$$

These three equations define a curve Γ, tangent at its ordinary points to the characteristic $C(\lambda)$. If the functional determinant of the first members of these equations in x,y,z is non-zero at a point satisfying the equations, then x,y,z are deinfed as a function of λ, and by calculating x',y',z', we see that the point is ordinary if the third derivative of f with respect to λ is non-zero, which will be the case in general. We obtain an arc of the curve tangent to the characteristics. We are going to look at the disposition of the surface envelope in the neighbourhood of this curve.

62 - The cusp (Arête de rebrousrement)

We will just consider the most simple case. The surface $S(\lambda)$,

$$f(x,y,z,\lambda) = 0 \quad , \tag{38}$$

has an envelope Σ given by this equation to which we add

$$\frac{\partial}{\partial\lambda}(x,y,z,\lambda) = 0 \quad . \tag{39}$$

The contact occurs along the curve $C(\lambda)$ defined by these two equations. It is the *characteristic curve*. We assume that this curve $C(\lambda)$ has an envelope Γ, which is defined by the above equations to which we add

$$\frac{\partial^2 f}{\partial\lambda^2}(x,y,z,\lambda) = 0 \quad ; \tag{40}$$

the contact occurs at a point $M(\lambda)$. We shall assume that $C(\lambda)$ are ordinary at this point, i.e., we can solve the system of equations (38), (39), (40); the functional determinant is non-zero. The surfaces $S(\lambda)$ and $S'(\lambda)$ defined by (39) have distinct tangent planes at the point $M(\lambda)$. We can take the point $M(\lambda)$ as the origin, assume it corresponds to $\lambda = 0$, take as the Oxy plane the tangent to $S(o)$, and for the plane Oxy, the plane tangent to $S'(o)$. The plane Oxy will be an arbitrary plane distinct from the first two. We assume the existence of all useful derivatives. On expanding with respect to λ, we have

$$f \equiv f(x,y,z) + \lambda g(x,y,z) + \frac{\lambda^2}{2} h(x,y,z) + \frac{\lambda^3}{6} k(x,y,z) + \ldots ,$$

$$\frac{\partial f}{\partial\lambda} \equiv g(x,y,z) + \lambda h(x,y,z) + \frac{\lambda^2}{2} k(x,y,z) + \ldots$$

$$\frac{\partial^2 f}{\partial\lambda^2} \equiv h(x,y,z) + \lambda k(x,y,z) + \ldots$$

$$\frac{\partial^3 f}{\partial\lambda^3} \equiv k(x,y,z) + \ldots$$

For $\lambda = 0$, the equations (39) and (40) are satisfied, the tangent plane to 0 at $S(o)$ is Oxy, to $S'(o)$ is Oxy, and we can solve at x,y,z. Finally, as was said in No. 61, the third derivative is non-zero for λ zero. Hence

$$f(x,y,z) \equiv az + f_2(x,y,z) + \ldots \qquad a \neq 0$$

$$g(x,y,z) \equiv by + g_2(x,y,z) + \ldots \qquad b \neq 0$$

$$h(x,y,z) \equiv cx + \alpha y + \beta z + h_2(x,y,z) + \ldots \qquad c \neq 0$$

$$k(0,0,0) \neq 0$$

where f_2, g_2 and h_2 are homogeneous polynomials of degree equal to their index. The curve of intersection of Σ and of the plane Oxy is defined by the expressions of y and z as functions of λ deduced from the equations (28) and (39) when we set $x = 0$. We have

$$az + \lambda by + \ldots + \frac{\lambda^3}{6} k(o,y,z) + \ldots = 0 \quad ,$$

$$by + \lambda(\alpha y + \beta z) + \ldots + \frac{\lambda^2}{6} k(o,y,z) + \ldots = 0 \quad .$$

The derivatives of y and z with respect to λ are obtained by differentiation; these are given for $\lambda = 0$ by two equations which provide them separately step by step. We see that z' and y' are zero (for $\lambda = 0$) as well as z'', but we have

$$az'''_o - 2k(0,0,0) = 0 \quad , \quad by''_o + k(0,0,0) = 0 \quad .$$

Thus y''_o and z''_o are non-zero; we have for λ close to 0,

$$y = b_2\lambda^2 + \ldots \quad , \quad z = c_3\lambda^3 + \ldots$$

hence *a turning point of the first kind.* Consequently: *In general, a plane which passes through a point* M *of the cusp and which is not tangent to this cusp, intersects the surface* Σ *along a curve which presents a turning point of the first kind at* M. It is clear that the intersection of Σ with any surface passing through M, admitting this point as an ordinary point, and which is not tangent to M at the cusp, possesses the same property. Since instead of setting $x = 0$, we replace it by an expansion in y and z without first degree terms, then the terms of at least fourth order will be included in f_2, g_2, and h_2.

ORDER OF CONTACT

The order of contact of the cusp Γ with the surface $S(o)$ is obtained by replacing the variables x,y,z in the function $f(x,y,z,0)$ by their expansions in λ. These expansions are given by

$$az + \lambda by + f_2(x,y,z) + \frac{\lambda^2}{2} cx + \frac{\lambda^3}{6} k(x,y,z) + \ldots = 0 \quad ,$$

$$by + \lambda(\alpha y + \beta z) + \lambda cx + \ldots + \frac{\lambda^2}{2} k(x,y,z) + \ldots = 0 \quad ,$$

$$cx + \alpha y + \beta z + \lambda k(x,y,z) + \ldots = 0 \quad .$$

By effectively calculating the derivatives, we deduce that for $\lambda = 0$, we have $ax' = -k(0,0,0)$, $y' = 0$, $z' = 0$; $by'' = k(0,0,0)$, $az'' + 2Sx'^2 = 0$ where S is the coefficient of x^2 in the function f_2. If we then take the expansions of x,y,z in

$$az + f_2(x,y,z) \quad ,$$

we see that the terms in λ and λ^2, which have az' and $\frac{1}{2}az'' + Sx'^2$ respectively as their coefficients are zero. Hence:

At its ordinary points, the cusp has at least a second order contact with the enveloped surface.

THE CASE OF DEVELOPABLE SURFACES

We have seen (no. 59) the case of the surface envelope of a plane depending on one parameter. The characteristic line envelopes a curve Γ which is skew, unless all the given planes pass through a fixed point, the case where the curve reduces to this point. The contact of the plane with Γ having order of at least two, this plane is the osculating plane to Γ.

Conversely, the osculating planes of a curve Γ have as their envelope a developable surface (it depends on a single parameter). *The characteristics are the tangents to* Γ. For, if M is a point of Γ and P an arbitrary point, then the equation of the osculating plane is

$$[(\vec{P} - \vec{M}), \vec{M}', \vec{M}''] = 0 \quad .$$

The characteristic is defined by this equation and its derivative which reduces to

$$[(\vec{P} - \vec{M}), \vec{M}', \vec{M}''] = 0 \quad ,$$

which provides the result. Hence Γ is the cusp.

We also see that the tangents to a shew curve Γ engender a developable surface for which Γ is the cusp.

If $z = F(x,y)$ is the equation of a developable surface, then the tangent plane $Z - z = p(X - x) + q(Y - y)$ (notation due to Monge) need only depend on one parameter, hence these functions of x and y are functions of one of them, e.g., q. It follows that the functional determinants

$$\frac{D(p,q)}{D(x,y)} \quad , \quad \frac{D(z - px - qy, q)}{D(x,y)}$$

are zero, which implies the relation $rt - s^2 = 0$. Conversely, if $rt - s^2 = 0$, the tangent plane only depends on one parameter, and the surface is developable (1, 127).

Chapter V

ELEMENTARY THEORY OF DIFFERENTIAL EQUATIONS

The most simple differential equations, i.e., those for which the solutions are expressable in terms of elementary functions or integrals over such functions, were studied throughout the 18th century. Amongst the first mathematicians to have studied and solved them are: Clairaut, Bernouilli, Lagrange, Euler and Riccati. D'Alembent was interested in linear equations with constant coefficients, and Laplace considered more general equations. At the start of the 19th century, Jacobi and Abel introduced some special equations. Cauchy, as we have seen in Chapter III, described general methods which in certain cases lent themselves to integration. The linear equations which will be considered later have inspired some remarkable works, in particular those bearing the names of Gauss, Bessel, Fuchs, Picard, Mathieu and Halphen. Liouville first distinguished the equations which could be integrated by means of elementary functions and has shown (1841) that the Riccati equation cannot in general enter into this class. The works of Briot and Bouquet, Poincaré, Lie and of Painlevé have been the point of departure for vast theories of which several will be developed or outlined in the following chapters.

In this chapter there will be an emphasis on those differential equations whose solution may be obtained directly or which reduce to such equations by simple transformations. There will not be any appeal to the existence theorems in Chapter III.

Recall that the *order* of a differential equation is the order of the highest derivative of the unknown function y which remains when all the reductions have been made: A differential equation of order n is of the form $F(x,y,y',\dots,y^{(n)}) = 0$. The equation is said to be *algebraic* if the function F is a polynomial with respect to the $n+2$ variables $x,\dots,y^{(n)}$. The functions $y(x)$ satisfying the equation are the *solutions* or *integrals*. When x and y are real, we can consider the curves representing $y = y(x)$: these are the *integral curves*. We have seen (nos. 46, 48) that under certain conditions, the solutions depend on *arbitrary constants equal in number to the order*. We shall recover these directly. When we have found a solution depending on constants and one susceptible to yielding all of the solutions, save a few, by means of a suitable choice of these constants, we shall say that the *general integral* or *general solution* is known. An achieved solution will be a *particular integral* or *particular solution*.

I. FIRST ORDER EQUATIONS

63 - Equations with separable variables. Examples

There are equations which can be written in the form

$$y' = \frac{f(x)}{g(y)} \quad .$$

By separating the variables and using to the extent, the following differential notation, we may write it as

$$f(x)dx - g(y)dy = 0 \quad . \tag{1}$$

Let $F(x)$ and $G(y)$ be primitives of x and y respectively. *The solutions of (1) are the implicit functions* y *of* x *defined by*

$$F(x) - G(y) = C \quad , \tag{2}$$

where C *denotes an arbitrary constant.* Since if $y = \phi(x)$ is a solution of (1), then the function $F(x) - G(\phi(x))$ has as its derivation $f(x) - g(\phi(x))\phi'(x)$, hence a zero derivative since $\phi(x)$ satisfies equation (1). Consequently, $F(x) - G(\phi(x))$ is constant, $y = \phi(x)$ is a solution of (2), where C has a certain value. Conversely, if $y = \phi(x)$ is a differentiable function satisfying (2), then its derivative is zero, hence $\phi(x)$ is a solution of (1).

Having said this, we may sharpen the hypotheses somewhat. If x and y are real along with f and g, then matters are placed in an interval of values of x and one of y where f and g are continuous and, naturally, well defined; hence they are *uniform*. If x any y are complex, we assume that f and g are analytic. We will actually put this case aside.

SINGULAR INTEGRALS

What happens if $g(y)$ is infinite for $y = y_o$, hence $1/g(y)$ zero, in the case where the differential equation, taken in its original form, is preserved. Firtly, $y = y_o$ will be a solution, this is clear. Let us suppose that $g(y)$ keeps a constant sign for $y_1 < y < y_o$ for example; let $g(y) > 0$ to fix matters. Then

$$G(y) = \int_{y_1}^{y} g(y)dy$$

increases with y. If $y \to y_o$, two cases are possible: the integral either has or has not a sense. In the first case, the equation (2) yields y as a function of s:

$$G(y) = \int_{y_1}^{y} g(y)dy = F(x) - C \quad ,$$

and if C is chosen such that $F(x_o) - C = G(y_o)$ where x_o is a point belonging to an interval of continuity of $f(x)$, we see that we obtain an *integral curve* which is tangent at the point x_o, y_o to the line $y = y_o$. The line $y = y_o$ is tangent to the other integral curves, it is an envelope of integral curves depending on the constant C, and it is a solution which is not obtained for any value of C in (2). We say that $y = y_o$ is a *singular integral.*

Let us assume now that $G(y)$ has no sense for $y = y_o$, hence $G(y) \to \infty$ when $y \to y_o$. By writing the curves (2) under the form

$$\frac{1}{G(y)} = \frac{1}{F(x) - C} \quad ,$$

we see that if $C \to \infty$, $G(y) \to \infty$, $y \to y_o$. The line $y = y_o$ is the limit of the interval curves (2) when $C \to \infty$; we can consider this as the integral curve corresponding to $C = \infty$. The integral curves corresponding to C finite do no, moreover, attain to the line $y = y_o$.

As in equation (1), x and y play a symmetric role; we can make the same remarks as those concerning the isolated zero of $f(x)$. They will provide integral curves $x =$ const. which will be singular if the integral of $f(x)$ taken up to these values has a meaning, and will be ordinary in the contrary.

Example

If we take the equation

$$y'^2 = \frac{1 - y^2}{1 - x^2}$$

we obtain

$$\frac{dy}{\sqrt{1 - y^2}} = \pm \frac{dx}{\sqrt{1 - x^2}}$$

and

$$\text{arc sin } y = \pm \text{ arc sin } x + C \quad .$$

Extracting y and rationalizing, it becomes

$$y = \pm x \text{ cox } C \pm \sqrt{1 - x^2} \text{ sin } C \quad ,$$

since

$$x^2 + y^2 - 2\lambda xy = 1 - \lambda^2 \quad , \tag{3}$$

where λ is an arbitrary real number (since we are free to integrate in terms of general arc-sine function). The lines $x = \pm 1$ and $y = \pm 1$ are singular integral curves since the corresponding integrals have a meaning.

These are the envelopes of the integral curves, that can be easily shown by equation (3) of these curves. Those integral curves are ellipses, hyperbolae or lines whose axes of symmetry are $y = x$ and $y = -x$ (fig. 33).

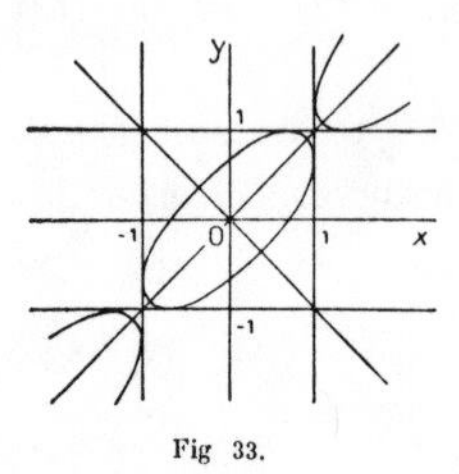

Fig 33.

Remarks. We note that the conditions for applying the Cauchy theorems of Chapter II imply, in the case where one of the domains is bounded by the four lines $y = \pm 1$ and $x = \pm 1$, that the constant of integration enters in all cases in a similar way as in equation (3). When we study the singular integrals in the general case, we will see what there is to say on the subject of the arbitrary constant (when it involves a first order equation) when the singular integral is crossed.

We might also remark that, in the actual case, we can find an infinity of integrals of the given equation that pass through a point M_o (fig. 34).

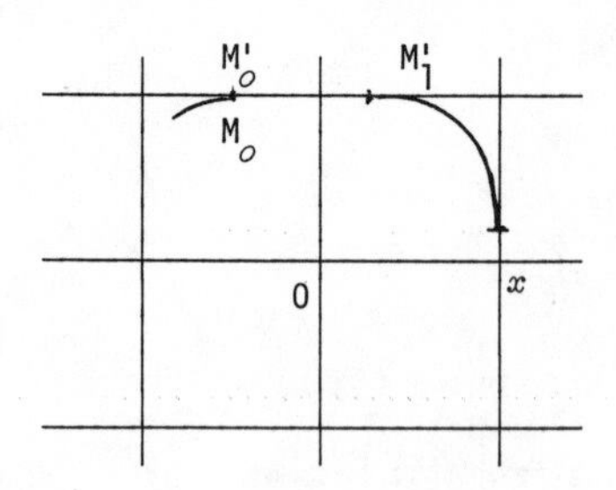

Fig. 34

Let us take, for example, M_o to be in the square with centre the origin, and one of the two ellipses passing through M_o (and which correspond to the double sign appearing in the differential equation). This ellipse is tangent to $y = 1$ at the point M'_o. Let us take on this line $y = 1$, a second point M'_1, then the ellipse will be tangent to this point. We thus form *a continuous integral curve*, if M'_1 is conveniently placed; this curve depends on the distance $M'_oM'_1$, i.e., *an arbitrary parameter*. These circumstances only arise because the boundary of the domain has been reached, where existence and uniqueness applies. They will be the same each time the integral curves have an envelope which they will touch at the ordinary points.

64 - Euler's equation

The equation

$$\frac{dx}{\sqrt{P(x)}} = \pm \frac{dy}{\sqrt{P(x)}} \tag{4}$$

where $P(x)$ is a polynomial of degree three or four, had been integrated by Euler by a beautiful generalisation of the result obtained in no. 63 in the case where $P(x)$ is of the second degree. Euler has sought a solution under the form

$$\Phi(x,y) = Ax^2y^2 + B(x+y)xy + C(x^2+y^2) + Dxy + E(x+y) + F = 0$$

where $A \ldots F$ are constants. The symmetric polynomial $\Phi(x,y)$ could be written as

$$\Phi \equiv F(y)x^2 + G(y)x + H(y) = F(x)y^2 + G(x)y + H(x)$$

with

$$F(x) \equiv Ax^2 + Bx + C\ , \quad G(x) \equiv Bx^2 + Dx + E\ , \quad H(x) \equiv Cx^2 + Ex + F$$

If x and y are related by the equation $\Phi = 0$, we have

$$\frac{\partial\Phi}{\partial x}\,dx + \frac{\partial\Phi}{\partial y}\,dy = 0 \quad . \tag{5}$$

Now

$$\frac{\partial\Phi}{\partial x} \equiv 2F(y)x + G(y) \quad ,$$

and $\Phi = 0$ is also written as

$$[2F(y)x + G(y)]^2 + 2F(y)H(y) - G(y)^2 = 0 \quad ,$$

such that the equation (5) is written under the form (4) if we set

$$P(x) \equiv G(x)^2 - 4F(x)H(x) \quad . \tag{6}$$

By means of condition (6), the polynomial Φ equal to zero, provides an integral of equation (4). Now $P(x)$ depends on five homogeneous coefficients, there are six in Φ and consequently six in the second member of (6). It is thus possible to determine A, B, C, D, E, and F by taking one of the curves to be arbitrary. The actual calculus of identifcation, which is tedious in general was systematically conducted by Jacobi.

THE RELATION WITH THE THEORY OF ELLIPTIC FUNCTIONS

We know that we can reduce the case where $P(x)$ is of the fourth degree to that where it is of the third degree (I, 96). If a is a zero of $P(x)$, we set $x = a + \frac{1}{X}$. We could then put the third degree polynomial in the normal Weierstrass form, by means of a linear transformation $X = k\xi' + k'$. If at the same time, we carry out these transformations on x and y, we see that we recover the Euler equation in the canonical form

$$\frac{dx}{\sqrt{4x^3 - g_2x - g_3}} = \pm\frac{dy}{\sqrt{4y^3 - g_2g - g_3}} \quad .$$

If we introduce the inverse functions of the elliptic integrals obtained by integrating both members, by setting

$$x = pt \ , \qquad y = pu \quad ,$$

then we have the relation

$$t = \pm u + C \quad ,$$

where C is an arbitrary constant. The relation between x and y is therefore given by

$$x = pt \ , \qquad y = p(t + C) \quad ;$$

the sign $\pm$ disappears since pt is even. Euler's result is hence equivalent to the fact that there exists an algebraic relation between pt and $p(t + C)$ (I, 228).

AN APPLICATION OF ABEL'S THEOREM

Let us assume that $P(x)$ and $P(y)$ have been reduced to the third degree. The integral of Euler's equation is written as

$$\int_{x_o}^{x} \frac{dx}{\sqrt{P(x)}} + \int_{y_o}^{y} \frac{dy}{\sqrt{P(y)}} = C \quad . \tag{7}$$

When x and y are the z of two variable points of the cubic

$$u^2 = P(z) \quad ,$$

then the sum of the abelian integrals calculated up to the points is therefore constant. Now the line joining the points

$$x\ \sqrt{P(x)}\quad ;\qquad y\ \sqrt{P(y)}$$

of the cubic cuts the cubic at a point

$$z,\ \sqrt{P(z)}\quad ,$$

and following Abel's theorem, the sum of the integrals of the first kind calculated up to these three points is also constant, consequently by (7),

$$\int_{z_o}^{z} \frac{dz}{\sqrt{P(t)}} = C' \quad .$$

The point $z, \sqrt{P(z)}$ is therefore fixed. Conversely, a line passing through a fixed point of the cubic cuts it at points whose z is equal to x and y respectively, and for which an equation of the form (7) is obtained. The integral of Euler's equation could then be obtained by indicating that the points

$$x,\ \sqrt{P(x)}\ ;\quad y,\ \sqrt{P(y)}\ ;\quad \lambda,\ \sqrt{P(\lambda)}\quad ,$$

where λ is an arbitrary constant are aligned. We then obtain the general integral under the form

$$\begin{vmatrix} x & P(x) & 1 \\ y & P(y) & 1 \\ \lambda & P(\lambda) & 1 \end{vmatrix} = 0 \quad .$$

65 - Homogeneous equations

These are equations which reduce to the form

$$y' = f\left(\frac{y}{x}\right) \quad . \tag{8}$$

An equation of this kind does not change if a homothety is made with respect to the origin: $x = kY$, $y = kY$, k = const. The integral curve other than the lines passing through the origin are therefore deduced from each other by a homothety with respect to the origin. This property is conducive to taking $u = y/x$ as the variable and to determine x as a function of u. We have

$$y = ux\ ,\quad \frac{dy}{dx} = u + x\,\frac{du}{dx}\quad ,$$

and the equation becomes

$$u + x\,\frac{du}{dx} = f(u) \quad .$$

This is an equation with variables separable that may be written as

$$\frac{dx}{x} = \frac{du}{f(u) - u} \tag{9}$$

and if we set

$$F(u) = \int \frac{du}{f(u) - u}$$

we obtain

$$x = Ce^{F(u)} , \quad y = Cue^{F(u)} , \tag{10}$$

where C is an arbitrary constant. We indeed see that the integral curves are homothetiques with respect to the origin. Under the form (9), we see that the equation is satisfied for $u = u_o$ if u_o is a root of $f(u) = u$. If $F(u)$ exists for $u = u_o$, then $u = u_o$ is a singular integral of (9). We also see that, in this case, the curve (10) passes through the point at a finite distance corresponding to $u = u_o$ and is tangent at this point to the line $y - u_o x = 0$. This line is an envelope of the integral curves; it is *a singular integral*. On the contrary, if the integral $F(u)$ increases indefinitely in absolute value when $u \to u_o$, then the corresponding point of the curve (10) tends towards the origin or moves away indefinitely. The line $y - u_o x = 0$ is tangent to the integral curves at the origin or is even an asymptotic direction of these curves. When the constant C tends to infinity or to zero, the points of the curve (10) corresponding to u close to u_o, tend towards those of $y - u_o x = 0$. This line enters into the general integral.

THE CASE OF NON-SOLVABLE EQUATIONS

In a general way, an equation of the form $F(x,y,y') = 0$ will provide a homogeneous equation after solving with respect to y' if it does not cease to be satisfied by kx, ky, y' for any k. Taking $k = 1/x$, we see that it must be of the form

$$F(1,y/x,x') = 0 \quad ;$$

we could then only have y' and y/x appearing.

Remark. If for any k, x, y, y', we have

$$F(kx,ky,y') \equiv k^{\alpha}F(x,y,y') \quad ,$$

then the equation is homogeneous. The condition that results, with k having any sign, is more restrictive than the condition of homogeneity given previously (I, 1116). We can distinguish the homogeneous functions in x and y, of degree α, satisfying the above identity and those functions that are *positively* homogeneous where it is assumed $k > 0$; these last functions are those considered previously.

PASSAGE TO POLAR COORDINATES

If we set $x = r \cos \theta$, $y = r \operatorname{Sin} \theta$ and if r is considered as the unknown and θ as the variable, then after a simple calculation, we obtain the equation with separated variables deduced from (8):

$$\frac{dr}{r} = \frac{1 + tg\theta f(ty\theta)}{f(tg\theta) - tg\theta} \quad d\theta \quad .$$

Remark. The equation will be homogeneous whenever we determine by means of a first order differential equation of the first order, the curves are homethic with respect to the origin.

Examples 1. The equation $4y^2 y'^2 + 4y^2 - (x + yy')^2 = 0$ is homogeneous. On integrating by the general method, we will find that the lines $x \pm\sqrt{3}\, y = 0$ are singular integrals. The general integral is given by

$$3(x^2 + y^2) = 4\lambda x + \lambda^2 = 0 \quad .$$

The integral curves are circles centred on Ox and tangent to the two lines above.

II. The homogeneous equation

$$y^2(x + yy')^2 - (x^2 + y^2)(xy' - y)^2 = 0$$

becomes in polar coordinates

$$r^2 \left(\frac{dr}{d}\right)^2 \operatorname{Sin}^2\theta \ - r^4 = 0$$

or

$$\frac{r'}{r} = \pm \frac{1}{\operatorname{Sin} \theta}$$

which may be integrated straight away. The integral curves are the strophoid lines

$$[x^2 + (y - \lambda)^2](y - \lambda) - \lambda\, [x^2 - (y - \lambda)^2] = 0 \quad .$$

III. The equation $xy'(2y - x) - y^2 = 0$ has for integral curves the hyperbolae $y(y - x) - \lambda x = 0$. For $\lambda = 0$, the curve reduces to its asymptotes.

EQUATIONS OF THE FORM $y' = f\left[\frac{ax + by + c}{a'x + b'y + c'}\right]$ WHERE a, b, c, a', b', c' ARE CONSTANTS

In general, we can recover a homogeneous equation by transporting to the origin at the point of intersection of the lines

$$ax + by + c = 0 \quad , \qquad a'x + b'y + c' = 0 \quad ,$$

which assumes that these lines are not parallel. We therefore set $x = X + \alpha$, $y = Y + \beta$ are constants; this gives

$$ax + by + c \equiv aX + aY, \quad a'x + b'y + c' \equiv a'X + b'Y \quad .$$

If we assume

$$a\alpha + b\beta + c = 0, \qquad a'\alpha + b'\beta + c' = 0 \quad ,$$

this determines α and β. The equation is then

$$\frac{dY}{dX} = f\left(\frac{aX + bY}{a'X + b'Y}\right)$$

it is homogeneous.

In the particular case that was left aside, where $a'b - b'a = 0$, the equation is written as

$$y' = f\left(\frac{ax + by + c}{k(ax+by) + c'}\right)$$

and on taking $Y = ax + by$ as the new unknown function, with x remaining as a variable, we have an equation with separable variables.

66 - Linear equations. The equations of Bernouilli and Jacobi

The linear equations are of the form

$$A(x)y' + B(x)y + C(x) = 0 \quad .$$

On dividing by $A(x)$, we obtain the canonical form

$$y' + A(x)y + X_1(x) = 0$$

or

$$y' + X(x)y = -X_1(x) \quad . \tag{11}$$

One such equation is said to be *without* a second member or is *linear and homogeneous* when $X_1(x) \equiv 0$. Then the equation is an equation with variables separable which can be integrated straight away. Its integral is

$$y = Ce^{U(x)}, \qquad U(\ ) = -\int X(x)dx \quad , \tag{12}$$

where C is an arbitrary constant.

To integrate the equation *with* the second member (11), we start by integrating the equation obtained by suppressing the second member. Its general integral is given by the formula (12). We then use *the method of the variation of constants* due to Lagrange. We then seek a solution of the form

$$y = \lambda(x)e^{U(x)}$$

$$y = \lambda(x)y_1$$

where y_1 denotes the integral $e^{U(x)}$ of the equation without the second member. By stating that y is a solution of equation (11), we obtain the new equation

$$y_1 \frac{d\lambda}{dx} + \lambda[y_1' + X(x)y_1] = -X_1(x) \quad ,$$

which reduces to

$$\frac{d\lambda}{dx} = \frac{-X_1}{y_1}$$

and gives λ by a quadrature

$$\lambda(x) = \left[-\int \frac{X_1}{y_1}\,dx + C\right]$$

where C is an arbitrary constant. It is obtained by two successive quadratures and depends linearly on one constant.

Remarks.

I. The solution is therefore equal to the general solution of the equation without the second member, plus a solution of the complete equation. If *one* solution of the complete equation is known, then it suffices to combine this solution with the general solution of the equation without the second member.

The form of the solution shows that the integral curves do not have any envelope and there is no singular integral.

II. The differential equation of the functions depending linearly on one constant is a linear equation. For, if $y = f(x) + Cg(x)$ are these functions, we have $y' = f'(x) + Cg'(x)$ and on eliminating C, we obtain the linear equation

$$[y' - f'(x)]g(x) - g'(x)\,[y - f(x)] = 0 \quad .$$

III. If y_1, y_2, y_3 are three solutions of a complete linear equation, corresponding to the values C_1, C_2, C_3 of the constant of integration, then we have

$$y_1 = f(x) + C_1 g(x), \quad y_2 = f(x) + C_2 g(x) \ , \quad y_3 = f(x) + C_3 g(x) \quad .$$

By eliminating $f(x)$ and $g(x)$, we have the relation

$$\frac{y_3 - y_1}{y_2 - y_1} = \frac{C_3 - C_1}{C_2 - C_1} = \text{const.}$$

Consequently, if y_2 and y_1 are two integrals, then every other integral is given by $y - y_1 = k(y_2 - y_1)$, where k is constant. We can then deduce a geometric property of the integral curves. If Γ, Γ_1 and Γ_2 are three such curves, and M, M_1 and M_2 are points of these curves situated on the parallel to Oy intersecting Ox at the point P, then we have $\overline{M_1M} = k\overline{M_1M_2}$. We can outline Γ pointwise when Γ_1 and Γ_2 are described and k is given.

If we consider another parallel to Oy, $P'M_1'M_2'M'$ (fig. 35)

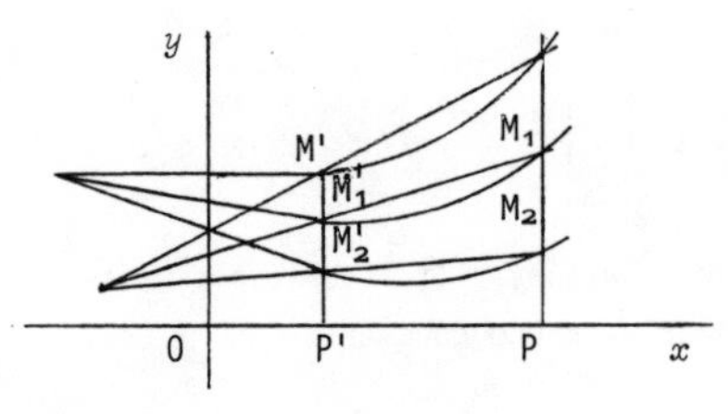

Fig. 35

$$\frac{M_1'M'}{M_2'M_1'} = \frac{\overline{M_1M}}{M_2M_1} \quad ,$$

and the lines M_1M_1', M_2M_2', MM' are concurrent. This gives the outline. In particular, the tangent to M passes through the point of intersection of the tangents to M_1 and M_2.

IV. To integrate the equation without the second member, we divided by y. We therefore assume $y \neq 0$. The formula (12), which gives y, shows moreover that if $C \neq 0$, y is non-zero for as long as $X(x)$ is continuous. But for certain x, $X(x)$ could be infinite and $U(x)$ could tend to $-\infty$ when x tends to such a value. Under these conditions, the function y defined by (12) can exist and is continuous for this x; it can continue to satisfy, even for $y = 0$, the equation without the second member. For example, the equation

$$xy' - 2y = 0$$

has for its integral $y = Cx^2$ which is well defined, continuous and differentiable, even for $x = 0$, a value for which $y = 0$.

BERNOUILLI'S EQUATION

This is an equation of the form

$$y' + X(x)y + X_1(x)y^n = 0 \quad ,$$

where n is any real constant, which implies $y > 0$ in general. If $n = 0$, the equation is linear; if $n = 1$, it is linear and homogeneous. These cases aside, the equation is transformed to a linear equation and on dividing both members by y^n and setting $y^{1-n} = Y$, we obtain

$$\frac{1}{1-n}\frac{dY}{dx} + X(x)Y + X_1(x) = 0 \quad .$$

By using the result relating to the linear equation, we see how the constant of integration enters into the expression of the solution.

Example. Take the equation $y'x^4 - yx^3 - y^2 = 0$. We set $y^{-2} = Y$, which yields the linear equation

$$\frac{1}{2}x^4Y' + x^3Y + 1 = 0 \quad .$$

The equation without the second member has as its general integral $Y = Cx^{-2}$, and by considering C as a function of x, we see that we must have

$$C' + \frac{2}{x^2} = 0 \quad ,$$

which yields

$$C = \frac{2}{x} + \lambda, \qquad Y = \frac{\lambda}{x^4} + \frac{2}{x^3}, \qquad y^2 = \frac{x^3}{\lambda x + 2} \quad .$$

AN APPLICATION

The equation

$$A\left(\frac{y}{x}\right)dx + B\left(\frac{y}{x}\right)dy + kx^m(xdy - ydx) = 0 \quad ,$$

where m is any real constant, reduces to a Bernouilli equation when we set $y = ux$. Since it then becomes

$$A(u) + B(u)\left(x\frac{du}{dx} + u\right) + kx^{m+2}\frac{du}{dx} = 0 \quad .$$

JACOBI'S EQUATION

The equation with constant coefficients

$$(ax + by + c)(xdy + ydx) + (a'x + b'y + c')dx$$

$$+ (a''x + b''y + c'')dy = 0$$

becomes of type (13) when $c = c' = c'' = 0$, since, in this case, the functions of y/x appear as coefficients of dx and dy on dividing both members by $ax + by$. (We then have $m = 0$.) We can then try to recover this case by setting $x = X + \alpha$, $y = Y + \beta$, where α and β are any suitably chosen constants. We obtain

$$(aX + bY)(XdY - YdX) + (AX + BY + C)dX + (A'X + B'Y + C')dY + 0$$

$$C = a'\alpha + b'\beta + c' - \beta(a\alpha + b\beta + c) \quad ,$$

$$C' = a''\alpha + b''\beta + c'' + \alpha(a\alpha + b\beta + c) \quad ;$$

as for A, B, A', B', these are constants which will be easy to calculate. It suffices to choose α and β in order that $C = C' = 0$. If we put

$$a\alpha + b\beta + c = \lambda \quad , \tag{14}$$

we will have

$$\begin{aligned} a'\alpha + b'\beta + c' &= \lambda\beta \\ a''\alpha + b''\beta + c'' &= -\lambda\alpha \quad . \end{aligned} \tag{15}$$

These three equations are only compatible if

$$\begin{vmatrix} a''+\lambda & b'' & c'' \\ a' & b'-\lambda & c' \\ a & b & c-\lambda \end{vmatrix} = 0 \quad .$$

We will then take for λ, a root of this equation of degree three, and α and β will then be given by the two equations (14) and (15), it could be indeterminante there.

67 - Riccati's equations

This is the equation

$$y' + X(x) + X_1(x) + X_2(x)y^2 = 0$$

where $X(x)$ and $X_2(x)$ are not identically zero. Liouville has shown that it is not integrable in general. But if, *a priori*, a solution y_1 is known, then changing it to the unknown function $y = y_1 + z$, reduces matters to

$$z' + (X_1(x) + 2y_1X_2(x))z + X_2(x)z^2 = 0 \quad ,$$

that is, a Bernoulli equation for which the transformation $z = 1/Z$ will reduce it to a linear equation. With respect to the arbitrary constant C, we will successively obtain

$$Z = f(x) + Cy(x), \quad z = \frac{1}{f(x) + Cy(x)}, \quad y = y_1 + z = \frac{h(x) + Ck(x)}{f(x) + Cy(x)} \quad .$$

The general solution is therefore a homographic function of the constant of integration. Conversely, if a function y depends homographically on a constant C, it is the integral of a Riccati equation. Since, as kind of inverse to the preceding calculation, we could set $y = y_1 + \frac{1}{Z}$, where Z is linear in C, and hence a solution of a linear equation. 1/Z will be a solution of a Bernouilli equation and finally y a solution of a Riccati equation.

It follows that if y_1, y_2, y_3, y_4 are four solutions correspondings to the values C_1, C_2, C_3, C_4 of the constant of integration, we have

$$\frac{y_4 - y_1}{y_4 - y_2} : \frac{y_3 - y_1}{y_3 - y_2} = \frac{C_4 - C_1}{C_4 - C_2} : \frac{C_3 - C_1}{C_3 - C_2}$$

since the anharmonic ratio or cross-ratio of the four numbers is preserved under a homography. *The anharmonic ratio of four solutions is therefore constant.* If three solutions y_1, y_2, y_3 are known, then every other solution y having a constant anharmonic ratio with y_1, y_2, y_3, will be given by

$$\frac{y - y_1}{y - y_2} : \frac{y_3 - y_1}{y_3 - y_2} = C \quad ,$$

where C is a constant. Hence, the equation is integrated if three particular solutions are known; this also results from the method given for integration when a solution y_1 is known. For Bernouilli's equation, which is recovered, we will know two solutions $y_2 - y_1$ and $y_3 - y_1$, therefore for the final linear equation, we will obtain two solutions, and hence all of the others. The same applies when two particular solutions are known; we will know a solution of the linear equation which is recovered and we will integrate by a single quadrature.

Examples

I. If we take a plane in which the curves intersect the circles of a one parameter family, in a constant angle V, then we are led to integrate a Riccati equation. Since if $a(\lambda)$ and $b(\lambda)$ are the coordinates of the centre of the circle, with radius $R(\lambda)$, its parametric equations (in rectangular axes) are

$$x = a(\lambda) + R(\lambda)\cos t, \quad y = b(\lambda) + R(\lambda)\sin t \quad .$$

We can determine t as a function of λ provided the curve as defined, intersects the circles in the angle V. For this curve, we will have

$$dx = (a' + R'\cos t)d\lambda - R\sin t\, dt$$

$$dy = (b' + R'\sin t)d\lambda + R\cos t\, dt \quad .$$

The angle of this direction with the tangent to the circle must have a given number for its tangent. We then have the equation

$$dy\cos t - dx\sin t = b(dx\cos t + dy\sin t), \quad k = \operatorname{cot}g\, V \quad .$$

On replacing dx and dy by their values, we obtain

$$(b'\cos t - a'\sin t)d\lambda + R dt = k(a'\cos t + b'\sin t + R')d\lambda$$

and on setting $u = tg\,\frac{t}{2}$, we arrive at Riccati's equation.

We will know the integral if we know the solution in advance. For example, if it entails orthogonal trajectories, $k = 0$ and if one of them are known, we look at the quadratures. For orthogonal circles to a line or to a circle, this line or this circle counts twice, and a single quadrature will suffice.

II. Euler studied the equation

$$y' + y^2 = ax^m$$

where a and m are given. The equation is integrated when

$$m = -\frac{4N}{2N+1} \quad \text{or} \quad m = \frac{-4N}{2N-1} \quad ,$$

where N is a positive integer. In the first case, the transformation

$$x = X'/(m+1) \quad , \quad y = \frac{a}{m+1}\,\frac{1}{Y} \quad ,$$

leads to

$$Y' + Y^2 = \frac{a}{(m+1)^2}X^n \quad , \quad n = \frac{4N}{2N-1} \quad ,$$

i.e., to the second case. In this second case, if we set

$$x = \frac{1}{X}, \quad y = X - X^2Y \quad ,$$

we obtain

$$Y' + Y^2 = aX^p \quad , \qquad p = \frac{-4(N-1)}{2(N-1) + 1} \quad ,$$

i.e., the first case, but with a decrease of 1 is the value of N. Iterating what we had before will reduce matters to the case $N = 0$, and then the integration is immediate.

68 - The integration of the total differentials. The integrating factor

We know (I, 144) that, if the continuous functions $P(x,y)$ and $Q(x,y)$ admit partial derivatives with respect to x and y that are continuous in a certain domain and if in this domain D we have

$$\frac{\partial P(x,y)}{\partial y} \equiv \frac{\partial Q(x,y)}{\partial y} \quad , \tag{16}$$

then the expression $P(x,y)dx + Q(x,y)dy$ is the differential of a function $U(x,y)$. The differential equation

$$P(x,y)dx + Q(x,y)dy = 0 \tag{17}$$

has as its integral

$$U(x,y) = C \quad , \tag{18}$$

where C is an arbitrary constant (which could take any value taken by $U(x,y)$ in D). The function U is obtained by two quadratures. It is clear that every solution of (17), $y = \phi(x)$, renders U constant, and is defined by the implicit equation (18). Conversely, a solution of (18) is differentiable and satisfies (17).

Given a first order differential equation

$$A(x,y)dx + B(x,y)dy = 0 \quad ,$$

we could follow Euler, by integrating by the above method after having multiplied by a factor $\lambda(x,y)$ such that the first member becomes a total differential

$$\lambda(x,y)[A(x,y)dx + B(x,y)dy] \equiv dU(x,y) \quad .$$

For this to be so, it is necessary to have

$$\frac{\partial(\lambda A)}{\partial y} = \frac{\partial(\lambda B)}{\partial x} \quad , \qquad \text{or}$$

$$A \frac{\partial\lambda}{\partial y} - B \frac{\partial\lambda}{\partial x} + \lambda \left(\frac{\partial A}{\partial y} - \frac{\partial B}{\partial x}\right) = 0 \quad . \tag{19}$$

The factors $\lambda(x,y)$ determined, which we call the *integrating factors*, must then satisfy the equation with the partial derivatives (19). We shall see that the complete integration of one such equation will imply that of the system of differential equations

$$\frac{dx}{-B} = \frac{dy}{A} = \frac{d\lambda}{\left(\frac{\partial B}{\partial x} - \frac{\partial A}{\partial y}\right)\lambda}$$

hence, in particular, the equation proposed. But it will suffice to know a *particular solution*. When we know an integrating factor λ and the corresponding function U, we obtain the other integrating factors μ in seeking then to be of the form $\mu = \lambda v$. On account of (19) we must have

$$A\lambda \frac{\partial v}{\partial y} - B\lambda \frac{\partial v}{\partial x} = 0 \quad .$$

Now A and B are proportional to $\frac{\partial U}{\partial x}$ and $\frac{\partial U}{\partial y}$ respectively. The functional determinant of U, v, will then be zero; v will be a function of U and we will have $\mu = \lambda\phi(U)$, where ϕ is any differentiable function. We can straight away verify that every function $\lambda\phi(U)$ is an integrating factor. It follows from this that:

If two integrating factors whose ratio is non-constant are known, then the general integral is obtained by equating this ratio with an arbitrary constant.

Effectively, the general integral is $U = \text{const}$, hence $\mu/\lambda = \text{cost}$.

Remark. If we can integrate the equation $Adx + Bdy = 0$, and if $U(x,y) = \text{const.}$ is the integral, then on differentiating we have

$$\frac{\partial U}{\partial x} dx + \frac{\partial U}{\partial y} = dy = 0,$$

hence $\frac{\partial U}{\partial x}/A$ is an integrating factor.

APPLICATION. If the given equation is of the form

$$(Adx + Bdy) + (Cdx + Ddy) = 0 \quad , \tag{20}$$

where A, B, C, D are functions of x and y, and if an integrating factor is known for each of the points in brackets, say λ for the first with U the corresponding function, and μ for the second with corresponding V; then the general integrating factors for the two parts will be $\lambda\phi(U)$ and $\mu\phi(V)$ respectively. We can determine ϕ and ψ to obtain a common integrating factor, and this common factor will clearly be an integrating factor of the complete equation (20).

Example

Consider the equation

$$aydx + bxdy + x^m y = 0 \quad ,$$

where a, b, α, β are constants, along with the real exponents m and n. The equations $aydx + bxdy = 0$ and $\alpha ydx + \beta xdy = 0$ are integrated to provide integrating factors for the two parts of the equation. They are $\phi(x^a y^b)x^{-1}y^{-1}$ and $\psi(x^\alpha y^\beta)x^{-1-m}y^{-1-n}$ respectively. We will have a common integrating factor by taking ϕ and ψ as monomials, and in determining p and q such that

$$(x^a y^b)^p x^m y^n = (x^\alpha y^\beta)^q \quad .$$

The numbers p and q are determined by the equations

$$ap - \alpha q + m = 0 \ , \qquad bp - \beta q + n = 0 \quad ,$$

which yields a solution if $a\beta - b\alpha \neq 0$. If $a\beta = b\alpha$, the given equation simplifies and can at once be integrated.

69 - Nonsolvable equations. The equations of Lagrange and Clairaut

Let us consider an equation $F(x,y,y') = 0$ whose solution with respect to y' is not easily obtainable. We will attempt to use the parametric representations.

1. IF THE EQUATION DOES NOT CONTAIN y, IS OF THE FORM $G(x,y') = 0$, and of it the coordinates of a point of the curve $G(x,y) = 0$ can be expressed as functions of a parameter t, $X = \phi(t)$, $Y = \psi(t)$ say, then the integral curves will satisfy the conditions

$$x = \phi(t) \quad , \quad \frac{dy}{dx} = \psi(t) \quad .$$

Only one such curve x and y will be functions of t [since $y = \theta(t)$ and $x = \phi(t)$] and we will have

$$dy = \psi(t)dx = \psi(t)\phi'(t)dt \quad ,$$

when we deduce the integrals under the form

$$x = \ (t) \quad , \quad y = \int \psi(t)\ \phi'(t)dt \ + \text{const.} \quad .$$

II. IF THE EQUATION DOES NOT CONTAIN x, IS OF THE FORM $H(y,y') = 0$ and if this curve can be parametrically represented by $y = \lambda(t)$, $y' = \mu(t)$ then the coordinates of the points of the integral curves could be expressed as functions of t, and we have

$$y = \lambda(t)\,, \quad dy = \mu(t)dx \quad , \quad dx = \frac{dy}{\mu(t)} = \frac{\lambda'(t)}{\mu(t)}\,dt \quad ,$$

hence the representation

$$x = \int \frac{\lambda'(t)}{\mu(t)}dt + C\,, \quad y = \lambda(t)$$

where C is an arbitrary constant.

IN THE GENERAL CASE WHERE $F(x,y,z) = 0$, let us assume that a parametric representation of the surface $F(x,y,z) = 0$ is known to be of the form $x = f(u,v)$, $y = g(u,v)$, $z = h(u,v)$, a proper representation that has already been considered. If $y = \theta(x)$ is an integral of the differential equation, then at the points of this curve C, there will correspond values of u and v, hence a curve Γ in the plane of the (u,v), and we will have along Γ:

$$x = f(u,v), \quad y = g(u,v), \quad dy = h(u,v)dx \quad , \tag{21}$$

hence u and v will satisfy the differential equation

$$\frac{\partial g}{\partial u}\,du + \frac{\partial g}{\partial v}\,dv = h(u,v) \left[\frac{\partial f}{\partial u}\,du + \frac{\partial f}{\partial v}\,dv\right] \quad . \tag{22}$$

Conversely, if Γ is an integral curve of this equation (22), the curve C of Γ, transformed by means of the first two equations (21), will satisfy the third equation (21) and hence the equation $F(x,y,y') = 0$. The equation (22) is solvable with respect to dv/du or du/dv.

Remark. We could say that the integration of the equation $F(x,y,y') = 0$ amounts to finding, on the surface $F(x,y,z) = 0$, curves along which $dy = zdx$.

Particular cases. If we have the equation

$$x = K(y,y') \quad ,$$

then the parametric representation is $x = K(u,v)$, $y = u$, $z = y' = v$. We will have to integrate

$$v\left(\frac{\partial K}{\partial u}\,du + \frac{\partial K}{\partial v}\,dv\right) = du \quad ,$$

and consequently we will have $x = K(u,v)$ $y = u$.

Similarly, if the equation is of the form

$$y = L(x,y') \quad ,$$

then following $y' = p$, the parametric representation will be $x = x$, $y = L(x,p)$, $y' = p$, and we will have integral curves under the form

$y = L(x,p)$, where p and x are related by the differential equation

$$\frac{\partial L}{\partial x} dx + \frac{\partial L}{\partial p} dp = p dx \quad .$$

These arise by differentiating the equation $L(x,p) = y$.

AN APPLICATION OF LAGRANGE'S EQUATION

Lagrange's equation is of the form

$$y = xA(y') + B(y') \quad . \tag{23}$$

It enters into the preceding type. By setting $y' = p$, we will have

$$y = xA(p) + B(p) \tag{24}$$

and the relation between p and x is given by the equation obtained on differentiating

$$p dx = A(p)dx + (xA'(p) + B'(p))dp \quad . \tag{25}$$

If we take x as the unknown function and p as the variable we see that *when* $A(p) - p$ *is not identically zero*, the equation becomes

$$[A(p) - p]\frac{dx}{dp} + xA'(p) + B'(p) = 0 \quad .$$

This is a linear equation which may be integrated by two quadratures. With x obtained as a function of p, the equation (24) gives y as a function of p. We have a parametric representation as a function of p, which is the gradient of the tangent at a point of the integral curve.

We remark that, in the form of (25), the number p' satisfying the equation $A(p) - p = 0$ provide solutions; the corresponding curves are straight lines. One such line is an asymptotic direction of the integral curves if when p tends to p', x does not remain finite; it is a singular integral if x does remain finite.

Example. Consider the equation $y = xy'^2 + y'^3$. We have

$$y = xp^2 + p^3 \,, \qquad (p - p^2)\frac{dx}{dp} = 2px + 3p^2 \quad ;$$

the lines $y = x + 1$, $y = 0$, corresponding to $p = 1$ and $p = 0$ respectively, are integral curves. Moreover, we see that

$$x = \frac{-2p^3 + 3p^2 + C}{2(p-1)^2} \quad , \quad y = p^2x + p^3 \quad ,$$

where C is an arbitrary constant; x remains finite and is equal to $C/2$ for $p = 0$. The line $y = 0$ is an envelope of integral curves; it is a singular integral. But, when p tends to 1, $|x|$ increases indefinitely if $C \neq -1$. For $C = -1$, we have $2x = -2p - 1$. The line $y = x + 1$ is an asymptotic direction for the integral curves corresponding to $C \neq -1$. If $C = -1$, the integral curve decomposes into the line $y = x + 1$ and a conic.

PROPERTY OF THE INTEGRAL AND THE INTEGRAL CURVES

The non-singular integrals of the linear equation (25) are of the form $x = \lambda(p) + C\mu(p)$, where C is an arbitrary constant and λ and μ are determined functions of p. We will also have, therefore, $y = \lambda_1(p) + C\mu_1(p)$. For various values of C and the same p, we hence obtain points of the line (24) situated on the various integral curves, and at these points the tangents, whose slopes are p, are parallel. Moreover, if M_o, M_1 and M are points of the curves Γ_o, Γ_1 and Γ corresponding to C_o, C_1 and C respectively and taken for the same p we have

$$\overrightarrow{M_oM} = \frac{C - C_o}{C_1 - C_o} \overrightarrow{M_oM_1} \quad ,$$

which provides a pointwise construction of various integral curves when two non-singular integral curves are known.

CLAIRAUT'S EQUATION

This is the case left aside above, where, in equation (23), $A(y') \equiv y'$. We have then

$$y = xy' + B(y') \quad . \tag{26}$$

Proceeding in the same way, we will have by setting $y' = p$,

$$[x + B'(p)] \frac{dp}{dx} = 0 \quad .$$

This equation could only be satisfied when:

1° for $\frac{dp}{dx} = 0$, $p = 0$, and we then have

$$y = p(x) + B(p), \quad p = \text{const.} \tag{27}$$

2° for $x = -B'(p)$; we then obtain the curve Γ:

$$x = -B'(p) \ , \quad y = -pB'(p) + B(p) \quad .$$

This curve (28) is the envelope of the lines represented by equation (27). It is a singular integral.

Remark. We would arrive at a Clairaut equation if we intended to obtain, by means of a differential equation, a plane curve by means of a property of its tangents where the point of contact is not involved. For if $y = f(x)$ is the equation of the curve, then the equation of the tangent is $Y = y'(X-x) + y = y'X + y - xy'$. We will have a relationship between y' and $y - xy'$. But it is clear that, in such a case, we must use the tangential equation directly: if $Y = mX + n$ is the tangent, then the property in question is expressed by $n = B(m)$. We have to determine the envelope of $Y = mX + B(m)$ and we have to combine this equation with its derivative $X + B'(m) = 0$. The envelope is the singular integral of Clairaut's equation $y - xy' = B(y')$.

70 - The singular integral. Lie's geometric interpretation

In the cases discussed up to now, we have regarded the singular integrals as particular integrals that do not enter into the general integral; these integrals are the envelopes of the general integral. Lagrange observed that, more generally, if $F(x,y,y') = 0$ is a differential equation of the first order, then the integral depends on an arbitrary constant. If $\Phi(x,y,C) = 0$ is the equation of the integral curves and if these curves have an evenlope Γ, at every point $M(x,y)$ of Γ, then the value of y' on Γ is the same as the enveloped curve passing through M, hence $F(x,y,y')$ is zero, and Γ is an integral. This integral does not enter into the family $\Phi(x,y,C) = 0$.

From the point of view of the theory of envelopes, we could try to determine, *a priori*, the singular integrals if there are any. Let $M(x,y)$ be a point of the curve envelope Γ, and M' a point of Γ near to M. The integral curves tangent to Γ at M and M' intersect at a point $P(x_1,y_1)$ which tends towards M when M' tends towards M (no. 58). If y'_1 and y'_2 are the values of y' at the point P on the two integral curves passing through this point, then they tend to the value of y' at the point M. Following the theorem on the continuity of the roots, the equation in y', $F(x,y,y') = 0$, therefore admits a double root when x and y are the coordinates of M. *Thus, in order for the point* $M(x,y)$ *to be belong to a singular integral, it is necessary that the equation in* y', $F(x,y,y') = 0$ *has a double root; we must then have* $\frac{\partial}{\partial y'} F(x,y,y') = 0$. If we eliminate y' from the equations

$$F(x,y,y') = 0 , \quad \frac{\partial F}{\partial y'} (x,y,y') = 0 \quad , \tag{29}$$

we obtain a relation $R(x,y) = 0$ which could define a curve Γ'. (In the case where F is a polynomial with respect to y', $R(x,y)$ is the discriminant, it defines a real or imaginary curve.) *The curve* Γ' *is not in general an integral curve of the differential equation.* Since y', the parameter appearing in the equations (29), is not *a priori* the value of $\frac{dy}{dx}$ along the curve $R(x,y) = 0$. To avoid all confusion, let us put a parameter m into the equations (29): the curve $R(x,y) = 0$ defined by

$$F(x,y,m) = 0\ , \qquad \frac{\partial}{\partial m} F(x,y,m) = 0 \quad , \tag{30}$$

is in general the envelope of the curves $F(x,y,m) = 0$. Along this curve, which in certain cases, will be a singular locus, the tangent is given by

$$dx \frac{\partial F}{\partial x} (x,y,m) + dy \frac{\partial F}{\partial y} (x,y,m) = 0 \quad ;$$

its gradient will equal m only when we have

$$\frac{\partial F}{\partial x} (x,y,m) + m \frac{\partial F}{\partial y} (x,y,m) = 0 \quad . \tag{31}$$

Thus: *In order for a singular integral to exist, it is necessary that the equations (30) and (31) are compatible all along a curve* Γ', *which will not occur in general.*

Nothing here allows us to say that this necessary condition is sufficient in general. To arrive at a conclusion, it will be necessary to study the branches of integral curves in the neighbourhood of the curve $R(x,y) = 0$ and consequently, to apply the existence theorems (see Chapter VIII). But we can show that, in the case of Lagrange's equation, the necessary condition is sufficient in general. We have the equations

$$y - xA(m) - B(m) = 0\ , \qquad -xA'(m) - B'(m) = 0\ ,$$

$$m - A(m) = 0$$

and if they are compatible, i.e., if $A'(m)$ and $B'(m)$ are zero, where m is the solution of the last equation, then we have a singular integral in general.

LIE'S GEOMETRIC INTERPRETATION

As in no. 69, let us take the surface S,

$$F(x,y,z) = 0 \quad .$$

To integrate the differential equation $F(x,y,y') = 0$ amounts to finding the curves c of this surface, along which $dy = z dx$. Let us consider the apparent contour to S when we take the point of view of infinity, i.e., the curve outlined on S along which the tangent plane is parallel to OZ, let γ' be this curve and Γ' its projection onto the plane Oxy (fig. 36).

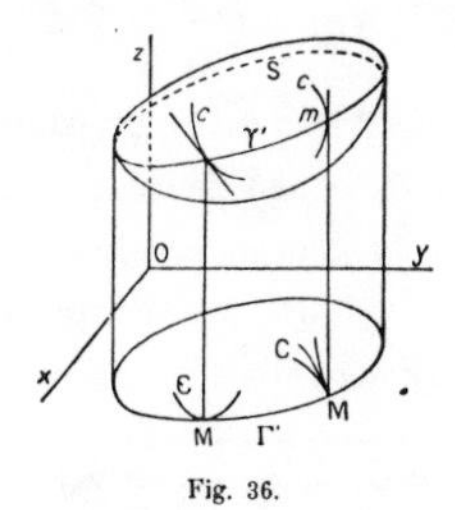

Fig. 36.

The curves c of S project onto the plane Oxy along the integral curves C of the differential equation. Let us assume that γ' is an ordinary curve on S and let us consider the curves c which intersect the curve γ'. At a point of intersection m of c with γ', the tangent to c is defined by the intersection of the tangent plane to S with the plane

$$Y - y = z(X-x) \quad ,$$

where x,y,z are the coordinates of m, since along c, $dy = zdx$. The tangent plane at m has as its equation $(X-x)F'_x + (Y-y)F'_y = 0$. Two cases are possible: these planes are either identified or not identified with each other. For them to be identified, it is necessary that $F'_x + zF'_y$ is zero, and since along γ', we have $F = 0$, $F'_z = 0$, it is necessary that the three equations

$$F(x,y,z) = 0 \ , \quad \frac{\partial F}{\partial z}(x,y,z) = 0 \ ,$$
$$\frac{\partial F}{\partial x}(x,y,z) + z\frac{\partial F}{\partial y}(x,y,z) = 0 \quad , \tag{32}$$

are compatible. *If this condition is not realised,* then the two planes in question, parallel to Oz, intersect along a parallel to Oz. The tangent at m to the curve c is parallel to Oz; the point M of the projection of m onto the plane Oxy, will in general be a turning point and the tangent at this point to C will be the trace of the plane Oxy, of the osculating plane to c at m. It will not in general be the tangent at M at Γ'. On the contrary, *if the equations (32) are compatible along* γ', then the tangent to c will not in general be parallel to Oz; it will project along the tangent at M to Γ' and C *will be a tangent to* Γ' *in general.*

We see that the above reasoning tends to show that if the equations (32), which are none other than the equations (30), (31), are compatible, then the curve Γ' defined by these equations is a singular integral, at the very least in general. In the opposite case, we will have a locus of turning points. But we have implicitly allowed those curves c intersecting γ'.

Remark. When we consider a 1-parameter family of plane curves $\Phi(x,y,C) = 0$ where Φ is a polynomial in x,y,C, then these curves have an envelope Γ *in general*, for if the polynomial is chosen arbitrarily, then these curves will have no singular points. If we form the differential equation of these curves by eliminating C from the equations

$$\Phi(x,y,C) = 0 , \quad \frac{\partial\Phi}{\partial x}(x,y,C) + y' \frac{\partial\Phi}{\partial y}(x,y,C) = 0 ,$$

then we obtain an algebraic differential equation $F(x,y,y') = 0$, whose general integral is formed by the given curves, and there exists a singular integral Γ. This reasoning lead Lagrange to think that an algebraic differential equation has, in general, a singular integral. Now we have seen that for this to be so, it is necessary that the equations (30) and (31) should be compatible along a curve, which is not the case if $F(x,y,m)$ is a polynomial taken at random. These two results do not contradict each other. An algebraic differential equation does not always have a general algebraic integral, and the confrontation of the two facts, allow us to say that, *in general*, an algebraic differential equation does not have its general integral algebraic with respect to x,y and the constant of integration C.

II. EQUATIONS OF ORDER GREATER THAN ONE

71 - The equation $y^{(n)} = f(x)$

The integration of this n-th order equation seems to involve n successive integrals. But we may observe that it involves determining the functions where n-th order derivative is known; the Lagrange form of Taylor's formula (I, 49) resolves matters straight away. We obtain the solution which takes the value y_o for $x = x_o$ and whose derivatives of order $p < n$ takes the values $y_o^{(p)}$ at this point, under the form

$$y = y_o + y_o'(x-x_o) + \dots + \frac{(x-x_o)^{n-1}}{(n-1)!} y_o^{(n-1)}$$

$$+ \frac{1}{(n-1)!} \int_{x_o}^{x} (x-t)^{n-1} f(t)dt$$

We assume of course that $f(x)$ is continuous from x_o to x.

The formula extends to the case of an equation of the form

$$F(x,y^{(n)}) = 0$$

when we know how to deduce x and $y^{(n)}$ as functions of a parameter t,

$$x = \phi(t) \ , \qquad y^{(n)} = \psi(t) \tag{33}$$

say. Let us imagine then that we have extracted t to be a function of x from the first of these relations and that this function is taken into the second equation (33). We will have $y^{(n)} = f(x)$, hence

$$y = P(x) + \frac{1}{(n-1)!} \int_{x_o}^{x} (x-u)^{n-1} f(u)du \quad ,$$

where $P(x)$ is an arbitrary polynomial of degree $n-1$. Let us replace x by $\phi(t)$; we will obtain y as a function of t:

$$y = P(\phi(t) + \frac{1}{(n-1)!} \int_{\phi(t_o)}^{\phi(t)} (\phi(t)-u)^{n-1} f(u)du \quad ,$$

since we are allowed to take $x_o = \phi(t_o)$. Finally, let us make the change of variable $u = \phi(v)$ in the integral. We will obtain

$$y = P(\ (t) + \frac{1}{(n-1)!} \int_{t_o}^{t} [\phi(t)-\phi(v)]^{n-1} \psi(v)\phi'(v)dv$$

since $f(\phi(v)) = \psi(v)$. This formula, combined with the first formula (33), gives the integral curves under the parametric form.

<u>A particular case</u>. If we have $x = G(y^{(n)})$, we will take $\psi(t) \equiv t$ and $\phi(t) \equiv G(t)$ and we shall have

$$x = G(\ ) \ , \quad y = P[G(t)] + \frac{1}{(n-1)!} \int_{t_o}^{t} [G(t)-G(v)]^{n-1} vG'(v)dv \quad .$$

Integrating by parts permits reducing the calculation of the last integral to that of the integral

$$\int_{t_o}^{t} [G(t)-G(v)]^{n} \, dv \quad .$$

<u>Example</u>. If the curve $F(x,y^{(n)}) = 0$, where x and $y^{(n)}$ are coordinates, is unicursal, then matters reduce to integrating a rational fraction.

72 - <u>Some simple cases for which the order is reduced</u>

In certain cases we can reduce the order of the equation.

1. THE EQUATION DOES NOT CONTAIN THE UNKNOWN FUNCTION

   It is of the form

$$F\left(x, \frac{d^k y}{dx^k}, \ldots, \frac{d^n y}{dx^n}\right) = 0 \quad .$$

This reduces to an equation of order $n-k$ by taking $z = \dfrac{d^k y}{dx^k}$ as an unknown function. If we can integrate the equation of order $n-k$ so obtained, then we have to apply the result of no. 71 to obtain y by a quadrature. If, in the integration of

$$F(x,z,z',\ldots,z^{(n-k)}) = 0 \quad ,$$

we had been lead to express x and z as a function of a parameter, then we proceed as in no. 71, since we have $x = \phi(t)$ and $y^{(k)} = \psi(t)$.

2. THE EQUATION DOES NOT CONTAIN THE VARIABLE x.

It is then of the form

$$F\left(y, \frac{dy}{dx}, \ldots, \frac{d^n y}{dx^n}\right) = 0 \quad .$$

We see that if we consider x as an unknown function and y as the variable, we will recover the previous case, since

$$\frac{dy}{dx} = \frac{1}{x'}, \quad \frac{d^2y}{dx^2} = \frac{-x''}{x'^3}, \ \ldots$$

where $x', x'', \ldots$ are the successive derivatives of x with respect to y. We would obtain an equation which would not contain the function x and we would be lead to take x' as the unknown function, where y is the variable. We simplify this procedure by taking y as the variable and $p = y'$ as the unknown function. We will have

$$p = \frac{dy}{dx}, \quad \frac{d^2y}{dx^2} = \frac{dp}{dx} = \frac{dp}{dy} : \frac{dy}{dx} = p\,\frac{dp}{dy} \quad ,$$

$$\frac{d^3y}{dx^3} = \frac{d}{dy}\left(p\,\frac{dp}{dy}\right) \cdot \frac{dy}{dx} \quad p\left(p\,\frac{d^2p}{dy^2} + \left(\frac{dp}{dy}\right)^2\right), \quad \ldots$$

We differentiate with respect to y, each expression successively obtained and multiply by p. For example, for $n = 4$, we obtain

$$F(y,p,pp',p^2p'' + pp'^2, p^3p''' + 4p^2p'p'' + pp'^3) = 0,$$

where p', p'' and p''' are the derivatives of $p = y'$ with respect to y. The equation in p and y being integrable, we will then have

$$\frac{dx}{dy} = \frac{1}{p} \quad ,$$

which will give x as a function of p, by a quadrature if y has been expressed as a function of p. If y and p have been expressed as functions of a parameter, $y = \psi(t)$, $p = \phi(t)$ say we again have

$$x = \int \frac{\psi'(t)}{\phi(t)}\, dt + \text{const.}$$

<u>A particular case</u>. In particular, a second order equation which does not contain the variable, hence of the form $F(y,y',y'') = 0$, reduces to a first order equation, by setting $y' = p$. If $p = f(y,C)$ is the integral of this equation, then we have

$$x = \int \frac{dy}{f(y,C)} + C' \quad .$$

<u>Remark</u>. The equation $y'' = G(y)$ enters into the above case. The method outlined leads to $pdp = G(y)dy$ and we have

$$y'^2 = p^2 = 2\int G(y)dy + C \quad . \tag{34}$$

This is what is obtained directly by multiplying both members of the given equation $y'' = G(y)$ by y'; this is what appears in the first member, $y''y'$ the derivative of $\frac{1}{2}y'^2$, and in the second member, $G(y)y'$, the derivative with respect to x of the primitive of $G(y)$. We then achieve the integration as it has been indicated above, since y' is known as a function of y.

<u>Example</u>. If $G(y)$ is a polynomial of degree m in y, then y'^2 will be a polynomial of degree $m+1$, $Q(y) + C$ say, and we will have

$$x = \int \frac{dy}{\sqrt{Q(y)+C}} + C' \quad .$$

x is given by a hyperelliptic integral if $m > 3$. For $m = 2$ or 3, $Q(y)$ is of degree 3 or 4, the integral is elliptic, and y is an elliptic function of x. This function degenerates if C is chosen in order that $Q(y)+C$ has a multiple zero.

3. A HOMOGENEOUS EQUATION IN $y, y', y'', \ldots, y^{(n)}$

If y *is a solution of one such equation, then* λy *is also a solution, for any constant* λ. If we take $\log y$ as the unknown function, the transformation of $\log y$ to $\log y + \mu$, where μ is an arbitrary constant, does not change the equation; $\log y$ does not occur there, which means we would take the derivative y'/y as the unknown function. We will then set $y'/y = u$ or $|y| = e^{\int u dx}$. We will have

$$y' = uy \ , \qquad y'' = uy' + u'y \qquad y(u^2 + u') \quad ,$$

$$y''' = y'(u^2 + u') + y(2u'u + u'') = y(u^3 + 3uu' + u'') \quad ,\ldots$$

and the equation, which can be written as

$$F\left(x, \frac{y'}{y}, \frac{y''}{y}, \ldots, \frac{y^{(n)}}{y}\right) = 0 \quad ,$$

will become

$$F(x, u, u^2 + u', u^3 + 3uu' + u'', \ldots) = 0 \quad ,$$

which is an equation of order $(n-1)$. Having integrated this equation, there remains a quadrature to be made, to obtain y as a function of x.

Example. The homogeneous linear, second order equation

$$y'' + ay' + by = 0 \quad , \tag{35}$$

where a and b are functions of x is transformed by the above method, to the Riccati equation $u' + u^2 + au + b = 0$. In the particular case where a and b are constants, this Riccati equation is an equation with separated variables which is easily integrated. On integrating and passing to y, we will find solutions of the linear, homogeneous, second order equations with constant coefficients.

Conversely, a general Riccati equation $u' + Au^2 + Bu + C = 0$, where A,B,C are functions of x, can first of all reduce to the cases where $A \equiv 1$ by making the transformation $u = \lambda v$, where λ as a function of x. It effectively takes the form

$$\lambda v' + A\lambda^2 v^2 + (B\lambda + \lambda')v + C = 0 \quad ,$$

and it suffices to take $\lambda A \equiv 1$, hence $\lambda = 1/A$, to reduce to the canonical case $u' + u^2 + au + b = 0$. If we make the transformation

$$y = e^{\int u dx}$$

then this canonical equation becomes the linear equation (35).

4. A HOMOGENEOUS EQUATION IN $x, y, dx, dy, \ldots, d^n y$

The equation does not change when x and y are transformed to kx and ky respectively, whatever the constant k. This transformation preserves $y/x, y', xy'', \ldots, x^{n-1}y^{(n)}$; the equation is then of the form

$$F\left(\frac{y}{x}, y', \ldots, x^{n-1}y^{(n)}\right) = 0 \quad .$$

The case $n = 1$ is that of the first order homogeneous equation.

If we take $\log|x|$ as the variable and y/x as the unknown function, then the transformation of $\log|x|$ to $\log|x| + \mu$ will not change the equation, $\log|x|$ will not occur and we will be in case II (lower order).

We will therefore set $|x| = e^t$, $y = ux$. The transformed equation will not contain the variable t. This is easily seen, since $y',\ldots,x^{n-1}y^{(n)}$ is expressed in terms of u and its derivatives. We have first of all

$$\frac{d^p y}{dx^p} = \frac{d^p u}{dx^p}\,x + p\,\frac{d^{p-1}u}{dx^{p-1}}$$

hence

$$x^{p-1}y^{(p)} = \frac{d^p u}{dx^p}\,x^p + p\,\frac{d^{p-1}u}{dx^{p-1}}\,x^{p-1} \quad .$$

Then on introducing derivatives with respect to t,

$$x\,\frac{du}{dx} = \frac{du}{dt}\,, \qquad x\,\frac{du}{dx} + x^2\frac{d^2u}{dx} = \frac{d^2u}{dt}\,, \ \ldots$$

We will arrive at an equation

$$F(u,u' + u,u'' + u',\ldots) = 0\ , \qquad u' = \frac{du}{dt}\,,\ldots,$$

in which we will take u as a variable and u' as the unknown function.

<u>Remark</u>. An equation can become homogeneous after a change of variable $x = X^\alpha$ (α constant). This does not change if we replace x and y by $k^\alpha x$ and ky whatever the constant k. We will set $y^\alpha = ux$, $|x| = e^t$, and the transformed equation will no longer contain the variable t.

5. A HOMOGENEOUS EQUATION IN x AND dx

The equation is of the form

$$F\left(y, x\,\frac{dy}{dx}\,,\ldots,x^n\,\frac{dy^n}{dx^n}\right) = 0 \quad .$$

It does not change when x is replaced by kx. Therefore, by setting $|x| = e^t$, we will have an equation not containing t, which will reduce to order $n-1$. The calculation is that of above.

73 - <u>Simplification by differentiation</u>

In certain cases it may be advantageous to increase the order of equation in order to simplify it. A differentiation may lead to a more simple equation. This procedure is applicable to Lagrange's equation and leads to the method previously employed. By differentiating both members of the equation

$$y = xA(y') + B(y') \tag{36}$$

we obtain

$$y' - A(y') = [xA'(y) + B'(y')] \frac{dy'}{dx} \quad ,$$

a linear equation which has x as the unknown function and $y' = p$ as the variable. But as this linear equation is integrable, it suffices to the value x' into equation (36), in order to achieve the integration.

Generally, the equation derived from differentiating an n-th order equation, is an equation of order $n+1$ whose general integral depends on $n+1$ constants; these are only certain solutions of this equation which will be solutions of the intended equation. By stating that the general solution of the equation of order $n+1$ satisfies the proposed equation, we have a relationship between the $n+1$ constants.

More often, we will combine the derived equation with the given equation; what was said still holds.

Monge's Example. If a, b, c are the semi-axes of an ellipsoid taken from its axes, with $a < b < c$, then the projections of the lines of curvature (see no. 242) onto the plane Oxy are given by the differential equation

$$\alpha xyy'^2 + (x^2 - \alpha y^2 - \beta)y' \quad xy$$

$$\alpha = \frac{a^2(b^2 - c^2)}{b^2(a^2 - c^2)} \qquad \beta = \frac{a^2(a^2 - b^2)}{a^2 - c^2} \quad .$$

We can integrate by changing the variables of the function; the equation does not change if we change x to $-x$ and y to $-y$. We can then set $x^2 = X$ and $y^2 = Y$, which, after multiplication by y/x, leads to the equation

$$\alpha Y'^2 X + (X - \alpha Y - \beta)Y' - Y = 0 \quad ,$$

which becomes

$$(XY' - Y)(\alpha Y' + 1) - \beta Y' = 0 \quad ,$$

and this is a Clairut equation. The general solution corresponds to $Y' = C$, where C is an arbitrary constant. The lines of curvature then have as their projection onto the plane Oxy, the conics

$$\alpha C^2 x^2 + C(x^2 - \alpha y^2 - \beta) - y^2 = 0 \quad .$$

To integrate (37), Monge differentiated equation (37) and in the equation so obtained, replaced $x^2 - \alpha y^2 - \beta$ by its value extracted from (37). We thus obtain

$$(\alpha y'^2 + 1)(xyy'' + xy'^2 - yy') = 0$$

and since α is positive, there remains the second order equation

$$x(y'^2 + yy'') - yy' = 0 \quad ,$$

which gives $yy' = \lambda x$, then $y^2 = \lambda x^2 + \mu$, where λ and μ are constants. On taking them into (37), we see that λ and μ must satisfy the relation $\alpha\lambda\mu + \mu + B\lambda = 0$. We thus arrive at the conics with equation

$$(y^2 - \lambda x^2)(\alpha\lambda + 1) + \beta\lambda = 0 \quad ,$$

in accordance with the preceding result.

DARBOUX'S METHOD FOR EULER'S EQUATION

Let us consider Euler's equation (no. 64) in the case where $P(x)$ has been reduced to the Legendre form (I, 97); $P(x) \equiv (1 - x^2)(1 - k^2x^2)$. Euler's equation may be written under the guise

$$\frac{dx}{\sqrt{P(x)}} = \frac{dy}{\sqrt{P(y)}} = dt$$

or

$$\left(\frac{dx}{dy}\right)^2 = (1 - x^2)(1 - k^2x^2), \quad \left(\frac{dy}{dt}\right)^2 = (1 - y^2)(1 - k^2y^2) \quad , \qquad (38)$$

and on differentiating with respect to t, we have

$$\frac{d^2x}{dt^2} = 2k^2x^3 - (1 + k^2)x \ , \qquad \frac{d^2y}{dt^2} = 2k^2y^3 - (1 + k^2)y \quad .$$

We deduce from these inequalities

$$y\,\frac{d^2x}{dt^2} - x\,\frac{d^2y}{dt^2} = 2k^2xy\,(x^2 - y^2)$$

and from (38)

$$y^2\left(\frac{dx}{dy}\right)^2 - x^2\left(\frac{dy}{dt}\right)^2 = (1 - k^2x^2y^2)(y^2 - x^2) \quad .$$

By dividing one member by the other, we obtain, where the primes denote the derivatives with respect to t,

$$\frac{yx'' - xy''}{y^2x'^2 - x^2y'^2} \quad - \frac{2k^2xy}{1 - k^2x^2y^2} \quad ,$$

which can be written as

$$\frac{(yx' - y'x')}{yx' - y'x} \quad - \frac{2k^2xy(xy)}{1 - k^2x^2y^2} \quad .$$

On integrating we see that

$$\frac{yx' - xy'}{1 - k^2x^2y^2} = \text{const.}$$

therefore

$$y\sqrt{P(x)} - x\sqrt{P(y)} \; = C(1 - k^2x^2y^2)$$

is the integral of Euler's equation when $P(x)$ has the normal Legendre form.

74 - An example of integration by elliptic functions. A plane elastic curve

We seek a plane curve whose curvature at each point $M(x,y)$ is proportional to its ordinate y. It is the point of equilibrium of an elastic sheet whose mean fibre is naturally rectilinear and on whose extremities the forces and couples situated in the plane of the sheet, act. We then have

$$\frac{y''}{(1 + y'^2)^{3/2}} = \frac{2y}{x^2}$$

where a is a constant and y' and y'' the first and second derivatives respectively of y, with respect to x. On applying the method on no. 72, we must set $y' = p$ and take y as the variable. We will have

$$\frac{pdp}{(1 + p^2)^{3/2}} = \frac{2ydy}{a^2}$$

then

$$- \frac{1}{\sqrt{1 + p^2}} = \frac{y^2 - C}{a^2}$$

where C is a constant, and on solving with respect to p,

$$p = \left[\frac{a^4 - (y^2 - C)^2}{(\ ^2 - C)^2}\right]^{1/2} \quad .$$

The value of x is given by an elliptic integral

$$x = \int \frac{dy}{y'} = \int \frac{\ ^2 - C}{\sqrt{a^4 - (y^2 - C)^2}} \; dy + C' \quad .$$

We can recover the Weierstrass elliptic functions by setting $y^2 = C - z$, which will transform the integral to

$$\int \frac{z\,dz}{\sqrt{4(z - C)(z^2 - a^4)}} \quad .$$

We will obtain the normal form by setting $z = \frac{C}{3} + u$. We obtain

$$y = \pm\sqrt{\frac{2C}{3} - u} \ , \qquad x = \int \frac{(C + 3u)du}{3\sqrt{4u^3 - g_2 u - g_3}} \quad ,$$

where the invariants g_2 and g_3 are expressed in terms of a^2 and C. If we set $u = pt$ (notation due to Weierstrass, 1, 97, 229), we have

$$y = \pm\sqrt{\frac{2C}{3} - pt} \ , \quad \pm x = \frac{C}{3}t + \int p(t)dt + C' = \frac{C}{3}t - \zeta t + C' \quad ,$$

where $\zeta(t)$ is the Weierstrass function (I, 231).

75 - A differential equation of a family of curves

Given a family of plane curves depending on p parameters $C_1, C_2, \ldots, C_p$, we can eliminate these parameters from the equation of these curves, $f(x, y, C_1, \ldots, C_p) = 0$, and the equations obtained by differentiating once, twice, ... p times, with respect to x, where y is the function of x defined by the equation of the curve. We will obtain in general a p-th order differential equation which admits all of the curves of the family as integral curves. These curves will provide the general integral of the differential equation obtained. This equation can, moreover, admit singular integrals (we can appeal to the existence theorem of no. 48). This assumes that the curves in question depend effectively on p parameters (without which the order of the equation would be lower).

Examples

I. The circles have the equation $x^2 + y^2 + 2Ax + 2By + C = 0$; by differentiating three times, we can eliminate the three parameters A,B,C. We obtain

$$x + yy' + A + By' = 0 \ , \quad 1 + y'^2 + yy'' + By'' = 0 \quad ,$$

$$3y'y'' + yy''' + By''' = 0 \quad ,$$

and on eliminating B from the last two equations, we obtain the differential equation of the circles

$$y'''(1 + y'^2) - 3y'y''^2 = 0 \quad . \tag{39}$$

This equation is satisfied by the circles and by *the straight lines.* Since if we had written the equation of the circle as $\alpha(x^2 + y^2) + 2Ax + 2By + C = 0$, the result would be the same, and for $\alpha = 0$, we obtain the lines. Indeed, for a line, $y'' = 0$ and hence also $y''' = 0$.

Equation (39) is with separable variables when we regard y' as the variable and y'' as the function, in accordance with no. 72. We can, moreover, write it under the form

$$\frac{y'''}{y''} = \frac{y'y''}{1+y'^2} \quad .$$

By integrating, we obtain $\log y'' = \frac{3}{2}\log(1 + y'^2) + \log C$ hence

$$y''(1 + y'^2)^{-3/2} = C \quad ,$$

and another integration yields, for $C \neq 0$,

$$y'(1 + y'^2)^{-1/2} = Cx + C' \quad ,$$

or

$$y' = \frac{Cx + C'}{\sqrt{1 - (C \quad + C')^2}} \quad .$$

A final integration provides

$$Cy = -\sqrt{1 - (Cx + C')^2} + C'' \quad ,$$

which is an equation of a circle under the solved form. For $C = 0$, we have $y'' = 0$, and $y = C_1x + C_2$,

II. A differential equation of curves of degree two. This has been obtained by Halphen by considering the equation

$$y = ax + b + \sqrt{cx^2 + 2dx + e} \quad ,$$

where $a, b, c, d,$ and e are constants. We obtain first of all

$$y'' = (ce - d^2)(cx^2 + 2dx + e)^{-3/2} \quad , \tag{40}$$

which eliminates a and b. It follows that $(y'')^{-2/3}$ is a trinomial of the second degree; its third derivative is zero. We thus have the required equation

$$[(y'')^{-2/3}]''' = 0 \quad . \tag{41}$$

On expanding the calculation, we find that

$$40(y''')^3 - 45y''y'''y^{IV} + 9(y'')^2y^v = 0 \quad .$$

Working backwards, we pass from (41) to (40), and then to the initial equation.

Remark. The parabolas correspond to the case $p = 0$ for these curves, the expression $(y'')^{-2/3}$ will be of the first degree. The differential equation of the parabolas is therefore

$$[(y'')^{-2/3}]'' = 0$$

or on expanding

$$5(y''')^2 - 3y''y^{IV} = 0 \quad .$$

III. LINEAR DIFFERENTIAL EQUATIONS

76 - Definitions. General properties. The Wronskian

A linear differential equation of order n is an equation of the form

$$A_o(x)\,\frac{d^n y}{dx^n} + A_1(x)\,\frac{d^{n-} y}{dx^{n-1}} + \ldots + A_{n-1}(x)\,\frac{dy}{dx} + A_n(x)y = B(x) \quad . \quad (42)$$

The functions $A_k(x)$ are the coefficients, $B(x)$ is the second member. If $B(x) \equiv 0$, we say that the equation is *homogeneous*, or as before, *without second member*. If the $A_k(x)$ are constants, then the equation is said to have constant coefficients. We shall assume that the $A_k(x)$ and the $B(x)$ are continuous.

If Y is a solution and if we set $y = Y + z$, we see at once, by substituting y by this expression that z will satisfy the equation

$$A_o(x)\,\frac{d^n z}{dx^n} + A_1(x)\,\frac{d^{n-1} x}{dx^{n-1}} + \ldots + A_{n-1}(x)\,\frac{dz}{dx} + A_n(z)x = 0 \quad , \quad (43)$$

i.e., the equation without its second member. Hence

Theorem 1. *The general integral of a linear equation with a second member is obtained by combining a particular integral of this equation with the general integral of the equation obtained by suppressing the second member.*

This proposition reduces the study of n-th order linear equations to those equations with a second member. For such an equation of the form (43), it is clear that if z is a solution, Cz is also a solution, where C is an arbitrary constant. Similarly, if z_1 and z_2 are two solutions, then $z_1 + z_2$ is a solution. Consequently:

Theorem 2. *If* $z_1, z_2, \ldots, z_p$ *are solutions of the equation (43) without second member, and if* $C_1, C_2, \ldots, C_p$ *are arbitrary constants, then the combination* $C_1 z_1 + C_2 z_2 + \ldots + C_p z_p$ *is also a solution of this equation.*

LOWERING OF ORDER

Let us assume that we know a particular solution (apart from zero), Z, of equation (43). Let us set $z = Zu$; we will have

$$z^{(p)} = u^{(p)}Z + pu^{(p-1)}Z' + \ldots + uZ^{(p)} \quad .$$

The equation will be transformed to a linear, homogeneous equation in u, and the coefficient of u will be zero when Z satisfies (43). We will then obtain an equation in u' of order $n-1$.

$$B_o u^{(n)} + B_1 u^{(n-1)} + \ldots + B_{n-1} u' = 0 \quad ,$$

where the B_k are functions of x, simultaneously continuous with the $A_k(x)$. If v is the general solution of this equation in u', then we have, where C is a constant,

$$u = v dx + C \ , \qquad z = Zu \quad .$$

We have obtained the general solution of equation (43).

FUNCTIONS THAT ARE STRICTLY LINEARLY INDEPENDENT

If $z_1, z_2, \ldots, z_p$ have derivatives with respect to x up to order $p-1$, then the determinant

$$\Delta(z_1, z_2, \ldots, z_p) = \begin{vmatrix} z_1 & z_2 & \cdots & z_p \\ z_1' & z_2' & \cdots & z_p' \\ \vdots & & & \\ z_1^{(p-1)} & z_2^{(p-1)} & \cdots & z_p^{(p-1)} \end{vmatrix}$$

is called *the Wronskian determinant* or simply, *the Wronskian* of these p functions. We shall say that the $z_1, z_2, \ldots, z_p$ are *strictly linearly independent* in an interval of their Wronskian exists and is non-zero in this interval, i.e., is different from zero at every point of this interval. (We assume here that x is real.)

Theorem 3. *If the functions* $z_1, z_2, \ldots, z_p$ *have derivatives up to order* p *in an interval* (α, β) *where they are strictly linearly independent and if, for* z *another function of* x *admitting a derivative of order* p, *we have*

$\Delta(z_1,z_2,\ldots,z_p,z) \equiv 0$ *in this interval, then* z *is a linear and homogeneous function of* $z_1,z_2,\ldots,z_p$ *with constant coefficients:*

$$z \equiv C_1z_1 + C_2z_2 + \ldots + C_pz_p$$

where the C_j *are constants.*

The determinant $\Delta(z_1,z_2,\ldots,z_p,z)$ is taken as above; its last line is $z_1^{(p)}z_2^{(p)}\ldots z_p^{(p)}$. The expansion of the determinant along the elements of this line is

$$\lambda_1z_1^{(p)} + \lambda_2z_2^{(p)} + \ldots + \lambda_pz_p^{(p)} + \lambda_z^{(p)} \equiv 0 \quad ,$$

λ, which is the Wronskian of $z_1,z_2,\ldots,z_p$ is non-zero in $(\alpha,\beta);\lambda_1,\ldots,\lambda_p,\lambda$, have derivatives with respect to x. On dividing the expression (44) by λ, we can then write

$$z^{(p)} \equiv \mu_1z_1^{(p)} + \ldots + \mu_pz_p^{(p)} \ ,$$

where the μ_k are differentiable. When we substitute the derivatives $z_j^{(p)}$ in (44) by derivatives of lower order or by the z_j, we then obtain an expression that is identically zero, an expansion of a determinant having two identical lines. Hence, we have

$$z^{(k)} \equiv \mu_1z_1^{(k)} + \ldots + \mu_pz_p^{(k)} \ , \qquad k = 0,1,\ldots,p \quad .$$

Let us differentiate successively the first p of these identities and take account of the identity following each derivation, we will have

$$0 \equiv \mu_1'z_1^{(k)} + \ldots + \mu_p'z_p^{(k)} \qquad k = 0,1,\ldots,p-1 \quad .$$

These identites can be considered as a system of equations giving for each value x of the interval (α,β), the values of $\mu_1',\ldots,\mu_p'$. This is a linear, homogeneous system of p equations with p unknowns whose determinant is $\Delta(z_1,z_2,\ldots,z_p) \neq 0$. Hence

$$\mu_1' = \mu_2' = \ldots = \mu_p' = 0 \quad .$$

The functions $\mu_1,\mu_2,\ldots,\mu_p$ are constants, which proves the theorem.

Theorem 4. *If* $z_1,z_2,\ldots,z_p$ *have derivatives up to order* $p-1$ *in the interval* (α,β) *and if there exists constants* $C_1,C_2,\ldots,C_p$, *not all zero, such that* $C_1z_1 + C_2z_2 + \ldots + C_pz_p \equiv 0$ *in* (α,β) *then the Wronskian of the* z_j, $\Delta(z_1,z_2,\ldots,z_p)$ *is identically zero in* (α,β).

Effectively, we have, by differentiating k-times the linear relation between the z_j,

$$C_1 z_1^{(k)} + C_2 z_2^{(k)} + \ldots + C_p z_p^{(k)} \equiv 0 \ , \qquad k = 0,1,\ldots,p-1$$

and the system of linear homogeneous equations in $C_1, C_2, \ldots, C_p$ with a non-zero solution, has its determinant, which is the Wronskian, zero at each point of (α,β).

Theorem 5. *If* $z_1, z_2, \ldots, z_p$ *have derivatives up to order* $p-1$ *in the interval* (α,β) *and if their Wronskian is identically zero in* (α,β), *then there exists an interval* (γ,δ) *contained in* (α,β) *in which we have*

$$C_1 z_1 + C_2 z_2 + \ldots + C_p z_p \equiv 0$$

where the $C_1, C_2, \ldots, C_p$ *are constants that are not all zero.*

Let us assume that the Wronskian of $p-1$ of the functions, e.g., $z_1, \ldots, z_{p-1}$, is not identically zero. As it is continuous, because it only has derivatives to order $p-2$, it will not be zero in an interval (γ,δ) and theorem 3 will apply. If all the $(p-1)$ by $(p-1)$ Wronskians of the z_k are identically zero, then we will consider the Wronskians' of $p-2$ of these functions, etc. It is clear that if all the functions z_j are identically zero, then the theorem still remains true, where the C_j are then arbitrary.

Remark. We will see in Chapter VI that, in the case of analytic functions, theorem 4 admits a single converse. In the real case, Fréchet has shown that the necessity of statements of the form III or V is similary imposed if the functions are indefinitely differentiables. He gives the following example. Let us take on the segment $(-1,+1)$ $z_1 = 0$ for $x = 0$ and

$$z_1 = e^{-1/x^2} \quad \text{for} \quad x \neq 0 \ .$$

This function is indefinitely differentiable, for its p-th order derivative is of the form

$$P(x)x^{-3p}e^{-1/x^2}$$

for $x \neq 0$, where $P(x)$ is a polynomial, and is equal to 0 for $x = 0$. The quotient of this expression by x tends to zero when $x \to 0$ by virtue of the properties of the exponential. Hence the derivative of order $p+1$ exists at the origin and is zero there. The function z_2 defined by

$$z_2 = -z_1 \quad \text{if} \quad -1 \le x \le 0, \quad z_2 = z_1 \quad \text{if} \quad 0 \le x \le 1 \ ,$$

has the same properties. The functional determinant of z_1 and z_2 is zero for $-1 \le x \le 0$ and $0 < x \le 1$ and is also zero at the origin; *it is identically zero.* There is no relation between z_1 and z_2 tenable for all $(-1,1)$.

For functions with only a finite number of derivatives, the analogous examples are immediate. Let us take for example

$$z_1 \equiv 0 \quad \text{if} \quad -1 \le x \le 0 \ , \quad z_1 \equiv x^2 \quad \text{if} \quad 0 \le x \le 1 \quad ,$$

$$z_2 \equiv x^3 \quad \text{if} \quad -1 \le x \le 0 \ , \quad z_2 \equiv 0 \quad \text{if} \quad 0 \le x \le 1 \quad .$$

The Wronskian is zero between -1 and 1 and the relation between z_1 and z_2 is $z_1 \equiv 0$ if $-1 \le x \le 0$ and $z_2 \equiv 0$ if $0 < x \le 1$. We may compare these considerations with those relative to the functional determinants (I, 126).

77 - A fundamental system of solutions. The general integral

Let us consider a linear homogeneous equation

$$A_o(x)y^{(n)} + \ldots + A_n(x)y = 0$$

where x is real and belongs to an interval, and where the coefficients $A_k(x)$ are continuous. *We will assume, moreover, that* $A_o(x)$ *is non-zero in this interval,* which allows us to write the equation in *the canonical form* obtained by dividing both members by $A_o(x)$

$$y^{(n)} + a_1(x)y^{(n-1)} + \ldots + a_k(x)y = 0 \quad . \tag{45}$$

The $a_1(x),\ldots,a_n(x)$ are continuous in the interval (α,β) in question. We shall say that the n solutions $y_1,y_2,\ldots,y_n$ *form a fundamental system of solutions in this interval, if the Wronskian* $\Delta(y\ ,y\ ,\ldots,y_n)$ *is not identically zero in this interval.*

Liouville's Theorem. *If* $y_1,y_2,\ldots,y_n$ *is a fundamental system of solutions in an interval* (α,β) *then the Wronskian* $\Delta(y\ ,\ldots,y_n)$ *is non-zero in this interval.*

By hypothesis, we have in the interval (α,β)

$$y_j^{(n)} + a_1(x)y_j^{(n-1)} + \ldots + a_n(x)y_j \equiv 0 \qquad j = 1,2,\ldots,n \quad .$$

The Wronskian determinant $\Delta(y_1,\ldots,y_n)$ has a derivative with respect to x since the y_j have derivatives up to order n. We shall always assume that the first line of this determinant is

$$y_1 y_2 \cdots y_n \quad .$$

Following the product rule of differentiation, its derivative is obtained by summing the n determinants obtained by successive differentiation of the elements of the first line, then the second, etc. Now if we differentiate the elements of the first line, we obtain a determinant having two identical lines, the first two, and it is zero. If we differentiate the elements of the line of rank $<n$, we obtain a zero determinant as having lines of rank k and $k+1$ identically. There only remains then, the determinant obtained by differentiating the elements of line of rank n; we have

$$\frac{d}{dx}\Delta(y_1,\ldots,y_n) \equiv \begin{vmatrix} y_1 & y_2 & \cdot & y_n \\ y_1' & y_2' & \cdot & y_n' \\ \vdots & \vdots & \cdot & \vdots \\ y_1^{(n-2)} & y_2^{(n-2)} & \cdot & y_n^{(n-2)} \\ y^{(n)} & y^{(n)} & \cdot & y_n^{(n)} \end{vmatrix} .$$

In the last line of this determinant, we can replace the $y_j^{(n)}$ by their values drawn from the identities (46). We will obtain the sum of n determinants obtained by successively placing the elements of the last line

$$\begin{matrix} -a_n(x)y_1 & -a_n(x)y_2 & \ldots & -a_n(x)y_n & , \\ -a_{n-1}(x)y_1' & -a_{n-1}(x)y_2' & \ldots & -a_{n-1}(x)y_n' & , \\ \cdot & \cdot & \ldots & \cdot & \\ -a_2(x)y^{(n-2)} & -a_2(x)y_2^{(n-2)} & \ldots & -a_2(x)y_n^{(n-2)} & , \\ -a_1(x)y_1^{(n-1)} & -a_1(x)y_1^{(n-1)} & \ldots & -a_1(x)y_n^{(n-1)} & . \end{matrix}$$

All the determinants thus formed are zero by having two lines proportional, except the last which is the product of $-a_1(x)$ by $\Delta(y_1,\ldots,y_n)$. By writing Δ in place of $\Delta(y_1,\ldots,y_n)$, we then have

$$\frac{d\Delta}{dx} \equiv -a_1(x)\Delta \quad .$$

Now Δ is not identically zero. It is continuous. Let (γ,δ) be an interval belonging to (α,β), where it is non-zero. By integrating the differential equation in Δ in this interval, starting from a value x_o at which Δ takes the value $\Delta_o \neq 0$, we will have

$$\Delta = \Delta_o e^{D(x)} \quad , \qquad D(x) = -\int_{x_o}^{x} a_1(x)dx \quad .$$

$D(x)$ is continuous throughout (α,β). *It follows that* Δ *is non-zero in this interval.* For if, for example, δ is the first value to the right of x_o for which Δ is zero, then the equation giving Δ would be in the interval (x_o,δ). The first member tends to zero when x tends to δ whilst the second member tends to a non-zero value

$$\Delta_o e^{D(\delta)} \quad .$$

The theorem is thus proved.

<u>A Fundamental Theorem</u>. *If* $y_1,y_2,\ldots,y_n$ *is a fundamental system of solutions of the linear homogeneous equation (45) in the interval* (α,β), *then every solution* y *of this equation is in this interval, a linear homogeneous function with constant coefficients of these solutions:*

$$y = C_1y_1 + C_2y_2 + \ldots + C_ny_n \quad ,$$

where the $C_1,C_2,\ldots,C_n$ *are constants.*

If $y,y_1,\ldots,y_n$ are solutions, then in (α,β) we have the identities (46), to which it is necessary to include

$$y^{(n)} + a_1(x)y^{(n-1)} + \ldots + a_n(x)y \equiv 0 \quad . \tag{47}$$

The $a_k(x)$ are determined in a unique way by the equations (46), since $\Delta(y_1,\ldots,y_n)$ is non-zero at each point of the interval (α,β). For these quantities to satisfy (47), it is necessary that the total determinant of the system should be zero, for any x in (α,β). We must have

$$\Delta(y_1,y_2,\ldots,y_n,y) \equiv 0 \quad .$$

Now $y_1,y_2,\ldots,y_n$ are strictly linearly independent following Liouville's theorem. Theorem 3 of no. 76 applies and we have

$$y = C_1y_1 + \ldots + C_ny_n$$

where the C_j are constants.

<u>Conclusion</u>. *If a fundamental system of solutions is known in an interval* (α,β), *hence a system of* n *solutions* $y_1,y_2,\ldots,y_n$ *whose Wronskian is not identically zero, then the general integral of the equation in this interval is a linear homogeneous combination of* $y_1,y_2,\ldots,y_n$ *with arbitrary constant coefficients.*

78 - <u>Equations with second members. The method of the variation of constants</u>

The method of varying the constants, due to Lagrange, allows us to obtain, via quadratures, an integral of differential linear equation with second member when we know a fundamental system of solutions of the linear homogeneous equation obtained by suppressing the second member. Let

$$y^{(n)} + a_1(x)y^{(n-1)} + \dots + a_n(x)y = b(x) \tag{48}$$

be the given equation, written under the canonical form by division of both members by the coefficient of $y^{(n)}$ assumed to be everywhere non-zero in the interval (α,β) in question. The coefficients $a_j(x)$ and $b(x)$ are assumed to be continuous. Let us assume that we know a fundamental system of solutions $y_1, y_2, \dots, y_n$ of the homogeneous equation deduced from (48), by replacing $b(x)$ by 0. We seek to satisfy the equation (48) by taking

$$y \equiv C_1 y_1 + C_2 y_2 + \dots + C_n y_n \quad , \tag{49}$$

where the C_k are functions of x, differentiable and subjected to n conditions in all.

We shall assume that

$$C_1' y_1 + C_2' y_2 + \dots + C_n' y_n \equiv 0 \quad . \tag{50}$$

Under those conditions, on differentiating the identity (49), we obtain only

$$y' \equiv C_1 y_1' + C_2 y_2' + \dots + C_n y_n' \quad . \tag{51}$$

In the same way, we will have

$$y'' = C_1 y_1'' + C_2 y_2'' + \dots + C_n y_n'' \tag{52}$$

if we subject the C_k *to satisfy the identity*

$$C_1' y_1' + C_2' y_2' + \dots + C_n' y_n' \equiv 0 \quad . \tag{53}$$

In a more general way, we will again have for $p < n$

$$y^{(p)} = Cy^{(p)} + \dots + C_n y_n^{(p)} \tag{54}$$

if we assume that up to this value p, *we have*

$$C_1' y_1^{(p-1)} + \dots + C_n' y_n^{(p-1)} \equiv 0 \quad . \tag{55}$$

The identity (54) occurs for $p = 1$, and we deduce that

$$y^{(n)} = C_1 y_1^{(n)} + \dots + C_n y^{(n)} + C_1' y_1^{(n-1)} + \dots + C_n' y_n^{(n-1)} \quad . \tag{56}$$

By taking the values of y and its derivatives obtained by the identities (49), (51), ..., (56), into equation (48), by means of conditions (50), (53) and (55), we see that in the first member, the coefficient of C_k will be zero for $k = 1,\ldots,n$ since y_k satisfies equation (48) less its second member. There remains then, the new condition

$$C_1'y_1^{(n-1)} + \ldots + C_n'y_n^{(n-1)} \equiv b(x) \quad . \tag{57}$$

The system of conditions (55) satisfied for $p = 1$ [which gives (50)], $2,\ldots,n-1$ and (57) determined $C_1',C_2',\ldots,C_n'$ since it is a system of n equations in n unknowns whose determinant $(y_1,\ldots,y_n)$ is non-zero. The values of the C_k are given by Cramer's formula. We will have

$$C_k' = (-1)^{n-k}b(x) \; \frac{\Delta(y_1,\ldots,y_{k-1},y_{k+1},\ldots,y_n)}{\Delta(y_1,\ldots,y_k,\ldots,y_n)} \quad .$$

By integrating, we will obtain each C_k, defined to within an additive constant. When these constants appear, we obtain the general integral of equation (48). But it suffices, which is equivalent, to take for each C_k any well-determined primitive of C_k'. We will then have by formula (49), a particular integral of equation (48), to which we will add the general integral of the equation without second member (theorem 1 of no. 76). Cauchy has outlined a method conducive to the same calculations.

In certain cases, and in the case of equations with constant coefficients, we can more quickly find a particular integral of the equation for certain forms of the function $b(x)$. We can then utilise the following proposition:

Theorem. *If the second member $b(x)$ of equation (48) is the sum of several functions: $b(x) = \Sigma b_m(x)$ and if Y_m is a particular integral of the equation deduced from (48) by replacing its second member by $b_m(x)$, then the sum ΣY_m is a particular solution of equation (48).*

It clearly suffices to prove this for a sum of two terms only. Now if we have

$$Y_1^{(n)} + a_1(x)Y_1^{(n-1)} + \ldots + a_n(x)Y_1 \equiv b_1(x) \quad ,$$

$$Y_2^{(n)} + a_1(x)Y_2^{(n-1)} + \ldots + a_n(x)Y_2 \equiv b_2(x) \quad ,$$

then by addition, we see at once that $(Y_1 + Y_2)^{(n)} + a_1(x)(Y_1 + Y_2)^{(n-1)} + \ldots + a_n(x)(Y_1 + Y_2) \equiv b_1(x) + b_2(x)$.

79 - An application to equations with constant coefficients

Let us consider first of all, a homogeneous linear with constant coefficients

$$y^{(n)} + a_1y^{(n-1)} + \ldots + a_ny = 0 \tag{58}$$

where the a_j, $j = 1,2,\ldots,n$ are constants. We can obtain particular solutions of the form $y = e^{rx}$, where r is a constant. For if we have

$$y = e^{rx}\ , \qquad y^{(p)} = r^p e^{rx}\ , \qquad p = 1,2,\ldots$$

and on carrying this into the equation we see that we have a solution if, on dividing by the factor e^{rx} which is never zero, r is a solution of the algebraic equation

$$f(r) \equiv r^n + a_1 r^{n-1} + \ldots + a_n = 0 \quad . \tag{59}$$

This equation $f(r) = 0$, which is called *the characteristic equation of equation* (58), will have in general, n distinct roots $r_1, r_2, \ldots, r_n$. Under these conditions we will obtain n particular solutions

$$e^{r_1 x}, e^{r_2 x}, \ldots, e^{r_n x} \quad .$$

These solutions are strictly linearly independent. For their Wronskian contains as a factor, the product of these solutions, that is never zero. This product is multiplied by the determinant

$$\delta = \begin{vmatrix} 1 & 1 & . & . & 1 \\ r_1 & r_2 & . & . & r_n \\ \vdots & \vdots & . & . & \\ r_1^{n-1} & r_2^{n-1} & . & . & r_n^{n-1} \end{vmatrix} \quad ;$$

δ is a Vandermonde determinant, equal to the product of the differences r_j pairwise; it is non-zero.

We have then, for any x, a fundamental system of solutions and the general solution, tenable for any x, is

$$y = \sum_{1}^{n} {}_k e^{r_k x} \quad ,$$

where the C_k are arbitrary constants.

THE CASE WHERE THE CHARACTERISTIC EQUATION HAS MULTIPLE ROOTS

Let us denote by $D(y)$, the first member of equation (58). $D(y)$ is an operator symbol. For each function, n times differentiable, $D(y)$ is a determined function. *For a constant* r, *let us calculate*

$$D(ze^{rx}) \quad .$$

Following the Leibniz formula, the derivative of order p of ze^{rx} is

$$(ze^{rx})^{(p)} = e^{rx}[z^{(p)} + C_p z^{(p-1)}r + \dots + C_p^q\, z^{(p-q)}r^q + \dots + zr^p]$$

where the C_j are, here, the bionomial coefficients. Let us multiply this identity by a_{n-p}, give p the values $0,1,\dots,n$ and add member to member. e^{rx} will be a factor and the coefficient of $z^{(p)}$ will be

$$a_{n-p} + C'_{p+1}\, a_{n-p-1}r + \dots + C_{p+q}^q\, a_{n-p-q}r^q + \dots + C_n^{n-p}r^{n-p},$$

$$\text{i.e.,}\ \frac{1}{p!} f^{(p)}(r) \quad .$$

Consequently, we have

$$D(ze^{rx}) \equiv e^{rx} \sum_{o}^{n} \frac{f^{(p)}(r)}{p!}\, z^{(p)} \quad . \tag{60}$$

Remark. We can get to formula (60) more quickly by observing that the first member is clearly of the form $e^{rx}D_1(z)$, where D_1 is another linear operator symbol of order n. We will have then, where $f_1(r)$ is the characteristic equation associated to D_1,

$$D(e^{(r+r')x}) \equiv e^{rx}D_1(e^{r'x}) \equiv e^{(r+r')x}\, f_1(r')$$

and, as the first member is $e^{(r+r')x}f(r+r')$, we have

$$f_1(r') \equiv f(r+r') \quad .$$

On bringing in the characteristic equation with symbol D_1, we recover equation (60).

Solutions corresponding to a multiple root - If r' is a multiple root of order q of the characteristic equation (59), we have

$$f(r') = f'(r') = \dots = f^{(q-1)}(r') = 0 \quad .$$

Hence by setting

$$y = e^{r'x}z \quad ,$$

the transformed equation

$$D(ze^{r'x}) = 0$$

is an equation

$$\sum_{q}^{n} \frac{f^{(p)}(r')}{p!}\, z^{(p)} = 0$$

which admits as a solution $z^{(q)} = 0$ or $z = P(x)$, where $P(x)$ is a polynomial of degree $q - 1$ in x, with arbitrary coefficients. *The equation (59) admits as a solution*

$$e^{r'x}P(x)$$

and, in particular

$$e^{r'x}, \quad e^{r'x}x, \quad e^{r'x}x^2, \quad \ldots, \quad e^{r'x}x^{q-1} \quad .$$

To each multiple root of the characteristic equation there so corresponds a number of solutions equal to the order of multiplicity of this root. The number of solutions so obtained is equal to the order n of the equation.

These solutions are again strictly linearly independent. To see this, we can again appeal directly to the Wronskian. But it is more simple to confine matters to showing that the Wronskian is not identically zero. The fact it is not zero will be a consequence of Liouville's theorem of no. 77. Now to show that it is not identically zero, it suffices, following theorem 5 of no. 76, to prove that these exists no linear relation with constant coefficients, not all zero, between these solutions. It suffices then to establish this lemma.

Lemma. *If the numbers* $\alpha, \beta, \ldots, \lambda$ *are all different and if the polynomials* $P_\alpha(x), \ldots, P_\lambda(x)$ *are not all identically zero, then we could not have an identity of the form*

$$P_\alpha(x)e^{\alpha x} + P_\beta(x)e^{\beta x} + \ldots + P_\lambda(x)e^{\lambda x} = 0 \quad . \tag{61}$$

Let us assume then, that we have such an identity, where the coefficients of the polynomials are evidently different from zero. It is clear that this identity cannot occur if there is only one term, for example, the first. Since an exponential is never zero and a non-identically zero polynomial has only a finite number of zeros, then their product is not identically zero. There is, therefore, at least two terms. Let us multiply both sides of the identity by $e^{-\alpha x}$ and differentiate $m+1$ times, if m is the degree of $P_\alpha(x)$. We have then

$$P_\alpha(x) + P_\beta(x)e^{(\beta-\alpha)x} + \ldots + P_\lambda(x)e^{(\lambda-\alpha)x} = 0 \quad ;$$

the differences $\beta-\alpha, \ldots, \lambda-\alpha$, are non-zero. With each derivation, a term such as the second yields a new term of the same form, since

$$\frac{d}{dx}\{P_\beta(x)e^{(\beta-\alpha)x}\} = e^{(\beta-\alpha)x}\,[(\beta-\alpha)P_\beta(x) + P'_\beta(x)] \quad ,$$

where the degree of the bracket is the same as that of $P_\beta(x)$. Hence, at the end of the $(m+1)$ derivations, we recover an identity of the same form as (61), but with one term less. In the exponentials, the coefficients of x: $\beta-\alpha,\ldots,\lambda-\alpha$, are again all different. By continuing these operations, we will reduce the number of terms to a single, and we have seen that it is impossible to have an identity such as (61) with a single term.

Conclusion. The solutions obtained in the case where there are multiple roots again form a fundamental system and the general solution may be deduced. We have, then, the following rule:

Rule: *To obtain the general solution of the homogeneous linear equation with constant coefficients (58), we form the corresponding characteristic equation (59). A simple or multiple root* r *of order* q $(q \geq 1)$ *provides* q *solutions*

$$e^{rx}, \quad e^{rx}x, \quad e^{rx}r^{q-1} \quad . \tag{62}$$

The n *solutions thus obtained form a fundamental system of solutions* $y_1, y_2, \ldots, y_n$. *The general solution is*

$$y = C_1 y_1 + \ldots + C_n y_n \quad ,$$

where the C_j *are arbitrary constants.*

Real solutions. We have assumed x is real. But the C_j and the roots of the characteristic equation are real or complex. If the coefficients of equation (58) are real and if r is a complex root of the characteristic equation, then the imaginary conjugate number $\bar{r}$, is also a root with the same order of multiplicity. To the solutions (62) that are also complex, we may associate the conjugate solutions. If we set $r = t + i\Sigma$ we will have

$$e^{tx}(\cos \tau x + i \sin \tau x),\ldots,e^{tx}x^{q-1}(\cos \tau x + i \sin \tau x) \quad (i = 5\text{-}1)$$

and

$$e^{tx}(\cos \tau x + i \sin \tau x),\ldots,e^{tx}x^{q-1}(\cos \tau x - i \sin \tau x) \quad .$$

By taking the 1/2-sum in one case; and the quotient by 2_o of the difference of these expressions of the same rank of these two lines, in the other case, we obtain the real solutions

$$e^{tx} \cos \tau x,\ldots,e^{tx}x^{q-1}\cos \tau x$$

$$e^{tx} \sin \tau x,\ldots,e^{tx}x^{q-1}\sin \tau x$$

and conversely, we may pass from these solutions to the original ones. We may then uniquely apply these real solutions and express the general integral in terms of the solutions $y_1, y_2, \ldots, y_n$, which are all real. *By taking then the* C_j *to be real, we will obtain real solutions.* The proof of the fundamental theorem of no. 77 and that of theorem 3 of no. 76, show that *all the real solutions are obtained in this way.*

80 - Equations with constant coefficients with a second member

If the second member is arbitrary, we may apply the general method of Lagrange (no. 78).

Example. Let the equation by $y'' + y = tgx$. The general integral of the equation without second member, which admits $r^2 + 1 = 0$ for the characteristic equation, is $\lambda \cos x + \mu \sin x$, where λ and μ are constants. Let us assume that λ and μ are functions of x. We will obtain a solution of the equation with second member if

$$\lambda' \cos x + \mu' \sin x = 0 \quad ,$$

$$-\lambda' \sin x + \mu' \cos x = tgx \quad .$$

We must then have

$$\lambda' = \frac{-\sin^2 x}{\cos x} = \frac{1}{\cos x} + \cos x, \quad \mu' = \sin x \quad ,$$

hence

$$\lambda = \sin x - \log\left| tg\left(\frac{x}{2} + \frac{\pi}{4}\right)\right| \qquad \mu = -\cos x$$

and the general solution is

$$y = C \cos x + C' \sin x - \cos \quad \log\left| tg\left(\frac{x}{2} + \frac{\pi}{4}\right)\right| \quad ,$$

where C and C' are arbitrary constants.

THE CASE WHERE THE SECOND MEMBER IS A SUM OF TERMS OF THE FORM $e^{\alpha x}P(x)$, WHERE α IS A CONSTANT AND $P(x)$ A POLYNOMIAL

Following the theorem at the end of no. 78, we are drawn to the case where there is a single term of this kind, with the second member. Let us consider, then, the equation

$$D(y) = y^{(n)} + a_1 y^{(n-1)} + \ldots + a_n y = e^{\alpha x}P(x) \quad .$$

If we set

$$y = e^{\alpha x} z \quad ,$$

we are lead to the equation

$$D(e^{\alpha x}z) = e^{\alpha x}P(x)$$

and via formula, to

$$\sum_{o}^{n} \frac{f(p)(\alpha)}{p!}\ z^{(p)} = P(x) \quad ,$$

i.e., an equation whose equation member is a polynomial and for which we seek a particular solution. Let

$$z^{(n)} + b_1 z^{(n-1)} + \ldots + b_n z = P(z) \ , \qquad b_{n-p} = \frac{f^{(p)}(\alpha)}{p!}$$

be this equation. *If* $b_n \neq 0$, *there exists a solution which is a polynomial of the same degree as* $P(x)$ *and which can be obtained by the method of indeterminate coefficients.*

In effect, let

$$P(x) \equiv c_m x^m + \ldots + c_1 x + c_o$$

Let us take

$$z \equiv d_m x^m + \ldots + d_1 x + d_o \quad .$$

The d_j will be determined iteratively by some equations, which are always possible since $b_n \neq 0$:

$$b_n d_m = c_m, \qquad b_n d_{m-1} + b_{n-1} m d_m = c_{m-1}, \ldots$$

$$b_n d_p + (p+1) b_{n-1} d_{p+1} + (p+2)(p+1) b_{n-2}\, d_{p+2} + \ldots = c_p, \ldots$$

$$b_n d_o + b_{n-1} d_1 + 2b_{n-2} d_2 + \ldots = c_o \quad .$$

The hypothesis $b_n \neq 0$ corresponds to $f(\alpha) \neq 0$; α *is not a root of the characteristic equation. If* α *is a root of order* q *of the characteristic equation,* we will have $b_n = b_{n-1} = \ldots = b_{n-q+1} = 0$ and $b_{n-q} \neq 0$. Equation (63) will be a linear equation with constant coefficients in $z^{(q)}$. The previous argument shows that the equation in $z^{(q)}$ will have as a solution a polynomial of the same degree as $P(x)$, which will again be obtained by the method of indeterminate coefficients. We deduce that z will be the product of x_q and a polynomial of degree m.

<u>Conclusion</u>. *If the second member is of the form* $e^{\alpha x}P(x)$, *where* α *is a constant and* $P(x)$ *a polynomial of degree* m, *there exists a solution of the*

form $e^{\alpha x}Q(x)$ *where* $Q(x)$ *is a polynomial of degree* m *if* α *is not a root of the characteristic equation, and the product of* x^q *and a polynomial of degree* m *of* α *is a root of order* q *of this equation.*

The method of proof is by calculation.

Remark. If we have a second member of the form $P(x)\cos\beta x$, or $P(x)\sin\beta x$, where $P(x)$ is a polynomial and β a constant, then the Euler formulae reduce to the preceding case. In the general case where $i\beta$ is not a root of the characteristic equation, the solution will be of the form $Q(x)\cos\beta x + Q_1(x)\sin\beta x$, where Q and Q_1 are polynomials of the same degree as $P(x)$. We can calculate these polynomials directly by identification. If $i\beta$ is a root of order q of the characteristic equation, it will be necessary to multiply Q and Q_1 by x^q.

We can make the same observations in the case where the second member is $e^{\alpha x}P(x)\cos\beta x$, for example.

Example. Let us take the equation $y^{IV} - y = \cos x$. The characteristic equation $r^4 - 1 = 0$ has for its roots $-1, +1, -i, +i$. The solution of the equation without second member is $Ce^x + C'e^{-x} + C''\cos x + C'''\sin x$. The second member $\cos x$ is a solution of the equation without second member, it corresponds to the exponentials whose coefficients are $+i$ and $-i$. It is necessary therefore to find a solution of the form

$$y = \lambda x\cos x + \mu x\sin x \quad .$$

The Leibniz formula yields for the fourth derivative

$$y^{IV} = \lambda(x\cos x + 4\sin x) + \mu(x\sin x - 4\cos x) \quad .$$

It will be necessary to take $\lambda = 0$ and $-4\mu = 1$; $-1/4\,x\sin x$ is a particular solution and the general solution is obtained by combining this function with the general integral of the equation without second member written as above.

IV. SYSTEMS OF LINEAR EQUATIONS

81 - The canonical system

The assumption in no. 49, that the functions are analytic, clearly holds. A linear differential equation of order n could be replaced by a system of n linear equations of the first order: the equation

$$A_o(x)y^{(n)} + A_1(x)y^{(n-1)} + \ldots + A_{n-1}(x)y' + A_n(x)y = B(x)$$

is equivalent to the system

$$y' = y_1 , \quad y_1' = y_2, \ldots, (y_{n-2})' = y_{n-1} \quad ,$$

$$A_o(x)(y_{n-1})' + A_1(x)y_{n-1} + \ldots + A_{n-1}(x)y_1 + A_n(x)y = B(x) \quad .$$

It is the same for a system of linear equations of order greater than 1; it can be replaced by one linear system. For example, the system of second order equations

$$y'' = a_1 y' + b_1 z' + c_1 y + d_1 z + e_1 \quad ,$$

$$z'' = a_2 y' + b_2 z' + c_2 y + d_2 z + e_2 \quad ,$$

where $a_1,\ldots,e_2$ are functions of x, and where the derivatives of y and z are taken with respect to x, is equivalent to the system of four linear equations

$$y' = y_1 , \quad z' = z_1$$

$$y_1' = a_1 y_1 + b_1 z_1 + c_1 y + d_1 z + e_1$$

$$z_1' = a_2 y_1 + b_2 z_1 + c_2 y + d_2 z + e_2 \quad .$$

More generally, given a system of linear differential equations, we shall assume that it has been solved with respect to the derivatives of the unknown functions and that the coefficients of these derivatives have been equated to one, by divisions. The system will then be taken in the *canonical form*

$$y_j' = \sum_1^n a_{j,k}\, y_k + b_j \qquad j = 1,2,\ldots n$$

where the coefficients $a_{j,k}$ and b_j are functions of the real variable x, which we assume is continuous in the interval in question; the y_j are unknown functions of x.

What we have to say about these systems will apply in particular to the system equivalent system with a linear equation of order n, and will allow us to recover what was obtained in the last section.

82 - Fundamental systems of solutions

We will first of all consider a homogeneous system

$$y_j' = \sum_1^n a_{j,k} y_k \qquad j = 1,2,\ldots,n$$

To simplify matters, we shall restrict our attention to a system of three equations of the form

$$\begin{cases} y' = ay + bz + cu \\ z' = a_1y + b_1z + c_1u \\ u' = a_2y + b_2z + c_2u \end{cases} \tag{64}$$

It is clear that if y_1, z_1, u_1 is a system of solutions and C is a constant, then Cy_1, Cz_1, Cu_1 is also a system of solutions, and that if y_1, z_1, u_1 and y_2, z_2, y_2 are two systems of solutions, then the functions $y_1 + y_2$, $z_1 + z_2$ and $u_1 + u_2$, will again satisfy the system. Consequently, if

$$y_1, z_1, u_1 \ ; \quad y_2, z_2, u_2 \ ; \quad y_3, z_3, u_3 \tag{65}$$

are three systems of solutions and C_1, C_2, C_3 are any three constants.

$$C_1y_1 + C_2y_2 + C_3y_3, \quad C_1z_1 + C_2z_2 + C_3z_3$$

$$C_1u_1 + C_2u_2 + C_3u_3 \quad ,$$

is again a system of solutions.

A FUNDAMENTAL SYSTEM OF SOLUTIONS

Three systems of solutions such as (65), form a fundamental system of the determinant

$$\Delta = \begin{vmatrix} y_1 & y_2 & y_3 \\ z_1 & z_2 & z_3 \\ u_1 & u_2 & u_3 \end{vmatrix}$$

is not identically zero in the interval where it applied. Liouville's theorem of no. 77 has here an analogous statement.

Theorem. *If* Δ *is the determinant of a fundamental system of solutions, then this determinant is non-zero in an interval where the coefficients of the equations are continuous.*

The derivative of Δ is the sum of three determinants obtained by replacing each line in Δ successively by its derivative. The first of these determinants is

$$\begin{vmatrix} y_1' & y_2' & y_3' \\ z_1' & z_2' & z_3' \\ u_1 & u_2 & u_3 \end{vmatrix} .$$

If we replace y_1', y_2', y_3' in the first line of this determinant, by their values $ay_1 + bz_1 + cu_1 \ldots$ taken from (64), then we see that it is the sum of three determinants of which the first is $a\Delta$ and the rest zero, by virtue of having two of their lines proportional. By proceeding in the same manner for the two other determinants, we see that

$$\Delta' = (a + b_1 + c_2)\Delta \quad ,$$

whence we deduce as in no. 77 that Δ, being different from zero at a point x_o of the interval in question, is non-zero in this interval and is given by

$$\Delta = \Delta(x_o)e^{D(x)} \quad , \qquad D(x) = \int_{x_o}^{x} (a + b_1 + c_1)dx \quad .$$

The theorem is thus proved.

<u>A Fundamental Theorem</u>. *In an interval where the coefficients of the system (64) are continuous, and where a fundamental system of solutions of (65) is known, the general solution of the system is given by*

$$\begin{aligned} y &= C_1y_1 + C_2y_2 + C_3y_3 \ , \qquad z = C_1z_1 + C_2z_2 + C_3z_3 \\ u &= C_1u_1 + C_2u_2 + C_3u_3 \quad , \end{aligned} \tag{66}$$

where C_1, C_2, C_3 *are arbitrary constants.*

We know already that the functions defined by equations (66), form a system of solutions. Let us show that every solution can be put into this form. First of all, as Δ is non-zero in the interval in question, we see that if y, z, u is a system of solutions, we can determine the functions $\lambda_1, \lambda_2, \lambda_3$ of x, such that

$$y = \lambda_1y_1 + \lambda_2y_2 + \lambda_3y_3 \quad , \qquad z = \lambda_1z_1 + \lambda_2z_2 + \lambda_3z_3 \quad ,$$

$$u = \lambda_1u_1 + \lambda_2u_2 + \lambda_3u_3$$

and it is necessary to show that these functions, which are differentiable since the y, z, u are differentiable, are actually constants. Now, on differentiating the first of these equations and on replacing the y' by their values drawn from the first equation (64), we see that we successively obtain

$$y' = \lambda_1y_1' + \lambda_2y_2' + \lambda_3y_3' + \lambda_1'y_1 + \lambda_2'y_2 + \lambda_3'y_3 \ ,$$

$$y' = ay + bz + cu = \lambda_1(ay_1 + b_1z + cu_1)$$

$$+ \lambda_2(ay_2 + bz_2 + cu_2) + \lambda_3(ay_3 + bz_3 + cu_3)$$

$$= \lambda_1 y_1' + \lambda_2 y_2' + \lambda_3 y_3'$$

hence

$$\lambda_1' y_1 + \lambda_2' y_2 + \lambda_3 y_3' \equiv 0 \quad .$$

Similarly,

$$\lambda_1' z_1 + \lambda_2 z_2' + \lambda_3 z_3' \equiv 0 \ , \qquad \lambda_1' u_1 + \lambda_2' u_2 + \lambda_3' u_3 \equiv 0 \quad ,$$

which implies, since $\Delta \neq 0$, $\lambda_1' = 0$, $\lambda_2' = 0$. Hence λ_1, λ_2, λ_3 are constants.

TRANSFORMATION OF THE CONDITION $\Delta \equiv 0$.

The results here are different from those concerning the Wronskian determinant (no. 76).

<u>Theorem</u>. *If we consider the systems of solutions (65) of the differential system (64) and if their determinant* Δ *is identically zero in an interval* (α,β), *then there exists an interval* (γ,δ) *in which*

$$C_1 y_1 + C_2 y_2 + C_3 y_3 \equiv 0 \ , \qquad C_1 z_1 + C_2 z_2 + C_3 \equiv 0 \ ,$$
$$C_1 u_1 + C_2 u_2 + C_3 u_3 \equiv 0 \ , \tag{67}$$

where C_1, C_2, C_3 *are constants that are not all zero.*

Let us assume that one of the minors of Δ, e.g., $y_2z_3 - y_3z_2$, is not identically zero in (α,β). As this minor is continuous, there will exist an interval (γ,δ) of (α,β) in which this minor will not be zero at any point. There will exist differentiable functions λ_2 and λ_3 of x, such that

$$y_1 \equiv \lambda_2 y_2 + \lambda_3 y_3 \ , \quad z_1 \equiv \lambda_2 z_2 + \lambda_3 z_2 \ , \quad u_1 \equiv \lambda_2 u_2 + \lambda_3 u_3 \quad .$$

These functions are provided by the first two identities and satisfy the third, since $\Delta = 0$. Let us differentiate these equations and, as in the above proof, note that the y_1, z_1, u_1 etc., are systems of solutions. We will have

$$\lambda_2' y_2 + \lambda_3' y_3 \equiv 0 \quad , \quad \lambda_2' z_2 + \lambda_3' z_3 \equiv 0 \quad ,$$

$$\lambda_2' u_2 + \lambda_3' u_3 \equiv 0 \quad ,$$

and the first two identities show that $\lambda_2' \equiv \lambda_3' \equiv 0$ since $y_2z_3 - y_3z_2$ is non-zero at any point of (γ,δ). Hence λ_2 and λ_3 are constants. We have indeed a relation agreeing with the conditions in the above statement. In this case, one of the systems of solutions is a consequence of the other two.

If all the minors of Δ are identically zero in (α,β), we will consider one of the functions y_1, say, non-zero in (γ,δ), and we can write $\lambda_2 = \lambda y_1$, where λ is finite and differentiable in (γ,δ). Since the minors of Δ are zero, we will also have $z_2 = \mu z_1$, $u_2 = \mu u_1$ and on differentiating, we will show that μ is a constant. In this same interval (γ,δ), we will likewise have $y_3 = \mu y_1$, $z_3 = \mu z_1$, $u_3 = \mu u_1$, where μ is constant. We will have here, two systems of relations of the form indicated in the statement.

This generalises to the case of order n.

Corollary. *If the identities (67) are impossible in a part of the interval* (α,β) *(where the coefficients of the differential system are continuous), then the system of these solutions is a fundamental system.*

For, in this interval, Δ is non-identically zero, otherwise the identities (67) will occur anywhere in (α,β).

83 - Lowering of order

When we know a system of solutions of a homogeneous linear system, we can lower the order of the system by one: in place of a system of n equations, we will have a system of $(n-1)$ equations and a quadrature will have to be made.

More generally, if we know p systems of solutions, we can, *if these solutions are distinct,* reduce the order by p units. Let us now denote the unknown functions by $y_1,\ldots,y_n$. We assume that the solutions

$$y_1^q, y_2^q, \ldots, y_n^q \quad , \qquad q = 1,2,\ldots,p \quad ,$$

are known.

These solutions will be said to be distinct in an interval if there exists a determinant formed by p *particular values of the unknowns, e.g.,*

$$\begin{vmatrix} y_1^1 & y_2^1 & \cdots & y_p^1 \\ y_1^2 & y_2^2 & \cdots & y_p^2 \\ \cdot & \cdot & \cdots & \cdot \\ y_1^p & y_2^p & \cdots & y^p \end{vmatrix}$$

which is non-zero at any point of the interval in question.

We reduce the order by making the following change of the unknown functions

$$y_j = y_j^1 Y_1 + y_j^2 Y_2 + \dots + y_j^p Y_p \quad \text{if} \quad j = 1,2,\dots, \quad ;$$

$$y_j = y_j^1 Y_1 + y_j^2 Y_2 + \dots + y_j^p Y_p + Y_j \quad \text{if} \quad j = p+1,\dots, \quad .$$

The system

$$y_j' = \sum_1^n a_{j,k} y_k \qquad j = 1,2,\dots n \quad ,$$

then becomes, on taking account that the y_j^q, $j = 1,2,\dots,n$, are solutions:

$$y_j^1 Y_1 + \dots + y_j^p Y_p' = \sum_{p+1}^n a_{j,k} Y_k \quad \text{if} \quad j = 1,2,\dots,p \quad ,$$

$$Y_j' + y_j^1 Y_1' + \dots + y_j^p Y_p' = \sum_{p+1}^n a_{j,k} Y_k \quad \text{if} \quad j = p+1,\dots,n \quad .$$

We can extract the values of $Y_1',\dots,Y_p'$ from the first p of these n equations since the determinant of the coefficients is non-zero in the interval in question, and bring these values into the last $n-p$ equations, which then provide a linear system of order $n-p$ in $Y_{p+1},\dots,Y_n$. If this system is integrable, then the Y_k will be known in the second members of the first p equations, hence the Y_j', $j = 1,2,\dots,p$, will be known, and by p integrations, we will obtain the Y_j.

84 - Equations without second members

The method of Lagrange remains tenable and in fact simplifies. It has already been implicitly applied in the proof of the fundamental theorem of no. 82. Let us take the system with three equations only

$$\begin{cases} y' = ay + bz + cu + d \\ z' = a_1 y + b_1 z + c_1 u + d_1 \\ u' = a_2 y + b_2 z + c_2 u + d_2 \quad , \end{cases} \tag{68}$$

where the coefficients are functions of x. If we put all of the terms in y, z, u in the first members, then the functions d, d_1, d_2 will be the second members. If we know a fundamental system of solutions of the homogeneous system obtained by replacing d, d_1, d_2 by zero, and if this system is the system in (65) of no. 82, then we seek a solution of the complete system (68), under the form

$$y = C_1 y_1 + C_2 y_2 + C_3 y_3 \quad , \qquad z = C_1 z_1 + C_2 z_2 + C_3 z_3 \quad ,$$

$$u = C_1 u_1 + C_2 u_2 + C_3 u_3 \quad ,$$

where C_1, C_2, C_3 are not constants, but are functions of x. On bringing those into (68) and taking account of the fact that the coefficients of C_1, C_2, C_3 are equal in both members (since the $y_1, z_1, u_1, \ldots$ are solutions), there remains the system of linear equations

$$\begin{cases} C_1' y_1 + C_2' y_2 + C_3' y_3 = d \\ C_1' z_1 + C_2' z_2 + C_3' z_3 = d_1 \\ C_1' u_1 + C_2' u_2 + C_3' u_3 = d_2 \end{cases} ,$$

which is solvable, since the determinant Δ is non-zero in the interval in question. We obtain

$$C_1' = f_1(x) \ , \quad C_2' = f_2(x) \ , \quad C_3' = f_3(x),$$

hence

$$C_1 = \int f_1(x)dx + C_4 \ , \qquad C_2 = \int f_2(x)dx + C_5 \ ,$$

$$C_3 = \int f_3(x)dx + C_6 \ .$$

It again follows that the general integral of the complete equation is obtained by combining a particular integral of this equation (corresponding to $C_4 = C_5 = C_6 = 0$) with the general integral of the equation obtained by setting $d = d_1 = d_2 = 0$, which can be satisfied directly.

85 - An application to equations with constant coefficients. Homogeneous systems

Let us take a system of equations of the form

$$y_j' + \sum_1^n a_{j,k} y_k = 0 \ , \qquad j = 1,2,\ldots,n$$

where the coefficients $a_{j,k}$ are constants. There exists solutions of the form

$$y_j = \alpha_j e^{rx} \ , \qquad j = 1,2,\ldots,n$$

where the α_j and r are constants. Replacing the y_j and their derivatives by these functions and their derivatives, e^{rx} will be a factor in the first member of each equation. We can suppress these factors which are never zero, and there remains to determine r and the α_j, the system of equations

$$\alpha_j r + \sum_1^n a_{j,k} \alpha_k = 0 \ , \qquad j = 1,2,\ldots,n$$

which can be written

$$a_{j,1}\alpha_1 + \dots + a_{j,j-1}\alpha_{j-1} + (a_{j,j} + r)\alpha_j + a_{j,j+1}\alpha_{j+1}$$
$$+ \dots + a_{j,n}\alpha_n = 0 \quad , \qquad = 1,2,\dots,n \quad . \tag{69}$$

We may consider this as a system of n homogeneous equations in $\alpha_1,\alpha_2,\dots,\alpha_n$. There will only be verifiable solutions (α_j not all zero) if the determinant is zero. r is then determined by the condition

$$F(r) \equiv \begin{vmatrix} a_{1,1}+r & a_{1,2} & \dots & a_{1,n} \\ a_{2,1} & a_{2,2}+r & \dots & a_{2,n} \\ \cdot & \cdot & \dots & \cdot \\ a_{n,1} & a_{n,2} & \dots & a_{n,n+r} \end{vmatrix} = 0 \quad .$$

If r is a root of this equation of order n, then the system (69) is possible. There exists at least one system of numbers $\alpha_1,\dots,\alpha_n$ satisfying it. *To each root of the equation* $F(r) = 0$, *called the characteristic equation, there corresponds a system of solutions. If this equation has* n *distinct roots* $r_1,\dots,r_n$, *we then have* n *systems of solutions: They give a fundamental system.* The n values of y_1, for example, are products of constants and exponentials e^{rx} corresponding to distinct values of r. Following the lemma of no. 79, they could only be bound by a linear relation with constant coefficients if the coefficients of the exponentials are all zero. This cannot happen for all the y_j, since, for each r, at least one of the α_j is non-zero. The corollary of no. 82 indeed shows that the system is fundamental. In this case, we have therefore found the general solution, valid for any x.

<u>Example</u>. Consider the system

$$\frac{dx}{dt} + qz - ry = 0 \ , \quad \frac{dy}{dt} + rx - pz = 0 \ , \qquad \frac{dz}{dt} + py - qx = 0 \quad ,$$

where p, q, r are three real constants. Let us name s as the coefficient of t in the exponentials. The characteristic equation will be

$$\begin{vmatrix} s & -r & q \\ r & s & -p \\ -q & p & s \end{vmatrix} = s^3 + s(p^2 + q^2 + v^2) = 0 \quad .$$

Its roots are 0 and $\pm i\omega$ if we set $\omega = \sqrt{p^2 + q^2 + r^2}$. The values of the coefficients α_1, α_2, α_3 corresponding to one of these roots, are the minors of the first line of the determinant, say

$$s^2 + p^2 \ , \quad pq - rs \ , \qquad rp + qs \quad .$$

We have then the systems of solutions x, y, z given in the following three lines,

$$\begin{array}{ccc} p, & q, & r, \\ -(q^2+r^2)e^{i\omega t} & (pq - vi\omega)e^{i\omega t} & (pr + qi\omega)e^{i\omega t} \\ -(q^2+r^2)e^{-i\omega t} & (pq + vi\omega)e^{-i\omega t} & (pr - qi\omega)e^{-i\omega t} \end{array}$$

To obtain real solutions, we will multiply the terms of the first line by the real constant C, those of the second by $1/2(C' + iC'')$ and those of the third by $1/2(C' - iC'')$, where C' and C'' are real. We thus obtain

$$\begin{cases} x = Cp - (q^2 + r^2)(C' \cos \omega t - C'' \sin \omega t) \\ y = Cq + pq(C' \cos \omega t - C'' \sin \omega t) + r\omega(C' \sin \omega t + C'' \cos \omega t) \\ z = Cr + pr(C' \cos \omega t - C'' \sin \omega t) - q\ (C' \sin \omega t + C'' \cos \omega t) \end{cases} .$$

THE CASE WHERE THE CHARACTERISTIC EQUATION HAS MULTIPLE ROOTS

Generaly, if r' is any constant and if we put

$$y_j = e^{r'x} Y_j \ , \qquad j = 1,2,\ldots,n \quad ,$$

we obtain a new system with constant coefficients

$$Y_j' + r'Y_j + \sum_1^n a_{j,k} Y_k = 0 \quad , \qquad j = 1,2,\ldots,n \quad ,$$

whose characteristic equation is $F(r+r') = 0$, where $F(r)$ is the equation defined above.

In order to study the case of a multiple root of $F(r) = 0$, we must therefore assume this root to be zero. To this zero root there corresponds a system of solutions formed by constants of which one is non-zero, we may suppose that it is the value of y_1, and on dividing by this volume, we obtain the system of solutions

$$y_1 = 1 \ , \quad y_2 = \alpha_2, \ldots, y_n = \alpha_n \quad .$$

We can reduce the order by putting

$$y_1 = Y_1 \ , \quad y_2 = \alpha_2 Y_1 + Y_2, \ldots \qquad y_n = \alpha_n Y_1 + Y_n \quad ,$$

which yields the system

$$
\begin{aligned}
&Y_1' + a_{1,2}Y_2 + \dots + a_{1,n}Y_n = 0 \quad , \\
&Y_j' + \alpha_j Y_1' + a_{j,2}Y_2 + \dots + a_{j,n}Y_n = 0 \; , \qquad j = 2,\dots,n \quad .
\end{aligned} \tag{70}
$$

By replacing Y_1' by its value drawn from the first equation, we see that $Y_2,\dots,Y_n$ are given by the system of order $n-1$.

$$
Y_j' + (a_{j,2} - \alpha_j a_{1,2})Y_2 + \dots + (a_{j,n} - \alpha_j a_{1,n})Y_n = 0 \quad , \qquad j = 2,\dots,n \quad . \tag{71}
$$

Y_1 will then be given by integrating equation (70). The characteristic equation of the system (71) is

$$
F_1(r) \equiv \begin{vmatrix} a_{2,2} - \alpha_2 a_{1,2} + r & \dots & a_{2,n} - \alpha_2 a_{1,n} \\ \cdot & \dots & \cdot \\ \cdot & \dots & \cdot \\ a_{n,2} - \alpha_n a_{1,2} & \dots & a_{n,n} - \alpha_n a_{1,n} + r \end{vmatrix} = 0
$$

The product with r could be put into the form

$$
rF_1(r) \equiv \begin{vmatrix} r & a_{1,2} & \dots & a_{1,n} \\ 0 & a_{1,2} - \alpha_2 a_{1,2} + r & \dots & a_{1,n} - \alpha_2 a_{1,n} \\ \cdot & & \dots & \\ \cdot & & \dots & \\ 0 & a_{n,2} - \alpha_n a_{1,2} & \dots & a_{n,n} - \alpha_n a_{1,n} + r \end{vmatrix}
$$

Let us combine the elements of the second line with the product of the elements of the first line and α_2, to those of the third, the product of α_3 and the elements of the first, etc., to those of the last line, the product of α_n and the elements of the first line. We will have

$$
rF_1(r) \equiv \begin{vmatrix} r & a_{1,2} & & a_{1,n} \\ r\alpha_2 & a_{2,2} + r & \dots & a_{2,n} \\ \cdot & \cdot & \dots & \cdot \\ \cdot & \cdot & \dots & \cdot \\ \cdot & \cdot & \dots & \cdot \\ r\alpha_n & a_{n,2} & & a_{n,n} + r \end{vmatrix} \quad .
$$

We recover $F(r)$. In fact, if we multiply the column elements of rank j by $-\alpha_j$, $j = 2,\ldots,n$, and all these products are combined with the elements of the first column, the first element of the first column becomes $r - (\alpha_2 a_{1,2} + \ldots + \alpha_n a_{1,n})$, the element of the first column situated in the line of rank k, becomes $-(\alpha_2 a_{k,2} + \ldots + \alpha_n a_{k,n})$. Now $1,\alpha_2,\ldots,\alpha_n$ is a solution of the differential system, we have therefore

$$a_{k,1} + \alpha_2 a_{k,2} + \ldots + \alpha_n a_{k,n} = 0 \ , \qquad k = 1,2,\ldots,n \quad .$$

We see then that

$$rF_1(r) \equiv F(r) \quad .$$

$F_1(r) \equiv 0$ therefore admits the root $r = 0$ with order of multiplicity $q - 1$, if it was of order q in $F(r) = 0$. We can continue these operations. We will arrive at a differential system of order $n - q + 1$ which will admit a characteristic equation for which 0 will be a simple root. This system will then admit a system of solutions formed by constants, defined to within an arbitrary constant. The preceding system will result from an intergration of an analogous equation to (70), where the Y_k will be constants which happens to be the case, and then from one transformation similar to that transforming Y to y. It will then have a system of solutions formed by polynomials of degree one at most, whose coefficients are linear and homogeneous functions of two arbitrary constants. *It follows that the given system will have a system of solutions formed by polynomials in x of degree $q - 1$ at most whose coefficients are linear homogeneous functions of q arbitrary constants.* This is seen by recurrence. Given this case, it suffices to show that, if the property is true for the system (71) whose characteristic equation admits the root 0 to order $q - 1$, then it is true for the given system. Now if the system (71) admits a system of solutions where the Y_k, $k = 2,\ldots,n$ are polynomials of degree $q - 2$ at most, whose coefficients are linear homogeneous functions of $q - 1$ arbitrary constants, then integration of equation (70) will yield for Y_1, a polynomial of degree $q - 1$ at most, whose coefficients will be linear homogeneous functions of q constants. Now, on passing to the y_j, $j = 1,2,\ldots,n$, we see that the property is established. On returning finally to the case of any multiple root of $F(r)$ we have the following.

If the characteristic equation admits the root r with order of multiplicity q, then there exists a system of solutions of the form

$$y_1 = e^{rx}P_1(x),\ y_2 = e^{rx}P_2(x),\ldots,\ \ y_n = e^{rx}P_n(x) \quad , \tag{72}$$

where $P_1,\ldots,P_n(x)$ are polynomials of degree at most $q - 1$ whose coefficients are linear homogeneous functions of q arbitrary constants.

Otherwise, if the q *arbitrary constants are all zero, then the functions* $y_1, y_2, \ldots, y_n$ *are not all identically zero.*

The last part of this statement is again verified by recurrence. The Y_k depend on $q - 1$ constants. If it is impossible that all are identically zero when these constants are not all zero, then the y_j given by

$$y_1 = Y_1, \quad y_2 = \alpha_2 Y_1 + Y_2, \ldots, \quad y_n = \alpha_n Y_1 + Y_n \quad ,$$

will not be either. For it would necessitate that Y_1 would be, hence also $Y_2, \ldots, Y_n$. The $q - 1$ constants appearing in these Y_k, $k = 2, \ldots,$, will be zero, and (70) would reduce to $Y_1' \equiv 0$, $Y_1 = \text{const}$, and the constant of integration would equally be zero. This last part of the statement allows us to apply the lemma of no. 79, to the systems of solutions corresponding to various simple or multiple roots of the characteristic equation. The application of the above theorem to these roots provides a system of solutions of the form

$$y_1 = \sum e^{rx} P_1(x), \ y_2 = \sum e^{rx} P_2(x), \ldots, \ y_n = \sum e^{rx} P_n(x) \quad , \qquad (73)$$

which depends on n arbitrary constants appearing linear and homogeneous. This will be the general solution if it is impossible to choose these constants in such a way that $y_1 \equiv 0, \ldots, y_n \equiv 0$. Now, following the lemma, this condition would imply that $P_1(x) \equiv 0, \ldots, P_n(x) \equiv 0$, for each value of r.

Conclusion. *The general solution of a system of* n *linear homogeneous differential equations with constant coefficients, is obtained by solving the characteristic equation, and determining for each root* r *of this equation, the corresponding solutions of the form (72). The sums (73) of these systems of solutions yield the general solution.*

In practice, we can find the constants corresponding to the simple roots of $F(r) = 0$, by using the minors of $F(r)$ as we did in the examples discussed. When r is a multiple root, we will determine the corresponding polynomials $P_j(x)$ by the method of indeterminate coefficients. We could also apply D'Alembert's procedure which will be the case in no. 89.

Remark. We could apply the method of reduction and thus determine, independently of the general results of no. 82, the form (73) of the general solution.

Example. Consider the system

$$\frac{dx}{dt} + y + 2z = 0, \quad \frac{dy}{dt} - x - z = 0, \quad \frac{5dz}{dt} + 6x + 8y = 0 \quad .$$

The characteristic equation is

$$\begin{vmatrix} r & 1 & 2 \\ -1 & r & -1 \\ 6 & -8 & 5r \end{vmatrix} \equiv 5r^3 - 15r + 10 = 0 \quad .$$

The roots are $r = 1$, which is a double root, and $r = -2$. For $r = -2$. the minors of the determinant are 12, -16, 20; we thus have the solution

$$x = 3e^{-2t} \; , \quad y = -4e^{-2t} \; , \quad z = 5e^{-2t} \quad .$$

For $r = 1$, we have a solution of the form

$$x = e^t(\alpha t + \beta) \; , \quad y = e^t(\alpha' t + \beta') \; , \quad z = e^t(\alpha'' t + \beta'') \quad .$$

By stating that the equations are satisfied by these functions, we find, after suppressing the factor e^t, two systems of equations, obtained by equating the terms in t, and the independent terms:

$$\begin{cases} \alpha + \alpha' + 2\alpha'' = 0 \\ \alpha' - \alpha - \alpha'' = 0 \\ 5\alpha'' + 6\alpha - 8\alpha' = 0 \end{cases} ; \qquad \begin{cases} \beta + \beta' + 2\beta'' = \alpha \; , \\ -\beta + \beta' + \beta'' = -\alpha \; , \\ 6\beta - 8\beta' + 5\beta'' = -5\alpha'' \end{cases} .$$

The first system yields $\alpha = 3\alpha'$, $\alpha' = -2\alpha'$, where α' remains arbitrary. On taking this into the second, we obtain $\beta = 5\alpha' + 3\beta'$, $\beta'' = -4\alpha' - 2\beta'$, where β' remains arbitrary. On denoting α' by C' and β' by C", we therefore have

$$x = e^t(3C't + 5C' + 3C'') \; , \quad y = e^t(C't + C'')$$

$$z = e^t(-2C't - 4C' - 2C'') \quad ,$$

and the general solution is

$$x = 3Ce^{-2t} + e^t(3C't + 5C' + 3c'') \quad ,$$

$$y = -4Ce^{-2t} + e^t(C't + C'')$$

$$z = 5Ce^{-2t} + e^t(-2C't - 4C' - 2C'') \quad ,$$

where C, C', C" are arbitrary constants.

86 - Linear systems with constant coefficients with second members

When the second members are arbitrary, we employ the method of variation of constants. But if the second members are sums of products of exponentials and polynomials, then we can proceed as in the case of order n (no. 80). Firstly, given the system

$$y_j' + \sum_1^n a_{j,k}y_k = f_j(x) + g_j(x) , \quad j = 1,2,\dots,n \quad ,$$

we can (whether or not the $a_{j,k}$ are constants) integrate separately the two systems obtained by putting separately in the second members, $f_j(x)$ first and then $g_j(x)$, and add the values obtained for the various unknown functions. In the case examined here, we firstly return to the case of a system of the form

$$y_j' + \sum_1^n a_{j,k}y_k = e^{\alpha x}Q_j(x) , \quad a_{j,k} = \text{const}, \quad j = 1,2,\dots,n$$

where α is a constant and the $Q_j(x)$ are polynomials. Given, then, such a system

$$y_j' + \sum_1^n a_{j,k}y_k = Q_j(x) \quad . \tag{75}$$

We can determine termwise, the coefficients of the polynomials $y_k = R_k(x)$ whose degree will at most be equal to the largest degree of the $Q_j(x)$ and which satisfy this system, providing the determinant of the $a_{j,k}$ is non-zero. For if we assume that

$$Q_j(x) = b_{j,m}x^m + b_{j,m-1}x^{m-1} + \dots , \quad j = 1,2,\dots,n$$

where one of the $b_{j,m}$ is non-zero, and if we set

$$R_j(x) = c_{j,m}x^m + c_{j,m-1}x^{m-1} + \dots , \quad j = 1,2,\dots,n$$

then the $c_{j,m}$ will be given by the system

$$\sum_1^n a_{j,k}c_{k,m} = b_{j,m} , \quad j = 1,2,\dots,n \quad ,$$

which is a Cramer system. We then calculate the $c_{j,m-1}$ which will be given by

$$\sum_1^n a_{j,k}c_{k,m-1} = b_{j,m-1} - mc_{j,m} , \quad j = 1,2,\dots,n \quad ,$$

and so on. On returning to the system (74), we see that:

If α *is not a root of the characteristic equation, then the system (74) has a system of particular solutions of the form*

$$y_j = e^{\alpha x}R_j(x) , \quad j = 1,2,\dots,n \quad ,$$

where the degree of the polynomials $R_j(x)$ *is less than or equal to the highest degree of the polynomials* $Q_j(x)$.

When α is a root of the characteristic equation, zero is a root of the characteristic equation of the transformed system of the form (75). [In (75), the $a_{j,k}$ are not the same as in (74).] We can apply the method of reduction, the only difference from that of no. 85 being that there will be second members which are the polynomials $Q_j(x)$, then linear combinations of these polynomials. It follows that:

If α is a root of order q of the characteristic equation, then the system (74) has a system of particular solutions of the form

$$y_j = e^{\alpha x} R_j(x) \ , \qquad j = 1,2,\dots,n$$

where the degree of the polynomials $Q_j(x)$ is at most equal to the largest degree of the polynomials $Q_j(x)$, plus q units.

87 - The Jacobi equation

Consider the linear homogeneous system

$$\frac{dx}{ax+by+cz} = \frac{dz}{a_1x+b_1y+cz} = \frac{dz}{a_2x+b_2y+c_2z} = dt \quad , \tag{76}$$

with constant coefficients; it defines a system of curves of the space (x,y,z) in terms of the parameter t. The first three ratios give a system of two differential equations

$$\frac{dx}{dt} = \frac{ax+by+cz}{a_2x+b_2y+c_2z} \ , \quad \frac{dy}{dz} = \frac{a_1x+b_1y+c_1z}{a_2x+b_2y+c_2z} \tag{77}$$

which define these same curves. To integrate a system of the form (77), we can always write it in the form (76). The equations (77) are homogeneous in x,y,z. Let us set $x = zX$, $y = zY$, the system becomes

$$X + z\frac{dX}{dz} = \frac{aX+bY+c}{z_2X+b_2Y+c_2} \ , \qquad Y + z\frac{dY}{dz} = \frac{a_1X+b_1Y+c_1}{a_2X+b_2Y+c_2} \quad . \tag{78}$$

If we multiply the first equation by dY, the second by dX and subtract, then if we multiply the first by Y, the second by X and subtract again, we obtain the equivalent system to (78)

$$\begin{aligned}(YdX - XdY)(a_2X + b_2Y + c_2) + (aX + bY + c)dY \\ - (a_1X + b_1Y + c_1)dX = 0 \quad ,\end{aligned} \tag{79}$$

$$\frac{dz}{z} = \frac{(a_2X + b_2Y + c_2)(YdX - XdY)}{Y(aX + bY + c) - X(a_1X + b_1Y + c_1)} \quad . \tag{80}$$

The first of these equations is a Jacobi equation (no. 66). The integrals of the system (76) therefore lead to integrals of this equation (79). Conversely, we combine an integral of the equation (80) with an integral of equation (79), which is obtained by a quadrature; we then obtain an integral of the system (77), and hence that of the system (76). The problem of integration of the Jacobi system is therefore equivalent to the problem of integration of the system (76); it depends on the solution of the characteristic equation of this system, which only differs by notation from that of no. 66.

This allows us to put the integral of equation (79) in the form given by Jacobi. The integral of the system (76) is given by

$$x = C_1 e^{r_1 t} + C_2 e^{r_2 t} + C_3 e^{r_3 t}, \quad y = C_1\alpha_1 e^{r_1 t} + C_2\alpha_2 e^{r_2 t} + C_3\alpha_3 e^{r_3 t},$$

$$z = C_1\beta_1 e^{r_1 t} + C_2\beta_2 e^{r_2 t} + C_3\beta_3 e^{r_3 t},$$

in the general case where the roots of the characteristic equation are simple. Let us consider these equations as a system of linear equations in $C_1 e^{r_1 t}$, $C_2 e^{r_2 t}$, $C_3 e^{r_3 t}$. We can solve with respect to these quantities since the determinant of the fundamental system of solutions is non-zero, and put these solutions in the form

$$C_1 e^{r_1 t} = \alpha x + \beta y + \gamma z, \quad C_2 e^{r_2 t} = \alpha' x + \beta' y + \gamma' z,$$

$$C_3 = \alpha'' x + \beta'' y + \gamma'' z$$

where the $\alpha, \beta, \gamma, \ldots,$ are constants. [We remark that if we put the exponentials in the second members, then we would obtain the first in these second members (no. 47).] If we divide the first two equations by the last, then we obtain in the second member, the ratios in which X and Y appear,

$$\frac{C_1}{C_3} e^{(r_1 - r_3)t} = \frac{\alpha X + \beta Y + \gamma}{\alpha'' X + \beta'' Y + \gamma''}, \quad \frac{C_2}{C_3} e^{(r_2 - r_3)r} = \frac{\alpha' X + \beta' Y + \gamma'}{\alpha'' X + \beta'' Y + \gamma''}.$$

Finally, by raising both members of the first equation to the power $r_2 - r_3$, those of the second to the power $r_1 - r_3$; then on dividing each member by the other, we arrive at the form

$$(\alpha X + \beta Y + \gamma)^{r_2 - r_3}(\alpha' X + \beta' Y + \gamma')^{r_3 - r_1}(\alpha'' X + \beta'' Y + \gamma'')^{r_1 - r_2} = C$$

where C is an arbitrary constant. It is the form given by Jacobi.

We will proceed in a similar way to eliminate t, and introduce X and Y in the case where there will be a multiple root.

Chapter VI

LINEAR DIFFERENTIAL EQUATIONS IN THE COMPLEX DOMAIN

Having given, in the preceding chapter, the elementary methods which permit the direct integration of certain types of equations, we are going to consider in this chapter, and in the two following, the analytic differential equations in complex variables. As mentioned in no. 48, the first difficulty arises in the study of the solutions of the differential equations and the presence of moving singularities. This difficulty does not exist in the case of linear equations (no. 49); also, their study had been developed earlier. Besides, these equations lend themselves to numerous applications. In this chapter we shall apply the general theorems obtained by the method of majorant functions to linear equations (Chapter III, §III); they will allow us to present under a different guise the results obtained directly in Chapter V, and to complete them. We shall then consider the singularities of homogeneous equations whose coefficients are meromorphic; we shall discuss the general theorem on the permutations of integrals following Jordan's method. In the proof of Fuch's theorem, we shall apply a procedure due to Birkhoff. The general theorems concerning the Fuschian equations will be applied to the study of the equations of Halphen and Picard. We shall then study the method of integration of Laplace, in the case of Laplace type equations, and further consider, in the general case, the equations whose coefficients are rational fractions, those having been studied by Poincaré. Finally, we shall discuss some general theorems.

I. AN APPLICATION OF THE EXISTENCE THEOREMS

88 - Solutions in the general case

Following no. 49, consider a linear differential equation of order n,

$$A_o(x)y^{(n)} + \dots + A_n(x)y = B(x) \quad , \tag{1}$$

where the $A_j(x)$ and $B(x)$ are holomorphic in the circle $|x - x_o| < R$ and where $A_o(x)$ is non-zero at any point of this circle. It admits a holomorphic solution y in this circle, taking the value y_o at the point x_o, and whose derivatives $y^{(k)}$, $k = 1,2,\dots,n-1$ take, at this point, the values $y_o^{(k)}$ where y_o and $y_o^{(k)}$ are arbitrary. We have a solution depending on n arbitrary constants $y_o, y_o', \dots, y_o^{(n-1)}$. We have seen, moreover, that it is possible to extend these solutions to the whole domain where the coefficients remain holomorphic and where $A_o(x)$ is non-zero, the only possible singularities being those of the functions

$$\frac{A_1(x)}{A_o(x)},\ldots,\frac{A_n(x)}{A_o(x)},\frac{B(x)}{A_o(x)} .$$

If we set aside the case where the functions possess singular lines, we can, in general, make the continuation along a polygonal line starting from x_o and returning to this point after having passed around the singularities of the coefficients. We return to the point x_o with the values $Y_o, Y'_o, \ldots, Y_n^{(n-1)}$ of y and its $(n-1)$ first derivatives; these are, in general, distinct from those with which we had started. We could ask, in certain cases, if it is not possible, on starting from x_o with a determined solution, to choose a path, such that on returning to x_o, the values $Y_o,\ldots,Y_o^{(n-1)}$ coincide with the arbitrary numbers assigned in advance. The theorem of Poincaré-Volterra (I, 201) shows that this is not the case. The set of values Y_o is countable, and likewise the sets $Y'_o,\ldots,Y_o^{(n-1)}$ are countable. Consequently: *The general solution of the equation depends effectively on* n *arbitrary constants.*

This reasoning, on which we shall return in no. 136, clearly applies to the solutions of analytic differential equations more general than linear equations.

89 - Fundamental systems of solutions

Let us consider an equation without second member

$$A_o(x)y^{(n)} + A_1(x)y^{(n-1)} + \ldots A_n(x)y = 0 \quad , \tag{2}$$

inside a circle $|x - x_o| < R$ where $A_1(x)/A_o(x),\ldots,A_n(x)/A_o(x)$ are holomorphic. If $y_1, y_2, \ldots, y_n$ are n holomorphic solutions in this circle, then $y \equiv C_1y_1 + \ldots + C_ny_n$, where $C_1,\ldots,C_n$ are constants, is also a holomorphic solution in this circle. We shall say that these solutions form a *fundamental solution* if it is possible to determine the constants C_j, $j = 1,2,\ldots,n$, in such a way that $y \equiv C_1y_1 + \ldots + C_ny_n$ can coincide with any solution. For this to be the case, it is necessary and sufficient that at the point $x_o, y, \ldots, y^{(n-1)}$ can take any given values, hence that the system

$$\begin{cases} C_1y_1 + \ldots + C_ny_n = Y_o \quad , \\ C_1y_1' + \ldots + C_ny_n' = Y_o' \quad , \\ \vdots \\ C_1y_1^{(n-1)} + \ldots + C_ny_n^{(n-1)} = Y_o^{(n-1)} \quad , \end{cases}$$

where, in the first member, the values of the functions are taken at the point x_o, is possible whatever the second members. This necessitates that the Wronskian

$$\Delta(y_1,\ldots,y_n) = \begin{vmatrix} y_1 & y_2 & \cdots & y_n \\ y_1' & y_2' & \cdots & y_n' \\ \vdots & \vdots & \cdots & \\ y_1^{(n-1)} & y_2^{(n-1)} & \cdots & y_n^{(n-1)} \end{vmatrix} \tag{3}$$

is non-zero at the point x_o. This necessary condition is also *sufficient*. The definition of the fundamental system which was made is therefore equivalent to that of no. 77.

A Wronskian determinant (3) is, here, an analytic function of x. It is zero only at isolated points or else is identically zero. This is the case not only in the circle $|x - x_o| < R$, but in every domain obtained in the analytic continuation.

Liouville's theorem of no. 77 also applies here, since the proof results from the fact that, for holomorphic solutions $y_1, y_2, \ldots, y_n$ of (2) (there are no others, nos. 47-49), we have

$$\Delta = \Delta_o e^{D(x)} , \qquad D(\;) = -\int_{x_o}^{x} \frac{A_1(x)}{A_o(x)}\, dx \quad ,$$

where Δ is the Wronskian, and here we see that $D(x)$ remains finite for as long as $A_1(x)/A_o(x)$ is holomorphic. Hence, the *Wronskian determinant is non-zero* at any point of *the domain described by* x, *as long as the coefficients of the equation are holomorphic and* $A_o(x)$ *is non-zero.*

A fundamental system of solutions thus remains as such when any continuation of those solutions is made along the *same* path. *We might then speak of a fundamental system of solutions without specifying the circle* $|x - x_o| < R$ *in which it is considered.*

THE EXISTENCE OF FUNDAMENTAL SYSTEMS OF SOLUTIONS.

To have a fundamental system, it suffices to take at the point x_o, the solutions $y_1, \ldots, y_n$ corresponding to the initial conditions for which $\Delta(y_1, \ldots, y_n)$ is non-zero at the point x_o. In particular, we could take the solutions for which, at x_o, the determinant is

$$\begin{vmatrix} 1 & 0 & \cdots & 0 \\ 0 & 1 & \cdots & 0 \\ \cdot & \cdot & \cdots & \cdot \\ 0 & 0 & \cdots & 1 \end{vmatrix}$$

On knowing such a system, the general solution

$$y = C_1 y_1 + \ldots + C_n y_n$$

is known.

THE WRONSKIAN OF ANALYTIC FUNCTIONS

If $y_1, y_2, \ldots, y_n$ are given, *a priori*, and are analytic (uniform or multiform, but in the second case, it will be necessary to specify the branches), then the Wronskian is analytic. If it is not identically zero, then it is zero only at isolated points; the theorem relating to linearly independent functions (no. 76) simplifies: *if the Wronskian is identically zero in a domain, however small, or on a segment, then there exist constants* C_j, *not all zero, such that*

$$C_1 y_1 + C_2 y_2 + \ldots + C_n y_n = 0 \tag{4}$$

simultaneously, throughout the continuation of the y_j. This springs from the fact that a relationship such as (4) holds throughout this domain if it holds in one of the regions or segments belonging to it. Moreover, we remark that, as in the proof of Theorem 5 of no. 76, to assume that a Wronskian is not identically zero implies here that it is zero only at isolated points.

The circumstances indicated in the remark of no. 76 only arise for functions that are non-analytic in the interval in question. Thus, *in the analytic case:*

The necessary and sufficient condition for n *functions to satisfy an identity such as (4) is that their Wronskian is zero at non-isolated points.*

90 - The extension of the results obtained in the real case

If we take an equation with second member (1), then the existence theorem indicates directly the existence of a particular solution. The general solution will be obtained by combining this particular solution with the general solution of the homogeneous equation obtained by replacing the second member by zero.

We could also commence from a fundamental system of solutions and apply the method of the variation of constants, as in no. 78.

REDUCTION

Let us consider a homogeneous equation. If p solutions $y_1, \ldots, y_p$ are already known in a simple form, and they are *linearly independent*, i.e., such that we do not have the identity

$$C_1 y_1 + C_2 y_2 + \ldots + C_p y_p \equiv 0 \quad ,$$

where the $C_1, C_2, \ldots, C_p$ are constants and not all zero, then we can lower the order of p units. We shall firstly assume $y = y_1 z$, which will lead to an equation of order $n-1$ in z', as we have seen in no. 76. We will know $p-1$ linearly independent distinct solutions of this equation in z':

$$\left(\frac{y_2}{y_1}\right)', \left(\frac{y_3}{y_1}\right)', \ldots, \left(\frac{y_p}{y_1}\right) \quad ;$$

these are functions which are analytic; the only new singularities introduced, provide the zeros of y_1. These functions are distinct, for otherwise we would have

$$C_2\left(\frac{y_2}{y_1}\right)' + C_3\left(\frac{y_3}{y_1}\right)' + \ldots + C_p\left(\frac{y_p}{y_1}\right)' \equiv 0$$

and on integrating and multiplying by y_1, we obtain a linear, homogeneous relationship between the y_j, $j = 1,2,\ldots,p$, contrary to hypothesis. We could therefore pursue the reduction up to order $n-p$.

Remark. We can proceed in the same way, with some precaution, in the non-analytic case.

91 - D'Alembert's method

Let us consider a linear homogeneous equation with constant coefficients

$$y^{(n)} + a_1 y^{(n-1)} + \ldots + a_n y = 0 \quad , \tag{5}$$

in the case where the characteristic equation

$$f(r) \equiv r^n + a_1 r^{n-1} + \ldots + a_n = 0$$

admits a multiple root α. We therefore have $f(r) \equiv (r-\alpha)^p P(r)$, if p is a polynomial of degree $n-p$. The equation

$$(r-\alpha)(r-\alpha-t) \ldots (r-\alpha-t(p-1))P(r) = 0 \tag{6}$$

admits as its limiting equation, when the parameter t tends to zero, the equation $f(r) = 0$. This equation (6) is the characteristic equation of an equation

$$y^{(n)} + a_1(t) y^{(n-1)} + \ldots + a_n(t) y = 0 \quad , \tag{7}$$

whose coefficients, which are those of equation (6), are polynomials in t. These polynomials tend to the coefficients $a_1, \ldots, a_n$ of equation (5) when t tends to zero. The equation (7) admits as solutions, the functions

$$e^{\alpha x}, e^{(\alpha+t)x}, \ldots, e^{(\alpha+(p-1)t)x} ,$$

hence also the functions

$$e^{\alpha x} \frac{e^{tx}-1}{t}, \quad e^{\alpha x}\left(\frac{e^{tx}-1}{t}\right)^2, \ldots, e^{\alpha x}\left(\frac{e^{tx}-1}{t}\right)^{p-1} \tag{8}$$

which are linear combinations of it. When t tends to zero, these functions have as their respective limits

$$e^{\alpha x}x, e^{\alpha x}x^2, \ldots, e^{\alpha x}x^{p-1} .$$

These limiting functions are solutions of equation (5). This is more or less the procedure outlined by d'Alembert to obtain new solutions of equation (5), in this case. To make this more rigorous, it suffices to apply Poincaré's theorem relating to equations containing parameters.

The parameter t enters analytically into equation (7); the solutions (8), along with their first $n-1$ derivatives, take, for $x = 0$ say, values which are analytic functions of t, for

$$\frac{e^{tx}-1}{t} = x + \frac{tx^2}{2} + \ldots$$

They thus provide for $t = 0$, the solutions of equation (5).

92 - Euler's equations

These are equations of the form

$$x^n y^{(n)} + a_1 x^{n-1} y^{(n-1)} + \ldots. + a_{n-1} xy' + a_n y = 0 \quad ; \tag{9}$$

the coefficients $a_1, a_2, \ldots, a_n$ are constants. Homogeneous in x and dx, they amount to linear equations with constant coefficients if the variable $t = \log x$ is adopted, which will assume that x does not take the value 0 and does not turn about the origin. We can assume that the argument of x remains between $-\pi$ and $+\pi$; then the correspondence between t and x is bijective and t remains in the strip $|\mathfrak{I}t| < \pi$. But we could also assume that x is taken on the Riemann surface on which $\log x$ is uniform (I, 176). We will have

$$x\frac{dy}{dx} = \frac{dy}{dt}, \quad x^2\frac{d^2y}{dx^2} + x\frac{dy}{dx} = \frac{d^2y}{dt^2}, \ldots$$

which shows that the transformed equation is linear and with constant coefficients, e.g.,

$$\frac{d^n y}{dt^n} + b_1 \frac{d^{n-1} y}{dt^{n-1}} + \ldots + b_{n-1} \frac{dy}{dt} + b_n y = 0 \quad . \tag{10}$$

The fundamental solutions of this equation in t, are of the form e^{rt} if r is a simple root of the characteristic equation and $e^{rt}t^m$, $m = 0,1,\ldots,q-1$ if r is a root of order q of this equation. On returning to x, we see that the solutions of equation (9) are linear homogeneous equations with constant coefficients of solutions of the form x^r and $x^r(\log x)^m$, where it is understood that, if r is arbitrary, x^r signifies $e^{r\log x}$. These particular solutions will, therefore, in general, be multiform functions with an infinity of branches admitting the origin and the point at infinity as critical logarithmic points. However, if r is real and rational, non-integral, we will have as the corresponding solutions x^r, critical algebraic points and at infinity. Finally, if r is an integer, x^r is uniform and admits the origin on the point at infinity, as a pole. If equation (10) admits a characteristic equation having a multiple root, then there will be at least one logarithmic term, in a solution of equation (9): *the general solution of (9) is then multiform with an infinity of branches, except for certain values of the constants of integration.* We also see that:
In order that all the integrals of equation (9) are uniform, it is necessary and sufficient that the characteristic equation of equation (10) has all its roots real, integral and distinct.

To determine the numbers r, we can make a direct calculation without passing through the intervening t. We seek solutions of the form x^r. We have

$$y = x^r \; , \quad y^{(p)} = r(r-1)\ldots.(r-p+1)x^{r-p} \quad ,$$

and equation (9) is satisifed if

$$r(r-1)\ldots(r-n+1) + a_1 r(r-1)\ldots(r-n+2) + \ldots + a_{n-1}r + a_n = 0 \quad .$$

This equation, known as the *determinantal equation,* is clearly identical to the characteristic equation of the transformed equation (9). If r is a multiple root of order q of the determinantal equation, then

$$x^r, \; x^r \log x, \ldots, x^r(\log x)^{q-1}$$

are solutions.

THE CASE OF EQUATIONS WITH A SECOND MEMBER

If equation (9) has a second member, then we can determine a particular integral by the method of variation of constants. But if this second member is a sum of terms of the form $x^k(\log x)^h$, where k is a constant and h a positive or zero integer, then we see that the transformation in t will have a second member of the form $e^{kt}t^h$, hence a solution of the form $e^{kt}P(t)$, where P is a polynomial. Consequently, Euler's equation will then have a particular solution of the form $x^k(\log x)^q P(\log x)$, where P is a polynomial of degree h and q is the order of multiplicity of the determinantal equation if k satisfies this equation, and zero otherwise.

A REMARK ON THE REAL CASE

If the Euler equation (9) has real coefficients and if x is taken to be real, then the particular solutions corresponding to the imaginary roots $\alpha + i\beta$ of the determinantal equation will be of the form

$$x^{\alpha+i\beta}(\log x)^h = x^\alpha(\cos(\beta \log x) + i\sin(\beta \log x))(\log x)^h$$

if x is positive. If x is negative, it will be necessary to substitute $|x|$ for x in the second member. We can assign the corresponding solutions to the numbers $\alpha + i\beta$ for writing the particular real solutions, as we had done for the equation with constant coefficients (no. 79).

It will be necessary to note that when the real point of view is taken, there will not, in general, be any link between the solutions corresponding to $x > 0$ and those corresponding to $x < 0$.

Examples. The equation $x^3y'' - 3xy' + 3y = 0$ admits the general uniform integral

$$y = \frac{C}{x} + C'x + C''x^3 \quad ,$$

where C, C' and C" are constants. If we put x in the second member, in place of zero, then we have a particular solution which is $-\frac{1}{4}x \log x$. The general integral is then multiform. The equation $x^3y''' - xy' + y = 0$ admits, as a general integral,

$$y = Cx + C'x^{1+\sqrt{2}} + C''x^{1-\sqrt{2}}$$

which is a function that has an infinity of branches, except when $C' = C'' = 0$.

II. HOMOGENEOUS EQUATIONS WITH UNIFORM COEFFICIENTS

93 - The permutation of the solutions about isolated singularities

We shall take the equation in the form

$$y^{(n)} + a_1(x)y^{(n-1)} + \dots + a_{n-1}(x)y' + a_n(x)y = 0 \tag{11}$$

and we assume that the coefficients $a_j(x)$ are uniform in a domain D where their singularities are isolated; these are therefore poles on isolated singular points. We also assume that D does not contain the point at infinity. In the case where the coefficients are holomorphic about the point at infinity, we look at the solutions about this point and take it to be of finite distance by means of the transformation $x = 1/X$.

We intend to find the form of the solutions in the neighbourhood of the isolated singularities situated in D. It will clearly suffice to determine it for each of them. We consider then a circle C having as its centre an isolated singularity 0 of the coefficients and in which and outside of which these coefficients are holomorphic. To simplify matters, we shall assume that the centre 0 of the circle has been taken to be the origin. Thus $a_1(x)$, $\ldots, a_n(x)$ are holomorphic for $0 < |x| < R$, but at least one of those functions effectively admits $x = 0$ as a pole or an essential point.

Let us imagine that a radius Γ is drawn in C; the circle with Γ removed will be a simply connected region Δ (fig. 37).

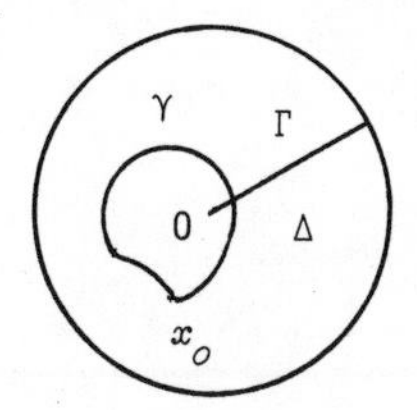

Fig. 37

Following the monodromy theorem (no. 1), any solution of the equation will be holomorphic in Δ since it can be analytically continued throughout the region. Let $y_1, y_2, \ldots, y_n$ be a fundamental system of holomorphic solutions in Δ. Let us start from a point x_o of Δ, with the solution y_1 and extend it along a simple closed curve around the origin 0, and hence intersecting Γ. When this course has been completed, we will return to the point x_o with a solution Y_1. This solution will be expressed in terms of y_1, y_2, $\ldots, y_n$. We will have

$$Y_1 \equiv \mu_{1,1} y_1 + \mu_{1,2} y_2 + \ldots + \mu_{1,n} y_n \quad ,$$

where the coefficients μ are constants. This identity will still hold if $y_1, y_2, \ldots, y_n$ and Y_1 are simultaneously extended from x_o arbitrarily; this shows that *the coefficients* μ *neither depend on* x_o, Γ, *nor* γ, *but uniquely on the fact that, on* γ, *a circuit about* 0 *has been once made, in a sense determined positively say*. Similarly, the rotation about in the positive sense, will change y_j and Y_j, and we will have

$$Y_j \equiv \mu_{j,1}y_1 + \mu_{j,2}{}_2 + \ldots + \mu_{j,n}y_n \qquad j = 2,3,\ldots,n \quad .$$

The solutions $Y_1, Y_2, \ldots, Y_n$ again form a fundamental system; their Wronskian is the product of the Wronskian of the y_j and the determinant of the $\mu_{j,k}$ as can easily be seen on writing this down. It follows that *the determinant of the* $\mu_{j,k}$ *is non-zero.*

Knowing the $\mu_{j,k}$ indicates how the solutions are permuted in the rotations about 0. We are going to look for a choice of the particular solutions y_j for which their permutations are particularly simple. A solution

$$\equiv \lambda_1 y_1 + \lambda_2 y_2 + \ldots + \lambda_n y_n$$

where the λ_j are constants, will be changed by a rotation about 0, in the direct sense, to

$$Z \equiv \lambda_1 Y_1 + {}_2 Y_2 + \ldots {}_n Y_n$$

$$\equiv \sum_1^n \lambda_j(\mu_{j,1}y_1 + \ldots + \mu_{j,n}y_n) \quad .$$

We will have

$$Z \equiv sz \equiv s(\lambda_1 y_1 + \ldots + \lambda_n y_n) \quad ,$$

where s is a constant, if the λ_j satisfy the system of equations

$$\lambda_1\mu_{1,k} + \lambda_2\mu_{2,k} + \ldots + \lambda_n\mu_{n,k} = s\lambda_k \, , \qquad k = 1,2,\ldots,n \quad ,$$

which may be written as

$$\left\{\begin{array}{l} \lambda_1(\mu_{1,1} - s) + \lambda_2\mu_{2,1} + \ldots + \lambda_n\mu_{n,} \\ \lambda_1\mu_{1,2} + \lambda_2(\mu_{2,2} - s) + \ldots + \lambda_n\mu_{n,2} = 0 \\ \lambda_1\mu_{1,n} + \lambda_2\mu_{2,} + \ldots + \lambda_n(\mu_{n,n} - s) = 0 \quad . \end{array}\right. \tag{12}$$

This system is compatible at least one system of λ_j, not all zero, if the equation in s

$$\Phi(s) \equiv \begin{vmatrix} \mu_{1,1} - s & \mu_{2,1} & & \mu_{n,1} \\ \mu_{1,2} & \mu_{2,2} - s & \ldots & \mu_{n,2} \\ \mu_{1,n} & \mu_{2,n} & & \mu_{n,n-s} \end{vmatrix} = 0 \tag{13}$$

is satisfied.

Let us assume that the equation in s *all its roots distinct.* To each there corresponds a solution z: if $s_1, s_2, \ldots, s_n$ are these roots, then we have n solutions $z_1, z_2, \ldots, z_n$ such that, by rotation about 0 in the direct sense, we have

$$Z_j = s_j z_j \qquad j = 1,2,\ldots,n \quad . \tag{14}$$

These solutions $z_1, \ldots, z_n$ *form a fundamental system.* Otherwise, there would exist constants v_j such that

$$\sum_1^n v_j z_j \equiv 0 \quad .$$

Following the permutation formula (14), we would have, after p rotations about the origin

$$\sum_1^n v_j s_j^p z_j \equiv 0 \qquad p = 1,2,\ldots,n-1 \quad .$$

Now these identities are impossible since the z_j are all different from zero at a suitably chosen point, and the determinant of the system of these equations in $v_j z_j$ is the Vandermonde determinant of the s_j.

THE FORM OF THE SOLUTIONS

In the case where the equation in s, $\Phi(s) = 0$, *has all its roots distinct, there exists a fundamental system of solutions* $z_1, \ldots, z_n$ *for which the permutations are defined by the identity (14).*

Let us determine the form of a function z admitting, in the cricle $|x| < R$, the origin as its only singular point and such that when a rotation about 0 is once made in the direct sense, then z changes to sz. The function

$$x^r \ , \quad r = \frac{\log s}{2i\pi}$$

has this property, whatever branch is considered, for when the argument of x increased by 2π, $r \log x$ becomes $r \log x + \log s$.

The ratio $z : x^r$ will not change by rotation about the origin, this will be a uniform function in the circle C, hence holomorphic in this circle, except at perhaps the point 0, which could be a pole or an essential point. It follows that z is of the form $x^r \phi(x)$, where ϕ has $x = 0$ as an essential point or is even holomorphic at this point. [For if $\phi(x)$ has a pole at the origin, then $\phi(x)$ will be the product of a holomorphic function and x^{-q} and we could replace r by $r-q$.] Consequently:
When the roots of the equation $\Phi(s) = 0$ *are distinct, there exists a fundamental system of solutions which are of the form*

$$x^{r_j}\phi_j(x) \qquad j = 1,2,\ldots,n \quad .$$

Each function $\phi_j(x)$ *is holomorphic in the circle* C *or may admit the origin as a singular isolated point.*

Definition. When the function $\phi_j(x)$ is holomorphic at the point 0, we say that the corresponding integral is *regular* at this point, if the point 0 is an essential point for $\phi_j(x)$, then the integral is said to be *irregular*.

94 - The case where the equation in s has multiple roots

We are going to prove that, in this case, we can find a fundamental system of solutions z_j *formed by groups of* p *solutions which, after a rotation about the origin in the positive sense, are transformed as follows:*

$$Z_1 = sz_1 \ , \quad Z_2 = s(z_1 + z_2) \ , \quad Z_3 = s(z_2 + z_3) \ , \ \ldots,$$

$$Z_p = s(z_{p-1} + z_p) \quad .$$

We shall say that such a group is a canonical group. A similar root s *of equation* $\Phi(s) = 0$ *can provide several groups.* In the case where $\Phi(s) = 0$ has its roots distinct, we have n groups with a single term.

The substitutions of the solutions are provided by the formulae

$$Y_j = \mu_{j,1}y_1 + \mu_{j,2}y_2 + \ldots + \mu_{j,n}y_n \ , \qquad j = 1,2,\ldots,n \ , \tag{15}$$

where the determinant of the $\mu_{j,k}$ is non-zero. The problem of the solution no longer involves the differential equation; this is a problem of reduction of the system of substitutions defined by (15), where we can regard the y_j as independent variables. When a homogeneous linear combination of the y_j is made, then the same combination is made for the Y_j, and it is necessary to choose these combinations in such a way as to form canonical groups. We shall pursue the proof by recurrence as given by Jordan. The theorem is true for a single variable (this is clear); let us show that, if it is true for $n - 1$, then it is true for n. We form the equation $\Phi(s) = 0$ and use one of the roots to give to the system, the substitutions of the form

$$Z_1 = s_1z_1 \quad , \tag{16}$$

$$Y_j = v_{j,1}z_1 + v_{j,2}y_2 + v_{j,3}y_3 + \ldots + v_{j,n}y_n \quad , \qquad j = 2,3,\ldots,n \quad . \tag{17}$$

As the total determinant of these substitutions (16) (17) is different from zero and is the product of s_1 and the determinant of the $v_{j,k}$,

$j = 2,3,\ldots,n$, $k = 2,\ldots,n$. The substitutions defined by those $v_{j,k}$

$$Y_j' = v_{j,2}y_2' + v_{j,3}y_3' + \ldots + v_{j,n}y_n', \quad j = 2,3,\ldots,n \quad , \tag{18}$$

can be reduced to substitutions formed by canonical groups since there are only $n - 1$ variables. We shall set

$$u_j' = \rho_{j,2}y_2' + \rho_{j,3}y_3' + \ldots + \rho_{j,n}y_n' \quad , \quad j = 2,3,\ldots,n \quad ;$$

the determinant of the $\rho_{j,k}$ is non-zero and the substitutions (18) will reduce to groups of the form

$$U_2' = su_2', \; U_3' = s(u_2' + u_3'), \ldots, \; U_p' = s(u_{p-1}' + u_p') \quad .$$

Having chosen the $\rho_{j,k}$, let us make the transformation

$$u_1 = z_1$$

$$u_j = \rho_{j,2}y_2 + \ldots + \rho_{j,n}y_n \quad , \quad j = 2,3,\ldots,n$$

on (16) and (17).

We will obtain new substitutions formed by the groups

$$U_1 = s_1u_1 \; ; \; U_2 = su_2 + \tau_2u_1 \; , \; U_3 = s(u_2 + u_3) + \tau u_1, \ldots,$$

$$U_p = s(u_{p-1} + u_p) + \tau_pu_1; \; \ldots$$

the other groups having the form of the second group which runs from U_2 to U_p. Let us form the new combinations with constant coefficients η

$$v_1 = u_1; \; v_2 = u_2 + \eta_2u_1 \; , \; v_3 = u_3 + \eta_3u_1, \ldots, \; v_p = u_p + \eta_pu_1; \; \ldots$$

which will provide the new system

$$V_1 = s_1v_1$$

$$V_2 = sv_2 + v_1(\tau_2 + \eta_2(s_1 - s)), \quad V_3 = s(v_2 + v_3)$$

$$+ v_1(\tau_3 + \eta_3(s_1 - s) - s\eta_2), \ldots$$

$$V_p = s(v_{p-1} + v_p) + v_1(\tau_p + \eta_p(s_1 - s) - s\eta_{p-1}), \ldots \; ;$$

the resulting substitutions form groups similar to the second, from V_2 to V_p. If $s - s_1 \neq 0$, then we can determine, step by step, the constant $\eta_2, \eta_3, \ldots, \eta_p$ in such a way as to make zero the coefficient of v_1 in the second members of the equations giving $V_2,\ldots,V_p$; the group is then canonical. If $s = s_1$, we can take $\eta_2,\eta_3,\ldots,\eta_{p-1}$ in such a way as to make zero the term in v_1 in the second member of $V_3,V_4,\ldots,V_p$; if this coefficient of v_1 is also zero in V_2, the group is canonical. We thus proceed in this way for each group. We will therefore obtain the canonical groups and there will remain with $V_1 = s_1 v_1$ a certain number of groups not yet reduced, for which $s = s_1$. Let us suppose there are two. We have then the substitutions

$$V_1 = sv_1 \quad (s = s_1)$$

$$V_2 = sv_2 + \varepsilon_1 v_1,\ V_3 = s(v_2 + v_3),\ldots,\ V_p = s(v_{p-1} + v_p),$$

$$V_{p+1} = sv_{p+1} + \varepsilon_2 v_1,\ \ V_{p+2} = s(v_{p+1} + v_{p+2}),\ldots,$$

$$V_{p+q} = s(v_{p+q-1} + v_{p+q})$$

starting from V_{p+q+1}; we obtain canonical groups. ε_1 and ε_2 are non-zero constants; it is permitted to assume that $p - 1 \geq q$. We form the combinations

$$w_1 = \frac{\varepsilon_1}{s} v_1 \ ; \ \ w_j \quad v_j \ , \qquad j = 2,\ldots,p,p+q+1,\ldots,n$$

$$w_j = v_j - \frac{\varepsilon_2}{\varepsilon_1} v_{j+1-p} \ , \qquad j = p+1,p+2,\ldots,p+q$$

which will yield,

$$W_1 = sw_1,\ W_2 = s(w_1 + w_2),\ W_3 = s(w_2 + w_3),\ldots,$$

$$W_p = s(w_{p-1} + w_p),\ \ W_{p+1} = sw_{p+1} \quad ,$$

$$W_{p+2} = s(w_{p+1} + w_{p+2}),\ldots,\ W_{p+q} = s(w_{p+q-1} + w_{p+q})$$

for which the first $p+q$ substitutions are the only ones changed.

All the groups are canonical. The solutions so obtained form a fundamental system, for all the successive transformations have been made by substitutions with non-zero determinant. The theorem is thus proved.

THE FORM OF THE SOLUTIONS

There exist solutions of equation (11) which may, in the circle C, divide themselves into groups such as $z_1,z_2,\ldots,z_p$, where for each of which, a rotation about the origin in the positive sense yields the substitutions

$$Z_1 = sz_1,\ \ Z_2 = s(z_1 + z_2),\ldots,\ Z_p = s(z_{p-1} + z_p) \ ; \qquad (19)$$

the set of these solutions provide a fundamental system. We have already seen that a function which is multiplied by s, is of the form $x^r\phi(x)$, where $\phi(x)$ is holomorphic at the origin or admits this point as an essential, isolated singular point. We have then

$$z_1 = x^r\phi(x) \quad . \tag{20}$$

To determine generally the z_q, we shall first of all set $z_q = x^r u_q$, where r is equal to $1/2\pi i$ logs. The substitutions (19) will become

$$U_1 = u_1,\ U_2 = u_1 + u_2, \ldots,\ U_p = u_{p-1} + u_p \quad .$$

The first equation signifies that u_1 is uniform and yields $u_1 = \phi_1(x)$, where the function $\phi_1(x)$ is holomorphic for $0 < |x| < R$. To obtain the solution of

$$U_q = u_q + u_{q-1} \quad ,$$

it suffices to know a solution v_q, hence $V_q = v_q + u_{q-1}$, and on curtailing, we see that $u_q = v_q + \phi_q(x)$, where $\phi_q(x)$ is holomorphic for $0 < |x| < R$. Let us consider first of all $U_2 = u_2 + u_1$. We know that $\log x/2\pi i$ increase by one when x turns about the origin in the direct sense and returns to the point of departure. Hence the product of this function and u_1, plus u_1, is a solution of $U_3 = u_2 + u_1$ and the general solution is

$$U_2 = \frac{\log x}{2\pi i}\phi_1(x) + \phi_2(x) \quad .$$

To solve $U_3 = u_3 + u_2$, it is necessary to find a function which increases by u_2 under the rotation of 2π about the origin. If we set $T = \frac{\log x}{2\pi i}$ to simplify matters, then $T\phi_2(x)$ will increase by $\phi_2(x)$; it will be necessary to find a function which increases by T and then multiply it by $\phi_1(x)$. We see that $1/2\ T(T-1)$ responds to our request.

We will have then a solution $\phi_1(x)\frac{T(T-1)}{2} + \phi_2(x)T$ and

$$u_3 = \frac{T(T-1)}{2}\phi_1(x) + T\phi_2(x) + \phi_3(x)$$

$$T = \frac{\log x}{2\pi i} \quad ,$$

where $\phi_3(x)$ is holomorphic for $0 < |x| < R$. We must then find a function $\psi(T)$ such that $\psi(T+1) - \psi(T) \equiv \frac{T(T-1)}{2}$. It is clearly a third degree polynomial which is determined by the method of indeterminate coefficients. Let us call it $P_3(T)$. We must then have a polynomial $P_4(T)$ such that $P_4(T+1) - P_4(T) \equiv P_3(T)$. Now $P_o(T) = 1$, $P_1(T) = T$, $P_2(T) = \frac{T(T-1)}{2}$. More generally, $P_n(T)$ will be integral for T integral equal to m for

m sufficiently large, for the coefficients of $P_n(T)$ are rational, and we will need to have $P_n(m+1) - P_n(m) \equiv P_{n-1}(m)$. This is the known relation

$$C^n_{m+1} = C^n_m + C^{n-1}_m \quad .$$

We necessarily have

$$P_n(T) \equiv \frac{T(T-1),\ldots,(T-n+1)}{1,2,\ldots,n} \tag{21}$$

and

$$u_4 = P_3(T)\phi_1(x) + P_2(T)\phi_2(x) + P_1(T)\phi_3(x) + \phi_4(x)$$

$$u_p = P_{p-1}(T)\phi_1(x) + P_{p-2}(T)\phi_2(x) + \ldots + P_1(T)\phi_{p-1}(x) + \phi_p(x) \quad ,$$

where the functions $\phi_q(x)$ are holomorphic for $0 < |x| < R$. The values of z_q, $q = 1,2,\ldots,p$ are obtained by multiplying the u_q by x^r. If some of the functions have a pole at the origin, than on replacing r by a number $r-h$ where h is a suitably chosen integer, we can replace these $\phi_q(x)$ by functions $x^h\phi_q(x)$ which are holomorphic at the origin. The definitive result so obtained has the following form: Each group of solutions *undergoing the permutations defined by the relationship in (19), is given by equations of the form*

$$\begin{aligned} z_1 &= x^r\phi_1(x), \quad z_2 = x^r\phi_1(x)\,\frac{\log x}{2\pi i} + x^r\phi_2(x),\ldots, \\ z_p &= x^r\phi_1(x)P_{p-1}\left(\frac{\log x}{2\pi i}\right) + x^r\phi_2(x)P_{p-2}\left(\frac{\log x}{2\pi i}\right) + \ldots + x^r\phi_p(x) \quad , \end{aligned} \tag{22}$$

where $P_n(T)$ *is the binomial polynomial (21), and each function* $\phi_q(x)$ *is holomorphic for* $|x| < R$ *or even for* $0 < |x| < R$, *where the origin is an essential point. The number* r *depends on the group in question; it could be the same for several groups.*

Definitions. An integral is said to be *regular* (at the origin) when the functions $\phi_q(x)$ which enter into it are holomorphic at the origin.

If all the functions $\phi_q(x)$ relative to various groups are holomorphic at the origin, then the differential equation (11) is said to be *regular at the origin.*

BIRKHOFF'S REGULARITY CRITERION

Let us consider a solution z that is regular at the origin. It is multiform in general and has an infinity of branches. Let us take one of these branches for the x whose argument belongs to the segment $(0,2\pi)$ and whose module is less than R. z is a sum of terms of the form (22).

For these terms, the $|\phi_q(x)|$ are uniformly bounded for $|x| \leq R < R$. The argument of x varies from $\lambda = 2n\pi$, n an integer, to $\lambda + 2\pi$, hence $|\log x| \leq |\log |x|| + h$, where h is a fixed number, $h = |\lambda| + 2\pi$. Similarly, if r' is the real part of r, and r'' the coefficient of i, then

$$|x^r| = |x|^{r'} e^{-r''}\psi \leq |x|^{r'} e^{h'} ,$$

where h' is fixed and ψ is the argument for x. It follows that:
For z regular at the origin, to every branch of z corresponding to an argument of x varying from 0 to 2π, there corresponds two members M and k positive, such that

$$|z| < M|x|^{-k} \quad . \tag{23}$$

(Where M and k depend on z and the branch chosen.)

Conversely, let us assume that *every* solution of equation (11) whose coefficients are holomorphic for $0 < |x| < R$, but at least one of the coefficients is not holomorphic at the origin, satisfies a condition of the form (23) when x has an argument describing the segment $(0,2\pi)$, and that we take *one* of the branches of z. *Under these conditions, the differential equation is regular at the origin.*

Once satisfied for z_1, this condition is also satisfied for $z_1 x^{-r}$, as can be easily seen. The function $\phi_1(x)$ *which is holomorphic about the origin*, thus satisfies uniformly the condition (23) for $|x| < R$. Following Liouville's theorem (I, 191), the origin is not an essential point for this function. Hence $\phi_1(x)$ is holomorphic at the origin, and z_1 is *regular*. Let us pass on to z_2. One of the branches of z_2 satisfies condition (23) for all x whose argument belongs to the segment $(0,2\pi)$ but the first term of the second member of z_2 also satisfies a similar condition, hence also the second term $x^r\phi_2$, and what had been said now applies and $\phi_2(x)$ is holomorphic at the origin. In the same way, we see that $\phi_3,\dots,\phi_p$ are holomorphic at the origin. Hence:
The necessary and sufficient condition for the equation (11) to be regular at the origin, taking z to be any of its solutions, is that if we take a branch of z corresponding to the x whose argument belongs to the segment $(0,2\pi)$ and whose modulus is less than R_o, then there exists two positive numbers M and k such that this branch satisfies condition (23).

95 - Fuch's theorem

Theorem. *The necessary and sufficient condition for the equation*

$$y^{(n)} + a_1(x) + \dots + a_{n-1}(x)y' + a_n(x) = 0 \tag{24}$$

to be regular at the point x_o, is that the functions $a_q(x)$, $q = 1,2,\dots,n,$

admit this point as a pole; the order of multiplicity of this pole being equal to its index q *at most, for each* $a_q(x)$.

To say that the equation is regular at the point x_o, is to say that if this point is taken to be the origin, then the equation is regular at the origin. Hence this means that we can find a fundamental system of solutions $y_1, y_2, \ldots, y_n$ which divide into groups $y_1, y_2, \ldots, y_1$, for each of which we have

$$y_1 = (x - x_o)^r \phi_1(x) \ , \quad z_2 = (x - x_o) \ \frac{\log(x - x_o)}{2\pi i} \ \phi_1(x)$$

$$+ (x - x_o)^r \phi_2(x), \ldots$$

$$y_p = (x - x_o)^r \left\{ \phi_1(x) P_{p-1}\left(\frac{\log(x - x_o)}{2\pi i}\right) + \phi_2(x) P_{p-2} \left(\frac{\log(x - x_o)}{2\pi i}\right) + \ldots + \phi_p(x) \right\} ,$$

where the functions $\phi_q(x)$ are holomorphic at the point x_o; r depends on the group as do the $\phi_q(x)$.

To establish the theorem, we now assume that $x_o = 0$.

Let us prove first of all that if the equation is regular, then the $a_q(x)$ only have poles at most at the origin. These functions are expressed in terms of the determinants formed by the solutions y_q and their derivatives (no. 77); $a_q(x)$ is the quotient of the minor of $y^{(n-q)q}$ in $\Delta(y_1, y_2, \ldots, y_n)$ by $\Delta(y_1, y_2, \ldots, y_n)$. Now $\Delta(y_1, \ldots, y_n)$ is a function that is holomorphic at all points $0 < |x| < R$. A rotation about the origin subjects the $y_1, y_2, \ldots, y_n$ to substitutions of the various canonical groups whose determinant is the product of the corresponding factors s^p. $\Delta(y_1, y_2, \ldots, y_n)$ is then multiplied by this determinant δ when a rotation of 2π is made about the origin. On the other hand, the derivatives of y_q are functions of the same form as y_q, being sums of products of the logarithm $(\log x)^k$ and the functions $x^r \phi_q^{(m)}(x)$. The product of such functions is again of the same form. $\Delta(y_1, y_2, \ldots, y_n)$ is therefore also of this form; it satisfies an inequality of the form (23) when x varies in a domain bounded by $|x| = R' < R$, and the segment $0 < x \leq R'$ of the real axis. It follows that, for $\Delta(y_1, \ldots, y_n)$ having the form $x^\lambda \psi(x)$, $\lambda = \frac{\log \delta}{2\pi i}$, where $\psi(x)$ is holomorphic about the origin; this function $\psi(x)$ does not admit the origin as an essential point. The same reasoning shows that the numerator of $a_q(x)$ is of the form $x^\lambda \psi_1(x)$ where ψ_1 is again holomorphic or meromorphic at the origin and consequently $a_q(x)$ has at most one pole at the origin.

Having established this point, we can show recursively that the condition in the statement of Fuch's theorem is necessary. It is the case for a first order equation $y' + a_1(x)y = 0$. In this case, a solution is of the form $y = Cx^r\phi(x)$, hence y'/y has a simple pole with residue r at the origin since we can assume that $\phi(o) \neq 0$. Let us suppose then, that the condition is necessary for an equation of order $n-1$. Having been given the n-th order equation (24), which by hypothesis is regular, then it has a solution of the form $y_1 = x^r\phi(x)$ where $\phi(x)$ is holomorphic and non-zero at the origin. The quotient of this solution by the other solutions forming the fundamental system composed of canonical groups, is again a fundamental system composed of canonical groups for the equation obtained by setting $y = y_1 Y$

$$\sum_{o}^{n} b_{n-q}(x)Y^{(q)} = 0 \quad , \tag{25}$$

say, where, following Leibniz's formula, we have

$$b_{n-q}(x) = a_{n-q}(x)y_1 + (q+1)a_{n-q-1}(x)y_1' + \dots + C_n^q y_1^{(n-q)} \tag{26}$$

$$y = 0,1,\dots,n \quad .$$

This equation (25), which admits the solution 1 has no term in Y, which indeed proves equation (26). This is an equation in $Y' = Z$ which can be written, on dividing both members by y_1, as

$$Z^{(n-1)} + B_1 Z^{(n-z)} + \dots + B_{n-1}Z = 0 \quad ,$$

with

$$B_q \equiv \frac{b_q(x)}{y_1} \equiv q_q(x) + (n-q+1)a_{q-1}\frac{y_1'}{y_1} + \dots + C_n^{n-q}\frac{y_1^{(q)}}{y_1} , \tag{27}$$

$$q = 1,2,\dots,n-1 \quad .$$

This differential equation in Z, of order $n-1$, admits as a fundamental system of solutions, the derivatives of y_j/y_1, $j = 2,\dots,n$. These derivatives, which are sums of derivatives of expressions of the form

$$x^8 \frac{\phi_q}{\phi_1} (\log x)^m$$

where $\phi_1(o) \neq 0$ and $\phi_q(x)$ is holomorphic at the origin, are again regular at the origin. The theorem therefore applies to the equation in Z; B_q has at most one pole of order q at the origin and consequently, after (27), $a_q(x)$ has also only a pole of order q at most, at the origin. *The condition is necessary.*

To show that it is sufficient, we shall appeal to this lemma.

Lemma. *If the function* $u(x)$ *is holomorphic in a neighbourhood of the point* x, $x \neq 0$, *and if we set* $|x| = \rho$, *then we obtain*

$$|u'(x)| = \left|\frac{du}{dx}\right| \geq \left|\frac{\partial|u(x)|}{\partial\rho}\right| .$$

To calculate $u'(x)$, we could give x an increment Δx whose argument is that of x or $-x$, then $|\Delta x| = |\Delta\rho|$. Now

$$|\Delta u| = |u(x + \Delta x) - u(x)| \geq ||u(x + \Delta x)| - |u(x)|| ,$$

hence when $\Delta x \to 0$,

$$\left|\frac{du}{dx}\right| = \lim \left|\frac{\Delta u}{\Delta x}\right| \geq \lim \left|\frac{\Delta|u(x)|}{\Delta(\rho)}\right| = \left|\frac{\partial|u(x)|}{\partial(\rho)}\right| .$$

We intend replacing the differential equation (24), (where we assume that the $a_q(x)$ are holomorphic about the origin, which is a pole of order q at most), by a differential system. Let us first of all write (24) in the form

$$y^{(n)}x^n + y^{(n-1)}x^{n-1}A_1(x) + \dots y'xA_{n-1}(x) + yA_n(x) = 0 .$$

The $A_j(x)$ will be holomorphic at the origin, hence $|A_j(x)| < M'$ for $j = 1,2,\dots,n$ if $|x| \leq R' < R$. Let us put $y = z_1$ and

$$xz_1' = z_2, \quad xz_2' = z_3, \dots, xz_{n-1}' = z_n ; \tag{28}$$

the equation will take the form

$$xz_n' = z_1D_1(x) + z_2D_2(x) + \dots + z_nD_n(x) \tag{29}$$

where the $D_j(x)$ are linear combinations with constant coefficients of the $A_j(x)$. We will also have $|D_j(x)| < M$, $j = 1,2,\dots,$ and we can assume that M is greater than one. From the system (28, (29), we deduce, *a fortiori*, that

$$|xz_n'| < M(|z_1| + |z_2| + \dots + |z_n|) ,$$

$$|xz_j'| < M|z_{j+1}| , \qquad j = 1,2,\dots,n-1 ,$$

and consequently if $|x| = \rho$ and $|z_1| + \dots + |z_n| = U(x)$,

$$\rho(|\ _1'| + |z_2'| + \dots + |z_n'|) < 2M\ U(x) .$$

On applying the lemma, we will have, assuming the argument of x is constant,

$$\rho \left|\frac{dU(x)}{d\rho}\right| < 2MU(x)$$

where if $d\rho > 0$

$$\frac{-2Md\rho}{\rho} \leq \frac{dU(x)}{U(x)} \leq \frac{2Md\rho}{\rho} \quad .$$

Let us integrate from $\rho = |x|$ to $|x_o| = \rho_o$, $\rho < \rho_o$, along the radius. We obtain

$$2M \log \frac{\rho}{\rho_o} < \log \frac{U(x_o)}{U(x)} < 2M \log \frac{\rho_o}{\rho} \quad .$$

The first part of this inequality yields, since $|y| \leq |U(x)|$,

$$|y|\,|x|^{2M} \leq U(x)|x|^{2M} < U(x_o)|x_o|^{2M} \quad . \tag{30}$$

Let us consider then a solution y of equation (24) and take x as describing a circle $|x| = |x_o|$. The argument of x varying on the segment $(0,2\pi)$, the moduli of y, $z_2 = zy'$, $z_3 = xz_2', \ldots, zn$, will have a finite bound K', and consequently, the second member of (30) will have, for all x_o having this modulus, a bound K. Birkhoff's regularity criterion will apply, and y will be regular at the origin. As this is the case for every solution, equation (24) will be regular at the origin. The condition of the statement of Fuch's theorem is sufficient and thus the theorem is proved completely.

96 - The determinant equation

Let us consider an equation (24) satisfying the conditions of Fuch's theorem, i.e., one that is regular at a singular point x_o. We will say that it is *of the Fuchsian type* at the point x_o. We shall say that the equation is *of the first type* if there exists a fundamental system of solutions of the form

$$y_j = (x - x_o)^{r_j} \phi_j(x) \, , \qquad j = 1,2,\ldots,n \quad ,$$

where the $\phi_j(x)$ are holomorphic at the point x_o. This is the case if the equation in s of no. 93 has all its roots simple. When the equation is not of the first type, it will be said to be of the *second type*.

The solutions of the equations of Fuchsian type can be calculated by its power series expansion. We shall always assume that x_o has been taken to be the origin and let us take the equation to be of the form

$$x^n y^{(n)} + x^{n-1} y^{(n-1)} A_1(x) + \dots + xy' A_{n-1}(x) + A_n(x) = 0 \quad . \tag{31}$$

The functions $A_q(x)$ which are holomorphic at the origin are given by the power series expansions

$$A_q(x) \equiv a_q + a_q' x + \dots + a_q^m x^m + \dots$$

convergent for $|x| < R$. A solution of the form

$$y = x^r \phi(x) \tag{32}$$

where $\phi(x)$ is holomorphic and non-zero at the origin (if it were zero at the origin, we would replace r by $r + 1$, or $r + 2, \dots$) can be expanded as a power series as

$$y = x^r(c_o + c_1 x + c_2 x^2 + \dots + c_m x^m + \dots)$$

or

$$y = c_o x^r + c_1 x^{r+1} + \dots + c_m x^{r+m} + \dots \quad .$$

The coefficients $c_o, c_1, \dots, c_m$ and the number r are determined by the fact that the $A_q(x)$ and y having been replaced by these expansions, the first member of (31) must be identically zero and the coefficients of the various powers of x must be zero. We see this by taking x^r as a factor. Let us order the first member by taking the increasing powers of $x: r, r+1, r+2, \dots$ (in fact, r is, in general, complex and we extend here the usual sense of word increasing). The coefficient of x^r must be zero; it contains c_o as a factor, which was anticipated. c_o *will be arbitrary and we can take it equal to* 1. We have then the condition

$$\begin{aligned} F(r) \equiv r(r-1)\dots(r - n + 1) + a_1 r(r-1)\dots(r - n + 2) + \dots \\ + r a_{n-1} + a_n = 0 \quad . \end{aligned} \tag{33}$$

This equation, $F(r) = 0$, *is the determinantal equation.* If the differential equation (31) is of the first Fuchsian type, there exists a fundamental system of solutions of the form (32). If two of these solutions correspond to the same value of r, then their difference contains at least x^{r+1} as a factor. We could then, step by step, see that the values of r corresponding to the solutions of the fundamental system, thus modified, are all different. Each must satisfy equation (33). Hence:

I. *In order that the equation is of the first Fuchsian type, it is necessary that the determinantl equation has all its roots distinct.*

Let us now suppose that the equation is of the second Fuchsian type. We will have two solutions of the form

$$y_1 \equiv x^r {}_1(x) , \quad y_2 \equiv x^r \frac{\log x}{2\pi i} \phi_1(x) + x^r \phi_2(x) \quad . \tag{34}$$

where r must be a root of the determinantal equation. It will be necessary, on the other hand, that y_2 or $2\, iy_2$, are solutions. Let us set

$$2\pi i y_2 = x^r \phi_1(x) \log x + x^r \psi(x) \equiv y_1 \log x + x^r \psi(x) \quad ,$$

$$\psi(x) \equiv d_\tau x^\tau + d_{\tau+1} x^{\tau+1} + \dots \quad \tau > 0, \ \ \tau < 0 \text{ or } \tau = 0 \quad .$$

The q-th order derivative of $2\pi i y_2$ is

$$y_1^{(q)} \log x + q y_1^{(q-1)} \frac{1}{x} + \frac{q(q-1)}{2} y_1^{(q-2)} \frac{-1}{x^2} + \dots$$

$$d_\tau (r+\tau)\dots(r+\tau-q+1) x^{r+\tau-q} + \dots \quad .$$

On putting in the first member of the equation, we see that the term in $\log x$ disappears; the term of the lowest degree in the terms arising from $y_1 \log x$ is then

$$x^r \sum_{q=0}^{n} a_{n-q} (qr(r-1)\dots(r-q+2) - \frac{q(q-1)}{2} r(r-1) \\ \dots(r-q+3) + \dots) \quad . \tag{35}$$

The coefficient of a_{n-q} is the quotient by x^r of the derivative of order q of $x^r \log x$ less the term in $\log x$. Now

$$\frac{d^q}{dx^q}(x^r \log x) = \frac{\partial^q}{\partial x^q} \frac{\partial}{dr}(x^r) = \frac{\partial}{\partial r} \frac{\partial^q}{\partial x^q}(x^r)$$

$$= \frac{\partial}{\partial r} [r(r-1)\dots(r-q+1\ x^r] \quad ;$$

the coefficient of a_{n-q} is hence the derivative with respect to r of

$$r(r-1)\dots(r-q+1)$$

and the expression (35) is equal to $x^r F'(r)$. On the other hand, the terms arising from $\psi(x)$ commence from

$$x^{r+\tau} d_\tau F(r+\tau) \quad .$$

We have therefore two cases to distinguish.

THE FIRST CASE

We have $\tau \geq 0$. If $\tau = 0$, $F(r) = 0$. The term of the lowest degree is, in all cases, $x^r F''(x)$; we have then $F'(r) = 0$, where r is a multiple root of the determinantal equation. The hypothesis $\phi_2(x) = 0$ enters into this case.

THE SECOND CASE

We have $\tau < 0$. The term of the lowest degree is that containing d_τ. We must then have $F(r+\tau) = 0$. It follows that:

II. *For the differential equation to be of the second Fuchsian type, it is necessary that the determinantal equation has a multiple root or has two roots whose difference is a real integer.*

From Propositions I and II, we see that:

III. *If the determinantal equation has all its roots distinct and if the difference of any two of its roots is not an integer, then the equation is of the first type.*

In this case, there are n solutions of the form (32) corresponding to the n roots of the determinantal equation. For each of them, the coefficients of the power series expansion is calculated step by step: if we take $c_o = 1$, we have, on equating to zero the coefficients of x^{r+1}, x^{r+2},...

$$c_1 F(r+1) = \Phi_1(a_q, a'_q)$$

$$c_2 F(r+2) = \Phi_2(a_q, a'_q, a^2_q)$$

$$\cdots\cdots$$

where the second members are linear functions of the coefficients of the $A_q(x)$. All the numbers $F(r+m)$ are different from zero. We know that the series obtained will converge for $|x| < R$, a circle in which the $A_q(x)$ has only the origin as a singular point. We will not pursue this general study any further in the case of equations of order n.

97 - <u>The necessary and sufficient condition for the equations to be of the first Fuchsian type. An application</u>

Let us assume that all the roots of the determinantal equation are distinct, but that a certain number of them have integral differences. Is it possible to have solutions with terms in $\log x$? We would have two solutions of the form (34) and, since the determinantal equation has its roots simple, the number τ defined above would be negative; $r+\tau$ would be a root of the determinantal equation. Let us assume that we have made the formal calculation of the coefficients of the expansion of $x^r\psi(x)$, where d_τ is arbitrary, since we can multiply y_2 by a constant which could be taken to be 1. We have then

$$d_{\tau+1}F(r+\tau+1) = \Phi_1(a_q, a_q^1) \quad ,$$

$$d_{\tau+1}F(r+\tau+2) = \Phi_2(a_q, a_q^1, a_q^2), \ldots$$

and the calculation is the same as if we have found a *solution* of the form $x^r\psi(x)$, once we have arrived at the term in x^r. But, for this term, the coefficient will be that providing the preceding terms of $\psi(x)$, e.g., $\Phi_{-\tau}(a_q, \ldots, a^{-\tau}q)$ plus $c_o F'(r)$, providing the term in $\log x$. We will have here

$$\Phi_{-\tau}(a_q, \ldots, a_q^{-\tau}) = c_o F'(r) \neq 0 \quad ,$$

as long as we have $\Phi_{-\tau}(a_q, \ldots, q_q^{-\tau}) = 0$, *if* $x^r\psi(x)$ *is a solution.* We will see then that a root r' of the determinantal equation that is equal to another root minus an integer, gives a solution for which the formal calculation of the coefficients of this solution is possible up to the term in x^r. Thus:

For the equation to be of the first type at the point x_o, *it is necessary and sufficient that the determinantal equation has all its roots distinct and that if a root* r' *is equal to* $r - m$, *where* r *is another root and* m *is a positive integer, then the calculation of the coefficients of the solution*

$$(x - x_o)^{r'} + g_{r'+1}(x - x_o)^{r'+1} + \ldots + g_r(x - x_o)^r + \ldots$$

is possible up to and including g_r.

Remark. It will be clearly possible for the coefficients g_j to be zero starting from a value less than r, or for some of these coefficients to be arbitrary if there are several roots whose pairwise differences are integers.

THE CONDITION FOR A GENERAL INTEGRAL TO BE MEROMORPHIC AT A SINGULAR POINT

In particular, for an integral of equation (24) to be meromorphic at a point x_o which is an isolated singular point of the coefficients and about which those coefficients are uniform, it is necessary and sufficient that the equation is of the first Fuchsian type, and that all the numbers r are real and integral. It will therefore be necessary for the determinantal equation to have all its roots distinct, real and integral, and that the formal expansions of the solutions formed from these various roots, are all possible up to the term in $(x - x_o)^r$, where r is the greatest root. This is the sufficiency.

THE CASE OF AN EQUATION OF THE SECOND ORDER

In this case, if the equation is not of the first type, we know its solution for which $\log(x - x_o)$ can only enter for degree 1. We have the following three cases:

1. The determinantal equation has two roots whose difference is not an integer. The equation is of the first type.

2. The determinantal equation has a double root. The equation is of the second type.

3. The determinantal equation has two distinct roots whose difference is an integer. We consider the series expansion from the "smallest" of these roots r, r',

$$(x - x_o)^r + c_1(x - x_o)^{r+1} + \dots$$

and we calculate the coefficients up to the term in $(x - x_o)^{r'}$. If the calculation is possible, then the equation is of the first type. If it is impossible to calculate this last term, then the equation is of the second type and it will be necessary to introduce the term in $\log(x - x_o)$.

THE POINT AT INFINITY

To study the behaviour about the point at infinity, we must make the transformation $x = 1/X$, as we have already seen.

III. APPLICATIONS

98 - Equations with two singularities

We can see at once that if we make a homographic transformation on x, with constant coefficients

$$x = \frac{\alpha X + \beta}{\gamma X + \delta} \quad , \qquad a\delta - \beta\gamma \neq 0$$

and if to x_o finite, there corresponds X_o finite, then we will have about these points

$$x - x_o = \eta(X - X_o) + \dots , \quad \eta \neq 0 \quad .$$

To a regular solution about x_o, there will correspond, in the transformed equation, a regular solution, and vice-versa. We have an analogous result if x_o finite corresponds to $X = \infty$, or to x_o infinite a finite value X_o. Moreover, in the case where we have obtained an equation of the first type, it will be the same after the transformation and the determinantal equation will not be changed.

To study the case of the differential equation with uniform coefficients only having two isolated singularities, we can assume that it has been reduced to the case where one of its singularities is at the origin and the other at the point at infinity. The equation will be of the form

$$x^n y^{(n)} + x^{n-1} y^{(n-1)} a_1(x) + \dots + xy' a_{n-1}(x) + y a_n(x) = 0 \quad ,$$

where the functions $a_q(x)$ are everywhere holomorphic, at a finite distance, except perhaps at the origin. If we require the equation to be of Fuchsian type at the origin, then it will suffice that the $a_q(x)$, $q = 1,2,\dots,n$ will again be holomorphic at the origin; these will be entire functions. The solutions will be of the form

$$x^r \phi(x) \; , \quad x^r \phi(x) \log x + x^{r'} \phi_1(x) \; ,\dots$$

where $\phi(x)$, $\phi_1(x)\dots$ are entire functions.

To express the equation at infinity as a Fuchsian type, we must set $x = 1/X$. We will have

$$\frac{dy}{dx} = \frac{-1}{x^2} \frac{dy}{dX} \, , \quad \frac{d^2y}{dx^2} = \frac{1}{x^4} \frac{d^2y}{dX^2} + \frac{2}{x^3} \frac{dy}{dX} \; ,\dots$$

$$\frac{d^py}{dx^p} = (-1)^p \frac{1}{x^{2p}} \frac{d^py}{dX^p} + (-1)^p \frac{d}{x^{2p-1}} \frac{d^{p-1}y}{dX^{p-1}} + \dots \quad ,$$

where dp is a positive constant. The transformed equation will be

$$X^n y^{(n)} - X^{n-1} y^{(n-1)} \left(a_1 \left(\frac{1}{X}\right) - d_n \right) + \dots + X^{n-q} y^{(n-q)} b_q \left(\frac{1}{X}\right) +$$

$$\dots + b_n \left(\frac{1}{X}\right) y = 0$$

where the $y^{(q)}$ are the derivatives with respect to X and where

$$b_q \left(\frac{1}{X}\right) = (-1)^q \left[a_q \left(\frac{1}{X}\right) - a_{q-1} \left(\frac{1}{X}\right) d_{n-q+1} + \dots \right] \quad .$$

For this equation to be regular, it is first of all necessary that $a_1 \left(\frac{1}{X}\right)$ is holomorphic at the origin. The entire function $a_1(x)$ must therefore be holomorphic at infinity; it is a constant. We see then that $b_2 \left(\frac{1}{X}\right)$ happens to be holomorphic at the origin, and since $a_1(x)$ is constant, $a_2(x)$ must be holomorphic at infinity, and hence constant, etc. All the coefficients $a_q(x)$ will be constants. Hence:

The only differential equations with uniform coefficients only admitting the origin and the point at infinity as singular points, and which are regular at these points, are the Euler equations.

Remark. It is important not to confuse the singularities of the differential equation and the singularities of its solutions. For example, the second order equation having the general integral $CP(x) + C'Q(x)$, where C and C' are arbitrary constants and $P(x)$ and $Q(x)$ are two given polynomials, which is

$$\begin{vmatrix} y'' & y' & y \\ P'' & P' & P \\ Q'' & Q' & Q \end{vmatrix} = 0 ,$$

will in general admit all the zeros of $P'Q - PQ'$ as singularities, points for which the general integral is holomorphic.

99 - Halphen's equation

The Halphen equations are of the form

$$P_o(x)y^{(n)} + P_1(x)y^{(n-1)} + \ldots + P_{n-1}(x)y' + P_n(x)y = 0 \tag{36}$$

whose coefficients $P_o, P_1, \ldots, P_n$ *are polynomials whose degree is at most equal to the degree of* $P_o(x)$ *and all of whose solutions are meromorphic at a finite distance.* It is therefore necessary that the equation should be of Fuschian type about each zero of $P_o(x)$, that the corresponding determinantal equation has all its roots real integral and distinct, and that the formal expansions corresponding to each of these roots, are possible. This can be seen to be the case providing we can solve $P_o(x) = 0$.

The condition relating to the degrees of the polynomials $P_j(x)$ can be interpreted and is equivalent to the fact that the rational fractions $P_j(x)/P_o(x)$, $j = 1,2,\ldots,n$ have expansions about the point at infinity, that commence from a term of degree 0, or, if we wish, are holomorphic at infinity. If we set

$$y = zR(x) ,$$

where $R(x)$ is a rational fraction, then equation (36) is transformed into an equation in z whose coefficients are clearly polynomials. The degrees of those polynomials again satisfy the same condition. For, if we assume that about the point at infinity, the expansion of $R(x)$ commences from a term in x^q, then we shall have

$$y = z(a_q x^q + \ldots) , \quad y' = q'(a_q x^q + \ldots) + z(qa_q x^{q-1} + \ldots), \ldots$$

$$y^{(p)} = z^{(p)}(a_q x^q + \ldots) + z^{(p-1)}p(qa_q x^{q-1} + \ldots) + \ldots$$

and on bringing this into equation (36) written in the form

$$y^{(n)} + (b_1 + b_1' x^{-1} + \dots) y^{(n-1)} + \dots + (b_n + b_n' x^{-1} + \dots) y = 0 \quad ,$$

then, on dividing by the coefficient of $z^{(n)}$, the apparent property is realised.

Let y be any integral of equation (36). Its only possible poles are the zeros of $P_o(x)$; their orders of multiplicity are at most equal to the absolute value of the smallest negative roots of the corresponding determinantal equations. A transformation of the form

$$y = \frac{z}{[P_o(x)]^k}$$

where k is a sufficiently large integer, hence transforms all of the solutions of equation (36) into entire functions which are solutions of the transformed equation

$$Q_o(x) z^{(n)} + Q_1(x) z^{(n-1)} + \dots + Q_n(x) z = 0 \quad , \tag{37}$$

an equation which, following what was said, is again of the same type; the degree of $Q_o(x)$ is at least equal to the degrees of the other polynomials. Let us set $z = e^{\alpha x} u$, where α is a constant. Following the calculations of the derivatives of this function that have already been used previously, we obtain an equation in u, which after suppression of the factor of $e^{\alpha x}$, is written as

$$R_o(x) u^{(n)} + R_1(x) u^{(n-1)} + \dots + R_n(x) u = 0 \quad . \tag{38}$$

$R_o(x) \equiv Q_o(x)$ is again of degree greater than or equal to that of the other polynomials $R_j(x)$, and we have

$$R_n(x) \equiv \alpha^n Q_o(x) + \alpha^{n-1} Q_1(x) + \dots + Q_n(x) \quad .$$

α is determined on stating that the highest term of this polynomial is zero; we will, in general have n possible values for α. We take one of these values. Then, in (38) *the degree of* $R_n(x)$ *is less than the degree* μ *of* $R_o(x)$. The solutions of (38) are again entire functions. The equation is of the first Fuchsian type about each zero of $R_o(x)$. Hence $R_1(x)/R_o$ is a rational fraction with simple poles. We have

$$\frac{R_1(x)}{R_o(x)} = A_o + \sum \frac{A_k}{x - \beta_k} \quad .$$

The determinantal equation relative to β_k is

$$r(r-1)\ldots(r-n+1) + A_k r(r-1)\ldots(r-n+2) + \ldots = 0 \quad ,$$

the sum of its roots is

$$\frac{n(n-1)}{2} - A_k \quad .$$

These roots are integers, positive or zero (only a few are zero) and unequal; their sum is at least equal to $0 + 1 + 2 + \ldots + n-1$, hence equal to $\frac{1}{2}n(n-1)$. A_k is therefore a negative integer or zero and we have $\sum A_k \leq 0$. Let us assume that $R_n(x)$ is not identically zero. Let us differentiate equation (38) with respect to x, giving

$$R_o u^{(n+1)} + (R_1 + R_o')u^{(n)} + (R_2 + R_1')u^{(n-1)} + \ldots + R_n' u = 0 \quad ,$$

and eliminate u from this equation and (38). We see that u' will be a solution of the equation

$$\begin{aligned} R_o R_n(x) + (R_o' R_n + R_1 R_n - R_o R_n')v^{(n-1)} + \ldots \\ + (R_p' R_n + R_{p+1} R_n - R_p R_n')v^{(n-p-1)} + \ldots = 0 \quad . \end{aligned} \tag{39}$$

This equation is again of the same type, for its last term is

$$R_n^2 + R_n R_{n-1}' - R_n' R_{n-1} \quad ;$$

it is of degree less than $R_o R_n$. The analogous sum to $\sum A_k$ is again negative or zero. Now $\sum A_k$ is the coefficient of $1/x$ in the expansion of R_1/R_o about the point at infinity. For the new equation, the ratio of the coefficient of $v^{(n-1)}$ at the coefficient of $v^{(n)}$ is

$$\frac{R_1}{R_o} + \frac{R_o'}{R_o} - \frac{R_n'}{R_n} \quad ,$$

the coefficient of $1/x$ in the expansion of this fraction about the point at infinity is that obtained for R_1/R_o, $\sum A_k$, plus μ, which is the coefficient of $1/x$ in the expansion of the logarithmic derivative R_o'/R_o since the degree of R_o is μ, and less $\mu - 1$ at most since R_n is of degree $\mu - 1$ at most. This coefficient is therefore at least $\sum A_k + 1$, and we must have $\sum A_k + 1 \leq 0$, hence $\sum A_k \leq -1$.

If the coefficient of v in equation (39) is not identically zero, then we re-commence the same transformation; u'' will be a solution of an equation of the same form and we will have $\sum A_k + 2 \leq 0$ or $\sum A_k \leq -2$. If the last coefficient of the equation is not identically zero, then we can start again; u''' will satisfy a similar equation and $\sum A_k$ will be less than or equal to

-3. As $\sum A_k$ is a determined number, we see that at the end of a number of operations, at most $-\sum A_k$, we will stop and we will thus have obtained an equation satisfied by $u^{(m)}$ and whose last coefficient will be identically zero. This equation will admit a solution which will be a constant, hence the equation in u admits a solution which is a polynomial; equation (37) admits a solution of the form

$$z_1 = e^{\alpha x}S(x) \quad ,$$

where $S(x)$ is a polynomial.

To start from such a solution, we can reduce the order. We will first of all set $z = z_1 A$, which we give an equation which, following what we have seen, is of the same type [since we can initially set $z = e^{\alpha x}U$, then $U = S(x)Z$]. This equation does not have a term in Z; it is an equation in Z'. We can apply the same argument to this equation. We will have a solution Z' of the form $e^{\beta x}T_1(x)$, where $T_1(x)$ is a rational fraction. But Z happens to become uniform and the logarithmic terms in the integration of Z' must disappear, Z is of the form $e^{\beta x}T(x)$, where $T(x)$ is rational. Returning to z, we will have a second solution

$$e^{(\beta+\alpha)x}S(x)T(x) \quad .$$

The fact presumes that equation (37) admits a fundamental system of solutions

$$z_p = e^{\alpha_p x}S_p(x) \qquad p = 1,2,\ldots,n \quad ,$$

where the α_p are constants, that may or may not be distinct, and where the $S_p(x)$ are polynomials. Let us prove this by recurrence. The property is true for a first order equation. We see this directly: we have

$$\frac{z'}{z} = \frac{-Q_1}{Q_o} = B_o + \sum \frac{B_j}{x-\beta_j} \quad ,$$

and the B_j are positive integers since, for each β_j, the determinantal equation has B_j as a root. Hence z is indeed the product of an exponential and a polynomial. If the property is true up to order $n - 1$, then the equation in Z' obtained by reduction, will admit a fundmental system of solutions of the form

$$Z'_p = e^{\gamma_p x}T_p(x) \qquad p = 1,2,\ldots,n-1 \quad ,$$

where the γ_p are constants and the $T_p(x)$ rational. The equation in Z will admit as solutions, forming a fundamental system, the primitives of these functions Z'_p, which are of the same form (with different fractions T_p),

and the constant 1. For, if these functions satisfy a linear identity with constant coefficients, then their derivatives, the Z'_p, also satisfy such an identity. By multiplying the solutions of this fundamental system by z_1, we obtain a fundamental system of solutions of the form indicated by the equation in z. By finally dividing by $[P_o(x)]^k$, we see that:
The general solution of the Halphen equation (36) (an equation satisfying the stated conditions), is of the form

$$y = \sum_1^n C_p e^{\alpha_p x} \rho_p(x) \tag{40}$$

where the α_p *are constants, the* $\rho_p(x)$ *rational fractions and the* C_p *arbitrary constants. Conversely, if an n-th order differential equation admits a general integral of the form (40), then this equation* which is obtained from the solutions of the fundamental system (as we have mentioned several times before) $e^{\alpha_p x}\rho_p(x)$ *is seen to be of the form*

$$P_o(x)y^{(n)} + P_1(x)y^{(n-1)} + \ldots + P_n(x)y = 0 \quad ,$$

where the $P_j(x)$ *are polynomials.*

We intend proving that the degree of $P_o(x)$ *is greater than or equal to the degree of the other polynomials.*

This is clear for $n = 1$. Since from

$$y = e^{\alpha x}\rho(x) \quad ,$$

we deduce

$$\frac{y'}{y} = \alpha + \frac{\rho'(x)}{\rho(x)} \quad ; \tag{41}$$

the second member of (41) is thus a rational fraction for which the degree of the numerator does not exceed that of the denominator.

Let us assume that the property is true up to order $n - 1$.

Let us consider the functions obtained by dividing the n given functions $y_p = e^{\alpha_p x}\rho_p(x)$ by one of them, y_1 say. Those functions

$$1, \frac{y_2}{y_1}, \ldots, \frac{y_n}{y_1}$$

are again the products of the exponentials and the rational fractions. They are solutions of an equation

$$Q_o(x)z^{(n)} + Q_1(x)z^{(n-1)} + \ldots + Q_{n-1}(x)z' = 0 \quad . \tag{42}$$

which has no term in z, since 1 is a solution. This equation is an equation of order $n - 1$ in z', whose functions

$$\frac{d}{dx}\left(\frac{y_p}{y_1}\right) \quad , \qquad p = 2,3,\ldots,n \quad ,$$

are solutions. The property holds for this equation of order $n - 1$; the degree of $Q_o(x)$ is at least equal to the degree of the other polynomials $Q_j(x)$. We pass from equation (42) to the differential equation in y by setting $y = y_1 z$ which, as we have seen, does not change the property of the degree of the coefficients. The proposition is thus proved.

Remarks

I. If a Halphen equation (36) admits an integral which is a rational fraction $R(x)$, we have, on dividing by the coefficient of $y^{(n)}$ as we did above, the form

$$y^{(n)} + (b_1 + b_1' \frac{1}{x} + \ldots)y^{(n-1)} + \ldots + (b_n + b_n' \frac{1}{x} + \ldots)y = 0 \quad ,$$

$$|x| > R \quad ,$$

in which we must replace y by $a_q x^q + a_{q-1}x^{q-1} + \ldots$. For this equation to be possible, it is necessary that $b_n = 0$. Hence if there exists a rational solution, the degree of $P_n(x)$ is (36) is less than that of $P_o(x)$. It follows that if

$$e^{\alpha x}R(x)$$

is a solution, if we set $y = e^{\alpha x}z$, we will obtain an equation in z in which the degree of the coefficient of z will be less than the degree of the coefficient of $z^{(n)}$. Hence, α *is necessarily one of the roots of the equation obtained by stating that the term of the highest degree of*

$$\alpha^n P_o(x) + \alpha^{n-1}P_1(x) + \ldots + P_n(x)$$

is zero.

II. In pursuing the method of reduction used in the first part of the proof, we evidently arrive at the general integral.

100 - The Picard equations

The Picard equations are the equations of the form

$$y^{(n)} + a_1(x)y^{(n-1)} + \ldots + a_n(x)y = 0 \tag{43}$$

whose coefficients are elliptic functions of x, *with the same periods* $2\omega, 2\omega'$ *and whose general integral is memorphic at every point of finite distance.*

This demands that the equation should be of Fuchsian type about each pole of the coefficients belonging to the same parallelogram of periods, that the determinantal equations have all their roots real, integral and distinct, and that the conditions for which there are no terms in $\log(x-x_j)$ are satisfied.

The integrals of these equations are expressed in terms of the functions of the theory of elliptic functions.

Let $y_1(x), y_2(x), \ldots, y_n(x)$ be a fundamental systems of solutions. As the equation does not change when we replace x by $x + 2$, $y_1(x+2\omega), \ldots, y_n(x+2\omega)$ also form a fundamental system and we have

$$y_j(x + 2\omega) = \lambda_{1,j} y_1(x) + \ldots + \lambda_{k,j}\, y_k(x) + \ldots + \lambda_{n,j}\, y_n(x)$$

$$j = 1, 2, \ldots, n \quad ,$$

where the $\lambda_{k,j}$ are constants and their determinant is non-zero. Following that which we have seen in no. 93, we can find a linear homogeneous combination of the $y_j(x)$, with constant coefficients,

$$Y_1(x) = \mu_1 y_1(x) + \ldots + \mu_n y_n(x) \quad ,$$

for which we will have $Y_1(x + 2\omega) \equiv sY_1(x)$, where s is a constant. The functions (x), $Y_2(x) \equiv Y_1(x + 2\omega')$, $Y_3(x) \equiv Y_2(x + 2\omega')$ ..., will again be integrals. For an index value m, at most equal to n, the function $Y_{m+1}(x)$ will be a linear homogeneous combination with constant coefficients of the m preceding functions, whereas this is not the case for $Y_m(x)$. We will have

$$Y_{m+1}(x) \equiv v_1 Y_1(x) + v_2 Y_2(x) + \ldots + v_m Y_m(x) \quad .$$

The functions $Y_1(x), \ldots, Y_m(x)$ therefore undergo the substitutions

$$Y_1(x + 2\omega') \equiv Y_2(x), \ldots, \quad Y_{m-1}(x + 2\omega') \equiv Y_m(x) \quad ,$$

$$Y_m(x + 2\omega') \equiv v_1 Y_1(x) + \ldots + v_m Y_m(x) \quad .$$

We can thus find a homogeneous linear combination with constant coefficients of these functions

$$z(x) = \tau_1 Y_1(x) + \ldots + \tau_m Y_m(x)$$

say, such that $z(x + 2\omega') \equiv s'z(x)$, where s' is constant. $z(x)$ is not identically zero, since the $Y_j(x)$, $j = 1, 2, \ldots, m$ are independent. On the other hand, following the property of $Y_1(x)$, we also have $Y_p(x + 2\ ) \equiv sY_p(x)$, hence $z(x + 2\ ) \equiv sz(x)$. Consequently:

There exists at least one integral of equation (43), $z(x)$, which is such that

$$z(x + 2\omega) \equiv sz(x) \ , \quad z(x + 2\omega') \equiv s'z(x) \quad ,$$

where x and s' are constant.

The meromorphic functions which have this property, for which ω and ω' are two numbers whose ratio is complex, have been studied by Hermite under the name of *doubly-periodic functions of the second kind;* they are also called *functions with constant multipliers.* The elliptic functions correspond to the case where $s = s' = 1$.

FUNCTIONS WITH CONSTANT MULTIPLIERS

It is clear that if ω, ω', s, s' are given and $z(x)$ and $u(x)$ are two functions with constant multipliers corresponding to these, then the quotient $u(x)/z(x)$ is an elliptic function with periods $2\omega, 2\omega'$. Conversely, if $z(x)$ is a function with constant multipliers, then all the functions $z(x)f(x)$, where $f(x)$ is elliptic with periods 2ω and $2\omega'$, are functions with multipliers. It suffices to construct one of these functions.

THE WEIERSTRASS NOTATION

We know that the function σx satisfies the equations (I, 232)

$$\sigma(x + 2\omega) = -\sigma x e^{2\eta(x+\omega)}$$

$$\sigma(x + 2\omega') = -\sigma x e^{2\eta'(x+\omega)} \quad .$$

The function

$$G(x) = \frac{\sigma(x+a)}{\sigma x} e^{bx} \quad ,$$

where a and b are constants, admit by virtue of the preceding formulae, the multipliers

$$e^{2\eta a + 2b\omega} \ , \qquad e^{2\eta' a + 2b\omega'} \quad .$$

We can choose a and b so that these multipliers are two given numbers s and s'. We must then have

$$2\eta a + 2b\omega = \log s, \qquad 2\eta' a + 2b\omega' = \log s' \quad ,$$

which determines a and b since the determinant $\eta\omega' - \eta'\omega = \frac{i\pi}{2} \neq 0$ (I, 231). We see that when $\omega \log s' - \omega' \log s = 0$, a is zero and $G(x)$ reduces to the exponential.

In all cases, *the functions with constant multipliers are of the form*

$$F(x) = \frac{\sigma(x+a)}{\sigma x} e^{bx} f(x) \tag{44}$$

where $f(x)$ *is an elliptic function with periods* 2ω, $2\omega'$. From the formula of decomposition into factors of $f(x)$ by means of the function σ, (I, 232) we straight away deduce that decomposition for $F(x)$. The formula (44) shows that $F(x)$ has the same number of zeros and poles in a parallelogram of periods, but they can only have one, when $f(x)$ = const. and none at all when $f(x)$ = const. and we are *in the special case* where $\omega \log s' - \omega' \log s = 0$. *The decomposition into simple elements. We assume that we are not in the special case.* If $F(x)$ is the given function with constant multipliers and if $G(x)$ is the function will the same multipliers as considered above, then the function $F(x)G(u-x)$, where u is a parameter, is an elliptic function, since the change from x to $x + 2\omega$ or $x + 2\omega'$ multipliers $G(u-x)$ by $1/s$ and $1/s'$ accordingly. The sum of the residues of the poles situated in a parallelogram of the periods (or relative to all the non-homologous poles, which is the same) is zero (I, 228). Let b_k be one of the poles of $F(x)$ and let

$$\frac{B_k^1}{x-b_k} + \frac{B^2}{(x-b_k)^2} + \dots + \frac{B_k^q}{(x-b_k)^q}$$

be the principal part of this pole. The residue of $F(x)G(u-x)$ at this point will be

$$B_k^1 G(u-b_k) - B_k^2 G'(u-b_k) + \dots + \frac{(-1)^{q-1}}{(q-1)!} B_k^q G^{(q-1)}(u-b_k) \quad .$$

On the other hand, $G(u-x)$ has $x = u$ as a simple pole and its residue is $-\sigma\alpha$; that of $F(x)G(u-x)$ will be $-\sigma a$ and we obtain Hermite's formula

$$F(u) = \frac{1}{\sigma a} \sum \left[B_k^1 G(u-b_k) + \dots + \frac{(-1)^{q-1}}{(q-1)!} B_k^q G^{(q-1)}(u-b_k) \right] \quad . \tag{45}$$

In the special case where $\omega \log s' = \omega' \log s$, we obtain a similar result. The function $F(x)e^{-bx}$ is then an elliptic function. We can apply the Hermite decomposition formula to it (I, 231). We will obtain an expression of the form

$$F(x) = e^{bx} \left\{ D_o + \sum \left[D_k^1 \zeta(x-b_k) + \dots + D_k^q \zeta^{(q-1)}(x-b_k) \right] \right\}$$

where the coefficients are simply related to the B_k^j. If here we set

$$G_1(x) = \zeta x e^{bx}$$

the formula becomes

$$F(x) = e^{bx}D_o + \sum\left[E_k^{'}G_1(x-b_k) + \ldots + E_k^{q}G_1^{(q-1)}(x-b_k)\right] \quad .$$

The sum of the residues of $F(x)e^{-bx}$ in a parallelogram of periods is zero.

THE JACOBI NOTATIONS

Here it is the function $H(x)$ (I, 237) which replaces σx. On account of the relation between $H(x)$ and σx, we will have

$$G(x) = K\frac{H(x+a)}{H(x)}e^{b'x} \quad ,$$

where a and b are constants determined as above, and K is also a constant which may be taken to equal 1 since $G(x)$ could be quite conveniently multiplied by a constant. We could also assign another value to K, such that the residue of $G(x)$ at the pole $x = 0$, is -1. Nothing changes from what was said previously, except that on taking K as indicated, the preceding decomposition formula will become, in the case of simple poles,

$$F(u) = BG_1(u-b) + \ldots + LG_1(u-\ell)$$

where G_1 is the function G corresponding to this value of K. In the special case, the function Z will be introduced in place of ζ.

THE INTEGRATION OF THE PICARD EQUATION

We have seen that the equation admits a doubly periodic solution of the second kind. This solution is expressed, as we will show, in terms of the functions σ or H, or ζ or , and an exponential. Let us call this new solution y_1. We can reduce the order by putting $y = y_1\int v dx$. We will obtain a solution of order $n - 1$ in v whose general integral will again be meromorphic. The coefficients will be linear functions of the $a_j(x)$ which are elliptic, the coefficients of these linear functions being y_1'/y_1, $y_1''/y_1, \ldots$. These functions are expressed in terms of y_1'/y_1 and its derivatives. Now, since y_1 is a function with constant multipliers, its logarithmic derivative is doubly periodic. The differential equation of order $n - 1$ is therefore a Picard equation. We can continue these operations. We thus see that the general integral of the Picard equation is expressed in terms of doubly periodic functions of the second kind, and their successive integrals. But, *in general*, the equation in s, which provides the solution $Y_1(x)$, will have distinct roots; we will have n integrals such as $Y_1(x)$ and we will find n integrals which are doubly periodic of the second kind.

THE CASE OF THE SECOND ORDER EQUATION

Following the method of reduction, we have two integrals: y_1 which is doubly periodic of the second kind and

$$y_2 = y_1 \int v dx$$

where v is also a doubly periodic function of the second kind. If the multipliers of v are not of the form s, $s^{\omega'/\omega}$, i.e., if they are not found in the special case, then the function v has the form of the second member of (45), all the terms $G^{(q-1)}(x - b_k)$ have primitives of the same form, $G^{(q-2)}(x - b_k)$ when $q - 1 \geq 1$; we have then to consider

$$\sum \int B_k' G(x - b_k) dx \quad ,$$

where the B_k' are constants. Now the primitive of $G(x - b_k)$ yields terms in $\log(x - b_k)$, since $G(x)$ has a simple pole at the origin. These terms cannot be reduced, unless the B_k' are not all zero, and this must be the case since y_2/y_1 is uniform. Hence

$$\int v dx = \sum \frac{1}{\sigma a} \left(-B_k^2 G(x - b_k) + \ldots + \frac{(-1)^{q-1}}{(q-1)!} B_k^q G^{(q-2)}(x - b_k) \right)$$

is a function with constant multipliers s, s' and its product with y_1 which is with constant multipliers (in general different from s and s'), is also with constant multipliers. *The general integral is expressed in terms of the function* σ *and its derivatives, hence in terms of* σ, ζ, p *and exponentials.*

If we are in the special case where $s' = s^{\omega'/\omega}$, v is expressed in terms of formula (46); in order that its derivative is uniform, it is necessary that the coefficients E_k^1 are all zero, and then we have, if $b \neq 0$,

$$\int v dx = \sum \left[E_k^2 G_1(x - b_k) + \ldots + E_k^q G_1^{(q-2)}(x - b_k) \right] + \frac{D_o}{b} e^{bx} \quad . \quad (47)$$

Now the condition relating to the residues of the product of the second member of (46) and e^{-bx}, gives the condition

$$\sum (E_k^1 + \ldots + E_k^q b^{q-1}) e^{-bb_k} = 0 \quad ,$$

hence, since the E_k^1 are zero,

$$\sum (E_k^2 + \ldots + E_k^q b^{q-1}) e^{-bb_k} = 0 \quad .$$

It follows that the expression (47) is again a function with constant multipliers. For the products of e^{bx} and the derivatives of $\zeta(x - b_k)$ are such functions, and the terms not containing these derivatives are

$$e^{bx}\sum\zeta(x-b_k)(E_k^2+\ldots+E_k^q\, b^{q-2})e^{-bb_k} \quad .$$

The change from x to $x+2\omega$ or $x+2\omega'$, which changes $\zeta(x-b_k)$ to $\zeta(x-b_k)+2\eta$ or $\zeta(x-b_k)+2\eta'$ respectively, will leave the $\sum$ invariant. *It follows that, in this case,* y_2 *is a function with constant multipliers. The general integral is expressed in terms of* e^{bx}, σ, ζ, p. If b was zero, i.e. if v was doubly periodic $G_1(x)$ would reduce to ζx, the $\sum$ of the second member of (47) would be doubly periodic, where all the $E_k^{\prime}$ are again zero, but the last term of the second member of (47) would be replaced by $D_o x$. y_2 is no longer a function with multipliers if $D_o \neq 0$. *But the general integral is again expressed in terms of the functions of the theory of elliptic functions, multiplied by some exponentials and by* x.

We can proceed in a similar fashion when the order of the equation is greater than 2.

101 - The Lamé equation

This equation, first integrated by Hermite in a particular case, reduces to the case of the Picard equations. It was introduced by Lamé in the theory of thermodynamics under the form

$$\frac{d^2y}{dx^2} = (n(n+1)k\ sn^2x+h)y \quad ,$$

where n is an integer and h are both constants, k is equally so and is the modulus and snx is the function from the Jacobi theory (I, 238). As sn^2x only has the (double) pole $x=\omega'$, in the parallelogram of periods and is even, we have $sn^2x(x+\omega') = \lambda^2 px+\mu$, where λ and μ are constants, and an appropriate change of variable shows that, in the Weierstrass notation we can give the Lamé equation the canonical form

$$\frac{d^2y}{dx^2} = (n(n+1)px + m)y \quad ,$$

where m is a constant.

The equation does not change if we replace n by $-n-1$; we can then assume that the integer n is positive. It is also clear that if we change x to $-x$, the equation does not change: if $y(x)$ is a solution, $y(-x)$ is also a solution. The equation is of Fuchsian type, since the poles of px are double. The determinantal equation for the pole $x=0$ is $r(r-1) = n(n+1)$, its roots are $-n$ and $n+1$. The root $n+1$ gives an integral y_1 that is holomorphic at the origin. The root $-n$ will provide a uniform solution if the formal expansion is possible. We can avoid this verification. y_1 is given by an expansion

$$y_1 = x^{n+1} + c_{n+2}x^{n+2} + \ldots$$

which is obtained by identification. As px is even, all the terms of y_1 have the same parity. The second solution y_2 is given by the formula relating to the reduction. We have here

$$y_1 \frac{d^2 y_2}{dx^2} - y_2 \frac{d^2 y_1}{dx^2} = 0 \quad ,$$

hence a solution

$$y_2 = y_1 \int \frac{dx}{y_1^2} \quad .$$

About the origin, all the terms of y_1^2 are even, the integral does not give any logarithmic term and y_2 is uniform. The equation is of the Picard type. It thus admits an integral with constant multipliers. By using the decomposition formula in terms of the function σ, this solution will be of the form

$$e^{bx} \frac{\sigma(x+a_1)\ldots\sigma(x+a_n)}{\sigma(x)^n}$$

since the origin is a pole of order n. On changing x to $-x$, we will obtain a second solution which, in general, will be distinct from the first. If we can calculate the a_j and b for an arbitrary value of m, we can apply the D'Alembert method to obtain a second solution in the case where the two solutions are identified.

THE CASE $n = 1$.

In this case, we have a solution of the form

$$y = e^{bx} \frac{\sigma(x+a)}{\sigma x} \quad .$$

Let us expand about the origin

$$e^{bx} = 1 + bx + \frac{b^2 x^2}{2} + \ldots , \quad \sigma(x+a) = \sigma a + x\sigma' a + \frac{x^2 \sigma'' a}{2} + \ldots$$

$$\frac{1}{x} = \frac{1}{x(1 + \alpha x^4 + \ldots)} = \frac{1}{x} - \alpha x^3 + \ldots \quad .$$

We then deduce

$$y = \frac{A}{x} + B + Cx + Dx^2 + \ldots , \quad y'' = \frac{2A}{x^3} + 2D + \ldots$$

with

$$A = \sigma a, \quad B = b\sigma a + \sigma' a, \quad C = \frac{b^2 \sigma a}{2} + b\sigma' a + \frac{\sigma'' a}{2}$$

The equation is

$$y'' = (2px + m)y = 0 , \quad px = \frac{1}{x^2} + c_2x^2 + \dots \quad .$$

By equating the coefficients of x^{-3}, x^{-2}, x^{-1} and that of the independent term, to zero, we obtain the conditions

$$2A - 2A = 0, \quad 2B = 0, \quad 2C + Am = 0, \quad 2D - 2D - Bm = 0$$

which reduce to $B = 0$, $2C + Am = 0$, hence to $b\sigma a + \sigma' a = 0$, $m\sigma a + b^2\sigma a + 2b\sigma' a + \sigma'' a = 0$, and in turn give

$$b = \frac{-\sigma' a}{\sigma a} = -\zeta a , \quad m = \frac{-\sigma'' a}{\sigma a} + \left(\frac{\sigma' a}{\sigma a}\right)^2 = -\left(\frac{\sigma' a}{\sigma a}\right)' = pa \quad .$$

The constant a is therefore determined by

$$pa = m \quad ,$$

which yields for a, up to a multiple of the periods, the two opposing values a and $-a$, to each there corresponds a value of b and these two values are opposite, i.e., $-\zeta a$ and ζa. If we replace a by $a + 2\omega$, for example, ζa will be increased by 2η and y will not be multiplied by a constant following the properties of σx. We then obtain two possible solutions

$$e^{-x\zeta a}\frac{\sigma(x+a)}{\sigma(x)} \quad , \quad e^{x\zeta a}\frac{\sigma(x-a)}{\sigma x} \quad .$$

As we pass from one equation to the other, by changing x to $-x$ (since σ and ζ are uneven), both of them agree. If they are distinct, then we have a general solution. They are distinct if a and $-x$ are not equal, to within the nearest period, for they then have distinct zeros. It is therefore necessary to assume that a is different from ω, ω', $\omega + \omega'$ if 2ω and $2\omega'$ are the periods. Hence: *If* m *is not equal to one of the numbers* $p\omega$, $p\omega'$, $p(\omega + \omega')$, *then the general solution to the Lamé equation* $y'' - (2px + m)y = 0$, *is*

$$y = Ce^{-x\zeta a}\frac{\sigma(x+a)}{\sigma x} + C'e^{x\zeta a}\frac{\sigma(x-a)}{\sigma x} \quad ,$$

where C *and* C′ *are arbitrary constants and* a *is determined, to within a sign, by the equation* $pa = m$.

If, for example, $m = p\omega$, we will first of all take $m = p(\omega + t)$, where t is a parameter that can be made to tend to zero. We will obtain the particular solution

$$\frac{e^{-x\zeta(\omega+\ )}\sigma(x+\omega+t) - e^{-x\zeta(\omega-t)}\sigma(x+\omega-t)}{2\sigma(x)t}$$

which is the quotient by $2t$ of the difference of the two integrals corresponding to $a = \omega + t$ and to $-\omega - t + 2\omega$. This function is analytic in t, the numerator is zero for $t = 0$; this is the solution to the equation $y'' - (2px + p(\omega + t))y = 0$ which takes, for $x = \omega$, a value which is an analytic function of t. For $t = 0$, it will be a solution of the equation $y'' - (2px + p\omega)y = 0$. We thus obtain the solution

$$\frac{1}{\sigma x}\frac{d}{d\omega}\left(e^{-x\zeta\omega}\sigma(x+\omega)\right) = e^{-x\eta}\frac{\sigma(x+\ )}{\sigma x}\left[\zeta(x+\omega) + xp\omega\right] ,$$

which when combined with the one already obtained, provides the general solution.

IV. THE LAPLACE METHOD OF INTEGRATION

102 - The Laplace equations

These are linear equations of the form

$$(a_o x + b_o)y^{(n)} + (a_1 x + b_1)y^{(n-1)} + \ldots + (a_n x + b_n)y = 0 , \quad (48)$$

where the a_j and b_j are constants. Laplace sought to obtain a solution under the integral form

$$y = \int_\alpha^\beta e^{zx}U(z)dz , \quad (49)$$

where the function $U(z)$ is an analytic function of z which has to be determined and α and β are suitably chosen limits along with the curve which joins these points and on which the integration takes place. This integral is a holomorphic function of x in every finite domain if the path of integration is finite and if $U(z)y$ is continuous (I, 240). Its derivatives are obtained by differentiating under the sign of integration; we have

$$y^{(p)} = \int_\alpha^\beta e^{zx}z^pU(z)dz .$$

If we take these values into the first member of equation (48), we obtain

$$\int_\alpha^\beta e^{zx}\ [P(t)x + Q(z)]\ U(z)dz , \quad (50)$$

by setting

$$P(z) \equiv a_o z^n + a_1 z^{n-1} + \dots + a_n, \quad Q(z) \equiv b_o z^n + b_1 z^{n-1} + \dots + b_n .$$

We can integrate by parts:

$$\int_\alpha^\beta e^{zx} x\, [P(z)U(z)]dz = [e^{zx}P(z)U(z)]_\alpha^\beta - \int_\alpha^\beta e^{zx} \frac{d}{dt} (PU)dz$$

and on replacement into (50), we see that this integral will be zero, and that the integral (49) will be a solution, if we have at the same time

$$[e^{zx}P(z)U(z)]_\alpha^\beta = 0 \quad ,$$
$$\int_\alpha^\beta e^{zx} [Q(z)U(z)] - \frac{d}{dz} [P(z)U(z)]dz = 0 \quad . \tag{51}$$

For the integral to be zero, it suffices that $U(z)$ satisfies the differential equation

$$\frac{d}{dz} (PU) - ()U) \frac{Q}{P} = 0 \quad ,$$

which gives

$$U(z) = \frac{1}{P(z)} e^{V(z)} , \quad V(z) = \frac{Q(z)}{P(z)} dz \quad . \tag{52}$$

The condition (51) is then written as

$$[e^{zx+V(z)}]_\alpha^\beta = 0 \quad . \tag{53}$$

Definitively, we can obtain an integral of the form (49) provided that the function $U(t)$ *defined (to within a constant) by equation (52) is continuous along the chosen path of integration and that condition (53) is satisfied.*

In general, the polynomial $P(z)$ will have its n zeros distinct, α_k say, $k = 1,2,\dots,n$. $V(z)$ will then be

$$V(z) = \int \left(A + \sum_1^n \frac{A_k}{z-\alpha_k}\right) dz = Az + \sum_1^n A_k \log(z-\alpha_k) \quad ,$$

where the A and the A_k are constants, and we will have

$$e^{V(z)} = e^{Az} \prod_1^n (z-\alpha_k)^{A_k}$$

a function which is multiform in general. Let us denote by Γ_k a simple closed contour starting from the point α and encircling α_k. When Γ_k is described in the direct sense, the function $e^{V(z)}$ is multiplied by $e^{2\pi i A_k}$. If we describe Γ_k in the opposite sense, then $e^{V(z)}$ will take its initial value and the condition (53) will be realised by taking the contour thus formed for the path of integration. If the integral (49), which is then equal, to within a factor, to

$$\int_{\alpha}^{\beta} e^{(A+x)z} \prod_{1}^{n} (z-\ _{k})^{A_k-1} \, dz \quad , \tag{54}$$

is not identically zero, then it yields an integral of the Laplace equation. By thus successively associating one of the points α_k, α_1 for example, to the remaining $n-1$, we will obtain $n-1$ solutions in the general case. If $I(\Gamma_k)$ denotes the integral (54) taken on Γ_k in the direct sense, then we see that these solutions are of the form

$$\gamma_k I(\Gamma_1) - \gamma_1 I(\Gamma_k), \qquad k = 2,\ldots,n \quad ,$$

where the γ_k are constants, and by associating α_k with α_j, we obtain an integral which is a linear combination of these. Hence, under the most favourable conditions, we thus obtain $n-1$ solutions.

When the n numbers A_k have their real part positive, then a path of integration joining α_k to α_j will permit a realisation of condition (53) since $(z-\alpha_k)^{A_k}$ is zero at the point α_k, and under these conditions, the integral (54) will have a meaning. We will again obtain in this case, $n-1$ solutions at most.

A REMARK ON THE CASE OF EQUATIONS WITH CONSTANT COEFFICIENTS

If the α_j are all zero, then equation (48) is with constant coefficients and the method breaks down. But. the integral (50), which is then

$$\int_{\alpha}^{\beta} e^{zx} Q(z) U(z) \, dz \quad ,$$

will again be zero and (49) will provide a solution if we for the path α, β, a simple closed curve, and for the product QU, a function that is holomorphic inside this curve and on this curve. Let $QU = f(z)$, hence $U = f/Q$. The integral (49) is then

$$\int_{\alpha}^{\beta} e^{zx} \frac{f(z)}{Q(z)} \, dz \quad .$$

It is equal to the product by $2\pi i$ of the sum of the residues of the poles of the function to be integrated, situated within the contour. The procedure, thus modified (this is due to Cauchy), a recovering of the known results. The equation $Q(z) = 0$ is the characteristic equation. If α is one of its roots, of order q, then we can take for U the function

$$U = \frac{C_q}{(z-\alpha)^q} + \frac{C_{q-1}}{(z-\alpha)^{q-1}} + \ldots + \frac{C_1}{z-\alpha}$$

where the C_j are arbitrary constants, and for the contour, we take a circle of centre α not containing any other roots of $Q(z) = 0$. The solution obtained is to within a factor, the residue of Ue^{zx} at the point α, which gives

$$e^{\alpha x}(C_1 + C_2 x + \dots + \frac{C_q}{(q-1)!}x^{q-1}) \quad .$$

We recover the q particular integrals corresponding to the root α.

103 - The integration of the Laplace equation in the general case

Let us return to the general case where $P(z)$ is of degree n and has its zeros distinct. We can obtain a system of n distinct solutions, tenable in a half-plane, by integrating on an infinite path. We must assume that the function

$$e^{z(A+x)} \prod^{n} (z-\alpha_k)^{A_k} \tag{55}$$

is zero at infinity when z is extended indefinitely in a given direction. As the product Π behaves as a power of $|z|$, it suffices to assume that the argument of $z(A+x)$ remains between $\pi/2 + \varepsilon$ and $3\pi/2 - \varepsilon$, where ε is positive provided the exponential behaves as $e^{-\varepsilon'|z|}$, $\varepsilon' > 0$, and that expression (55) tends to zero. If we assume that the argument of z has ϕ as its limit, then we will necessarily have

$$\frac{\pi}{2} - \phi < \arg(A+x) < \frac{3\pi}{2} - \phi \quad , \tag{56}$$

which indicates that the image of x will be found in a half-plane passing through the point $-A$. For example, if we take $\phi = \pi$, the point x must be found to the right of the parallel to the imaginary axis, taken through the point $-A$. Let us take through the points α_k, $k = 1,2,\dots,n$, the half-lines parallel to the half-line $\arg z = \phi$. In general, they will be without common points. If some of these half-lines have in common a half-line, e.g., if the half-line relative to α_2 contains the point α_3, then we replace this half-line by a small semi-circle in the neighbourhood of α_3 (fig. 38).

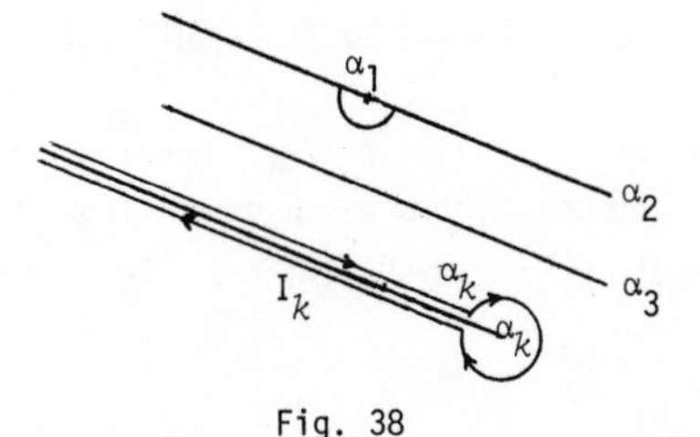

Fig. 38

The contour of integration Γ_k corresponding to α_k will be defined as follows: we pass along the half-line emanating from α_k from infinity, up to the point α'_k and when we return to the point α'_k, we return to infinity along the half-line. We assume that α'_k is sufficiently near to α_k such that the small circle does not contain in its interior or on its circumference, the points α_j, $j \neq k$. We take the integral (49) sur Γ_k; the condition (53), where α and β are replaced by the points at infinity of Γ_k, is satisfied. The integral (49) which is here

$$y_k = \int_{\Gamma_k} e^{(A+x)z} \prod_1^n (z-\alpha_j)^{A_j-1} dz \tag{57}$$

has a meaning and defines a holomorphic function of x in the half-plane defined by the inequalities (56). We thus obtain n solutions of the differential equation, but it is again necessary to check that they are not identically zero and that they are distinct. This results from the study (following Poincaré) of the behaviour of the solutions thus obtained for the greater values of $|x|$, where x is within an angle inside the half-plane defined by (56). We also appeal to the fact that the contour Γ_k can be modified without the value of y_k changing; we can always vary the position of the point α'_k. Since, between two contours corresponding to two different points α'_k, the function to be integrated, is holomorphic.

If we set $z = \alpha_k + Z$, we obtain

$$y_k = e^{(A+x)\alpha_k} \int_{\Gamma'_k} e^{(A+x)Z} Z^{A_k-1} \phi(Z)dZ \tag{58}$$

where $\phi(Z)$ denotes the product of the quantities $(\alpha_k - \alpha_j + Z)^{A_j-1}$, $j \neq k$ and where Γ'_k is the contour deduced from Γ_k by the translation which effects the passage from the point α_k, to the origin. The function $\phi(Z)$ is holomorphic at the origin (we take a branch of this function, is multiform in general). For $|Z|$ sufficiently small, $|Z| \leq \gamma$, it is series-expandable as $\phi(Z) = \beta_o + \beta_1 Z + \ldots B_q Z^q + \ldots$, $\beta_o \neq 0$. The terms following the term of degree q can be written as $Z^{q+1}\psi(Z)$, where $\psi(Z)$ is holomorphic for $|Z| \leq \gamma$, hence $|\psi(Z)| < K$. Consequently for $|Z| \leq \gamma$, we have

$$\phi(Z) = \beta_o + \beta_1 Z + \ldots + \beta_q Z^q + Z^{q+1}\psi(Z) \quad |\psi(Z)| < K \quad .$$

For $|Z| > \gamma$, we have on Γ'_k, $|\phi(Z)| < |Z|^s$, where S is fixed. We assume that the circle, of arbitrary radius, which starts from Γ'_k is of radius less than γ. On the part of Γ'_k on which $|Z| > \gamma$, Γ''_k say, we will have

$$\left| \int_{\Gamma_k''} e^{(A+\ )Z_A^{A\ -1}} \phi(Z)\ Z \right| > 2 \int_{\gamma} e^{-h|A+x||Z|} |Z|^s |dZ| \tag{60}$$

where h is positive, providing that

$$\frac{\pi}{2} - \phi + \eta < \arg(A + x) < \frac{3\pi}{2} - \phi - \eta, \qquad \eta > 0 \quad . \tag{61}$$

It follows that, once $|x|$ is sufficiently large, the first member of the inequality (60) will be less than $|x|^{-H}$, where H is a given to be positive and arbitrarily large. On the remainder of the contour Γ_k', we will set $Z(A+x) = T$. This part of Γ_k' will thus be transformed into a closed contour formed by a segment $\Omega\Omega'$ twice described, taken along a half-line emanating from 0 and whose argument is taken between $\pi/2 + \eta$ and $3\pi/2 - \eta$, and through a circle passing through Ω (fig. 39).

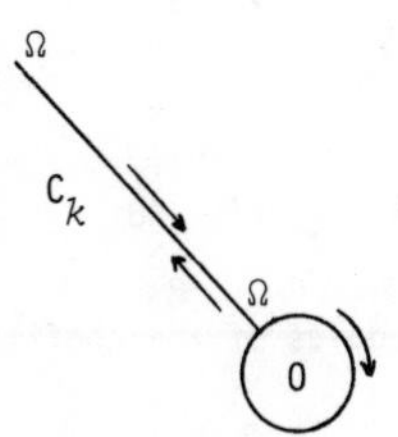

Fig. 39

It may be assumed that the radius of this circle is independent of x, for example equal to 1, since the point α_k' was arbitrary (we assume $|x|$ to be very large). As for Ω', its affix has $|A + x|\gamma$ as its modulus. On this contour, we apply equation (59). The integral along the remainder of Γ_k' will then be equal to

$$\begin{aligned} &\sum_{j=0}^{q} (A+x)^{-A_k-j} \beta_j \int_{C_k} e^T T^{A_k+j-1} dT \\ &\quad + (A+x)^{-A_k-q-1} \int_{C_k} e^T T^{A_k+q} \psi(Z) dT \quad . \end{aligned} \tag{62}$$

In these integrals, we have

$$T^{A_k} = e^{A_k \log T}$$

and the argument of T on the rectilinear, on one of the sides of C_k, is equal to $\phi + \arg(A + x)$; $\log T$ is therefore a function which is real on the positive real axis. We are now going to replace the contour C_k by another. First of all, we replace C_k by the contour Ω', a circle of radius 1 from Ω to Ω_1, and $\Omega_1\Omega_1'$, where the segment $\Omega_1\Omega_1'$ is symmetric to $\Omega\Omega'$ with respect to the real axis. The difference between the integrals taken on C_k and on this new contour is equal to the integral taken on the arc of the circle of centre $0, \Omega_1'N\Omega'$ (Fig. 40) since the functions to be integrated, occuring in expression (62) are holomorphic between these two contours.

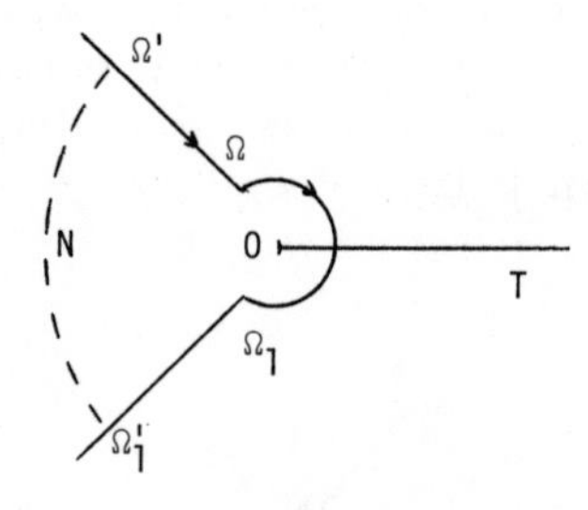

Fig. 40

On this circular arc, the modulus of the coefficient of β_j will be at most equal to

$$(1 + K)e^{-|T|\cos\phi'}\ |T|^{s} , \qquad |T| = \gamma|A + x|$$

$$|\phi'| < \frac{\pi}{2} - \eta ,$$

where S is fixed. By replacing C_k by this contour, we thus increase expression (62) by a quantity of the form $|x|^{-H}$, providing that $|x|$ is sufficiently large, where H is the given number introduced above. Similarly, we can replace, in all of the integrals, save the last, the contour $\Omega'\Omega\Omega_1\Omega_1'$ by the following contour C: the half-line $\Omega'\Omega$ taken from infinity to Ω, the circle of radius 1, and the half-line $\Omega_1\Omega_1'$ up to infinity. The last integral, for which $|\psi(Z)| < K$, is clearly bounded.

On gathering together all of these results, we see that within the angle (61), we have

$$y_k = e^{(A+x)\alpha_k}\left[\sum_{j=0}^{q}(A+x)^{-A_k-j}\beta_j\int_C e^{T}T^{A_k+j-1}dT + O(1)(A+x)^{-A_k-q-1}\right]$$

Following Hankel's formula (I, 240), the integral coefficient of β_j is

$$\frac{-2\pi i}{\Gamma(-A_k-j+1)} \quad ,$$

where $\Gamma(u)$ is the Euler function. It is non-zero if the A_k are not positive integers, which we shall assume henceforth. Be that as it may, we can write:

$$y_k = e^{(A+x)\alpha_k}(A+x)^{-A_k}\left[\sum_{j=0}^{q} \frac{-2\pi i\beta_j}{\Gamma(-A_k-j+1)(A+x)^j} + \frac{O(1)}{x^{q+1}}\right]$$

$$k = 1,2,\ldots,n \quad .$$

THE INDEPENDENCE OF THE SOLUTIONS

We can always assume that the origin has been shifted to the point -A ($A+x$ is replaced by X, the new origin A is, following the definition of A, the singular point $-b_o/a_o$ of the differential equation); this amounts to taking $A = 0$. It is impossible to have a relation of the form

$$C_1y_1 + C_2y_2 + \ldots + C_ny_n = 0 \quad , \tag{63}$$

where the C_j are constants, not all zero. Let us assume, in fact, that we have assigned to x an arbitrarily large modulus and an argument ψ such that the real parts of the numbers $e^{i\psi}\alpha_k$ are all different; this is impossible since all the points α_k are distinct. Now let us suppose that the y_j have been numbered in decreasing order of the real parts of $e^{i\psi}\alpha_k$. If we divide both members of (63) by y_1 (which is non-zero for as long as $|x|$ is sufficiently large), then the coefficients of $C_2,\ldots,C_n$ are as small as we wish once $|x|$ is suitably large since they contain exponentials of the form $e^{-h|x|}$, $h > 0$, when the other factors are of the order of a power of $|x|$. It follows that $C_1 = 0$. Applying the same argument, we see that $C_2 = C_3 = \ldots = C_{n-1} = 0$, and, since y_n is not identically zero, we also have $C_n = 0$. Hence: *The solutions $y_1,y_2,\ldots,y_n$ given by the integrals (57) are linearly independent.* We have thus obtained the general solution in a half-plane, except at -A and the point at infinity.

Remarks.

I. The Laplace equation (48) has for its singularities only the point $-b_o/a_o$ (or -A) and the point at infinity. At the point -A, it is of Fuchsian type. The determinantal equation has a root that could be arbitrary; the others are $0,1,\ldots,n-2$. By assuming that we have taken -A for the origin, we can write the equation in the form

$$xy^{(n)} + (b_1 + a_1x)y^{(n-1)} + \ldots + (b_n + a_nx)y = 0 \quad ; \tag{64}$$

the last root of the determinantal equation is $n - 1 - b_1$, when it is not integral, $n - 1$ of the solutions are holomorphic at the origin. This is the general case; this explains why, in no. 102, we had the possibility of having $n - 1$ holomorphic solutions at the origin.

The series expansions of the solutions about the point $x = 0$ will be convergent for all finite x, but the calculation of the general term of these expansions is impossible for the most part of these cases.

104 - Asymptotic expansions

Let us again take the equation in the reduced form (64). The solutions taken in the form (57) (with $A = 0$) are given, for $|x|$ sufficiently large by

$$y_k = e^{x\alpha_k} x^{-A_k}(\delta_o + \frac{\delta_1}{x} + \ldots + \frac{\delta_q}{x^q} + \frac{0(1)}{x^{q+1}}) \quad , \tag{65}$$

where the δ_j are constants, as are α_k and A_k. This relationship is valid in a half-plane. The expansion within the brackets, that arises from that of

$$(Z) = \Pi'(\alpha_k - \alpha_j + Z)^{A_j - 1} \equiv \beta + \beta_1 Z + \ldots + \beta_q Z^q + \ldots , \quad j \neq k$$

with

$$\delta_j = \frac{-2\pi i\beta_j}{\Gamma(-A_k - j + 1)} \quad ,$$

or, in general, susceptible to having as many terms as is desired. *We shall set aside the special case where, for the* A_j *positive integers,* $\phi(Z)$ *is a polynomial; this is a case where everything simplifies:* $0(1)$ *is replaced by zero in (65) as long as* q *is sufficiently large.*

We are thus in the situation where there is an infinity of non-zero δ_q.

THE FORMULA (65) IS DIFFERENTIABLE

Effectively, y_k is given by

$$y_k = \int_{\Gamma_k} e^{xz}(z - \alpha_k)^{A_k - 1} \; \Pi'(z - \alpha_j)^{A_j - 1} \; dz$$

y_k has a first derivative which is obtained by multiplying by z, the quantity to be integrated which preserves the same form. If then we show that this derivative has an expansion which is deduced from (65) by direct differentiation, then the property will extend to the following derivatives. Now we can apply y_k' what we had done for y_k; when we set $z = \alpha_k + Z$, we will introduce the supplementary term $\alpha_k + Z$; $\phi(Z)$ will be replaced by

$$\phi(Z)(\alpha_k + Z) = \beta_o\alpha_k + (\beta_1\alpha_k + \beta_o Z) + \ldots + (\beta_q\alpha_k + \beta_{q-1})Z^q + \ldots$$

and we will obtain

$$y_k' = e^{x\alpha_k} x^{-A_k}\left(\delta' + \frac{\delta_1'}{x} + \ldots + \frac{\delta_q'}{x^q} + \frac{O(1)}{x^{q+1}}\right)$$

with

$$\delta_j' = \frac{-2\pi i(\beta_j\alpha_k + \beta_{j-1})}{\Gamma(-A_k - j + 1)} = \delta_j\alpha_k + \delta_{j-1}\frac{\Gamma(-A_k - j + 2)}{\Gamma(-A_k - j + 1)}$$

$$= \delta_j\alpha_k - (A_k + j - 1)\delta_{j-1} ,$$

a formula in which we must regard δ_{-1} as being equal to zero. This is the quantity obtained by directly differentiating equation (65). Thus equation (65) is differentiable, as often as is required. It follows that these coefficients $\delta_o, \delta_1, \ldots, \delta_q$ can be calculated step by step by substituting these expansions into the equation, which will give another means of obtaining them. This type of direct calculation has been made by Thome. The calculations are those that would have to be made to produce a series expansion:

$$y_k e^{-x\alpha_k} = \delta_o x^{-r} + \delta_1 x^{-r-1} + \ldots + \delta_q x^{-r-q} + \ldots, \quad r = A_k .$$

This formal calculation is indeed possible, but the series obtained is divergent. *To establish that the series*

$$\delta_o + \frac{\delta_1}{x} + \ldots + \frac{\delta_q}{x^q} + \ldots \tag{66}$$

is divergent, where the δ_q *are calculated by the above formula, it suffices to assume the opposite.* Let us suppose that this entire series converges for certain x. It will define a holomorphic function for $|x| > R$; let $g(x)$ be this function. By substituting $g(x)$ into the differential equation obtained by setting

$$y = e^{x\alpha_k} x^{-A_k}\gamma ,$$

all the coefficients of the powers of $1/x$ will be zero, and we will have a solution. Thus, equation (64) will admit the solution

$$y = e^{x\alpha_k} x^{-A_k} g(x) \tag{67}$$

for $|x| > R$. This solution could be extended along the radii $\arg x = \text{const.}$ in the direction of the origin since the only singular point of equation (64) is $x = 0$. As $e^{-x\alpha_k}$ is holomorphic everywhere and x^{A_k} holomorphic on the radii $\arg x = \text{const}$, except at the origin, we see that

$$g(x) = e^{-x\alpha_k} x^{A_k} y$$

will be extendable up to the origin; the radius of convergence of the entire series (66) will be infinite, and (67) will provide a solution of equation (64) for x close to zero. *This solution cannot be regular, for the product*

$$h(x) \equiv e^{x\alpha_k} g(x) \tag{68}$$

will admit the origin as an essential point, and this will contradict Fuch's theorem, following which the solutions of equation (64) are regular at the origin. The fact that the product (68) necessarily admits the origin as an essential point, where $g(x)$ is the series (66) with infinite radius of convergence, proceeds from the fact that, if $h(x)$ has at most one pole at the origin, then it will be the same for $h(x)e^{-x\alpha_k}$, and hence $g(x)$.

<u>Conclusion</u>. *The expansions (65) obtained for the solutions in the half-planes where they are defined, cannot be replaced by power series expansions, where the series with general term* $\delta_q x^{-q}$ *is divergent. Poincaré asserted that the expansion*

$$\delta_o + \frac{\delta_1}{x} + \dots + \frac{\delta_q}{x^q} + \dots$$

asymptotically represents

$$y_k e^{-x\alpha_k} x^{A_k} \quad ,$$

which signifies that the difference between the sum of the first $q+1$ *terms of the series is of order less than that of* $\frac{1}{|x^q|}$ *as long as* $|x|$ *is sufficiently large.*

These asymptotic expansions, or asymptotic series, analogous to those encountered in the theory of the Euler function (see Stirling's formula, I, 106), had been likewise applied by Poincaré to the general theory of linear differential equations whose coefficients are polynomials. Poincaré studied the operations on these expansions from a more general view-point. We refer to *Leçons sur les séries divergentes* by Borel, for an account of this subject.

105 - Equations whose coefficients are polynomials

Poincaré applied the Laplace integral in the general case of equations of the form

$$P_o(x)y^{(n)} + P_1(x)y^{(n-1)} + \dots + P_n(x)y = 0 \quad , \tag{69}$$

where the $P_j(x)$ are polynomials. On stating that the integral

$$y = \int_\alpha^\beta e^{zx} U(z) dz \tag{70}$$

provides a solution, it will be necessary to calculate the products

$$x^m y^{(q)} = \int_\alpha^\beta e^{zx} z^q x^m U(z) dz \quad .$$

On integrating by parts m times, we obtain

$$x^m y^{(q)} = \sum_{j=0}^{m-1} \left[(-1)^j e^{zx} x^{m-j-1} (z^q U(z))^{(j)} \right]_\alpha^\beta + (-1)^m \int_\alpha^\beta e^{zx} \left[z^q U(z) \right]^{(m)} dz \tag{71}$$

If we assume that the polynomial $P_o(x)$ appearing in (69) has a degree at least equal to the degrees of the following polynomials and if we set

$$P_o(x) \equiv c_j x^p + \sum_{q=1}^{p} c^{q_j} x^{p-q}$$

where p is the degree of $P_o(x)$, we see that equation (69) will be satisfied by the integral (70) if $U(z)$ is a solution of the p-th order differential equation

$$\sum_o^n c_j (z^{n-j} U)^{(p)} + \sum_{q=1}^{p} \sum_o^n (-1)^q c^{q_j} [z^{n-j} U]^{(p-q)} = C \tag{72}$$

and moreover we have

$$[e^{zx} (z^q U)^{(j)}]_\alpha^\beta = 0 \ , \quad q = 0,1,\dots,n, \ j = 0,1,\dots,p-1 \quad . \tag{73}$$

The differential equation (72) is of order p, the coefficients of $U^{(p)}$ is the polynomial

$$c_o z^n + c_1 z^{n-1} + \dots + c_n \quad ; \tag{74}$$

the other coefficients are polynomials. We shall only consider the general case where this polynomial (74) has all its roots distinct. Each of these zeros $\alpha_1, \alpha_2, \dots, \alpha_n$ is a singular point of equation (72) which is of Fuchsian type about this point. Following the remark I of no. 103, about each point α_k, we will have, in the general case, $n - 1$ holomorphic solutions and one non-holomorphic solution of the form:

$$U(z) = (z - \alpha_k)^{r_k}\phi_k(z) \quad , \tag{75}$$

where the function $\phi_k(z)$ is holomorphic in the circle with centre α_k whose radius is the shortest distance from the point α_k to the other points α_j. It is this solution, extended to infinity along a half-line going from α_k to infinity, that was utilised by Poincaré. Having chosen this half-line, we take x in such a way that the real part of zx behaves as $-\mu|z|$, where μ is positive. This will at once guarantee the conditions (73) and the existence of the integral (70), where α and β are points at infinity on this half-line in question, from one part of this half-line to the other, and where x is sufficiently large, which implies that μ is sufficiently large. We have, in effect, the following proposition:

Liapounoff's Theorem. *Let us consider a homogeneous linear differential system*

$$\frac{dy^j}{dx} = \sum_1^n a_{j,k}y_k$$

where the $a_{j,k}$ *are functions of* x *that are holomorphic for* x *real and positive and satisfy an inequality of the form*

$$|a_{j,k}(x)| < K \,, \qquad j = 1,2,\ldots,n, \qquad k = 1,2,\ldots,n; \quad x \geq 0$$

Under these conditions, we can find a positive number μ *such that the solutions* y_j, $j = 1,2,\ldots,n$ *whose modulus is less than* M *at the origin, extended along* $0x$, *satisfy the inequalities*

$$|y_j| < 3\sqrt{n}\,Me^{\mu x} \qquad j = 1,2,\ldots,n \quad .$$

To prove this, let us produce the real part and the coefficient of i, by setting

$$y_j = u_j + iv_j \,, \qquad a_{j,k} = \gamma_{j,k} + i\delta_{j,k} \quad .$$

The system will be replaced by a system of $2n$ homogeneous linear equations with real solutions; we have, for example,

$$\frac{du_j}{dx} = \sum_1^n (\gamma_{j,k}u_k - \delta_{j,k}v_k) \,, \qquad j = 1,2,\ldots,n \quad . \tag{76}$$

The absolute values of the coefficients are bounded by K and at the origin, the solutions u_j, v_j have an absolute value less than M. Let us put $u_j = U_je^{\mu x}$, $v_j = V_je^{\mu x}$, $Y_j = U_j + iV_j$, $y_j = Y_je^{\mu x}$, $j = 1,2,\ldots,n$, where μ is real. Equation (76) becomes

$$\frac{dU_j}{dx} = \sum_1^n (\gamma_{j,k}U_k - \delta_{j,k}V_k) - \mu U_j \qquad j = 1,2,\ldots,n \quad , \tag{77}$$

and we obtain similar equations for the V_j. If we multiply each equation (77) by U_j, the equations in V_j' by V_j, and add them, we obtain

$$\frac{1}{2}\frac{d}{dx}\sum_1^n (U_j^2 + V_j^2) = \Phi(U_j, V_j) \quad ,$$

where the second member is a quadratic form in the $2n$ variables of U_j, V_j. In this form, the coefficients of the products U_jV_k have an absolute value less than 2K, and the coefficients of the squared terms U_j^2 or V_j^2 are of the form $\gamma_{j,j-\mu}$, hence less than $K - \mu$. We therefore have

$$\Phi(U_j, V_j) < K \left[\sum_1^n (|U_j| + |V_j|)\right]^2 - \mu \sum_1^n (U_j^2 + V_j^2) \quad ,$$

for which the maximum of the second member occurs for the equality of $|U_j|$ and $|V_j|$ and is negative as long as $\mu > 2nK$. Under these conditions we will have,

$$\frac{d}{dx}\sum_1^n (U_j^2 + V_j^2) < 0$$

and consequently, on integrating from 0 to x,

$$\sum_1^n (U_j^2 + V_j^2) < 2nM^2 \quad ,$$

since the U_j or V_j have an absolute value less than M for $x = 0$. From this, we have $|U_j| < \sqrt{2n}\,M$, $|V_j| < \sqrt{2n}\,M$, $|Y_j| < 2\sqrt{2n}\,M$, $j = 1,2,\ldots,n$ which proves the theorem.

A CONSEQUENCE

By using this result of Liapounoff, we see that we can apply the arguments in nos. 103 and 104 to the integral in (70). For, in a small circle of centre α_k, the function $U(z)$ has the same properties as the function considered in no. 103. Outside of this circle, along the line of integration, $U(z)$ will behave at most like an exponential $e^{\mu|z|}$ and for as long as $|x|$ is sufficiently large, the exponential e^{zx} will be dominant.

The result of no. 103 relating to the independence of the solutions so obtained in a half-plane, will stand, along with what was said in no. 104: we can make an asymptotic expansion of the solutions so obtained.

In his mémoire, (*American Journal of Mathematics*, 1885), Poincaré examined, in a like manner, the particular cases that we have left aside.

V. VARIOUS RESULTS

106 - Common solutions to two linear equations

Let us consider two linear homogeneous equations

$$P(y) \equiv y^{(n)} + a_1 y^{(n-1)} + \dots + a_n y = 0 \quad ,$$

$$Q(y) \equiv y^{(m)} + b_1 y^{(m-1)} + \dots + b_m y = 0 \quad ,$$

where the a and b are functions of x and where $m \leq n$. We can determine functions of $x, c_1, \dots, c_p$, with $p = n - m$ such that the expression

$$P(y) - \left[[Q(y)^p]^{(p)} + c_1 [Q(y)]^{(p-1)} + \dots + c_p Q(y) \right] \quad ,$$

where the derivatives are taken with respect to x and where y is an arbitrary function of x, no longer contains the derivative of y of order greater than $m - 1$. The functions c_j are obtained step by step: we have $c_1 = a_1 - b_1$, etc. We thus obtain

$$P(y) \equiv [Q(y)]^{(p)} + c_1 [Q(y)]^{(p-1)} + \dots + c_p Q(y) + c_{p+1} R(y) \tag{78}$$

with

$$R(y) \equiv y^{(q)} + d_1 y^{(q-1)} + \dots + d_q y \quad , \qquad q < m \quad ,$$

where the c_{p+1} and the d_j are functions of x. The identity in (78) is similar to the division identity. If the given equations have a common solution other than 0, then this function is a solution of $R(y) = 0$. Conversely, the common solutions of $R(y) = 0$ and $Q(y) = 0$, are solutions of $P(y) = 0$. *Determining the common solutions of* $Q(y) = 0$ *and* $P(y) = 0$ *is reduced to determining the common solutions of* $Q(y) = 0$ *and* $R(y) = 0$. *If* $R(y) = 0$, *then all the solutions of* $Q(y) = 0$, *satisfy* $P(y) = 0$. *If* $q = 0$, *hence if* $R(y) \equiv y$, *then the two equations have no other common solution other than* $y = 0$. If one of these two cases do not occur, then we can recommence the same operation with Q and R. After a number of operations, at most m, we will arrive at a remaining $R_k(y)$ which will be identically zero or will reduce to y. *The common solutions of* $P(y) = 0$ *and* $Q(y) = 0$ *will not have any common solutions if* $R_k(y)$ *reduces to* y. This method, due to Brassinne, is not far from the method of finding the greatest common divisor of two numbers or of two polynomials.

These considerations lead on to the notion of a *reducible* or *irreducible* linear equation. In the special case where the coefficients of the equation are rational fractions of the variable x, Picard asserted that *the equation is reducible if one of its solutions is a solution of a linear equation of a lower order, with rational coefficients.* If this is so, then the given equation will admit, on account of the preceding discussion, all the solutions of a linear equation of a lower order with rational coefficients. We can recognise this to be the case when we have an n-th order differential equation (see Picard, *Traité d'Analyse*, t. III). We shall confine ourselves to the case of a second order equation

$$y'' + a_1 y' + a_2 y = 0 \quad , \tag{79}$$

whose coefficients a_1, a_2 are rational fractions. If it is reducible, it admits solutions of a first order equation $y' + by = 0$, where b is a rational fraction. This amounts to saying that equation (79) admits an integral whose logarithmic derivative is rational. Now, if we set $y' = uy$, equation (79) is transformed into the Riccati equation (no. 72)

$$u' + u^2 + a_1 u + a_2 = 0 \quad , \tag{80}$$

and we must check that this equation admits a rational solution, where a_1 and a_2 are rational fractions. If c is a pole of a_1 or a_2, then the principal part of u about c,

$$u = \frac{A_\gamma}{(x-c)^\gamma} + \ldots + \frac{A_1}{x-c} + \ldots$$

can be calculated on substituting into equation (80); this will provide the value of γ and the possible values of $A_\gamma, \ldots, A_1$. This is the same for the principal part of u about the point at infinity. If d is a pole of u distinct from the points c, then it is necessarily a simple pole since equation (80) must be satisfied about this point; hence, u' and u^2 must be of the same order. We must have

$$u = \frac{D}{x-d} + \ldots \quad ,$$

and on substituting, we see that $D = 1$. Thus, we must also have

$$u = \frac{y'}{y} = R(x) + \sum \frac{1}{x-d} = R(x) + \frac{U'(x)}{U(x)}$$

where $R(x)$ is a known rational function (there are several possible solutions), and $U(x)$ is a polynomial. We will obtain

$$y = e^z U \quad , \qquad z = \int R(x)dx$$

and, on substituting into (79), we see that the polynomial U will satisfy an equation of the form

$$U''A_1 + U'A_2 + UA_3 = 0 \quad ,$$

where A_1, A_2, A_3 are polynomials; this will determine U if the problem is possible.

The theory relating to the symmetric functions of the roots of an algebraic equation have been extended by Appell to linear differential equations. The transformation theory and the Galois theory have been extended by Picard and Vessiot.

107 - The group of a linear equation

Given a linear equation of order n whose coefficients are holomorphic functions in all of the plane except at a finite number of points (we assume that the coefficient of $y^{(n)}$ is equal to 1) consider a fundamental system of solutions

$$y_1, y_2, \ldots, y_n$$

taken to be in a simply connected region containing no singularities of the coefficients. When the variable x describes a closed path starting from a point of this domain, and returning there after having gone round the singular points, the solutions y_j take the values Y_j which are given by

$$Y_j = \lambda_j^1 y_1 + \lambda_j^2 y_2 + \ldots + \lambda_j^n y_n \qquad j = 1,2,\ldots,n \tag{81}$$

where the λ are constant. The set of the various substitutions (81) relative to all of the paths considered, *constitutes the group of the differential equation in question*. It is a group of substitutions. For if we consider an initial substitution S, corresponding to a circuit S' which induces a passage from the system y_j to Y_j, then a new substitution T is induces on the set, corresponding to a circuit T'. It will induce a passage of the Y_j to solutions Z_j which will be those obtained by commencing from the y_j and successively circuiting S', then T'. The Z_j will then again be given in terms of a function the y_j by a substitution from the set. Hence, *the product* ST *of two substitutions of the set, belongs to the set,* with the product given as above. The inverse substitution of (81), obtained by solving for y_j, also belongs to the set; it corresponds to the circuit S' made in the opposite sense. *The substitutions* (81) *indeed form a group* G.

As there are only a finite number of singular points, then the paths S' can reduce to a certain number of braids described about these points. If $S_1, S_2, \ldots, S_m$ are the substitutions corresponding to these various braids, then every substitution S will be a product of these substitutions taken in a certain order. S will depend on this order.

If we replace the fundamental system $y_1, y_2, \ldots, y_n$ by another fundamental system, then the group is changed. But if we introduce the substitution τ which induces a passage from the first system to the second (τ does not form part of the group G, in general), then every substitution of the new group G' will deduce from a substitution S of G: it will be of the form $\tau S \tau^{-1}$, where τ^{-1} is the inverse of τ. The operation inducing a passage from one of these groups to the other is called a *transmutation*.

If the equation in question is reducible, then there will exist a system of $m < n$ solutions of the form

$$\mu_j' y_1 + \ldots + \mu_j^n y_n \quad , \qquad j = 1,2,\ldots,m$$

which will be solutions of an equation of order m. These solutions permute between themselves when substitutions of the group G are applied. We shall say that G is not a *primary* group. To determine the irreducibility of an equation reduces to determining whether its group is primary or not.

108 - The adjoint equation

The theory of the integrating factor was extended by Lagrange to homogeneous linear equations. If

$$P(y) \equiv a_o y^{(n)} + \ldots + a_n y = 0$$

is such an equation, then the a_j are functions of the variable x. We look for a function $u(x)$ such that the product uP is the derivative with respect to x of a linear function

$$b_o y^{(n-1)} + \ldots \, b_{n-1} y \quad ,$$

where the b are functions of x. By applying the general formula for integration by parts (I, 49), we have

$$a_{n-p} u_y^{(p)} = \frac{d}{dx}\left[a_{n-p} u y^{(p-1)} - y^{(p-2)} \frac{d}{dx}(a_{n-p}u) + \ldots + (-1)^{(p-1)} y \frac{d^{p-1}}{dx^{p-1}}(a_{n-p}u)\right] + (-1)^p y \frac{d^p}{dx^p}(a_{n-p}u) \; .$$

We must then take for u a solution of the equation

$$Q(u) \equiv \sum_{p=0}^{n} (-1)^p \frac{d^p}{dx^p}(a_{n-p}u) = 0 \quad ,$$

and we will then have

$$uP(y) \equiv \frac{d}{dx}[b_o y^{(n-1)} + \ldots + b_{n-1} y] \quad .$$

The coefficients b_j are deduced from the expression (82), we have

$$b_j = \sum_{k=0} (-1)^{j-k} \frac{d^{j-k}}{dx^{j-k}} (a_k u) \quad . \tag{83}$$

The linear equation of order n, $Q(u) = 0$ is called the *adjoint equation of* $P(y) = 0$. If one solution is known, then matters are reduced to the equation of order $n - 1$.

$$b_o y^{(n-1)} + \ldots + b_{n-1} y = \text{const.}$$

To each solution of the adjoint equation, there thus corresponds an equation of order $n - 1$. If we know p distinct solutions of $Q(u) = 0$, then y will satisfy p linear equations of order $n - 1$. On eliminating $y^{(n-1)}$, $\ldots, y^{(n-p+1)}$ between these equations, we will reduce matters to an equation of order $n-p$ alone. In particular, if we can integrate the adjoint, then we have the general integral of the given equation, without quadratures.

Remark. The adjoint equation of $Q(u) = 0$, is $P(y) = 0$. It can be satisfied by commencing with the expressions of the coefficients of the $u^{(k)}$ in $Q(u)$. But it is simpler to note that, with the b_j defined by (83), we have, for any y and e,

$$uP(y) - yQ(u) \equiv \frac{d}{dx} [b_o y^{(n-1)} + \ldots + b_{n-1} y] \quad .$$

If $R(y)$ is the adjoint to $Q(u)$, then likewise we have for any x, y, u

$$yQ(u) - uR(y) \equiv \frac{d}{dx} [c_o u^{(n-1)} + \ldots + c_{n-1} u] \quad .$$

By adding these equations, we see that

$$u[P(y) - R(y)]$$

is the derivative with respect to x of a function of x, y, u, linear with respect to the $u, u', \ldots, u^{(n-1)}$. This function is identically zero. For, if one of the terms is non-zero, then by differentiating, we obtain a linear expression with respective to certain $u^{(k)}$, $P(y) - R(y)$ will depend on u. We therefore have $R(y) \equiv P(y)$.

109 - Equations with algebraic coefficients

The theory of Fuchsian functions, of which a few words were said in no. 31, was developed by Poincaré with the aim of solving the following problem: *to integrate in terms of other transcendental functions, every linear differential equation with algebraic coefficients.* It concerns equations of the form

$$y^{(n)} + a_1(x,u)y^{(n-1)} + \ldots + a_n(x,u)y = 0 \ , \quad y = y(x) \quad , \tag{84}$$

where the $a_j(x,u)$ are rational fractions of x and u, and where u is determined as a function of x by the algebraic equation $P(x,u) = 0$. Poincaré firstly proved that every Fuchsian function provides the inversion of the quotient of two solutions of a second order equation with algebraic coefficients. The Fuchsian functions thus admit the integration of certain classes of second order differential equations of the type envisaged. To pass to the general case of equation (84), Poincaré introduced some other transcendental functions, which are no longer Fuchsian functions; these are called *zeta-Fuchsian functions*. The general integral of the equations of the form (84) can be expressed in terms of these new functions.

Chapter VII

SOME SECOND ORDER LINEAR DIFFERENTIAL EQUATIONS

In this chapter, we intend applying the general results obtained in the theory of linear equations to some second order equations which are frequently encountered in applications: the equations of Gauss, Legendre and Bessel. Some of these equations had been studied prior to the development of the general theory. The Bessel equation corresponding to integer values of the parameter was integrated by Euler as long ago as 1764. The Legendre polynomials were defined by Legendre in 1785 and put into their present day form by Rodrigues in 1815; they amount to a particular case of the class of spherical functions introduced by Laplace (1785). The Bessel functions were studied by Bessel (1819) and Fourier (1822). The hypergeometric series, which contain nearly all the elementary series as particular cases, was studied by Gauss in 1812. But it was only towards the middle of the 19th century that these equations and functions were studied in greater depth and were subjected to complex variable methods, in particular, by Neumann, Riemann, Schläfli, Heine and Hankel.

I. THE GAUSS EQUATION AND HYPERGEOMETRIC FUNCTIONS

110 - A second order equation with three regular singularities

Following what was said in no. 98, if we make a homographic transformation on the variable x, then a Fuchsian equation is transformed into a Fuchsian equation of the same type. Through such a transformation, we can replace a second order equation admitting three singularities about a, b, c about which the solutions are regular, by another equation for which the singularities will be $0, 1, \infty$. There are six ways of doing this. Let us consider then, a second order equation admitting the points $0, 1, \infty$ as singularities; the solutions will be regular about these points. About $x=0$ and $x=1$, we will have at least one solution of the form

$$x^r\phi(x) \ , \quad (x-1)^{r'}\phi_1(x) \quad ,$$

where r and r' are roots of the determinantal equation relative to these points $x = 0$, $x = 1$ and where ϕ and ϕ_1 are holomorphic about these points. If we make the change of the unknown function

$$y = x^r(x - 1) \quad ,$$

then z will satisfy a new second order equation whose singularities will again be $x = 0$, $x = 1$, $x = \infty$ and which will admit a holomorphic solution about $x = 0$ and $x = 1$. The determinantal equations relative to these points will have a zero root. This equation in z will be of the form

$$x^2(x-1)^2 z'' + x(x-1)P(x)z' + zQ(x) = 0 \quad , \tag{1}$$

where $P(x)$ and $Q(x)$ are holomorphic at every point at finite distance. If we change x to $1/X$, we obtain the equation

$$(1-X)^2 z'' + \frac{1-X}{X}\left[2(1-X) - XP\left(\frac{1}{X}\right)\right]z' + Q\left(\frac{1}{X}\right)z = 0$$

which is of Fuchsian type about $X = 0$. It follows that

$$XP\left(\frac{1}{X}\right) \; , \quad X^2Q\left(\frac{1}{X}\right)$$

must be holomorphic for $X = 0$; $P(x)$ is a polynomial of the first degree and $Q(x)$ is a polynomial of the second degree. The determinantal equations of equation (1) for $x = 0$ and $x = 1$ are

$$r(r-1) - rP(0) + Q(0) = 0 \; , \qquad r(r-1) + rP(1) + Q(1) = 0 \quad ,$$

and must have a zero root. We have $Q(0) = Q(1) = 0$; $Q(x)$ is of the form $Ax(x-1)$, where A is constant, and equation (1) is written as

$$x(x-1)z'' + (Bx + C)z' + Az = 0 \quad ,$$

where A, B, C are constants. By changing the notation and now putting y in place of z, we see that *every second order equation having only three regular singularities, reduces to the form*

$$x(1-x)y'' + [\gamma - (\alpha+\beta+1)x]y' - \alpha\beta y = 0 \quad , \tag{2}$$

where α, β, γ *are arbitrary constants.* Following what came before, there are at least six ways of obtaining this reduced form.

A PARTICULAR CASE

When the primitive equation is of the first Fuchsian type about singular points a, b, c and the three determinantal equations relative to these points have roots whose difference is not an integer (which implies that it is of the first type), then the roots will be preserved under a homographic transformation which assigns 0, 1, ∞ to a, b, c respectively. In the transformation $y = x^r(x-1)^{r'}z$, we can take for r, one or other of the roots of the determinant equation relative to $x = 1$. There will then be 24 ways of reducing the equation to the form in (2); but here α, β, γ will not be completely arbitrary, the roots of the determinantal equation of (2), for $x = 0$, $x = 1$ $x = \infty$, are distinct and their difference is non-integral.

THE EQUATION (2) IS THE GAUSS EQUATION

The roots of the determinantal equations relative to the three singular points $0, 1, \infty$ *are respectively the roots of*

$$r(r-1) + \gamma r = 0, \quad r(r-1) - (\gamma - \alpha - \beta - 1)r = 0 \quad ,$$

$$-r(r+1) + (\alpha + \beta + 1)r - \alpha\beta = 0 \quad ,$$

hence

$$0, 1-\gamma, \quad 0, \gamma-\alpha-\beta, \quad \alpha, \beta \quad .$$

111 - The hypergeometric function $F(\alpha,\beta,\gamma,x)$

If γ is not integral, or if γ is integral, $1-\gamma$ is negative or zero, the Gauss equation (2) admits a holomorphic solution at the origin (since 0 is the greatest root of the determinantal equation in the case where the difference of the roots is integral). If we take the solution to be equal to one at the origin, its series expansions given by

$$y = 1 + c_1 x + \dots + c_n x^n + \dots \quad , \tag{3}$$

then the recurrence relation between the c_n, obtained by equating the coefficient of x to be zero in (2) when y, y', y'' are replaced by their expansions, is

$$(n+1)(n+\gamma)c_{n+1} - [n(n-1) + n(\alpha+\beta+1) + \alpha\beta]c_n = 0$$

(where c_o is equal to 1). The coefficient of c_n is zero for $n = -\alpha$ and $n = -\beta$, and we have

$$c_{n+1} = \frac{(n+\alpha)(n+\beta)}{(n+1)(n+\gamma)} c_n \quad . \tag{4}$$

When neither α, nor β are negative integers, we obtain a function defined by the series (3) whose radius of convergence following (4), is equal to 1. Moreover, we know that the solution is holomorphic in the circle $|x| < 1$. *We call this series a hypergeometric series, and denote it by* $F(\alpha,\beta,\gamma,x)$. We have

$$F(\alpha,\beta,\gamma,x) = 1 + \sum_{n=1}^{\infty} \frac{\alpha(\alpha+1)\dots(\alpha+n-1)\beta(\beta+1)\dots(\beta+n-1)x^r}{n!\gamma(\gamma+1)\dots(\gamma+n-1)} \tag{5}$$

By introducing the function $\Gamma(u)$, we know that

$$\Gamma(u+n) = (u+n-1) \ldots (u+1)u\Gamma(u)$$

and we can write

$$\frac{\Gamma(\alpha)\Gamma(\beta)}{\Gamma(\gamma)} F(\alpha,\beta,\gamma,x) = \sum_{0}^{\infty} \frac{\Gamma(\alpha+n)\Gamma(\beta+n)}{\Gamma(\gamma+n)} \frac{x^n}{n!} \quad .$$

The analytic function defined by the analytic continuation of the hypergeometric series is called a *hypergeometric function*; we shall continue to denote it by $F(\alpha,\beta,\gamma,x)$.

Remark. We have

$$(1-x)^{-\alpha} = F(\alpha,\beta,\beta,x), \quad \log(1-x) = -xF(1,1,2,x) \quad .$$

THE GENERAL SOLUTION

For $|x| < 1$. *If we assume that* γ *is not an integer*, then the Gauss equation admits a second solution

$$y = x^{1-\gamma} u \quad , \tag{6}$$

where u is holomorphic for $|x| < 1$. Changing the unknown function (6) transforms the Gauss equation to another equation of the same type, following what was said above. We see that u satisfies the equation

$$x(1-x)u'' + [2-\gamma-(\alpha+\beta-2\gamma+3)x]u' - (\alpha+1-\gamma)(\beta+1-\gamma)u = 0 \quad .$$

We therefore have

$$u = F(\alpha-\gamma+1, \beta-\gamma+1, 2-\gamma, x)$$

and the general solution is, in this case, for $|x| < 1$.

$$y = \lambda F(\alpha,\beta,\gamma,x) + \mu x^{1-\gamma} F(\alpha-\gamma+1,\beta-\gamma+1,2-\gamma,x) \tag{7}$$

where λ and μ are arbitrary constants. *When* γ *is integral,* there are two cases to be distinguished. If $\gamma \neq 1$, then the difference of the roots of the determinantal equation is integral with the greatest root corresponding to a solution of the form obtained; the form of the second solution will depend on the values of α,β. When $\gamma = 1$, the determi equation has a double root, and there will be a logarithmic term in the second solution. We can obtain this second solution by d'Alembert's method. Let us assume $\gamma \neq 1$ and is near to 1. The general solution is of the form (7) obtained above. We have, in particular, the solution

$$\frac{1}{\gamma-1} [F(\alpha,\beta,\gamma,x) - x^{1-\gamma} F(\alpha-\gamma+1,\beta-\gamma+1,2-\gamma,x)] \tag{8}$$

which for $x = 1/2$, for example, along with its first derivative, takes well determined values when the branch of $x^{1-\gamma}$ has been chosen. These values y_o and y'_o are clearly continuous with respect to γ, and γ enters analytically into the Gauss equation. Following the Poincaré theorem, the solution taking the value y_o for $x = 1/2$ and whose derivative is y'_o, is continuous with respect to γ. The limit of the expression (8) for $\gamma = 1$, therefore yields a solution. As $F(\alpha-\beta+1,\beta-\gamma+1,2-\gamma,x)$ tends towards $F(\alpha,\beta,1,x)$, we can write (8) in the form

$$\frac{1-x^{1-\gamma}}{-1} F(\alpha-\beta+1,\beta-\gamma+1,2-\gamma,x) + \frac{1}{\gamma-1}[F(\alpha,\beta,\gamma,x)$$

$$- F(\alpha-\gamma+1,\beta-\gamma+1,2-\gamma,x)]$$

and when γ tends towards one, the first term tends towards $F(\alpha,\beta,1,x)\log x$. As for the second, its limit is obtained by writing the bracketed term as $[F(\alpha,\beta,\gamma,x) - F(\alpha,\beta,1,x)] + [F(\alpha,\beta,1,x) - F(\alpha,\beta,2-\gamma,x)] + [F(\alpha,\beta,2-\gamma,x) - F(\alpha,\beta-\gamma+1,2-x,\gamma)] + F(\alpha,\beta-\gamma+1,2-\gamma,x) - F(\alpha-\gamma+1,\beta-\gamma+1,2-\gamma,x)$; this is then

$$2\frac{\partial}{\partial\gamma}F(\alpha,\beta,1,x) + \frac{\partial}{\partial\beta}F(\alpha,\beta,1,x) + \frac{\partial}{\partial\alpha}F(\alpha,\beta,1,x) \quad .$$

Since, for example,

$$\frac{\partial}{\partial\alpha}[\alpha(\alpha+1)\dots(\alpha+n-1)] = \alpha(\alpha+1)\dots(\alpha+n-1)\sum_{j=1}^{n}\frac{1}{\alpha+j-1} \quad ,$$

we obtain as a second solution

$$F(\alpha,\beta,1,x)\log x + \sum_{1}^{\infty}\left[c_n\sum_{j=0}^{n-1}\left(\frac{1}{\alpha+j} + \frac{1}{\beta+j} - \frac{2}{j+1}\right)x^n\right]$$

where the c_n are the coefficients of the x^n in $F(\alpha,\beta,1,x)$. This last expansion converges for $|x| < 1$; the solution obtained goes into this circle.

THE GENERAL SOLUTION FOR $|x-1| < 1$.

Henceforth, let us confine ourselves to the general case where γ, $\beta-\alpha$, $\alpha+\beta-\gamma$ are not integers, and hence where the determinantal equations have distinct roots with non-integral differences; the equation is of the first type about the three singular points. If we set $x = 1 - X$, we recover a Gauss equation. We obtain

$$X(1-X)y'' - [\gamma(\alpha+\beta+1)(1-X)]y' - \alpha\beta y = 0 \quad .$$

It is a Gauss equation where α and β have retained the same value, but where γ is replaced by $\alpha + \beta - \gamma + 1$. We deduce that the Gauss equation (2) admits the general solution

$$y = \lambda' F(\alpha,\beta,\alpha+\beta-\gamma+1,1-x) + \mu'(1-x)^{\gamma-\alpha-\beta} F(\gamma-\beta,\gamma-\alpha, \\ 1+\gamma-\alpha-\beta,1-x) \quad , \tag{9}$$

which is valid for $|x-1| < 1$, where λ and λ' are arbitrary constants.

THE GENERAL SOLUTION FOR $|x| > 1$

Likewise, by setting $x = 1/X$, we transform the Gauss equation to another Gauss equation, if we simultaneously set $y = X^{\alpha}Y$ (or $y = X^{\beta}Y$) in such a way so as to annihilate one root of the determinantal equation. We obtain the equation

$$(1-X)XY'' + [\alpha+1-\beta-(2\alpha+2-\gamma)X]Y' - \alpha(\alpha+1-\gamma)Y = 0 \quad ,$$

whose general solution, for $|X| < 1$, is deduced from the expression in (7). We extract, for $|x| > 1$, the general solution of equation (2), in the form

$$y = \lambda'' x^{-\alpha} F(\alpha,\alpha+1-\gamma,\alpha+1-\beta, \frac{1}{x}) + \mu'' x^{-\beta} F(\beta,\beta+1-\gamma,1+\beta-\alpha, \frac{1}{x})$$

where λ'' and μ'' are arbitrary constants.

Remarks

I. We could likewise consider the equation transformed by the 18 other transformations envisaged in no. 110. We would obtain 18 other solutions that could be associated pairwise, as the six considered above, to obtain the other forms of the general solution. The solutions thus obtained are evidently expressed in terms of those given.

II. The solutions (7), (9), (10) respectively valid in the domains $|x| < 1$, $|x-1| < 1$, $|x| > 1$ provide the solution in the entire plane since the second of these circles has a portion common with the first and with its exterior. But it is fitting to find out how one passes from one solution to another.

112 - The continuation of the solutions

We will consider the case where at least one of the differences between the roots of the characteristic equations has a non-zero real part. Since we can permute the roles of the points $0, 1, \infty$ and that of the roots in the transformations providing the twenty-four forms of the equation, we can assume that the real part of $\gamma - \alpha - \beta$ is positive

$$R(\gamma-\alpha-\beta) > 0 \quad . \tag{11}$$

Under these conditions, i.e., (11) satisfied, the hypergeometric series

$$F(\alpha,\beta,\gamma,x),\quad F(\alpha-\gamma+1,\beta-\gamma+1,2-\gamma,x),$$

$$F(\alpha,\alpha-\gamma+1,\alpha-\beta+1,\tfrac{1}{x}),\quad F(\beta,\beta-\gamma+1,\beta-\alpha+1,\tfrac{1}{x}) \quad ,$$

is again absolutely convergent for $|x| = 1$.

In $F(\alpha,\beta,\gamma,x)$, we have

$$\frac{c_{n+1}}{c_n} = 1 + \frac{\alpha+\beta-\gamma-1}{n} + \frac{0(1)}{n^2}$$

where c_n is the coefficient of x^n, hence (I, 18),

$$\left|\frac{c_n}{c_{n+1}}\right| = 1 + \frac{R(1+\gamma-\alpha-\beta)}{n} + \frac{0(1)}{n^2}$$

and following Duhamel's rule (I, 18), the series with general term $|c_n|$ is convergent since $R(1+\gamma-\alpha-\beta) > 1$. In the three other series, the quantity playing the role of $1+\gamma-\alpha-\beta$ retains the same value; the hypothesis of (11) thus implies the same conclusion. We assume that the series exist: γ and $\alpha-\beta$ are not integers.

CONTIGUOUS FUNCTIONS

Gauss asserted that the hypergeometric functions $F(\alpha,\beta,\gamma,\delta)$ and $F(\alpha',\beta',\gamma',x)$ are contiguous when two of the parameters are equal, e.g., $\alpha' = \alpha$, $\gamma' = \gamma$, whilst the third parameters are different by one unit, $\beta' = \beta \pm 1$. There exist six functions contiguous with $F(\alpha,\beta,\gamma,x)$.

Theorem. *The function* $F(\alpha,\beta,\gamma,x)$ *and any two contiguous with it, satisfy a linear relationship whose coefficients are polynomials of the first degree in* x.

We will obtain a number of relations equal to $\frac{1}{2}\cdot 6\cdot 5 = 15$. In order to prove the theorem, it suffices to prove it for $|x| < 1$. Let us consider for example $F(\alpha,\beta,\gamma,x)$ and $F(\alpha-1,\beta,\gamma,x)$, $F(\alpha+1,\beta,\gamma,x)$. If c_n is the coefficient of x^n in the function, then this coefficient will be

$$c_n\frac{\alpha-1}{\alpha+n-1} \quad , \quad c_n\frac{\alpha+n}{\alpha}$$

in the other two, and in the product of these functions and x, the coefficient of x^n will be respectively

$$c_n \frac{n(n+\gamma-1)}{(n+\alpha-1)(n+\beta-1)} \quad , \quad c_n \frac{\alpha-1}{\alpha+n-2} \; \frac{n(n+\gamma-1)}{(n+\alpha-1)(n+\beta-1)}$$

$$c_n \frac{n(n+\gamma-1)}{\alpha(n+\gamma-1)} \quad .$$

In the sum

$$(A + Bx)F(\alpha+1,\beta,\gamma,x) + (C + Dx)F(\alpha,\beta,\gamma,x) + EF(\alpha-1,\beta,\gamma,x)$$

where A, B, C, D, E are constants, the coefficient of x^n will be the product of x^n and a rational fraction in n whose numerator will be of the third degree. The coefficients of this numerator are linear and homogeneous in A, B, C, D, E and we can determine these five numbers via four linear homogeneous equations and equation all the terms in zero.

We proceed in the same way to obtain the relationships corresponding to the values $\beta-1$, β, $\beta+1$ or $\gamma-1$, γ, $\gamma+1$ (α and γ in the first case, α and β in the second case, are interchanged). Likewise, we find

$$(\alpha-\beta)F(\alpha,\beta,\gamma,x) \equiv \alpha F(\alpha+1,\beta,\gamma,x) - \beta F(\alpha,\beta+1,\gamma,x),$$

$$(\alpha-\gamma+1)F(\alpha,\beta,\gamma,x) \equiv \alpha F(\alpha+1,\beta,\gamma,x) + (1-\gamma)F(\alpha,\beta,\gamma-1,x) \quad ,$$

relations which, when combined with the other three, lead to the general result. Amongst these relations, we shall utilise the following:

$$\begin{aligned}\gamma[\gamma-1-(2\gamma-\alpha-\beta-1)x]F(\alpha,\beta,\gamma,x) - \gamma(\gamma-1)(1-x)F(\alpha,\beta,\gamma-1,x) \\ = -(\gamma-\alpha)(\gamma-\beta)xF(\alpha,\beta,\gamma+1,x) \quad ,\end{aligned} \tag{12}$$

which is obtained as we have just said.

<u>Corollary I</u>. *If condition (11) occurs, we have*

$$\frac{F(\alpha,\beta,\gamma,1)}{F(\alpha,\beta,\gamma+1,1)} = \frac{(\gamma-\alpha)(\gamma-\beta)}{\gamma(\gamma-\alpha-\beta)} \quad . \tag{13}$$

Let us assume first of all that $R(\gamma-\alpha-\beta) > 1$. The three hypergeometric series occuring in relation (12) are then convergent for $x = 1$, and by setting $x = 1$, we obtain equation (13). But in this equality where we have assumed α and β fixed and γ variable, and which can be written as

$$\gamma(\gamma-\alpha-\beta)F(\alpha,\beta,\gamma,1) - (\gamma-\alpha)(\gamma-\beta)F(\alpha,\beta,\gamma+1,1) = 0 \quad , \tag{14}$$

the functions F, regarded as functions of γ, are holomorphic for γ distinct from negative integers or zero and for $R(\gamma-\alpha-\beta) \geq \eta > 0$, since the inequality above can then be written as

$$\left|\frac{c_n}{c_{n+1}}\right| > 1\,\frac{\eta'+1}{n} \quad , \qquad 0 < \eta' < \eta \qquad n > n'$$

and implies that $F(\alpha,\beta,\gamma,1)$ has its terms less in modulus than those of a numerical series. It follows that the first member of (14) is holomorphic for $R(\gamma-\alpha-\beta) > 0$, $\gamma \neq p$, $p = 0,1,\ldots$, and is zero for $R(\gamma-\alpha-\beta) > 1$, hence everywhere. This proves equation (13) via (11).

Corollary II. *By means of (11), we have*

$$F(\alpha,\beta,\gamma,1) = \frac{\Gamma(\gamma)\Gamma(\gamma-\alpha-\beta)}{\Gamma(\gamma-\alpha)\Gamma(\gamma-\beta)} \quad . \tag{15}$$

Since it is deduced from (13) by assigning the values $\gamma,\gamma+1,\ldots,\gamma+p-1$ to γ and taking the product of the equations obtained

$$\frac{F(\alpha,\beta,\gamma,1)}{F(\alpha,\beta,\gamma+p-1)} = \frac{(\gamma-\alpha)(\gamma+1-\alpha)\ldots(\gamma+p-\alpha-1)}{\gamma(\gamma+1)\ldots(\gamma+p-1)}\;\frac{(\gamma-\beta)\ldots(\gamma+p-\beta-1)}{(\gamma-\alpha-\beta)\ldots(\gamma-\alpha-\beta+p-1)} \quad .$$

When $p \to \infty$, each term of $F(\alpha,\beta,\gamma+p,1)$, except the first, tends towards zero and since this series has its terms less than that of a convergent numerical series, $F(\alpha,\beta,\gamma+p,1) \to 1$ (the Weierstrass rule).

On the other hand, we have, on account of the properties of the function $\Gamma(u)$, (I, 83)

$$\Gamma(u) = \lim_{p=\infty} \frac{(p-1)!\,(p-1)^u}{u(u+1)\ldots(u+p-1)} \quad . \tag{16}$$

This equality likewise holds for u complex [since it also results from the decomposition formula into factors of $\Gamma(u)$, (I, 83) which has been extended to the case of u complex]. As $(\gamma-\alpha) + (\gamma-\beta) = \gamma + (\gamma-\alpha-\beta)$, the equation (15) is deduced directly from (16).

AN APPLICATION TO THE EMBEDDING OF SOLUTIONS

Condition (11) having been satisfied, the functions $F(\alpha,\beta,\gamma,x)$, $F(\alpha-\gamma+1,\beta-\gamma+1,2-\gamma,x)$, $F(\alpha,\alpha-\gamma+1,\alpha-\beta+1,1/x)$ and $F(\beta,\beta-\gamma+1,$ $\beta-\alpha+1, 1/x)$ will take at the point 1 (when x tends to this point on remaining in the circle $|x| \leq 1$ for the other two), the respective values

$$A = \frac{\Gamma(\gamma)\Gamma(\gamma-\alpha-\beta)}{\Gamma(\gamma-\alpha)\Gamma(\gamma-\beta)} \quad , \qquad A' = \frac{\Gamma(2-\gamma)\Gamma(\gamma-\alpha-\beta)}{\Gamma(1-\alpha)\Gamma(1-\beta)} \ ,$$

$$A'' = \frac{\Gamma(\alpha-\beta+1)\Gamma(\gamma-\alpha-\beta)}{\Gamma(1-\beta)\Gamma(\gamma-\beta)} \quad , \qquad A''' = \frac{\Gamma(\beta-\alpha+1)\Gamma(\gamma-\alpha-\beta)}{\Gamma(1-\alpha)\Gamma(\gamma-\alpha)}$$

Let us consider the general solution taken in the form (7). Let us assume that the x-plane intersects along the portion of the real axis joining the point 0 to infinity $(0 \leq x < \infty)$ and take in $x^{1-\gamma}$ the branch which is defined by e^u with $u = (1-\gamma)\log x$, where $\log x$ is taken to be real above the intersection. The solution (7) then has the form

$$y = \lambda y_1 + \mu y_2 \quad ,$$

where y_1 and y_2 are uniform and defined for $|x| \leq 1$. Likewise, the solution (10) will be uniform in the plane so cut out, and will be determined if we take

$$x^{-\alpha} = e^{-\alpha \log x} \quad , \qquad x^{-\beta} = e^{-\beta \log x} \quad ,$$

where the branch of $\log x$ is the same as above. This solution (10) can be written as

$$y = \lambda'' y_3 + \mu'' y_4 \quad ,$$

where y_3 and y_4 are uniform and defined for $|x| \geq 1$. This second solution will be the extension of the first if, for $|x| = 1$, $x \neq 1$, we have

$$\lambda y_1 + \ y_2 = \lambda'' y_3 + \mu'' y_4 \quad .$$

When x, with modulus 1, tends toward $x = 1$ above the intersection, we have

$$\lambda A + \mu A' = \lambda'' A'' + \mu'' A'''$$

and when $x \to 1$, $|x| = 1$, below the intersection, we likewise have

$$\lambda A + \mu A' e^{-2\pi i \gamma} = \lambda'' A'' e^{-2\pi i \alpha} + \mu'' A''' e^{-2\pi i \beta} \quad .$$

These two equations yield λ'' and μ'' as a function of λ and μ, where the determinant of λ'' and μ'' is, to within a factor of positive modulus, equal to

$$e^{2\pi i(\alpha - \beta)} - 1 \quad ,$$

a quantity that is non-zero since $\alpha - \beta$ is not an integer.

113 - The Jacobi polynomials

When β is a negative integer, $-n$ say, the function $F(\alpha, -n, \gamma, x)$ is a polynomial of degree n. *The polynomials*

$$\Phi_n(x) = F(\alpha + n, -n, \gamma, x)$$

were introduced by Jacobi. They generalise the Legendre polynomials.

More generally, if y is a solution of the Gauss equation (2), its derivative $z = y'$ satisfies the equation obtained by differentiating (2):

$$x(1-x)z'' + [\gamma + 1 - (\alpha+\beta+3)]z' - (\alpha+1)(\beta+1)z = 0 \quad .$$

This is the Gauss equation where α, β, γ are replaced by $\alpha + 1, \beta + 1, \gamma + 1$. It follows that, for u demoting the derivative of order $n - 1$ of y, we shall obtain

$$x(1-x)u'' + [\gamma + n - 1 - (\alpha+\beta+2n-1)]u'$$
$$- (\alpha+n-1)(\beta+n-1)u = 0 \quad .$$

If we multiply both members of this equation by $x^{\gamma+n-2}(1-x)^{\alpha+\beta-\gamma+n-1}$, then we can write

$$\frac{d}{dx}[x^n(1-x)^n Ku'] = (\alpha+n-1)(\beta+n-1)x^{n-1}(1-x)^{n-1}Ku \tag{17}$$

with

$$K = K(x) = x^{\gamma-1}(1-x)^{\alpha+\beta-\gamma} \quad . \tag{18}$$

By differentiating both members of (17) $n - 1$ times, we obtain the reduction formula.

$$\frac{d^n}{dx^n}[x^n(1-x)^n Ky^{(n)}] = (\alpha+n-1)(\beta+n-1)\frac{d^{n-1}}{dx^{n-1}}[x^{n-1}(1-x)^{n-1}Ky^{(n-1)}].$$

Applying this formula for $n, n-1, \ldots, 1$, we arrive at the equation

$$y = \frac{\Gamma(\alpha)\Gamma(\beta)}{\Gamma(\alpha+n)\Gamma(\beta+n)}\, x^{1-\gamma}(1-x)^{\gamma-\alpha-\beta}\frac{d^n}{dx^n}[x^n(1-x)^n K(x)y^{(n)}] \quad . \tag{19}$$

If we take $\beta = -n$ and replace α by $\alpha + n$, then formula (19) applied to the Jacobi polynomial as defined above, yields

$$\Phi_n(x) = \frac{\Gamma(\gamma)}{\Gamma(\gamma+n)}\, x^{1-\gamma}(1-x)^{\gamma-\alpha}\frac{d^n}{dx^n}\left[x^{\gamma+n-1}(1-x)^{\alpha+n-\gamma}\right]$$

for the derivative of order n of $F(\alpha,\beta,\gamma,x)$ is then

$$\frac{\Gamma(\alpha+n)\Gamma(\beta+n)}{\Gamma(\alpha)\Gamma(\beta)}\quad\frac{\Gamma(\gamma)}{\Gamma(\gamma+n)} \quad .$$

We see that for $\gamma = 1$, $\alpha = 1$, $\Phi_n(x)$ is, to within a linear transformation on x, the Legendre polynomial of degree n, $P_n(x)$, (I, 100). We have

$$P_n(x) = F(n+1, -n, 1, \frac{1-x}{2}) \quad .$$

THE CONDITION OF ORTHOGONALITY

The Jacobi polynomials cannot form an orthogonal family between 0 and 1, since this condition defines the Legendre polynomials. But if we assume the real parts of γ and $\alpha+1-\gamma$ to be positive, then we have

$$\int_0^1 x^{\gamma-1}(1-x)^{\alpha-\gamma}\Phi_m(x)\Phi_n(x)dx = 0 \quad , \quad m \neq n \quad . \tag{21}$$

We can take m or n zero, $\Phi_0(x) \equiv 1$. In effect, the equation

$$x(1-x)\Phi_n'' + [\gamma-(\alpha+1)x]\Phi_n' + n(n+\alpha)\Phi_n = 0$$

can be written as

$$\frac{d}{dx}(x^\gamma(1-x)^{\alpha+1-\gamma}\Phi_n') = -n(n+\alpha)x^{\gamma-1}(1-x)^{\alpha-\gamma}\Phi_n \quad ,$$

and on multiplying by Φ_m and integrating from 0 to 1, we see that the product of $n(n+\alpha)$ and the first member of (21), is equal to

$$-\int_0^1 \Phi_m d(x^\gamma(1-x)^{\alpha+1-\gamma}\Phi_n') \quad .$$

On integrating by parts, the completely integrated part is zero and we obtain

$$n(n+\alpha)\mathrm{I} = \int_0^1 x^\gamma(1-x)^{\alpha+1-\gamma}\Phi_n'\Phi_m'dx \tag{22}$$

where I is the first member of (21). By permuting n and m, we will have

$$n(n+\alpha)\mathrm{I} = m(m+\alpha)\mathrm{I} \quad , \quad m \neq n \quad ,$$

hence $\mathrm{I}=0$ if α is not an integer and consequently for any α, considering the continuity. For $m=n$, we can operate in the same way by starting from the second member of (22) since Φ_n' also satisfies a similar Gauss equation. By continuing these reductions, we find that for $m=n$, the first member of (21) is equal to

$$\frac{1}{\alpha+2n}\,\frac{\Gamma(n+1)[\Gamma(\gamma)]^2\,\Gamma(\alpha+n-\gamma+1)}{\Gamma(\alpha+n)\Gamma(\gamma+n)} \quad .$$

If we assume that γ and $-\gamma$ are real and positive, then we can change the variable by setting

$$X = \int_0^x t^{\gamma-1}(1-t)^{\alpha-\gamma}dt \quad ,$$

and the transformed functions of the $\Phi_n(x)$ will form an orthogonal family. But we observe that when a family of functions $\psi_m(x)$, $n = 0,1,2,\ldots,$ continuous on a segment $a, b,$ satisfies the condition

$$\int_a^b K(x)\psi_n(x)\psi_m(x)dx = 0 , \qquad m \neq n \qquad ,$$

where $K(x)$ is given to be continuous on $a, b,$ the determination of the coefficients of an expansion of a function $f(x)$ under the form

$$f(x) = c_o\psi_o(x) + \ldots + c_n\psi_n(x) + \ldots \qquad ,$$

can be achieved by multiplying both members by $\psi_n(x)K(x)$ and integrating from a to b. Assuming the operation is legitimate, we will have

$$c_n \int_a^b [\psi_n(x)]^2 K(x)dx = \int_a^b K(x)f(x)\psi_n(x)dx \qquad .$$

114 - The representation of $F(\alpha,\beta,\gamma,x)$ by an integral

By introducing the Euler function of the first kind (I, 148)

$$B(u,v) = \int_o^1 t^{u-1}(1-t)^{v-1}dt , \qquad Ru > 0 , \quad Ru > 0 \qquad ,$$

that defines a holomorphic function of u and v [following (I, 240)], we can write

$$\frac{\Gamma(\beta+n)}{\Gamma(\gamma+n)} = \frac{B(\beta+n,\gamma-\beta)}{\Gamma(\gamma-\beta)} \qquad , \qquad n = 0,1,\ldots$$

if $R\beta > 0$, $R(\gamma-\beta) > 0$. Since the equation holds for β and $\gamma-\beta$ real and positive, and consequently it is extendable. It follows that, for $|x| < 1$,

$$\frac{\Gamma(\alpha)\Gamma(\beta)}{\Gamma(\gamma)} F(\alpha,\beta,\gamma,x)$$

$$= \frac{1}{\Gamma(\gamma-\beta)} \sum_{n=0}^{\infty} \int_o^1 \Gamma(\gamma+n)t^{\beta+n-1}(1-t)^{\gamma-\beta-1} \frac{x^n}{n!} dt \qquad .$$

As the series

$$\sum_o^{\infty} \Gamma(\alpha+n) \cdot \frac{t^n x^n}{n!}$$

converges uniformly for t taken between 0 and 1 when $|x| < 1$, we can reverse the order of the integration and summation to obtain

$$\frac{\Gamma(\alpha)\Gamma(\beta)}{\Gamma(\gamma)} F(\alpha,\beta,\gamma,x) = \frac{1}{\Gamma(\alpha-\beta)} \int_0^1 t^{\beta-1}(1-t)^{\gamma-\beta-1}\left(\sum_0^\infty \Gamma(\alpha+n) \frac{t^n x^n}{n!}\right)dt$$

$$= \frac{\Gamma(\alpha)}{\Gamma(\gamma-\beta)} \int_0^1 t^{\beta-1}(1-t)^{\gamma-\beta-1}(1-tx)^{-\alpha}\, dt \quad .$$

Finally,

$$F(\alpha,\beta,\gamma,x) = \frac{\Gamma(\gamma)}{\Gamma(\beta)\Gamma(\gamma-\beta)} \int_0^1 t^{\beta-1}(1-t)^{\gamma-\beta-1}(1-tx)^{-\alpha}\, dt \quad ,$$

by assuming $R\beta > 0$, $R(\gamma-\beta) > 0$ and $|x| < 1$. But the integral defines a holomorphic function of x for any x in a finite domain not containing the part of the real axis taken between 1 and $+\infty$. This formula thus gives the values of $F(\alpha,\beta,\gamma,x)$ in all of this domain, i.e., at every point other than the real points, $1 \leq x \leq \infty$.

115 - Hypergeometric integrals

More generally, let us consider an integral

$$y = \int_A^B (t-a)^{b-1}(t-a')^{b'-1}(t-x)^{k-1}\, dt \tag{23}$$

whose limits A, B are two of the numbers a, a', ∞. We take b, b', k such that the integral exists and admits a derivative with respect to x. It will then define a holomorphic function of x, as long as x does not cross the path of integration. By putting

$$V(t) = (t-a)^{b-1}(t-a')^{b'-1}$$

we have

$$y = \int_A^B (t-x)^{k-1}V(t)dt \ , \qquad y' = -(k-1)\int_A^B (t-x)^{k-2}V(t)dt \quad ,$$

and

$$y'' = (k-1)(k-2)\int_A^B (t-x)^{k-3}V(t)dt \quad .$$

We can determine the constants D, E, G such that we have

$$y''(x-a)(x-a') + (D+Ex)y' + Gy \equiv 0 \quad .$$

In effect, the first member is equal to

$$\int_A^B V(t)(t-x)^{k-3}[(k-1)(k-2)(x-a)(x-a') - (k-1)(t-x)(D+Ex) + G(t-x)^2]dt \quad ,$$

and we can choose D, E, G in order that the coeffficient of dt is, to within a constant factor, the derivative of

$$U(t) \equiv (t-a)^b (t-a')^{b'} (t-x)^{k-2} \quad ,$$

such that the integral will be $U(B) - U(A) = 0$. We have

$$U'(t) \equiv V(t)(t-x)^{k-3}\Big[[b(t-a') + b'(t-a)](t-x) + (k-2)(t-a)(t-a')\Big] \quad ,$$

and necessarily, we have

$$(k-1)(k-2)(x-a)(x-a') - (k-1)(t-x)(D+Ex) + G(t-x)^2$$

$$\equiv \lambda[(b(t-a') + b'(t-a))(t-x) + (k-2)(t-a)(t-a')] \quad .$$

We will have $\lambda = k-1$, and

$$D = a'b + ab' + (k-2)(a+a') \ , \qquad G = (k-1)(b+b'+k-2) \ ,$$

$$E = 4 - 2k - b - b' \quad .$$

The integral (23) is thus a solution of the Gauss equation having the canonical form (2) if we take $a = 0$, $b' = 1$. Then $D = -\gamma$, $G = \alpha\beta$ and $E = \alpha + \beta + 1$, such that

$$k = 1 - \alpha \ , \quad b = 1 + \alpha - \gamma, \quad b' = \gamma - \beta \quad . \tag{24}$$

The Gauss equation (22) thus admits an integral of the form (23), with $a = 0$, $a' = 1$, $k\,b$ and b' are given by (24), and the limits A and B are such that the integral has a meaning and is differentiable as we claimed. Moreover, the function $U(t)$ must have values equal to A and B, and that suffices. *It will not be necessary for* A *and* B *to be one of the numbers* 0, 1, ∞. As in the case of the Laplace integral (no. 102), we can take a closed path about the points 0 and 1, for example.

The integrals of the form (23) are called hypergeometric integrals.

We can also take the integral (23) between a or a' and x providing it exists along with its derivatives. In the derivatives, the terms arising from the limit x will be zero.

116 - An application to the conformal representation

The integral occuring in the expression giving the conformal representation of a half-plane on a quadrilateral (I, 220) is a hypergeometric integral.

We have seen, on the other hand (I, 221), that the function $Z = f(x)$ which gives the conformal representation of a half-plane $(y > o)$ on a domain bounded by arcs of circles, is defined by a Riccati equation in f''/f':

$$\left(\frac{f''}{f'}\right)' - \frac{1}{2}\left(\frac{f''}{f'}\right)^2 = \sum\left(\frac{1}{2}\ \frac{1 - \beta^2}{(z-b)^2} + \frac{\beta''}{z-b}\right) \quad , \tag{25}$$

where the b are real along with the β''. The angles at the vertices of the contour of the domain are the numbers $\beta\pi$; the $\beta\pi$ will also satisfy certain conditions. The first member of (25) is the Schwartzian derivative; if $f(x)$ is a solution, the others are obtained by taking

$$\frac{\lambda f(z) + \mu}{\lambda' f(z) + \mu'}$$

where λ, μ, λ' and μ' are arbitrary constants.

We have seen (no. 72) that a Riccati equation reduces to a second order equation. We shall go about this transformation in another way. Given the linear equation

$$y'' + a_1 y' + a_2 y = 0 \quad , \tag{26}$$

let us denote two particular integrals by y_1 and y_2. Let us put

$$u = \frac{y_1}{y_2} \quad .$$

We will have

$$u' = \frac{y_1' y_2 - y_1 y_2'}{y_2^2} \quad ,$$

hence for y_2 and y_1 solutions of the linear equation,

$$\frac{u''}{u'} = \frac{y_1'' y_2 - y_2'' y_1}{y_1' y_2 - y_2' y_1} - \frac{2y_2'}{y_2} = -a_1 - \frac{2y_2'}{y_2} \quad ,$$

and

$$\left(\frac{u''}{u'}\right)' - \frac{1}{2}\left(\frac{u''}{u'}\right)^2 = -a_1' - \frac{1}{2}a_1^2 - 2\left(\frac{y_2''}{y_2} + a_1\frac{y_2'}{y_2}\right) = 2a_2 - \frac{1}{2}a_1^2 - a_1' \quad .$$

The ratio of two integrals of equation (26) hence satisfies an equation of type (25). Conversely, the solutions of (25) will be the quotient of two integrals of an equation (26) such that

$$2a_2 - \frac{1}{2}a_1^2 - a_1' = \sum\left(\frac{1}{2}\frac{1 - \beta^2}{(z-b)^2} + \frac{b''}{z-b}\right) \quad .$$

We can choose a_1 arbitrarily and a_2 will then be determined. We can take a_1 to be identically zero. Equation (26) will be of Fuchsian type about each of the points b.

THE CASE OF THE DOMAIN BOUNDED BY THREE ARCS OF CIRCLES

We can assign the points $z = 0, 1, \infty$ to the three vertices. If $\lambda\pi$, $\mu\pi$, $\nu\pi$ are the angles at the vertices corresponding to the three points, then the second member of equation (25) is

$$\frac{1}{2}\,\frac{1-\lambda^2}{z^2} + \frac{1}{2}\,\frac{1-\mu^2}{(z-1)^2} + \frac{\lambda'}{z} + \frac{\mu'}{z-1} ,$$

and the expansion of this expression about the point at infinity must commence from $(1 - \nu^2)/2z^2$, which determines λ' and μ'. We obtain

$$\frac{1}{2}\,\frac{1-\lambda^2}{z^2} + \frac{1}{2}\,\frac{1-\mu^2}{(z-1)^2} + \frac{1}{2}\,\frac{1-\lambda^2-\mu^2+\nu^2}{z(1-z)} .$$

In equation (26) we shall take

$$a_1 = \frac{A}{z} + \frac{B}{1-z}$$

where A and B are constants such that a_2 admits $z = 0$ and $z = 1$ as simple poles. We have

$$4a_2 = \frac{1-\lambda^2+A^2-2A}{z^2} + \frac{1-\mu^2+B^2+2B}{(z-1)^2} + \frac{1-\lambda^2-\mu^2+\nu^2+2AB}{z(1-z)}$$

we must therefore take

$$(A-1)^2 = \lambda^2 \quad , \quad (B+1)^2 = \lambda^2 \quad ,$$

and then equation (26) is a Gauss equation:

$$(1-z)zy'' + [A - z(A-B)]y' + \frac{1}{4}(1-\lambda^2-\mu^2+\nu^2+2AB)y = 0 \quad .$$

By writing it in the form (2), we have $A = \gamma$, $A - B = \alpha + \beta + 1$, $1 - \lambda^2 - \mu^2 + \nu^2 + 2AB = -4\alpha\beta$, hence

$$\gamma = A, \quad \alpha = \frac{1}{2}(A - B - 1 + \nu), \quad \beta = \frac{1}{2}(A - B - 1 - \nu) \quad ,$$

or, by taking $A = 1 - \lambda$, $B = -1 + \mu$;

$$\gamma = 1 - \lambda, \quad \alpha = \frac{1}{2}(1-\lambda-\mu+\nu) \quad \beta = \frac{1}{2}(1-\lambda-\mu-\nu) \quad . \tag{27}$$

Having defined α, β, γ this way, the required function $Z = f(x)$, will be the ratio of two solutions of the Gauss equation. The variable z must be taken in the upper half-plane. We can apply the ratio of the solutions $F(\alpha,\beta,\gamma,z)$, $z^{1-\gamma}F(\alpha+1-\gamma,\ \beta+1-\gamma,\ 2-\gamma,z)$, extended in the upper half-plane as in no. 112. But we can also take the ratio of the two hypergeometric integrals calculated on the real axis between 0 and between 1 and ∞, for example.

A PARTICULAR CASE

If we take the triangle bounded by the lines $RZ = 0$, $RZ = 1$, and by the circle $|Z - \frac{1}{2}| = \frac{1}{2}, \lambda,\mu,\nu$ are zero, hence $\gamma = 1$, $\alpha = \beta = \frac{1}{2}$. Following the equations (24) of no. 115, the corresponding hypergometric integral is

$$\int_A^B \frac{dt}{\sqrt{t(t-1)(t-z)}} ,$$

where we can take $A = 0$, $B = 1$, and $A = 1$, $B = \infty$. We thus make the desired conformal representation via the ratio of the periods of this elliptic integral. If $Z = f(z)$ is the function giving this conformal representation on the upper half-plane $\mathfrak{I}_z > 0$ on this triangle, with the points $z = 0,1,\infty$ representing the vertices, *then the inverse function* $z = \phi(z)$ *is a modular function.* If we reduce the above elliptic integral to the normal Weierstrass form, with the quantity under the root taking the form $4v^3 - y_2v - g_3$, we find that

$$\frac{g_2^3}{g_2^3 - 27g_3^2} = \frac{4(1-z+z^2)^3}{z^2(1-z)^2} .$$

The first member is the modular function $J(\tau)$ considered previously (I, 239), where τ is the ratio of the periods, thus equal to Z. Therefore, between the functions J and ϕ we have the relation

$$J(Z) = \frac{4(1-\phi(Z)+\phi(Z)^2)^3}{\phi(Z)^2(1-\phi(Z))^2} .$$

117 - Degeneracies of the Gauss equation. Laguerre polynomials

If, in the Gauss equation (2), we change x to x/β and write the transformed equation in the form

$$x\left(1-\frac{x}{\beta}\right)y'' + \left(\gamma - x - \frac{\alpha+1}{\beta}x\right)y' - \alpha y = 0 ,$$

then on letting β tend to ∞, we obtain the degenerate equation

$$xy'' + (\gamma - x)y' - \alpha y = 0 , \tag{28}$$

which no longer admits $x = 1$ as a singularity. The origin is always a singularity of Fuchsian type, but this is no longer the same for the point at infinity. Following the Poincaré theorem, this equation must admit the limits of the solutions of (2) relative to the neighbourhood of the origin, as its solutions.

Moreover, we can show directly that when $1-\gamma$ is not a positive integer, the equation admits the solution

$$F_1(\alpha,\gamma, \ ) = 1 + \sum_1^\infty \frac{\alpha(\alpha+1)\dots(\alpha+n-1)}{\gamma(\gamma+1)\dots(\gamma+n-1)} \frac{x^n}{n!} \quad ,$$

which is an entire function. If $1-\gamma$ is not an integer, then the general solution is

$$\lambda F_1(\alpha,\gamma,x) + \mu x^{1-\gamma} F_1(\alpha-\gamma+1\ ,\ 2-\gamma,x) = 0 \quad .$$

LAGUERRE POLYNOMIALS

When α is a negative integer, equation (28) admits a polynomial as a particular solution. These polynomials were studied by Laguerre in the case $\gamma = 1$ and by Sonine in the general case. If we set

$$L_n(x) = \frac{\Gamma(\gamma+1+n)}{n!\Gamma(\gamma+1)} F_1(-n,\gamma+1,x) \ , \qquad n = 0,1,2,\dots$$

we have

$$L_n(x) = \frac{e^x x^{-\gamma}}{n!} \frac{dx}{dx^n} (e^{-x} x^{n+\gamma}) \quad .$$

When $R\gamma > -1$, we have for $m \neq n$,

$$\int_0^\infty e^{-x} x^\gamma L_m(x) L_n(x) dx = 0 \quad .$$

For $|t| < 1$, we have

$$(1-t)^{-\gamma-1} e^{-xt/(1-t)} = \sum_0^\infty t^n L_n(x) \quad .$$

118 - Hermite polynomials

The Hermite polynomials $H_n(z)$ are associated with a second order differential equation which has no singularity at a finite distance:

$$y'' - 2zy' + 2ny = 0 \quad , \tag{29}$$

where n is an integer, positive or zero. They are defined directly by the equation

$$H_n(z) = (-1)^n e^{z^2} \frac{d^n}{dz^n} e^{-z^2}, \quad H_o(z) = 1 \quad . \tag{30}$$

We deduce from this equation that

$$H_{n+1}(z) = (-1)^{n+1} e^{z^2} \frac{d^n}{dz^n} (-e^{z^2} 2z) = 2zH_n(z) - 2nH_{n-1}(z) \quad . \tag{31}$$

This recurrence relation allows us to compute step by step, and shows that $H_n(z)$ has the parity of n. Differentiating (30), we obtain, on account of (31),

$$H'_n(z) = 2zH_n(z) - H_{n+1}(z) = 2nH_{n-1}(z) \quad . \tag{32}$$

It follows that

$$H''_n(z) = 2zH'_n(z) + 2H_n(z) - H'_{n+1}(z) = 2zH'_n(z) - 2nH_n(z) \quad ;$$

$H_n(z)$ is thus a solution of equation (29). If we consider the function

$$e^{2tz-t^2} = \sum_o^\infty \frac{t^n}{n!} \mathbb{H}_n(z) \quad , \tag{33}$$

then the $\mathbb{H}_n(z)$ are clearly polynomials of degree n, and the equation is valid for any z and t since the double series

$$\sum^\infty \frac{(|2tz|+|t|^2)^n}{n!}$$

converges. By differentiating (33) with respect to t, we obtain

$$2(z-t)\sum_o^\infty \frac{t^n}{n!} \mathbb{H}_n(z) = \sum_1^\infty \frac{t^{n-1}}{(n-1)!} \mathbb{H}_n(z) \quad .$$

We therefore have

$$\mathbb{H}_{n+1}(z) = 2z\mathbb{H}_n(z) - 2n\mathbb{H}_{n-1}(z) \quad ,$$

and on comparison with (31) we see that $\mathbb{H}_n(z) \equiv H_n(z)$. The function e^{2tz-t^2} is called a *generating function* for the $H_n(z)$. Such functions were obtained above for the Laguerre polynomials and previously for those of Legendre (I, 101) and Bernouilli (I, 103).

The Hermite polynomials satisfy a generalised orthogonality relationship,

$$\int_{-\infty}^{+\infty} e^{-z^2} H_n(z)H_m(z)dz = 0 \; , \qquad n \neq m \quad , \tag{34}$$

where the integral is taken along the real axis and clearly has a meaning.

Effectively, by writing the differential equation of $H_n(z)$,

$$(e^{-z^2}H_n'(z))' + 2nH_n(z)e^{-z^2} = 0 \quad ,$$

then multiplying by $H_m(z)$ and integrating by parts, we see that the first member of (34) is equal to

$$\frac{1}{2n}\int_{-\infty}^{+\infty} e^{-z^2} H_n'H_m' \, dz$$

and on interchanging the roles of m and n, we obtain the result. We also see that, on account of (32),

$$\int_{-\infty}^{+\infty} e^{-z^2}(H_n(z))^2 dz = 2n\int_{-\infty}^{+\infty} e^{-z^2}(H_{n-1}(z))^2 dz$$

$$= 2^n n! \int_{-\infty}^{+\infty} e^{-z^2} dz = 2^n n! \sqrt{\pi} \quad .$$

The relation in (31) allows us to establish by recurrence that $H_n(z)$ has its zeros real, distinct and separated from those of $H_{n-1}(z)$. Let us assume that this is so up to the value and that n is odd; this is the most complicated case. The positive zeros of $H_n(z)$ are $\alpha_1,\alpha_2,\ldots,\alpha_q$, $2q + 1 = n$. Let us substitute the numbers $0,\alpha_1,\alpha_2,\ldots,\alpha_q$ into $H_{n+1}(z)$. Following (31), we obtain the values $H_{2q}(0),H_{2q}(\alpha_1),\ldots,H_{2q}(\alpha_q)$ multiplied by $-(2q + 1)$. We obtain the signs + and - in alternation. On the other hand, if q is even, $H_{2q}(0)$ is positive; it is negative if q is odd. We have then the signs of $(-1)^{q+1},(-1)^{q+2},\ldots,(-1)^{2q+1}$, and since we have the sign + for $H_{2q+2}(\infty)$,$H_{2q+2}(z)$ indeed has $q + 1$ positive zeros separated from those of $H_{2q+1}(z)$. We shall verify that the Hermite polynomials are tied to the Leguerre polynomials. We have

$$H_{2n}(z) = (-1)^n 2^{2n} n! \; L_n(z^2) \quad ,$$

where L_n corresponds to $\gamma = -\frac{1}{2}$, and $H_{2n+1}(z) = (-1)^n 2^{2n+1} n! z \, L_n(z^2)$, where L_n here corresponds to $\gamma = \frac{1}{2}$.

II. LEGENDRE EQUATIONS AND FUNCTIONS

119 - The Legendre equation. The Schlaefli and Laplace formulae

We have seen (I, 100, 101, 102) that the Legendre polynomial defined by the formula (due to Rodrigues)

$$P_n(x) = \frac{1}{2.4\ldots(2n)} \frac{d^n}{dx^n}(x^2 - 1)^n \tag{35}$$

is a solution of the differential equation

$$(1-x^2)y'' - 2xy' + n(n+1)y = 0 \quad , \tag{36}$$

which is a Gauss equation. It reduces to the canonical form (2) by taking the variable to be $X = (x+1)/2$ and corresponds to the values $\alpha = n+1$, $\beta = -n$, $\gamma = 1$; it occurs in the particular case examined in no. 111. We call this the Legendre equation.

SCHAEFLI'S FORMULA

The Cauchy formula giving the n-th order derivative of a holomorphic function $f(z)$ in the form

$$f^{(n)}(z) = \frac{n!}{2\pi i} \int_{C^+} \frac{f(u)}{(u-z)^{n+1}}\, du \quad ,$$

where C is a simple contour containing the point z in its interior, shows, as a consequence of (35), that

$$P_n(x) = \frac{1}{2\pi i 2^n} \int_{C^+} \frac{(u^2-1)}{(u-x)^{n+1}}\, du \quad ,$$

where C is an arbitrary simple closed curve about the point x. This is Schaefli's formula.

LAPLACE'S FORMULA

In particular, if we take C to be the circle

$$|u-x| = \sqrt{|x^2-1|} \quad ,$$

we can put

$$u = x + \sqrt{x^2-1}\; e^{i\phi} \quad .$$

We have

$$\begin{aligned} u^2 - 1 &= (x^2-1)(e^{2i\phi}+1) + 2x\sqrt{x^2-1}\; e^{i\phi} \\ &= 2(u-x)\,[x + \sqrt{x^2-1}\,\cos\phi] \end{aligned}$$

and we obtain

$$P_n(x) = \frac{1}{2\pi} \int_{-\pi}^{\pi} [x + \sqrt{x^2-1}\,\cos\phi]^n\, d\phi \quad .$$

The branch chosen for $\sqrt{x^2-1}$ is arbitrary. As the function to be integrated is even, we can take

$$P_n(x) = \frac{1}{\pi} \int_o^\pi [x + \sqrt{x^2-1} \cos \phi]^n d\phi \qquad (37)$$

This is the Laplace formula.

120 - The general solution of Legendre's equation

Knowing the solution $P_n(x)$ of equation (36), we obtain the general solution by the method of reducing the order (nos. 76, 90) by setting $y = P_n(x)u$. We thus obtain the general solution under the guise of

$$y = \lambda P_n(x) + \mu P_n(x) \left[\frac{1}{2} \log \frac{(x+1)}{(x-1)} - \sum_1^n \frac{1}{(x-x_j)(1-x_j^2)P'^2_n(x_j)}\right]$$

where the x_j are zeros, all real and distinct from $P_n(x)$ (I, 102), λ and μ are arbitrary constants. The coefficient of μ is of the form

$$Q_n(x) = \frac{1}{2} P_n(x) \log \frac{x+1}{x-1} - U_{n-1}(x) \quad , \qquad (38)$$

where $U_{n-1}(x)$ is a polynomial of degree at most equal to $n - 1$. *$Q_n(x)$ is the Legendre function of the second kind.* It admits the points $x = 1$ and $x = -1$ as critical logarithmic points, but it is uniform if x is restricted to not crossing the segment (-1,1), since after a rotation about the points -1 and +1, the logarithm will take its initial value. *We fix the chosen branch for $Q_n(x)$ by taking the real logarithm for x real and positive and greater than* 1.

If we assume that $|x| > 1$, we will have

$$\log \frac{x+1}{x-1} = \log (1 + \frac{1}{x}) - \log (1 - \frac{1}{x}) = 2 (\frac{1}{x} + \frac{1}{3x^3} + \ldots)$$

and

$$Q_n(x) = x^{n-1} \left[c_o + c_1 \frac{1}{x} + \ldots + c_p \frac{1}{x^p}\right] \quad .$$

The point at infinity is therefore of the first Fuchsian type, the solutions $P_n(x)$ and $Q_n(x)$ are uniform at this point. $Q_n(x)$ can therefore be expanded by a direct calculation. The roots of the determinantal equation about the point at infinity are $-n$ and $n + 1$; the first, $-n$, yields the solution $P_n(z)$, the second, $n + 1$, provides $Q_n(x)$. We will have

$$Q_n(x) = \frac{C_n}{x^{n+1}} \left[1 + \frac{d_1}{x} + \frac{d_2}{x} + \ldots\right] \quad ,$$

where C_n is a constant. We obtain $d_1 = d_3 = \ldots = 0$,

$$d_p = d_{p-2}\,\frac{(n+p)(n+p-1)}{p(2n+1+p)}$$

and we can write

$$Q_n(x) = \frac{C_n}{x^{n+1}}\; F\left(\frac{n+1}{2}, \frac{n}{2}+1, n+\frac{3}{2}, \frac{1}{x^2}\right) .$$

The constant C_n *is equal to the coefficient of* x^{-n-1} *in the expansion of*

$$\frac{1}{2}\,P_n(x)\,\log\frac{x+1}{x-1} ,$$

and the polynomial $U_{n-1}(x)$ *is equal to the sum of the first* n *terms of this expansion.* If we write

$$P_n(x) = \sum \gamma_{n-2p}\, x^{n-2p} ,$$

we have

$$C_n = \sum \frac{\gamma_{n-2p}}{2n-2p+1} = \int_0^1 x^n\, P_n(x)\, dx .$$

As P_n has the parity of n, we can write, on account of the orthogonality relationship (I, 100),

$$C_n = \frac{1}{2}\int_{-1}^{1} x^n P_n(x) dx = \frac{2^n (n!)^2}{(2n)!}\;\frac{1}{2}\int_{-1}^{1} (P_n(x))^2\, dx .$$

THE CALCULATION OF $\int_{-1}^{1} (P_n(x))^2 dx$.

The preceding equation reduces this calculation to that of

$$C_n = \int_0^1 x^n P_n(x) dx = \frac{1}{2^n n!}\int_0^1 x^n\,\frac{d^n}{dx^n}(x^2-1)^n\, dx$$

and on integrating by parts (I, 49), we obtain

$$C_n = \frac{1}{2^n}\int_0^1 (1-x^2)^n dx = \frac{1}{2^n}\int_0^{\pi/2} \sin^{2n+1} t\, dt = \frac{2^n (n!)^2}{(2n+1)!} .$$

It follows that

$$\int_{-1}^{1} (P_n(x))^2 dx = \frac{2}{2n+1} . \qquad (39)$$

We have, on the other hand,

$$Q_n(x) = \frac{2^n(n!)^2}{(2n+1)!}\frac{1}{x^{n+1}} \; F\left(\frac{n+1}{2}, \frac{n}{2}+1, n+\frac{3}{2}, \frac{1}{x^2}\right) \quad . \tag{40}$$

Remark. The legendre polynomials satisfy other relations apart from that already obtained between three successive polynomials (I, 101). On differentiating (35), and writing

$$[(x^2-1)^n]^{(\;+1)} = [2nx(x^2-1)^{n-1}]^{(n)} \quad ,$$

we obtain the identity

$$P_n'(x) = xP_{n-1}'(x) + nP_{n-1}(x) \quad .$$

On combining this identity with that obtained by differentiating

$$nP_n(x) = (2n-1)\,P_{n-1}(x) - (n-1)P_{n-2}(x) \quad ,$$

we see, after changing n to $n+1$, that

$$xP_n'(x) - P_{n-1}'(x) = nP_n(x) \quad .$$

We can combine these relations, which can also be deduced from Legendre identity (I, 101)

$$\frac{1}{\sqrt{1-2\alpha x+\alpha^2}} = P_o + \alpha P_1 + \ldots \alpha^n P_n + \ldots$$

121 - Heine's formula

The expression (40) of $Q_n(z)$ can be written explicitly, on simplifying C_n:

$$Q_n(x) = \frac{2.4,\ldots.2n}{3.5,\ldots(2n+1)}\frac{2}{(2x)^{n+1}}\sum_{p=0}^{\infty}\frac{(n+1)(n+2)\ldots(n+2p)}{2.4\ldots 2p(2n+3)(2n+5)\ldots(2n+2p+1)}$$

$$\cdot\frac{1}{x^{2p}} = \frac{1}{(2x)^{n+1}}\sum_{o}^{\infty}\frac{(n+2p)!}{(2p)!}\frac{3.5\ldots(2p-1)2^{n+1}}{3.5\ldots(2n+2p+1)}\frac{1}{x^{2p}}$$

$$= \frac{1}{(2x)^{n+1}} = \sum_{p=0}^{\infty}\frac{(n+2p)!}{(2p)!}\frac{\Gamma(p+\frac{1}{2})}{\Gamma(n+p+\frac{3}{2})}\frac{1}{x^{2p}} \quad .$$

Now on introducing the Euler function $B(u,v)$ we have

$$\frac{n!\Gamma(p+\frac{1}{2})}{\Gamma(n+p+\frac{3}{2})} = B(n+1, p+\frac{1}{2}) = \int_o^1(1-u)^n u^{p-1/2}du = \int_{-1}^{+1}(1-t^2)^n t^{2p}dt \quad .$$

It follows that, assuming always that $|x| > 1$,

$$Q_n(x) = \frac{1}{(2x)^{n+1}} \sum_{p=0}^{\infty} \frac{(n+2p)!}{n!(2p)!} \int_{-1}^{1} (1-t^2)^n (\frac{t}{x})^2 dt \quad .$$

By adding the terms

$$\frac{(n+2p+1)!}{n!(2p+1)!} \int_{-1}^{+1} (1-t^2)^n (\frac{t}{x})^{2p+1} dt$$

which are zero, and on reversing the order of integration and summation, which is legitimate since the series

$$\sum_{q=0}^{\infty} \frac{(n+q)!}{n!q!} (\frac{t}{x})^q = (1 - \frac{t}{x})^{-n-1}$$

converges uniformly, we obtain

$$Q_n(x) = \frac{1}{(2x)^{n+1}} \int_{-1}^{1} (1-t^2)^n (1-\frac{t}{x})^{-n-1} dt$$

or

$$Q_n(x) = \frac{1}{2^{n+1}} \int_{-1}^{1} \frac{(1-t^2)^n}{(x-t)^{n+1}} dt \quad . \tag{41}$$

This is analogous to the Schaefli formula of no. 119. As both members are holomorphic outside the segment $(-1,+1)$, the formula established for $|x| > 1$ is valid outside of this segment.

Let us assume that x is real and greater than 1, and in (41) make the change of variable

$$t = x - \sqrt{x^2-1}\, e^{\theta} \quad ;$$

when t varies from -1 to +1, θ varies from $-\alpha$ to $+\alpha$, with

$$\alpha = \frac{1}{2} \log \frac{x+1}{x-1} \quad .$$

We obtain

$$Q_n(x) = \frac{1}{2} \int_{-\alpha}^{+\alpha} [x - \sqrt{x^2-1}\; ch\theta]^n \, d\theta \quad ,$$

which can be written, since the integrand is even,

$$Q_n(x) = \int_{o}^{a} [x - \sqrt{x^2-1}\; ch\theta]^n \, d\theta \quad .$$

Finally by putting

$$[x + \sqrt{x^2 - 1}\ ch\psi][x - \sqrt{x^2 - 1}\ ch\theta] = 1 \quad ,$$

we find that

$$Q_n(x) = \int_0^\infty [x + \sqrt{x^2 - 1}\ ch\psi]^{-n-1}\, d\psi \quad .$$

This is Heine's formula, analogous to the Laplace formula relating $P_n(x)$. It is established for x real and greater than 1. But by assuming that x remains outside of the segment (-1,+1), we see that the integral defines an analytic function of x; the convergence is clearly uniform. *The identity is valid in this domain* since $Q_n(x)$ is holomorphic there.

122 - Neumann's Theorem

Let us assume that x is restricted to not going beyond the segment (-1,+1), and consider the function $Q_n(x)$ as defined in no. 120. It is holomorphic in the plane with this segment removed. On account of its expression (38), when x tends towards a point x_o of the interval -1, +1 by remaining in the upper half-plane, $Q_n(x)$ tends towards

$$Q_n(x_o + i\theta) = \frac{1}{2} P_n(x_o) \left(\log \left|\frac{x_o+1}{x_o-1}\right| - i\pi\right) - U_{n-1}(x_o) \quad .$$

In contrast, if x tends towards x_o below the real axis, we have as the limit

$$Q_n(x_o - i\theta) = \frac{1}{2} P_n(x_o) \left(\log \left|\frac{x_o+1}{x_o-1}\right| + i\pi\right) - U_{n-1}(x_o) \quad .$$

Hence

$$Q_n(x_o - i\theta) - Q_n(x_o + i\theta) = \pi i P_n(x_o) \quad . \tag{42}$$

On account of its expression (40), $Q_n(x)$, holomorphic at infinity, is zero there. Following the Cauchy theorems, if x belongs to the domain bounded by the segment (-1,+1) and if C is a simple closed curve encircling this segment and leaving the point x in its exterior, then we shall have

$$Q_n(x) = \frac{1}{2\pi i} \int_{C^-} \frac{Q_n(u)}{u-x}\, du \quad .$$

We may assume that C reduces to the segment $(-1+\varepsilon, 1-\varepsilon)$, where ε is arbitrarily small, twice passing over, above and below the intersection, and with two small circles with centres 1 and -1 and radius ε. With respect to the logarithmic form of the singularities of $Q_n(x)$ [formula (38)] at the points -1 and +1, we can let ε tend towards zero, and we obtain

$$Q_n(x) = \frac{1}{2\pi i} \int_{-1}^{+1} \frac{Q_n(u+i\ ) - Q_n(u-i\ )}{u - x}\, du \quad ,$$

and, taking (42) into account, we see that

$$Q_n(x) = \frac{1}{2} \int_{-1}^{+1} \frac{P_n(u)}{x-u}\, du \quad .$$

From this formula (established by Neumann in 8248), we easily deduce that the recurrence relationship relating the three functions $P_n(x)$ is valid for the $Q_n(x)$. Effectively, we have

$$nQ_n - (2n-1)\, x\, Q_{n-1} + (n-1)Q_{n-2} = \frac{1}{2} \int_{-1}^{+1} \frac{\phi(u)}{x-u}\, du \ , \qquad n \geq 1$$

with

$$\phi(u) \equiv nP_n(u) - (2n-1)xP_{n-1}(u) + (n-1)P_{n-2}(u)$$

$$\equiv -2(n-1)(x-u)P_{n-1}(u)$$

and the integral of $P_{n-1}(u)$ between -1 and +1 is zero on account of the orthogonality condition.

THE EXPANSION OF $1/(x-u)$ IN TERMS OF A SERIES OF LEGENDRE POLYNOMIALS

On consideration of certain hypotheses, we have

$$\frac{1}{(x-u)} = \sum_{o}^{\infty} (2n+1)Q_n(x)P_n(u) \quad . \tag{43}$$

This result, due to Heine, was proved by Christoffel as follows. For $n \geq 1$, we have

$$(2n+1)uP_n(u) = (n+1)P_{n+1}(u) + nP_{n-1}(u) \quad ,$$

$$(2n+1)xQ_n(x) = (n+1)Q_{n+1}(x) + nQ_{n-1}(x) \quad ,$$

hence

$$(2n+1)(x-u)Q_n(x)P_n(u) = (n+1)\ [Q_{n+1}(x)P_n(u) - Q_n(x)P_{n+1}(u)]$$

$$- n[Q_n(x)P_{n-1}(u) - Q_{n-1}(x)P_n(u)] \quad .$$

For $n = 1$, we have

$$(x-u)Q_o(x)P_o(u) = [Q_1(x)P_o(u) - Q_o(x)P_1(u)] + 1 \quad ,$$

since

$$Q_o = \frac{1}{2}\log\frac{x+1}{x-1}\ ,\quad Q_1 = \frac{1}{2}x\log\frac{x+1}{x-1} - 1\ ,\quad P_o = 1\ ,\quad P_1 = u \quad .$$

On adding these equations, we obtain

$$\sum_o^m (2n+1)Q_n(x)P_n(u) = \frac{1}{x-u} + \frac{n+1}{x-u}\left[Q_{n+1}(x)P_n(u) - Q_n(x)P_{n+1}(u)\right]$$

and we see that (43) holds as long as

$$(n+1)\left[Q_{n+1}(x)P_n(u) - Q_n(x)P_{n+1}(u)\right] \tag{44}$$

tends to zero when $n \to \infty$.

Let us put $x = ch(\alpha + i\beta)$ and apply Heine's formula (we assume that x is outside the segment (-1,+1). For ψ real and $\alpha > 0$, which is possible, we have

$$|x + \sqrt{x^2 - 1}\ ch\psi|^2 = |ch(\alpha + i\beta) + sh(\alpha + i\beta)ch\psi|^2$$

$$= \frac{1}{2}(ch2\alpha + \cos 2\beta) + sh2\alpha ch\psi + \frac{1}{2}(ch2\alpha - \cos 2\beta)ch^2\psi$$

$$\geq ch2\alpha + sh2\alpha ch\psi \geq e^{2\alpha} \quad ,$$

hence

$$|Q_n(x)| \leq \int_o^\infty [ch2\alpha + sh2\alpha ch\psi]^{-(n+1/2)}d\psi$$

$$\leq e^{-(n-1)\alpha}\int_o^\infty [ch2\alpha + sh2\alpha ch\psi]^{-1}d\psi = e^{-(n-1)\alpha}Q_o(ch2\alpha) \quad .$$

Similarly, we take $u = ch(\gamma + i\delta)$, $\gamma > 0$, and if we apply the Laplace formula (37), since

$$|u + \sqrt{u^2 - 1}\cos\phi|^2 = |ch(\gamma + i\delta) + sh(\gamma + i\delta)\cos\phi|^2$$

$$= \frac{1}{2}(ch2\gamma + \cos 2\delta) + sh2\cos\phi + \frac{1}{2}(ch2\gamma - \cos 2\delta)\cos^2\phi$$

$$\leq ch2\gamma + sh2\gamma = e^\gamma \quad ,$$

we find that

$$|P_n(u)| \leq e^{n\gamma} \quad .$$

From these two inequalities it may be deduced that expression (44) has a modulus at most equal to

$$(n+1)e^{n(\gamma-\alpha)}(1+e^{\alpha+\gamma})Q_o(ch2\alpha) \quad ;$$

it tends uniformly towards zero when $n \to \infty$, for any β and δ, provided α and γ are fixed and $\gamma > \alpha$.

Now given α and β variable, the point $x = ch(\alpha + i\beta)$ describes an ellipse with focii $x = 1$ and $x = -1$ and major axis $2ch\alpha$. The condition $\gamma < \alpha$ indicates that the point u is in a homofocal ellipse inside this first ellipse. We thus establish

Heine's Theorem. *The expansion (43) is valid for all points* u *inside the ellipse* E *with focii* +1 *and* -1, *and which passes through the point* x. *The convergence is uniform with respect to* u *and* x *if the points* u *are in an ellipse with focii* +1 *and* -1 *inside* E, *and if* x *is on* E.

Neumann's Theorem. *If the function* $f(z)$ *is holomorphic inside and on an ellipse* E *with focii* $z = \pm 1$, *then this function can be expanded in* E *as a series of Legendre polynomials*

$$f(z) = \sum_{o}^{\infty} a_n P_n(z) \tag{45}$$

which converges uniformly in E.

In effect, $f(z)$ is holomorphic on and within an ellipse E_1 homofocal to E and containing E in its interior. For z belonging to E or its interior, we have

$$f(z) = \frac{1}{2\pi i} \int_{E_1^+} \frac{f(x)dx}{x-z} \quad .$$

Now, we have for x belonging to E_1, z either within or on E,

$$\frac{1}{x-z} = \sum_{o}^{\infty} (2n+1)Q_n(x)P_n(z) \quad .$$

We can replace $1/(x-z)$ by this series in the integral and integrate term by term, giving the equation (45) valid for the conditions indicated with

$$a_n = \frac{2n+1}{2\pi i} \int_{E_1^+} f(x)Q_n(x)dx \quad .$$

By a deformation of the contour of integration E_1, we can reduce this integral to another form as we did above to establish the Neumann formula with respect to $Q_n(x)$. But we know that (45) occurs uniformly on the segment -1, +1; we can integrate the second member term-wise after multiplying by $P_n(z)$, and with respect to the orthogonality condition, we will have

$$a_n \int_{-1}^{+1} (P_n(x))^2 dx = \int_{-1}^{+1} f(x)P_n(x)dx \quad .$$

Hence, following (39), we have

$$a_n = \frac{2n+1}{2} \int_{-1}^{+1} f(x)P_n(x)dx \quad .$$

123 - Legendre functions regarded as spherical functions

Assuming the variables x, y, z are real, a homogeneous polynomial in x, y, z which satisfies the Laplace equation

$$\Delta V \equiv \frac{\partial^2 V}{\partial x^2} + \frac{\partial^2 V}{\partial y^2} + \frac{\partial^2 V}{\partial z} = 0 \quad ,$$

is called *a harmonic polynomial.* If we pass to spherical coordinates defined by

$$x = r \sin\theta \cos\phi, \quad y = r \sin\theta \sin\phi, \quad z = r \cos\theta \ ,$$

a harmonic polynomial of degree n takes the form $r^n S_n(\theta,\phi)$. The function $S_n(\theta,\phi)$ is called *a spherical function.* By considering the Laplace equation in spherical coordinates (I, 119), we see that S_n satisfies the equation

$$\frac{1}{\sin\theta} \frac{\partial}{\partial\theta}\left(\sin\theta \frac{\partial S_n}{\partial\theta}\right) + \frac{1}{\sin^2\theta} \frac{\partial^2 S_n}{\partial\phi^2} + n(n+1)S_n = 0 \quad .$$

If we take $\cos\theta = u$ as the variable, then the equation becomes

$$\frac{\partial}{\partial u}\left[(1-u^2)\frac{\partial S_n}{\partial u}\right] + \frac{1}{1-u^2} \frac{\partial^2 S_n}{\partial\phi^2} + n(n+1)S_n = 0 \quad . \tag{46}$$

If we apply the Euler formulae (as did Poincaré) we have

$$x = r\sqrt{1-u^2}\ \frac{e^{i\phi}+e^{-i\phi}}{2} \qquad y = r\sqrt{1-u^2}\ \frac{e^{i\phi}-e^{i\phi}}{2i}$$

$$z = ru \ ;$$

and S_n is of the form

$$S_n = \sum_{-n}^{n} e^{ip\phi} R_{p,n}(u)$$

where $R_{p,n}$ is a polynomial of degree n, homogeneous in u and $\sqrt{1-u^2}$.

If p is even, it is a polynomial in u; if p is odd, it is the product of $\sqrt{1-u^2}$ and a polynomial in u. By taking this expression for S_n into equation (46), we obtain a function of u and ϕ which must be identically zero. The $R_{p,n}(u)$ must therefore satisfy the equations obtained by equating the coefficients of $e^{ip\phi}$ to zero:

$$\frac{d}{du}\left[(1-u^2)\frac{dR}{du}\right] + \left[n(n+1) - \frac{p^2}{1-u^2}\right] R = 0\ , \qquad p \leq n \tag{47}$$

For $p = 0$, we recover the Legendre equation; the polynomials $R_{p,n}(u)$ are then Legendre polynomials. The functions $R_{p,n}(u)$, known as *the fundamental spherical functions or the associated Legendre functions,* are deduced from the Legendre polynomials. The differential equation (47) again enters into the class of Gauss equations. If we set $R = (u^2-1)^{p/2}Y$, we obtain the equation

$$(1-u^2)\frac{d^2Y}{du^2} - 2(p+1)u\frac{dY}{du} + (n-p)(n+p+1)Y = 0 \quad ,$$

which can be written, following Leibnitz's formula, as

$$\frac{d^p}{du^p}\left[(1-u^2)\frac{d^2Z}{du^2} - 2u\frac{dZ}{du} + n(n+1)Z\right] = 0 \quad ,$$

when we set

$$Y = \frac{d^pZ}{du^p} \quad .$$

It follows that, when $Z(u)$ is a solution of the Legendre equation

$$(1-u^2)Z'' - 2uZ' + n(n+1)Z = 0 \quad ,$$

the function

$$R = (u^2-1)^{p/2}\frac{d^pZ}{du^p}$$

is a solution of the associated equation (47). This provides the general solution of equation (47) and the fundamental spherical functions.

III. BESSEL EQUATIONS AND FUNCTIONS

124 - The differential equation and functions of Bessel

When the Laplace equation $\Delta V = 0$ is transformed by taking polar coordinates r, ϕ, z or *cylindrical coordinates,* then the equation becomes (I, 119),

$$\frac{d^2u}{dr^2} + \frac{1}{r}\frac{du}{dr} + \left(k^2 - \frac{h^2}{r^2}\right)u = 0 \quad .$$

By replacing r by x/k and u by y, we arrive at the equation

$$\frac{d^2y}{dx^2} + \frac{1}{x}\frac{du}{dx} + \left(1 - \frac{v^2}{x^2}\right) y = 0 \quad . \tag{48}$$

This is Bessel's equation; its solutions are known as *Bessel functions* or *cylindrical functions*. When the parameter v is an integer, we usually denote it by n. This equation, which may be regarded as a degeneracy of the Gauss equations, is of Fuchsian type at the origin: the roots of the determinantal equation are $+v$ and $-v$, with difference $2v$. Infinity is an irregular singularity.

THE BESSEL FUNCTION $J_v(x)$

Even when $2v$ is an integer, which may be assumed to be positive, the root v provides a solution of the form

$$y = x^v(c_o + c_1x + \dots + c_px^p + \dots) \quad ,$$

where the bracket is an entire function. By showing that this series satisfies equation (48) we obtain the relation

$$(v+p+2)(v+p+1)c_{p+2} + (v+p+2)c_{p+2} + c_p - v^2c_{p+2} = 0$$

which shows that p is even, $p = 2_q$ say, and yields

$$c_{2q+2} = \frac{-c_{2q}}{(v+p+2)^2 - v^2} = \frac{-c_{2q}}{4(q+1)(v+q+1)} \quad .$$

Thus we obtain

$$y = x^v c_o \sum_o^\infty \frac{(-1)^q x^{2q}}{2^{2q}q!(v+1)(v+2)\dots(v+q)} \quad .$$

By introducing the function $\Gamma(u)$, we proceed to take $c_o = 1/2^v\Gamma(v+1)$ and consider the solution

$$J_v(x) = \left(\frac{x}{2}\right)^v \sum_o^\infty \frac{(-1)^q \left(\frac{x}{2}\right)^{2q}}{\Gamma(q+1)\Gamma(v+q+1)} \quad .$$

$J_v(x)$ is know as *the Bessel function. When* $2v$ *is not an integer,* the second root of the determinantal equation, $-v$, equally provides a solution which is obtained by changing v to $-v$ in the preceding calculation.

We thus obtain the solution

$$J_{-v}(x) = \left(\frac{x}{2}\right)^{-v} \sum_{o}^{\infty} \frac{(-1)\left(\frac{x}{2}\right)^{2q}}{\Gamma(q+1)\Gamma(-v+q+1)} \quad ,$$

which remains valid and clearly independent of $J_v(x)$ *provided that* v *is not an integer.*

When v *is not an integer, the general solution of the Bessel equation is*

$$\lambda J_v(x) + \mu J_{-v}(x) \quad ,$$

where λ *and* μ *are arbitrary constants. When* v *is an integer,* the first n terms of $J_{-n}(x)$, $(x{=}v)$ vanish since $1/\Gamma(u)$ is zero for $u = 0,-1,\ldots,$ $J_{-n}(x)$ is therefore an entire function which, on setting $q = q' + n$, can be written as

$$\sum_{q=n}^{\infty} \frac{(-1)^q}{\Gamma(q+1)\Gamma(-n+q+1)} \left(\frac{x}{2}\right)^{2q-n}$$

$$= (-1)^n \sum_{q=0}^{\infty} \frac{(-1)^{q'}}{\Gamma(q'+1)\ (n+q'+1)} \left(\frac{x}{2}\right)^{2q'+n} = (-1)^n J_n(x) \quad .$$

To obtain a second integral in this case, we can apply d'Alembert's method, which is justified by Poincaré's theorem. We look for the limit of

$$\frac{(-1)^n J_{-n+\varepsilon} - J_{n-\varepsilon}}{\varepsilon}$$

when ε tends to zero. We must set apart the first n terms of $J_{-n+\varepsilon}$ which tend towards zero, and, by changing the index of summation q to $q + n$ in those remaining, we will have to determine the limit of

$$\sum_{o}^{n-1} (-1)^{n+q} \frac{1}{\varepsilon\Gamma(q+1)\Gamma(-n+\varepsilon+q+1)} \left(\frac{x}{2}\right)^{-n+\varepsilon+2y}$$

$$+ \sum_{o}^{\infty} (-1) \left(\frac{x}{2}\right)^{n+2q} \frac{1}{\varepsilon} \left[\frac{(\ /2)^{\varepsilon}}{\Gamma(n+q+1)\Gamma(\varepsilon+q+1)} - \frac{(\ /2)^{-\varepsilon}}{\Gamma(q+1)\Gamma(n-\varepsilon+q+1)}\right] \quad .$$

We know that the residue of the point $q + 1 - n$ of $\Gamma(u)$ is $(-1)^{n-q-1}/\Gamma(n-q)$ (I, 82), hence $\varepsilon\Gamma(-n+q+1+\varepsilon) \to (-1)^{n-q-1}/\Gamma(n-q)$.

In the second sum, we have to take the limit of $g(\varepsilon)/\varepsilon$, where $g(\varepsilon)$ is a holomorphic function of ε about $\varepsilon = 0$, and $g(o) = 0$, hence it is $g'(o)$; let it be

$$b_q = \frac{1}{\Gamma(q+1)\Gamma(n+q+1)} \left[2 \log \frac{x}{2} - \frac{\Gamma'(q+1)}{\Gamma(q+1)} - \frac{\Gamma'(n+q+1)}{\Gamma(n+q+1)}\right] .$$

We thus obtain the solution

$$y = - \sum_0^{n-1} \frac{\Gamma(n-q)}{\Gamma(q+1)} \left(\frac{x}{2}\right)^{-n+2q} + \sum_0^{\infty} b_q \, (-1)^q \left(\frac{x}{2}\right)^{n+2q}$$

which, along with $J_n(x)$, will provide the general solution.

125 - Recurrence formulae

The formulae relating the contiguous hypergeometric functions (no. 112) simplify in this case. We have

$$J_{v-1}(x) + J_{v+1}(x) = \left(\frac{x}{2}\right)^{v-1} \sum_0^{\infty} \frac{(-x^2/4)}{q!\Gamma(v+q)} + \left(\frac{x}{2}\right)^{v+1} \sum_0^{\infty} \frac{(-x^2/4)^q}{q!\Gamma(v+q+2)}$$

$$= \left(\frac{x}{2}\right)^{v-1} \left[\sum_0^{\infty} \frac{(x^2/4)^q}{q!\Gamma(q+v)} - \sum_0^{\infty} \frac{(-x^2/4)^{q+1}}{q!\Gamma(q+v+2)}\right]$$

$$= \left(\frac{x}{2}\right)^{v-1} \left[\frac{1}{\Gamma(v)} + \sum_1^{\infty} \left(\frac{1}{q!\Gamma(q+v)} - \frac{1}{(q-1)!\Gamma(q+v+1)}\right) \left(\frac{-x^2}{4}\right)^q\right]$$

$$= v \left(\frac{x}{2}\right)^{v-1} \sum_0^{\infty} \frac{(-x^2/4)^q}{q!\Gamma(v+q+1)} = \frac{2v}{x} J_v(x) .$$

Thus

$$J_{v-1}(x) + J_{v+1}(x) = \frac{2v}{x} J_v(x) . \tag{49}$$

Similarly, we see that

$$J_{v-1}(x) - J_{v+1}(x) = 2J'_v(x) . \tag{50}$$

From this we deduce

$$J_{v-1}(x) = \frac{v}{x} J_v(x) + J'_v(x) , \quad J_{v+1}(x) = \frac{v}{x} J_v(x) - J'_v(x) .$$

Remark. We see straight away from the expression for J_v that

$$J_{1/2}(x) = \sqrt{2/\pi x} \sin x ; \quad J_{-1/2}(x) = \sqrt{2/\pi x} \cos x ;$$

it follows that, when $v = q + 1/2$, where q is an integer, J_v is expressed in terms of $\cos x$ and $\sin x$.

126 - Schaefli's formula. The generating function of $J_n(x)$

By applying Hankel's formula (I, 240), we can write

$$\frac{1}{\Gamma(v+q+1)} = \frac{1}{2\pi i}\int_C e^t\, t^{-(v+q+1)}dt \quad ,$$

where C is a closed contour formed by the real axis of the t-plane, from $-\infty$ to $-\rho$, $\rho > 0$, by a circle of centre 0 and radius ρ and by the real axis from $-\rho$ to $-\infty$. We assume that the absolute value of the argument of t is less than or equal to π and C is described by firstly following the real axis below Ot. We deduce from this that

$$\left(\frac{x}{2}\right)^{-v} J_v(x) = \sum_o^\infty \frac{1}{2\pi i}\int_C e^t\, t^{-v-1}\,\frac{1}{q!}\left(\frac{-x^2}{4t}\right)^q dt \quad .$$

We can reverse the order of integration and summation, since this is possible when the integral is taken over a finite of C. On the other hand, the integral

$$\int e^t |t|^K \sum_o^\infty \frac{1}{q!}\left|\frac{x^2}{4t}\right|^q dt \ , \quad K = |v|$$

converges uniformly on C. Thus we have

$$J_v(x) = \frac{1}{2\pi i}\left(\frac{x}{2}\right)^v \int e^t\, e^{-x^2/4t}\, t^{-v-1}\, dt \quad ,$$

which is Schaefli's formula.

THE CASE WHERE v IS AN INTEGER

If v is an integer n, positive, negative or zero, then the integral (51) depends on a uniform function of t which is holomorphic, except at infinity and at the origin which are essential points. The contour C can then be replaced by a circle with centre the origin, taken in the direct sense. If x is non-zero, we can take the circle C', $|t| = \frac{1}{2}|x|\,R$; we put

$$t = \frac{1}{2}xu \ , \qquad |u| = R \quad ,$$

and we have

$$J_n(x) = \frac{1}{2\pi i}\int_{C'+} e^{x\left(\frac{u}{2}-\frac{1}{2u}\right)}\,\frac{du}{u^{n+1}} \quad . \tag{52}$$

In particular, if $R = 1$, we can put $u = e^{i\theta}$ and obtain

$$J_n(x) = \frac{1}{2\pi}\int_{-\pi}^{+\pi} e^{-in\theta+ix\sin\theta}\,d\theta$$

$$= \frac{1}{2\pi}\int_{0}^{+\pi} (e^{-in\theta+ix\sin\theta} + e^{in\theta-ix\sin\theta})d\theta$$

which yields the Bessel formula

$$J_n(x) = \frac{1}{\pi}\int_0^{\pi} \cos(n\theta - x\sin\theta)d\theta \quad .$$

THE GENERATING FUNCTION

The identity (52) shows that the $J_n(x)$ are the coefficients of the Laurent expansion of

$$e^{x\left(\frac{u}{2}-\frac{1}{2u}\right)} \quad .$$

For any x and u finite and non-zero,

$$e^{x\frac{u}{2}-\frac{1}{2u}} = \sum_{-\infty}^{+\infty} u^n J_n(x) \quad .$$

From this formula (due to Schölmilch) on changing u to $1/u$ and taking the product of the identities obtained, we deduce that

$$1 = (J_o(x))^2 + 2\sum_{1}^{\infty} (J_n(x))^2$$

(on account of $J_{-n}(x) = (-1)^n J_n(x)$). It follows that *for* x *real*,

$$|J_o(x)| < 1 \ , \qquad |J_n(x)| < \frac{1}{\sqrt{2}}$$

The second of these inequalities is more concise than that deduced from the Bessel formula.

LIMITS OF $|J_v(x)|$

The entire function

$$J_v(x)\left(\frac{x}{2}\right)^{-v} = \sum_{o}^{\infty} \frac{(-1)^q}{\Gamma(q+1)\Gamma(v+q+1)}\left(\frac{x}{2}\right)^{2q} \tag{53}$$

is of order 1, as can be seen by applying the method previously given (I, 211). The inequality which occurs for $J_n(x)$ when x is real, shows moreover that, in the case v is an integer, the order with respect to x^2 is at least 1/2 (I, 242). We can obtain a limit of $|J_v(x)|$ for $x = X + iY$ by putting the second member of (53) in the form of an integral. When the real part of $v + 1/2$ is positive, we have

$$\frac{1}{\Gamma(v+q+1)} = \frac{1}{\Gamma(\frac{1}{2}+v)\Gamma(q+\frac{1}{2})} \mathrm{B}\left(v+\frac{1}{2}, q+\frac{1}{2}\right) \quad ,$$

where $\mathrm{B}(u,v)$ is the Euler function of the first kind, hence (I, 148)

$$\mathrm{B}\left(v+\frac{1}{2}, q+\frac{1}{2}\right) = 2\int_o^{\pi/2} \cos^{2v}\theta \sin^{2q}\theta \, d\theta \quad .$$

As

$$\Gamma(q+\tfrac{1}{2}) = \Gamma(\tfrac{1}{2})\, 2^{-}, 3.5\ldots(2q-1) \quad ,$$

we see that the second member of (53) is written as

$$\frac{2}{\Gamma(v+\frac{1}{2})\Gamma(\frac{1}{2})} \sum_o^\infty \int_o^{\pi/2} (\cos^{2v}\theta)(-1)^q \frac{(x \sin\theta)^{2q}}{(2q)!} \, d\theta \quad ,$$

which shows that when the real part of is greater than $-1/2$, we have

$$J_v(x) = \frac{2(x/2)^v}{\Gamma(\frac{1}{2})\Gamma(v+\frac{1}{2})} \int_o^{\pi/2} \cos(x \sin\theta)\cos^{2v}\theta \, d\theta \quad .$$

Let us assume that v is real, positive or zero and put $x = X + iY$, $X = \Re x$. On account of the Euler formulae and the formula relative to $\mathrm{B}(u,v)$ we have

$$|\cos(x \sin\theta)| \quad e^{|Y \sin\theta|} \leq e^{|Y|} \quad ,$$

$$2\int_o^{\pi/2} \cos^{2v}\theta \, d\theta = \frac{\Gamma(\frac{1}{2})\Gamma(v+\frac{1}{2})}{\Gamma(v+1)}$$

hence

$$|J_v(x)| \leq \left|\frac{x}{2}\right|^v \frac{e^{|Y|}}{\Gamma(v+1)} \quad , \qquad Y = \Im x \quad . \tag{54}$$

Remark. For $v = n$, the Bessel formula yields the inequalities

$$|J_n(x)| < e^{|Y|} \quad , \qquad Y = \Im x \quad .$$

127 - Neumann's theorem

In studying the expansion of a holomorphic function as a series of Bessel functions, Neumann introduced the polynomials $O_n(t)$ in $1/t$, defined by

$$\frac{1}{t-x} = J_o(x)O_o(t) + 2\sum_1^\infty J_n(x)O_n(t) \quad .$$

To prove this identity and to determine the conditions under which it is valid, we start with the Schlömilch identity in which we will put

$$\frac{1}{2}\,(u-\frac{1}{u}) = v, \qquad u = v + \sqrt{v^2+1}\ , \qquad v \geq 0 \quad ,$$

in such a way that

$$e^{xv} = \sum_{-\infty}^{+\infty} (v + \ v^2+1)^n \ J_n(x)$$

$$= J_o(x) + \sum_{1}^{\infty}\Big[\Big((v+\sqrt{v^2+1})^n + (-1)^n(v+\sqrt{v^2+1})^{-n}\Big)\Big]J_n(x)$$

$$= J_o(x) + \sum_{1}^{\infty} R_n(v)J_n(x) \quad ,$$

where we put

$$R_n(v) = (v+\sqrt{v^2+1})^n + (v - \sqrt{v^2+1}) \quad ;$$

$R_n(v)$ is a polynomial of degree n in v known as Tchebycheff polynomial. On multiplying by e^{-vt}, we deduce from this that

$$e^{(x-t)v} = J_o(x)e^{-vt} + \sum_{o}^{\infty} R_n(v)e^{-vt}\ J_n(x) \quad .$$

By integrating from 0 to ∞ with respect to v, we obtain

$$\frac{1}{t-x} = J_o(x)\,\frac{1}{t} + \int_o^{\infty}\Big(\sum_{1}^{\infty} R_n(v)J_n(x)e^{-vt}\Big)\,dv \quad ; \tag{55}$$

this is possible if the real part of t is positive and greater than that of x. On account of the inequality (54), we have

$$|J_n(x)| \leq \frac{e^{|x|}|x|^n}{2^n n!} \quad .$$

On the other hand, for $v \geq 0$, $R_n(v)$ is positive, $R_n(v) < 2^{n+1}(v+1)^n$, hence

$$\int_o^{\infty} R_n(v)|e^{-vt}|dv < 2^{n+1}\int_o^{\infty} e^{-v\text{R}t}\ (v+1)^n dv$$

$$= 2^{n+1}e^{\text{R}t}\int_o^{\infty} e^{-(v+1)\text{R}t}(v+1)^n dv < \Big(\frac{2}{\text{R}t}\Big)^{n+1} e^{\text{R}t}\ n! \quad ,$$

and consequently

$$|J_n(x)|\int_o^{\infty}|R_n(v)e^{-vt}|dv < K\Big(\frac{|x|}{\text{R}t}\Big)^{n+1} e^{|x|+\text{R}t} \quad , \tag{56}$$

where K is a numerical constant (independent of n). We can then reverse the order of summation and integration in (55) when $\mathrm{R}t > |x|$, which gives, in this case, the Neumann formula with

$$2O_n(t) = \int_o^\infty R_n(v)e^{-vt}dv \, , \qquad \mathrm{R}t > 0 \quad .$$

This relationship defines a polynomial of degree $n + 1$ in $1/t$ with positive coefficients. We have, in effect,

$$R_n(v) = 2 \sum C_n^{2q} v^{n-q}(v^2+1)^q \quad ,$$

$$\int_o^\infty v^k e^{-vt} dv = \frac{\Gamma(k+1)}{t^{k+1}} \quad .$$

The limit obtained above for t positive is hence valid for any t by replacing t by $|t|$. We have a similar inequality to that of (56),

$$2|J_n(x)O_n(t)| < K \left|\frac{x}{t}\right|^{n+1} e^{|x|+|t|} \quad .$$

The series with general term $O_n(t)J_n(x)$ therefore converges absolutely and uniformly for $|t| \geq |x| + \varepsilon$ and defines a holomorphic function of t. Its sum coincides with $1/(t-x) - J_o(x)O_o(t)$ for $\mathrm{R}t \geq |x| + \varepsilon$, and so it is equal to this function for $|t| \geq |x| + \varepsilon$. Consequently, *the relationship*

$$\frac{1}{t-x} = J_o(x)O_o(t) + 2\sum_1^\infty J_n(x)O_n(t)$$

where

$$O_o(t) = \frac{1}{t} \, , \quad O_n(t) = \frac{1}{2}\int_o^\infty e^{-vt}\left[v + \sqrt{v^2+1})^n + (v - \sqrt{v^2+1})^n\right] dv$$

occurs for $|t| > 0$, $|t| > |x|$. *For given* x, *the convergence is absolute and uniform for*

$$|t| \geq |x| + \varepsilon, \qquad \varepsilon > 0 \quad .$$

By considering the Cauchy formula

$$f(x) = \frac{1}{2\pi i}\int_{C^+} \frac{f(t)}{t-x}\, dt \quad ,$$

where $f(x)$ is taken to be holomorphic in the circle C, $|x| < R'$, and on its circumference and where $|x|$ is taken to be less than R', we can replace $1/(t-x)$ by its value (57) and integrate term by term. Consequently:

Neumann's Theorem. *A holomorphic function* $f(x)$ *for* $|x| < R$, *is expandable in this circle in a series of the form*

$$f(x) = \sum_{o}^{\infty} c_n J_n(x) \quad ,$$

uniformly and absolutely convergent for $|x| \leq R' < R$. *We have*

$$c_o = f(o)\,, \quad c_n = \frac{1}{\pi i}\int_{C^+} f(t)O_n(t)dt, \quad n = 1,2,\ldots$$

where C *denotes a circle of radius less than* R.

Remark. By applying the formula to $J_p(x)$, $p \neq n$, we see that

$$\int_{C^+} J_p(t)O_n(t)dt = 0 \quad .$$

128 - An application of the Laplace method

By means of an elementary transformation, the Bessel equation reduces to a type to which the Laplace method may be directly applied (no. 102).

Let us set $y = Yx^v$ in equation (48), we obtain the equation

$$xY'' + (2v + 1)Y' + xY = 0 \quad ,$$

which, on account of the result of no. 102, admits solutions of the form

$$Y = \int_{\alpha}^{\beta} e^{zx}U(z)dz\,, \quad U(z) = \frac{e^{V(z)}}{z^2 + 1}$$

$$V(z) = \int \frac{(2v+1)z}{z^2+1}\,dz \quad ,$$

that is

$$Y = \int_{\alpha}^{\beta} e^{zx}(z^2 + 1)^{v - \frac{1}{2}}\,dz \quad ,$$

providing

$$\left[e^{zx}(z^2 + 1)^{v + \frac{1}{2}}\right]_{\alpha}^{\beta} = 0 \quad .$$

This is the case when the polynomial $P(z)$ of no. 102 has distinct roots; we can obtain the general integral by the Poincaré method as outlined in no. 103. The integrals that were introduced in this particular case had actually been studied by Hankel in 1869.

By pursuing the technique of no. 104, we will obtain the asymptotic expansions; the first asymptotic relationships relating to these functions were formulated by Poincaré, the general relationships were due to Hankel.

As an application, let us assume that the real part of v is greater than -1/2, and consider the integral taken along the segment joining $\alpha = -i$ to $\beta = i$. It satisfies the conditions and provides a solution

$$Y = \int_{-i}^{i} e^{zx}(z^2+1)^{v-\frac{1}{2}}\,dz = i\int_{-1}^{1} e^{itx}(1-t^2)^{v-\frac{1}{2}}\,dt \quad , \tag{58}$$

which is defined and is holomorphic for any x.

It therefore provides the solution $J_v(x)$ to the Bessel equation, to within a factor, since we have set $y = Yx^v$; we have

$$J_v(x) = Cx^v Y \quad ,$$

where C is a constant. By setting $x = 0$, we obtain the value of C, given by

$$Ci\int_{-1}^{1} (1-t^2)^{v-\frac{1}{2}}\,dt = \frac{1}{2^v\Gamma(v+1)} \quad .$$

The integral appearing in this last identity is the double of that taken between 0 and 1, which on changing the variable $t^2 = u$, is equal to

$$\frac{1}{2}\,B\left(v+\frac{1}{2}, \frac{1}{2}\right) = \frac{\Gamma\left(\frac{1}{2}\right)\Gamma\left(v+\frac{1}{2}\right)}{2\Gamma(v+1)} \quad .$$

It follows that

$$Ci = \frac{1}{2^v\sqrt{\pi}\,\Gamma(v+\frac{1}{2})} \quad .$$

We obtained this result by taking in the integral of the second member of (58), the argument of $1+z^2$ equal to zero for $z = 0$. This integral can be replaced by another taken along a contour L turning in an opposite sense about the points i and $-i$ (fig. 41). We will thus have

$$Y = \frac{1}{1-e^{2\pi i(v-\frac{1}{2})}}\quad I = \frac{e^{-i\pi v}}{2\cos \pi v}\, I \quad , \tag{59}$$

where

$$I = \int_L e^{zx}(z^2+1)^{v-\frac{1}{2}}\, dz \quad .$$

We can deform L be replacing it by two contours L_1 and L_2 about two parallels ending at the points i and $-i$ and whose direction will be that of the negative real axis when we wish to study Y for the x whose real part is positive. The method of no. 103 will apply to these integrals taken along L_1 and L_2. By setting $z = Z + i$, we will have on L,

$$e^{ix} \int e^{Zx}\left[Z(2i + Z)\right]^{v-\frac{1}{2}} dZ \quad ,$$

but it will be necessary to state clearly the signification of the second factor under the sign of integration, a factor which is written as

$$(2Z)^{v-\frac{1}{2}} \left(1 - i\, \frac{Z}{2}\right)^{v-\frac{1}{2}} e^{i\tau} \tag{60}$$

and for which, to apply the arguments of no. 103, we must assume that when Z is real positive, its argument is taken to equal 0.

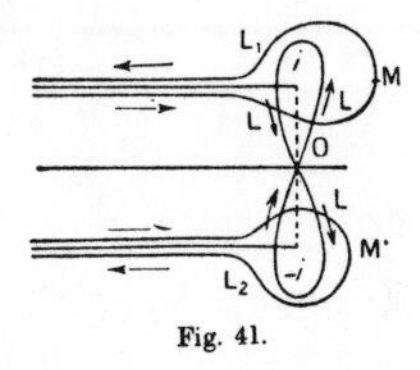

Fig. 41.

Now, at the point at infinity on the real negative axis, z^2+1 has argument zero. When we arrive at the point M, $i + \rho$, $\rho > 0$ (fig. 41) by following L_1 in the direction of the arrow, the argument of $1+z^2$ will be equal to $\pi/2$ and on multiplying by $v - \frac{1}{2}$, we see that in the formula in (60), τ is equal to $\frac{\pi}{2}(v-\frac{1}{2})$. For L_2, we will set $z = -i + Z$. But here the value of $z^2 + 1$ at infinity, no longer has zero argument, since for $z = 0$, its value is that obtained after rotating about the point i, $2\pi(v-\frac{1}{2})$ say. Therefore, at the point $M'(-i + \rho, \rho > 0)$, we will obtain the argument $\tau' = 2\pi(v-\frac{1}{2}) - \frac{\pi}{2}(v-\frac{1}{2})$, and on L_2, we will obtain

$$e^{-ix} \int e^{Zx}(2Z)^{v-\frac{1}{2}} \left(1 + i\frac{Z}{2}\right)^{v-\frac{1}{2}} e^{i\tau'} dZ \quad ,$$

where the integral is here taken in the opposite direction to that of the integral corresponding to L_1. We thus see that

$$e^{-i\,v}\,I = \int e^{Zx}(2Z)^{v-\frac{1}{2}}\left[e^{i\xi}\left(1 - i\frac{Z}{2}\right)^{v-\frac{1}{2}} + e^{-i\xi}\left(1 + i\frac{Z}{2}\right)^{v-\frac{1}{2}}\right] d \tag{61}$$

with

$$\xi = x - \frac{\pi v}{2} - \frac{\pi}{4} \quad ,$$

where the integral is taken in the $Z = X + Y$ plane, on a path about the real negative axis $(-\infty, -X_o, |Z| = X_o, X_o, +\infty)$ in the same sense as on L_1. The bracket appearing in the integral (61) is equal to

$$2\cos\xi\left(1 - \frac{(2v-1)(2v-3)}{32}Z^2 + \ldots\right) + \frac{1}{2}(2v-1)\sin\xi(Z + \ldots) \quad ,$$

and on setting $Zx = T$, with $|\arg x| < \frac{\pi}{2}$, we obtain, for large $|x|$, the asymptotic expansion of expression (61), (no. 103):

$$\frac{2^v}{x^v}\left(\frac{2}{x}\right)^{1/2}\left\{2\pi i\cos\xi\left(\frac{1}{\Gamma(\frac{1}{2}-v)} - \frac{(2v-1)(2v-3)}{32\Gamma(-\frac{3}{2}-v)}\frac{1}{x^2} + \ldots\right) + 2\pi i\sin\xi\left(\frac{2(v-1)}{4\Gamma(-\frac{1}{2}-v)}\frac{1}{x} - \ldots\right)\right\} \quad .$$

To pass to $J_v(x)$, we must multiply by

$$\frac{x^v}{2i\cos(\pi v)2^v\sqrt{\pi}\,\Gamma(\frac{1}{2}+v)}$$

and since, on account of the formula of the complements,

$$\Gamma(\tfrac{1}{2}-v)\Gamma(\tfrac{1}{2}+v) = \frac{\pi}{\sin\pi(v+\frac{1}{2})} = \frac{\pi}{\cos(\pi v)}$$

we obtain

$$J_v(x) = \sqrt{\frac{2}{\pi x}} \left\{ \cos\left(x - \frac{\pi v}{2} - \frac{\pi}{4}\right) \left[1 - \frac{(4v^2 - 1)(4v^2 - \ )}{128} \frac{1}{x^2} + \ldots\right] \right.$$

$$\left. - \sin\left(x - \frac{\pi v}{2} - \frac{\pi}{4}\right) \left[\frac{4v^2 - 1}{8} \frac{1}{x} - \ldots\right]\right\} \quad . \tag{62}$$

Following the proof, this asymptotic formula holds for $Rv > -\frac{1}{2}$, *and* $|\arg x| \le \frac{\pi}{2} - \varepsilon$, *where* ε *is an arbitrary positive no.* But we were able to take the contours L_1 and L_2 as going to infinity in a direction making an angle θ (less than $\pi/2$) with the negative real axis. We would then have a formula valid for $|\arg x - \theta| \le \frac{\pi}{2} - \varepsilon$ which would coincide with that about in the region common to the two angles of validity. *This shows that formula (62) remains valid for* $|\arg x| \le \pi - \varepsilon$, $\varepsilon > 0$.

Chapter VIII

NONLINEAR DIFFERENTIAL EQUATIONS IN THE COMPLEX DOMAIN

The study of the solutions of nonlinear equations presents difficulties which furnish the existence of moving singularities. Briot and Bouquet established the first results relating to the behaviour of the solutions in the neighbourhood of certain singular points of the coefficients of the equation (1856). Their results were extended to differential systems by Poincaré and Picard. The study of equations whose solutions are uniform or admit fixed singular points of a determined type was undertaken for first order equations by Briot and Bouquet, Hermite, Fuchs and Poincaré. From 1888, Painlevé introduced some new techniques, which for the first order case, were mainly developed by Boutroux and Malmquist (1913). In 1900, Painlevé devised a powerful method of investigation which allowed him to determine all the rational equations of the second order with fixed critical points, and to discover some new functions. His works were pursued in particular by Gambier, Chazy and Garnier.

In this chapter, we will first of all establish the most simple propositions of Briot and Bouquet, the theorem of Painlevé on the singularities of first order rational equations, the most simple theorems of Boutroux and Malmquist, and the theorem of Briot and Bouquet and Hermite on first order algebraic equations where the variable does not figure. We shall then discuss the results of Painlevé concerning the most simple of the second order equations with fixed critical points. Finally, we shall then return to the singular integrals of first and second order equations.

I. THE LOCAL STUDY OF SINGULARITIES. THE THEOREMS OF BRIOT AND BOUQUET

129 - The case where the differential coefficient becomes infinite

Given a first order differential equation

$$\frac{dy}{dx} = f(x,y) \quad , \tag{1}$$

we know that it admits a unique solution which for $x = x_o$, takes the value y_o when $f(x,y)$ is analysed at the point (x_o,y_o) (nos. 45, 46). This solution is holomorphic in a circle of centre x_o; the coefficients of its expansion can be calculated term by term. *Let us suppose that,* for $x = x_o$, $y = y_o$, *the function* $f(x,y)$ *is not analytic, but* $\frac{1}{f(x,y)}$ *is analytic.*

Let us put

$$g(x,y) = \frac{1}{f(x,y)} \quad .$$

We shall have $g(x_o,y_o) = 0$, otherwise $f(x,y)$ would be analytic. We can then say that $f(x,y)$ is infinite at the point (x_o,y_o). To study the solutions of equation (1) in the neighbourhood of the point (x_o,y_o), we can take this point to be the origin, and consider the equation

$$\frac{dx}{dy} = g(x,y) = a_{1,o}x + a_{o,1}y + \ldots \quad . \tag{2}$$

The existence theorem applies to this equation. It admits a unique solution which is zero for $y = 0$ and which is holomorphic for $|y|$ sufficiently small. As its derivative is zero for $y = 0$, this solution will be of the form

$$x = c_p y^p + \ldots \tag{3}$$

unless it is not identically zero. *In the general case* where the solution has the form (3), we obtain, on inverting,

$$y = \gamma_p x^{1/p} + \ldots \quad .$$

There exists a unique solution taking the value for $x = 0$; *this is an analytic function admitting the origin as a critical algebraic point.*

In order that equation (2) admits zero as a solution, it is necessary that the second member should be zero for $x = 0$, hence a factor that is a power of x, and this suffices. *In this case, there does not exist a solution of equation (1) taking the value* y_o *for* $x = x_o$.

Examples. The equations

$$\frac{dy}{dx} = \frac{1}{2y} , \quad \frac{dy}{dx} = \frac{1}{2xy}$$

have for solutions

$$y = \sqrt{x + C} , \quad y = \sqrt{\log(Cx)} \quad ,$$

respectively.

For $C = 0$, the first curve passes through the origin, whereas the second never passes through it.

130 - The case where the differential coefficient is indeterminate The general case

Briot and Bouquet have studied the case where the function $f(x,y)$ appearing in equation (1) is indeterminate at the point (x_o,y_o) and analytic about this point. We shall assume that *this point is the origin.* Let us consider the equation

$$x\frac{dy}{dx} = f(x,y) \tag{4}$$

where $f(x,y)$ is analytic about the origin and zero at this point. We have

$$f(x,y) = a_{1,o}x + a_{o,1}y + \dots$$

and by changing the notation, we can write (4) as:

$$xy' - \lambda y = ax + \phi(x,y) \quad , \tag{5}$$

where $\phi(x,y)$ is a power series in x and y commencing from the second degree terms. Let us see if there exists a holomorphic solution, necessarily zero at the origin, given by

$$y = c_1x + c_2x + \dots \tag{6}$$

The formal calculation of the coefficients of this series is possible *if* λ *is not a positive integer.* We will have

$$(-\lambda + 1)c_1 = a,\dots,(-\lambda + n)c_n = P_n(c_1,c_2,\dots,c_{n-1})\dots$$

P_n is a polynomial in $c_1,c_2,\dots,c_{n-1}$ and with respect to the coefficients of the function $\phi(x,y)$; the coefficients of this polynomial are positive. We can prove the convergence of the series (6) thus obtained by applying the method of the majorant functions. Let us denote the minimum of the numbers $|1-\lambda|$, $|2-\lambda|,\dots$ by H. This minimum exists since these numbers are not zero and have $+\infty$ as the only accumulation value; this is a positive number. Let

$$F(X,Y) = \frac{M}{(1-\frac{X}{r})(1-\frac{Y}{R})} - M - \frac{MY}{R} = AX + \Phi(X,Y)$$

be a majorant of the second member of (5). Let us consider the implicit equation

$$HY - F(X,Y) = 0 \quad . \tag{7}$$

If we look for a solution in the form of an entire series $Y = C_1X + C_2X^2 + \ldots$, then the coefficients will be given by the identities

$$HC_1 = A, \ldots, \quad HC_n = P_n\,(C_1, C_2, \ldots, C_{n-1}) \quad ,$$

where the polynomial P_n in $C_1, \ldots, C_{n-1}$ and with the coefficients of $\Phi(X,Y)$, is the same as that above. The series Y hence majors that of (6) since $H \leq |n - \lambda|$. Now, following the existence theorems of implicit functions in the analytic case (no. 39), equation (7) admits a holomorphic solution which is zero for $X = 0$ since the derivative of the first member with respect to Y, is $H \neq 0$ for $X = 0$. The series $Y = C_1X + C_2X^2 + \ldots$ converges in a certain circle, hence (6) does also. Thus we have:

The Theorem of Briot and Bouquet. *If λ is not a positive integer, then the differential equation (5), where $\phi(x,y)$ is analytic about the origin and has neither terms independent of x and y, nor terms of the first degree in x and y, admits a unique holomorphic solution which is zero for $x = 0$.*

A glance at the most simple cases shows that this holomorphic solution is not necessarily the only one zero at the origin.

Example. Consider the equation

$$xy' - \lambda y = ax$$

(where a and λ are constants). It is homogeneous and easily integrated (no. 65). For $\lambda - 1 \neq 0$, we have

$$y = \frac{a}{1-\lambda}\,x + Cx^{\lambda} \quad ,$$

where C is an arbitrary constant. We assume that λ is not a positive integer; the holomorphic solution at the origin corresponds to $C = 0$. If λ has a positive real part, then all solutions tend to zero if x tends to zero along a suitable path. If r is the modulus and ψ the argument of x (defined by continuity), and if $\lambda = \alpha + i\beta$, we have

$$|x^{\lambda}| = r^{\alpha}\,e^{-\psi\beta} \quad .$$

As α is positive, it suffices to have r tneding towards zero and ψ to remain fixed in order that y tends towards zero.

If α is negative or zero and $\beta \neq 0$, it suffices to assume that ψ is a function $\psi(r)$ such that

$$\lim_{r \to 0}\,(\alpha \log r - \beta\psi(r)) = -\infty$$

in order for y to tend to zero. We can assume that x tends to zero on a logarithmic spiral $\psi(r) = k \log r$, where k is a constant such that $\alpha - k\beta > 0$. Thus, the only case where the holomorphic integral is the only one which tends to zero when x tends to zero is that where λ is a negative real number or zero.

NONHOLOMORPHIC INTEGRALS AT THE ORIGIN TENDING TOWARDS ZERO WITH x.

In the general case of equation (5), where λ is not a positive integer, Briot and Bouquet obtained integrals that could tend towards zero as $x \to 0$, along a suitable path. As before, it is necessary to assume that λ is not a negative real number or zero.

Let z denote the holomorphic solution for $x = 0$ and $y = z + u$. Equation (5) is transformed into the equation

$$x \frac{du}{dx} = u(\lambda + \alpha_{1,0}x + \alpha_{0,1}u + \ldots) \quad .$$

Taking account of the preceding example, let us set

$$u = x^{\lambda}v \quad ,$$

where v is another unknown function which will be a solution of the equation

$$x \frac{dv}{dx} = v[\alpha_{1,0}x + \alpha_{0,1}x^{\lambda}v + \ldots] \quad . \tag{8}$$

We can look for a solution in the form of a double series in x and x^{λ}

$$v = \sum c_{m,n}x^{m+\lambda n}, \qquad n \geq 0, \ \ m \geq 0 \quad . \tag{9}$$

Calculating the successive coefficients is made step by step. We will have

$$(m + \lambda n)c_{m,n} = Q_{m,n}\ (\alpha_{p,q}, c_{\mu,v}) \ , \qquad m+n > 0 \quad ,$$

where $Q_{m,n}$ is a polynomial in $\alpha_{p,q}c_{\mu,v}$ with $\mu + v < m + n$, with positive coefficients. The coefficient $c_{0,0}$ is arbitrary. We will replace $c_{m,n}$ by a positive number $C_{m,n}$ greater than its modulus if we replace the $\alpha_{p,q}$, in $Q_{m,n}$, by the positive numbers $A_{p,q}$ greater than their moduli; $c_{\mu,v}$ will be replaced by $C_{\mu,v} \geq |c_{\mu,v}|$, and if $m + \lambda n$ is replaced by a positive number H less than $|m + \lambda n|$, $m \geq 0$, $n \geq 0$. For the $A_{p,q}$, we can take the coefficients of a majorant of the bracket of the second member of (8), taken to be of the form

$$\sum A_{p,q}X^{p+\Lambda q}V^{q} = \frac{M}{\left(1 - \frac{X}{r}\right)\left(1 - \frac{VX^{\Lambda}}{R}\right)} - M \ , \qquad \Lambda \geq |\lambda|$$

The coefficients of the expansion of that of the solutions of the equation

$$H(V - C_{o,o}) = \frac{MV}{\left(1 - \frac{X}{}\right)\left(1 - \frac{VX^{\Lambda}}{R}\right)} - MV \quad , \tag{10}$$

which is equal to $C_{o,o}$ for $X = 0$, are given by the identities

$$HC_{m,n} = Q_{m,n}(A_{p,q}, C_{\mu,\nu}) \ , \qquad m + n > 0 \quad .$$

This solution will be a majorant of the series (9). It is permissible to assume that Λ is integral; we know that equation (10) admits a holomorphic solution equal to $C_{o,o}$ for $X = 0$, provided that $C_{o,o}$ is sufficiently small. The series (9) converges when $|x^{\lambda}|$ and $|x|$ are sufficiently small since

$$\sum |c_{m,n}| A^{m+\Lambda n}$$

converges for $X < X_{\nu}$. We thus obtain the second theorem of Briot and Bouquet relative to equation (5):

<u>Theorem</u>. *When* λ *is neither a positive integer, a negative real number or zero, then every solution of equation (5) which takes a value sufficient near to that of the holomorphic solution at the origin for sufficiently small* $|x|$, *tends to zero when* x *tends to zero on suitably chosen paths.*

131 - <u>The case where</u> λ <u>is a positive integer</u>

If we take equation (5) and first of all assume that $\lambda = 1$, then the equation giving the coefficient c_1 of the series (6) reduces to $a = 0$; *there are no holomorphic solutions at the origin if* $\lambda = 1$, $a \neq 0$.

ASSUME $\lambda = 1$, $a = 0$. If we set $y = xz$ we obtain the equation

$$\frac{dz}{dx} = \frac{\phi(x, zx)}{x^2}$$

whose second member is analytic at the origin. There exists a holomorphic solution at the origin, $z = \psi(x)$, taking for $x = 0$ an arbitrary value z_o provided that $|z_o|$ is sufficiently small. *Equation (5) admits an infinity of holomorphic solutions at the origin and zero at this point.*

If $\lambda = 1$, $a \neq 0$, the situation is different. Let us consider the particular case of the equation

$$xy' - y = ax \qquad a = \text{const.} \quad ,$$

a homogeneous Euler equation whose general integral is

$$y = x[C + a \log x] \quad ,$$

where C is the arbitrary constant. There is no holomorphic integral at the origin, but all the solutions tend to zero when x tends to zero in a suitable way (if r is the modulus of x and θ its argument, then it suffices that θr tend towards zero).

This is a general fact. When $\lambda = 1$, $a \neq 0$, we can show that we obtain solutions depending on an arbitrary constant and which are given by a power series in x and $x \log x$; they tend towards zero with x under the conditions stated.

THE CASE WHERE $\lambda = n$, n A POSITIVE INTEGER

We make the transformation

$$y = \frac{ax}{1 - \lambda} + xz$$

which leads to the equation

$$xz' - (\lambda - 1)z = a_1 x + \phi_1(x,z)$$

which takes the same form, but with λ replaced by $\lambda - 1$. A finite number of these transformations reduces matters to the case $\lambda = 1$ which has just been discussed. Consequently: For all cases where λ is a positive integer there are two possible outcomes: there is either no holomorphic solution at the origin, or else there is an infinity of them. There always exists an infinity of solutions, depending on a parameter which tends towards zero when x tends towards zero on the suitably chosen paths.

132 - The general case

If we assume that we have an equation of the form

$$\frac{dy}{dx} = \frac{Y}{X} \quad , \tag{11}$$

where X and Y are power series in x and y convergent in the neighbourhood of the origin and zero at the origin:

$$Y = ax + by + \dots , \quad X = a'x + b'y + \dots \quad ,$$

then we may assume (since X and Y reduce to their terms of the first degree) that

$$y = xz \quad .$$

We obtain the equation

$$x\frac{dz}{dx} = \frac{a + (b-a')z - b'z^2 + x\psi(x,z)}{a' + b' \quad + x\phi(x,z)} \tag{12}$$

where ϕ and ψ are complete series. If $|a| + |a'| \neq 0$ matters are reduced to the cases previously studied: if $a \neq 0$, the differential coefficient is infinite at the origin, if $a = 0$, it is indeterminate. When we pass to $y = xz$, a solution z admitting a simple pole at the origin will yield a holomorphic solution at the origin; a solution z nonzero and holomorphic at the origin will yield a solution y of (11), holomorphic and zero at the origin.

In order to obtain the solutions of (12) which are holomorphic at the origin, we can set $z = z_o + Z$, where z_o is the value of this solution at the origin. Equation (12) becomes

$$x\frac{d}{dx} = \frac{a + (b-a')z_o - bz_o^2 + \ldots}{a' + b'z_o + \ldots} \quad .$$

As the first member is zero for $x = 0$ (if Z is holomorphic), we must have

$$a + (b-a')z_o - bz_o^2 = 0 \quad , \tag{13}$$

and if $a' + b'z_o \neq 0$, it is reduced to an equation of the type studied in no. 130, which, in general, admits a holomorphic solution. Equation (12) will have, in general, two holomorphic solutions at the origin, corresponding the two roots of equation (13).

If a and a' are zero, but b and b' nonzero simultaneously, then we can permute the roles of x and y, which will provide the solutions $y(x)$ tending to zero and x tends towards zero.

If a, a', b, b' are zero, then the above procedure shows that, in general, there will exist three integrals of equation (11) holomorphic and zero at the origin.

Remark. In the case where a root z_o of equation (13) forces $a' + b'z_o$ to be zero, then the equation in Z can be taken in the form

$$x^2\frac{dZ}{dx} = \alpha x + \beta Z + \ldots \quad .$$

There cannot exist any holomorphic solution at the origin. This will be so for the equation

$$x^2\frac{dZ}{dx} = \beta Z + \alpha x \quad ,$$

whose general integral

$$Z = e^{-\beta/x}\left[C + \alpha\int e^{\beta/x}\,\frac{dx}{x}\right], \qquad C = \text{const.}$$

admits the origin as an isolated singularity, for any C.

133 - The case of differential systems

The preceding results were extended to differential systems by Poincaré and Picard. Let us consider such a system

$$\frac{dx_1}{X_1} = \frac{dx_2}{X_2} = \ldots = \frac{dx_n}{X_n} \quad . \tag{14}$$

We assume that $X_1, X_2, \ldots, X_n$ are zero for $x_1 = x_2 = \ldots = x_n = 0$ and are analytic in a neighbourhood of this point. We shall assume that

$$X_j = a_j^1 x_1 + a_j^2 x_2 + \ldots + a_j^n x_n + \ldots, \qquad j = 1,2,\ldots,n \quad .$$

We can make a substitution for the x_k to reduce it to the system in (14) in a canonical form. If we set

$$y_k = \alpha_k^1 x_1 + \alpha_k^2 x_2 + \ldots + \alpha_k^n x_n, \qquad k = 1,2,\ldots,n$$

$$Y_k = \alpha_k^1 X_1 + \alpha_k^2 X_2 + \ldots + \alpha_k^n X_n, \qquad k = 1,2,\ldots,n$$

where the α_k^j are constants. The ratios in (14) are equal to the ratios

$$\frac{dy_1}{Y_1} = \frac{dy_2}{Y_2} = \ldots = \frac{dy_n}{Y_n} \tag{15}$$

and for the Y_k, we have the expansions

$$Y_k = \sum_{j=1}^{n} \alpha_k^j X_j = \sum_{j=1}^{n} \alpha_k^j \,(a_j^1 x_1 + \ldots + a_j^n x_n) + \ldots$$

in which we can replace the x_j as functions of y_k. The first degree term in the expansion of Y_k will be reduced to $\lambda_k y_k$ when we have

$$\sum_{j=1}^{n} \alpha_k^j a_j^m = \lambda_k \alpha_k^m \qquad m = 1,2,\ldots,n \quad . \tag{16}$$

If the equation in λ

$$\begin{vmatrix} a_1^1 - \lambda a_2^1 & \cdots & \cdots & a_n \\ a_1^2 & a_2^2 - \lambda & \cdots & a_n^2 \\ a_1 & a_2^n & & a_n^n - \lambda \end{vmatrix} = 0$$

has n distinct roots, then we can take the λ_k equal to these roots; to each λ_k there will correspond a system of numbers α_k^j, $j = 1,2,\ldots,n$ satisfying the equations (16). We know that the determinant of the α_k^j will be nonzero; we will indeed have a substitution. The system (14) will be replaced by the canonical system (15), in which we will have expansions of Y_k of the form

$$Y_k = \lambda_k y_k + \ldots \qquad k = 1,2,\ldots,n \quad .$$

If we denote by

$$\frac{dt}{t}$$

the common value of the ratios (15), we obtain the differential system

$$ty_k - \lambda_k y_k = \phi_k(y_1,\ldots,y_n), \qquad k = 1,2,\ldots,n \quad ,$$

where the ϕ_k are power series commencing from the terms of the second degree.

Proceeding as in no. 130, we will look for solutions in the form of power series in

$$t^{\lambda_1},\ldots,t^{\lambda_n} \quad .$$

The coefficients will be determined step by step; in each y_k, the coefficient of t^{λ_k} will be arbitrary and there will be no term independent of t. The formal calculation is possible and convergence results after calculating the limits, if the sums

$$m_1\lambda_1 + m_2\lambda_2 + \ldots + m_n\lambda_n - \lambda_j \qquad j = 1,2,\ldots,n \quad ,$$

where the m_k are arbitrary integers that are positive or zero (but $m_1^2 + \ldots + m_n^2 > 1$ when $m_j = 1$), have their modulus greater than a fixed number, and if

$$|t^{\lambda_1}| + \ldots + |t^{\lambda_n}|$$

is sufficiently small.

In certain cases, we can obtain a system of solutions tending towards zero when t tends towards zero or to infinity. The conditions in which the aforementioned result is valid, are, in fact, broader than those stated.

A number of works on this subject have been published, in particular, the works of Bendixon, Boutrous, Chazy, Dulac, Garnier, and Malmquist.

II. SINGULARITIES OF FIRST ORDER EQUATIONS

134 - The theorem of Painlevé

Let us consider a differential equation of the form

$$\frac{dy}{dx} = \frac{P(x,y)}{Q(x,y)} \tag{17}$$

where P and Q are polynomials in x and y not having a common factor. The curves $P(x,y) = 0$, $Q(x,y) = 0$ have a certain number of common points at a finite distance, call these (a_j,b_j) $(j = 1,2,\ldots,p)$. If we perform the transformation $y = 1/z$, equation (17) takes the form

$$\frac{dz}{dx} = \frac{P_1(x,z)}{Q_1(x,z)} \tag{18}$$

where P_1 and Q_1 are polynomials in x and z. We again consider the common points (a'_k,b'_k) to the two curves $P_1 = 0$, $Q_1 = 0$, $(k = 1,2,\ldots,)$. The points a_j and a'_k for which the solutions of equation (17) (take the values b_j or ∞) could have singularities, are known. These are *the singularities of the differential equation*. They are *fixed*. At the points a_j we shall add the points γ such that $x = \gamma$ satisfies (17). Let us consider a solution of equation (17) which, for $x = x_o$, takes a value such that $Q(x_o,y_o) \neq 0$. This solution is holomorphic in a circle of centre x_o, and its analytic continuation defines the solution in question (no. 48), which is uniform or multiform. A point α is a singular point for a branch $y = \phi(x)$ of this solution if α is a point of accumulation of points x' for which the circle of convergence of the Taylor expansion of $\phi(x)$ about x' does not contain the point α within its interior (I, 202).

Painlevé's Theorem. *The only singular points of the solutions other than the fixed points a_j, a'_k, are poles and critical algebraic points.*

Let α be a singular point (at a finite distance of a branch $y = \phi(x)$, distinct from the points a_j, a'_k. By β we denote a *finite* limiting value of $y = \phi(x)$ when x tends towards α. If such a value β does not exist, the $|y|$ tends towards $+\infty$ when x tends towards α. We shall discuss this case later. *A priori*, the following two cases are possible. 1) $Q(\alpha,\beta)$

is nonzero; 2) $Q(\alpha,\beta)$ is zero, but $P(\alpha,\beta)$ is not, since α is not a point a_j.

Let us show that it is possible that $Q(\alpha,\beta) \neq 0$. In effect, the second member of (17) will be analytic at the point α,β. Provided that $|x' - \alpha| < \varepsilon$, $|y' - \beta| < \varepsilon$ where ε is taken to be sufficiently small, the solution taking the value y' for $x = x'$ will be holomorphic in a circle containing the point α in its interior (no. 46, Theorem I). The branch $y = \phi(x)$ will be holomorphic at the point and y, moreover, will take the value β; α will not be a singularity for this branch.

Assume then that $Q(\alpha,\beta) = 0$, $P(\alpha,\beta) \neq 0$. Let us consider the equation

$$\frac{dx}{dy} = \frac{Q(x,y)}{P(x,y)} \tag{19}$$

which is satisfied by the inverse function of $y = \phi(x)$, $x = \psi(y)$ say. By hypothesis, there exists a sequence of points x_n tending towards α and such that $y_n = \phi(x_n)$ tends towards β. Equation (19), whose second member is analytic for $x = \alpha$, $y = \beta$, admits a unique solution, holomorphic for $y = \beta$ and taking the value β at this point, $x = \theta(y)$ say. This solution $\theta(u)$ is not reduced to the constant α, for $\alpha \neq \gamma$. When y_n and x_n are sufficiently near to β and α, then the solution $x = \psi_n(y)$ which for y_n takes the value x_n, is holomorphic in a circle $|y - \beta| < \eta$ (no. 46). It is also analytic with respect to x_n and y_n (no. 46), and when $x_n \to \alpha$ $y_n \to \beta$. It tends uniformly towards $\theta(y)$, in the circle $|y - \beta| < \eta$, provided that η is sufficiently small. Now $\theta(y)$, which is equal to α for $y = \beta$, which is not constant, takes all the values x such that $|x - \alpha| < \varepsilon$, providing ε is sufficiently small, in a circle $|y - \beta| < \eta$, whilst $|\theta(y) - \alpha| \geq 2\varepsilon$ for $|y - \beta| = \eta_1$. This results from the fact that when y describes a small circle $|y - \beta| \leq \eta_1$, $x - \alpha$ describes a domain with p sheets (if $\theta^{(q)}(\beta) = 0$ for $q = 1,\ldots,p$) bounded by a curve on which $|\theta(y) - \alpha| \geq 2\varepsilon$. It follows that, for n sufficiently large, the function $x = \psi_n(y)$, holomorphic for $|y - \beta| < \eta$, also takes the values x in this circle, such that $|x - \alpha| < \varepsilon$ (I, 193). The inverse function $y = \phi_n(x)$ of $x = \psi_n(y)$, which is the solution of equation (17), coincides with the branch of the solution $y = \phi(x)$ considered about x_n, admits the point α as a critical algebraic point. It follows that it is the same for $y = \phi(x)$, which is identical to $\phi_n(x)$.

The theorem is thus proved when, for x tending towards α, $|\phi(x)|$ does not tend towards $+\infty$. If $|\phi(x)| \to \infty$, then we consider the transformed equation (18) for which, when $x \to \alpha$, the solution tends towards 0. What has just been said will apply: α will be an ordinary point or a critical algebraic point for the solution $z = 1/y$, hence a pole or a critical algebraic point for y. The theorem is thus established.

THE CASE OF THE POINT AT INFINITY

In order to study the solutions when x tends towards infinity, we make the transformation $x = 1/X$. Matters are reduced to an equation of the same form. The point at infinity could then be a fixed singular point; otherwise, when it is a singular point for a branch of the solution, it will be a pole or critical algebraic point.

Remark. Painlevé's theorem is again valid when the equation is of the form (17), for which P and Q are polynomials in y and are analytic in x.

135 - Rational equations whose solutions do not have a mobile critical algebraic point

As an application of Painlevé's theorem, let us determine the form of an equation of the type in (17) for which the mobile singular points are all poles.

It is impossible for the denominator $Q(x,y)$ *to contain* y. Let us assume that y appears in $Q(x,y)$. Taking α to be arbitrary, distinct a_j, a_k', let us take for β a root of $Q(\alpha,y) = 0$, such that $P(\alpha,\beta)$. There exists a solution $y = \phi(x)$ such that y tends towards β when x tends towards α and which admits the point α as a critical algebraic point (no. 129).

The equation must then take the form $y' = P(x,y)$, where P' is a polynomial in y where coefficients are rational fractions in x. If we make the transformation $y = 1/z$, we obtain the equation

$$z' = -z^2 P(x, \frac{1}{z})$$

whose second member must be a polynomial in z. It is therefore necessary, and sufficient, that the polynomial P should be of the second degree in y (or of a lower degree). Thus we have the following:

Theorem. *In order that the solutions of a first order equation solved with respect to* y', *whose second member is a rational fraction in* x *and* y, *only admit poles as mobile singularities, it is necessary and sufficient that it should be a Riccati equation*

$$\frac{dy}{dx} = R_1(x)y^2 + R_2(x)y + R_3(x) \quad , \tag{20}$$

where R_1, R_2, R_3 *are rational fractions.*

The theorem remains true if we consider solvable equations whose second member is a rational fraction in y, with analytic coefficients in x. The fixed singularities are then the singularities of the coefficients and the analogous points to the points a_j, a_k'. In order that the mobile singularities are poles, it is necessary and sufficient that the equation should be of the form (20), where R_1, R_2 and R_3 are then analytic functions. The fixed singularities are the singularities of R_1, R_2 and R_3.

Remarks.

I. It is possible to determine the equation of type (17), whose mobile singularities are uniquely algebraic critical points. If we take it transformed as in (18), then it must not be a solution that is not identically zero, taking the value 0 for an arbitrary x; z must be a factor of the second member. *The degree of* $P(x,y)$ *with respect to* y *must be less than the degree of* $Q(x,y)$ *with respect to* y, *increased by* 2.

It follows that the solvable equations whose solutions do not have any mobile singularities are the linear equations.

II. A fortiori, the only solvable equations, rational in y and x, whose general integral is uniform, are necessarily of the form (20), but this condition is not sufficient. The supplementary conditions to be imposed on the coefficients R_1, R_2, R_3, which are not known, will involve the arithmetic properties of the coefficients of R_1, R_2, R_3; we see this in the simple case of $xy' = ay$ where a is a constant.

136 - A remark on the constants of integration

We have seen in no. 46 that the solutions of a solvable differential equation

$$y' = f(x,y) \quad ,$$

where the function $f(x,y)$ is analytic at the point x_o,y_o, depend analytically on an arbitrary constant. We obtain a sheaf of solutions in the neighbourhood of the point x_o and of the values of y close to y_o. When we analytically extend these elements of the solutions, we can, in certain cases, make x describe a path returning to the point x_o and it is possible to have the solutions extended to the point x_o. The solution initially equal to y_o, will in general take a new value y_1 on return to the point x_o. It is possible that the solution (x_o,y_1) thus obtained forms part of the sheaf of solutions defined about the point (x_o,y_o). For example, if we take the equation $xy' = ay$, where a is a real irrational constant very close to 1, then the general solution is $y = Cx^a$, where C is an arbitrary constant. If we consider the neighbourhood of the point 1 then the solution equal to 1 at the point 1 takes the value $e^{2i\pi ap}$ after p rotations about the origin; this is again close to 1 provided that p is not too large. But following the theorem of Poincaré-Volterra (I, 201), the values takes at a point by analytic function, form a countable set. For as long as the initial value y_o is arbitrary in a domain, we see *that there always exists an infinity of distinct solutions*. This remark justifies the distinction that was made between the fixed singularities and the moving singularities.

THE CASE WHERE THE SOLUTION DEPENDS RATIONALLY ON THE CONSTANT OF INTEGRATION

If $y = R(x,C)$ is an analytic function of x and a rational function of the constant C, then the elimination of C from this equation and its derivative $y' = R'_x(x,C)$ leads to the differential equation

$$F(y,y',x) = 0 \tag{21}$$

which is algebraic with respect to y and y'; F is a polynomial in y and y'. If x is considered as a constant, then the curve defined by equation (21), where y and y' are the coordinates, is of *genus zero*.

Let us suppose that we are given, *a priori*, a differential equation (21) algebraic in y and y', of *genus zero* and analytic in x. By expressing y and y' rationally as a function of a parameter u, then equation (21) may be written as

$$y = R(x,u) \ , \qquad \frac{dy}{dx} = R_1(x,u) \quad , \tag{22}$$

where R and R_1 are rationals in u and analytic in x, and we have $u = S(x,y,y')$ rational in y, y'. The solutions of (21) will be obtained by replacing u in (22), by the solution of the equation

$$R_1(x,u) = \frac{\partial R}{\partial x}(x,u) + \frac{du}{dx}\frac{\partial R}{\partial u}(x,u) \quad , \tag{23}$$

which is of the form $u' = G(x,u)$, where G is rational in u and analytic in x. If the solution $y = \phi(x,C)$ of (21) depends rationally on the constant C, then the solution of (23) can be written as $u = S(x, \phi(x,C), \phi'_x(x,C))$, its mobile singularities depend on C. Now, when C varies, the critical algebraic points of a rational function of C, are not displaced. The mobile singularities of the solutions of (23) will be poles; equation (23) will become a Riccati equation. Conversely, if equation (23) is of Riccati type, then we know that its general solution will depend homographically on the constant of integration; hence $y = R(x,u)$ will be a rational function of this constant.

137 - The theorem of Boutroux on the growth of solutions of $y' = P(y)$.

Let us consider an equation of the form

$$\frac{dy}{dx} = a_o(x)y^p + a_1(x)y^{p-1} + \ldots + a_p(x) \quad , \tag{24}$$

where the $a_j(x)$ are rational fractions. The fixed singularities are the poles of the $a_j(x)$ and, in general, the point at infinity. Let us consider a branch of the solution which has no singularity at a finite distance outside the circle $|x| > R$; it is then indefinitely extendable outside this circle and is multiform in general. We shall establish the hypothesis that one such solution $y = \phi(x)$ exists. We assume that R is sufficiently large in order that we can expand the $a_j(x)$ as a Laurent series

$$a_j(x) = \alpha_j x_j^{\beta} [1 + \psi_j(x)], \qquad \psi_j(x) = \sum_1^{\infty} \alpha_j^k x^{-k}$$

with

$$|\psi_j(x)| < \frac{1}{2}|x|^{\varepsilon} \qquad j = 0,1,2,\dots,p \quad ,$$

where ε is a given positive number less than 1. The second member of (24) could therefore be written as

$$x^{\beta} y^{p} [1 + \psi(x,y)] \qquad \alpha = \alpha_o \qquad \beta = \beta_o$$

with

$$\psi(x,y) \equiv \psi_o(x) + \sum_1^p a_j(x) \frac{1}{\alpha y^j x^{\beta}} \quad .$$

As the $\psi_j(x)$ are less than 1 if R is sufficiently large, then we will have

$$|\psi(x,y)| < |x|^{-\varepsilon} \tag{25}$$

provided that R is sufficiently large and that

$$|y| < |x|^{\sigma+\varepsilon} \tag{26}$$

where σ is greater than the maximum of the two numbers $(\beta_j - \beta)/j$. The number R only depends on p, the $a_j(x)$ and on ε.

Let us assume that $p \geq 2$ and consider a branch of the solution $y = \phi(x)$ having the above properties. Equation (24) can be written as

$$y^{-p} dy = \alpha x^{\beta} [1 + \psi(x,y)] dx$$

and can be integrated on a segment (X,x) under the form

$$y^{1-p} = Y^{1-p} + (1-p) \left\{ \int_X^x \alpha x^{\beta} dx + \int_X^x \alpha x^{\beta} \psi(x,\phi)\, dx \right\} \tag{27}$$

where y and Y are the values of $\phi(x)$ corresponding to x and X. The binomial formula (if $\beta \neq -1$), and the logarithmic expansion (if $\beta = -1$), show that the first integral of the second member of (27) is equal to

$$\alpha(x - X)X^{\beta}(1 + \theta(x)) \quad ,$$

with $|\theta(x)| < \frac{1}{4}$, provided that $|x - X| < \gamma|X|$ for which the number γ, less than 1, only depends on β. On the other hand, if on the segment (X,x), with $|X| > 2R$, $y = \phi(x)$ satisfies the inequality (26), then $\psi(x,y)$ will satisfy the inequality (25). The modulus of the second integral of the second member of (27) is less than

$$\frac{1}{4}\,|\alpha(x - X)X^{\beta}|$$

provided that R is sufficiently large. Under these conditions, we have then

$$y^{1-p} = Y^{1-p} + (1 - p)\alpha(x - X)X^{\beta}(1 + \omega(x)), \qquad |\omega(x)| < \frac{1}{2} \quad . \tag{28}$$

Let us assume that at a point X, $|X| > 2R$, the value $Y = \phi(X)$ satisfies the inequality

$$|Y| \geq |X|^{\sigma+\eta} , \qquad \eta > \varepsilon \tag{29}$$

The inequality (28) will hold on a segment (X,x) such that $|x - X| < \gamma|X|$ as long as $y = \phi(x)$ satisfies (26); this implies that

$$|y|^{1-p} < |Y|^{1-p} + 2(p - 1)|\alpha(x - X)X^{\beta}| \quad .$$

The first value x (with $|x - X| < \gamma|X|$) for which (26) is not satisfied, is then such that

$$2(p - 1)|\alpha||x - X| \geq [|x|^{(\sigma+\varepsilon)(1-p)} - |Y|^{1-p}]|X|^{-\beta}$$

$$\geq K|X|^{(\sigma+\varepsilon)(1- \ )-\varepsilon}$$

where K is a fixed number. In the last member of this inequality, the exponent of $|X|$ is greater than $(\sigma+\eta)(1-p) - \beta$. Assuming that η is chosen in such a way that

$$\tau = (\sigma+\eta)(1-p) - \beta < 1 - \varepsilon \quad , \qquad (n-\varepsilon)(p-1) > \varepsilon$$

we see that equation (28) is valid in the circle

$$|x - X| \leq |X|^{\tau+\varepsilon}$$

and it follows that on the circumference of this circle,

$$|y| < k|X|^{\sigma+\eta-\varepsilon/(p-1)}, \qquad k \text{ fixed} \quad .$$

The value of $|\phi(x)|$ at any point of this circumference would be less than the value at the centre, which satisfies the inequality (29), contradicting the Cauchy theorem on the maximum modulus hypothesis (29) is absurd. Thus we have

Boutroux's theorem. *If* p *is greater than* 1 *and if equation (29) admits a solution* $y = \phi(x)$ *which does not have any singularity at a finite distance outside a circle* $|x| > R$, *then this solution satisfies, at every point of* x *of modulus greater than* R', *an inequality of the form*

$$|y(x)| < |x|^k \quad ,$$

where the positive numbers R' *and* k *only depend on the coefficients* $a_j(x)$.

138 - The theorem of Malmquist

Let us consider another equation of the form

$$\frac{dy}{dx} = \frac{P(x,y)}{Q(x,y)} \quad , \tag{31}$$

where P and Q are polynomials in x and y. We can assume that the degree p of P with respect to y is tied to the degree q of Q with respect to y by the formula $p = q - 2$.

If this is not so, then we can always make a transformation $y = b + \frac{1}{Y}$ where b is such that $P(x,b)$, $Q(x,b)$ are not identically zero; the transformed equation will have the property as indicated. We intend, first of all, to look at certain properties of the solutions about a fixed singular point. Via a transforamtion $x = a + \frac{1}{X}$, we may assume that this point is at infinity. Hence we assume that in equation (31), $p = q + 2$ and that we consider matters outside a circle, $|x| > R$. We can write

$$P(x,y) = a_o(x)y^p + \ldots + a_p(x) \ , \quad Q(x,y) = b_o(x)y^q + \ldots b_q(x)$$

and expand the $a_j(x)$ and $b_k(x)$ about the point at infinity. The branches $y = \zeta_m(x)$ of the algebraic function defined by $Q(x,y) = 0$, can equally be expanded about the point at infinity. We will have

$$\zeta_m(x) = c_m x^{\lambda_m} + \ldots \ , \qquad m = 1,2,\ldots,q \qquad q \geq 1$$

whereby some of the branches may coincide: the numbers λ_m are in general fractional, positive or negative. If we set

$$y = \zeta_m(x) + \frac{1}{z} \quad ,$$

z verifies an equation of the form

$$\frac{dz}{dx} = \frac{\gamma_o(x)z^p + \ldots}{\delta_o(x)z^{q-\mu} + \ldots} \quad ,$$

where μ is the order of multiplicity of the branch $\zeta_m(x)$. In this expression, the coefficients of z are expandable about the point at infinity, as a series with fractional exponents. Proceeding as in no. 137, we shall put the equation in the form

$$\frac{dz}{dx} = \alpha z^2 + \mu x^\beta [1 + \psi(x,z)] \quad ,$$

and $\psi(x,y)$ will again satisfy condition (25), via condition (26), where σ is defined in an analogous way. We can then apply the same argument when we assume that z has no singularity at a finite distance, hence if $y = \phi(x)$, the solution of (31) does not have any algebraic singularities and if $\phi(x) - \zeta_m(x)$ is nonzero. We thus obtain the lemma:

Lemma. *If a solution* $y = \phi(x)$ *of equation (31), where* $p = q + 2$, $q \geq 1$, *has no critical algebraic points outside a circle,* $|x| < R$, *and if* $Q(x,\phi(x))$ *is nonzero outside this circle, then we have*

$$\left|\frac{1}{y - \zeta_m(x)}\right| < |x|^{k'} \quad \textit{if} \quad |x| > R \quad ,$$

where k' *and* R' *are positive numbers determined by the coefficients of (31) and* $\zeta_m(x)$ *is any one of the branches of the algebraic function defined by* $Q(x,y) = 0$.

Then let us assume that equation (31), with $p = q + 2$, has *a uniform integral* $y = \phi(x)$. With the exception of the fixed singular points, the function $\phi(x)$ only has poles. The function $1/Q(x,\phi)$ is a uniform function which does not admit any singular point outside the fixed singular points. For if x_o is a pole of $\phi(x)$, we see, on passing to the equation in $z = 1/y$, that $1/Q(x,\phi)$ is zero at the point x_o. In effect, the equation becomes

$$\frac{dz}{dx} = -\frac{a_o(x) + \ldots + a_p(x)z^p}{b_o(x) + \ldots + b_q(z)z^q}$$

(where the notations are the same as above), and, on setting $z = 0$, we see that $b_o(x_o) \neq 0$. The result is again valid when x_o is infinite. If x_o is a point of holomorphy of $\phi(x)$, then $Q(x_o,\phi(x_o))$ is nonzero, since P/Q must be finite and $P(x_o,\phi(x_o)) \neq 0$ if $Q(x_o,\phi(x_o)) = 0$. Thus the function

$$\frac{1}{Q(x,y)} \quad , \quad y = \phi(x) \quad ,$$

is holomorphic, except perhaps when x is one of the fixed singularities. In order to study this function about a fixed singularity we can assume that this singularity is at infinity. Then $y = \phi(x)$ has no critical algebraic points for $|x| > R$ and $Q(x,\phi(x))$ is nonzero. We may then apply the lemma. We have

$$\left|\frac{1}{y - \zeta_m(x)}\right| < |x|^{k'}, \qquad m = 1,2,\ldots,q \qquad q:$$

and consequently

$$\left|\frac{1}{Q(x,y)}\right| = \left|\frac{1}{b_q(x)\Pi(y - {}_m(x))}\right| < |x|^k$$

where k is a fixed number. Following the Liouville theorem in its general form (I, 191), the point at infinity is a pole of $1/Q(x,y)$. It follows that $1/Q(x,\phi)$ only has poles, up to infinity; it is a rational fraction (I, 194). Hence $Q(x,\phi)$ is a rational fraction $R(x)$; $y = \phi$ is defined by $Q(x,y) = R(x)$; it is an algebraic function. Now $\phi(x)$ is uniform, it is therefore a rational fraction. In the proof we have utilised the fact that $q \geq 1$, and we have the following result (since we can reduce matters to the case $p = q + 2$, as we have seen):

The Theorem of Malmquist. *If equation (31) in which* P *and* Q *are polynomials in* x *and* y, *is not a Riccati equation, then every uniform integral of this equation is a rational fraction.*

Malmquist established this other proposition the proof of which we shall leave aside: *If equation (31) does not transform into a Riccati equation by a transformation of the form*

$$z = \frac{y^n + \alpha_1(x)y^{n-1} + \ldots + \alpha_n(x)}{y^{n-1} + \beta_1(x)y^{n-2} + \ldots + \beta_n(x)} ,$$

where the $\alpha_j(x)$ *and* $\beta_k(x)$ *are rational fractions, then every integral of (31) which has only a finite number of branches is an algebraic function.*

For the study of these problems, we refer to the memoir of Malmquist (in *Acta Mathematica*, t. XXXVI) and to a note by Painlevé in *Lecons sur les fonctions dèfinies par les equations differentielles du premieur ordre* by Boutroux.

139 - Algebraic equations in y and y' with uniform solutions. The theorems of Hermite and Briot and Bouquet

Let us consider a differential equation of the form

$$P(y, \frac{dy}{dx}) = 0 \tag{32}$$

where P is a polynomial in y and $\frac{dy}{dx}$ is indecomposable. Let us determine the condition for which this equation admits a uniform solution and, consequently, has a solution that is generally uniform. This problem, posed by Briot and Bouquet, was solved by Hermite. If we set $u = \frac{dy}{dx}$, we have

$$P(y,u) = 0 , \quad dx = \frac{dy}{u} , \quad x = \int \frac{dy}{u} .$$

If y is a uniform function of x, then $u = y'_x$ is a multiform function of x. The coordinates of a point y,u of the algebraic curve $P(y,u) = 0$ are therefore uniform functions of x, i.e., of an abelian integral assigned to the curve. Following the theorem of no. 30, for which this is the case, it is necessary that the curve should have a genus equal to zero or one. We have seen in no. 30 that, if the genus is one, the abelian integral x must be of the first kind. In the case of genus zero, it will suffice to complete here what was said in no. 30. Retaining the notations of no. 30, we have seen that R(M) has at most $2m$ zeros; let $2m - \delta$, $\delta \geq 0$ be the number of these zeros. On the other hand, every point of ramification, of order r, is a pole of order $r - 1$ at least. The number of poles is then $\sum(r - 1) + \delta'$, $\delta' \geq 0$ and we have $\sum(r - 1) = 2m - \delta - \delta'$, hence following the Riemann formula (no. 9),

$$\frac{\sum(r-1)}{2} - m + 1 = 0 , \qquad \delta + \delta' = 2 .$$

The possibilities are then the following:

1. Each point of ramification of order r is a pole of order $r - 1$, and there are two other simple poles or another double pole ($\delta' = 2$).

2. One of the points of ramification of order r is a pole of order r; the other points of ramification are poles of order less by one unit of the order of ramification. There is one other simple pole ($\delta' = 2$).

3. Two of the points of ramification are poles of order equal to the order of ramification; the other points of ramification of order r are poles of order $r - 1$ ($\delta' = 2$).

4. A point of ramification of order r is a pole of order $r + 1$; the other ramification points are poles of order less by one unit of the order of ramification ($\delta' = 2$).

5. We have $\delta = \delta' = 1$. R(M) has a simple zero and $(m - 1)$ double zeros at infinity. It has a simple pole distinct from the ramification points, or one of the ramification points is a pole of order equal to that of the ramification.

6. We have $\delta = 2$, $\delta' = 0$. R(M) has two simple zeros at $m - 2$ double zeros at infinity or we have $(m - 1)$ zeros, all double, at infinity.

We see that the integral is of the third kind [1 (first hypothesis), 2, 3, 5, 6 (first hypothesis)] or of the second kind. In the first case, it has two critical logarithmic points; in the second case, it has a single simple pole. These conditions must be applied to the abelian integral x considered above.

Hermite's Theorem. *In order that equation (32) has its general solution uniform it is necessary and sufficient that either the curve* $P(y,u) = 0$ *is of genus one, and that the abelian integral*

$$x = \int \frac{dy}{u} \tag{33}$$

is of the first kind, or that the curve $P(y,u) = 0$ *is of genus zero and that the integral (33) is an integral of the second kind only having a single simple pole, or an integral of the third kind with two critical logarithmic points as its only singularities.*

We are about to see that the condition is necessary. Let us now suppose that this condition is satisfied. Let us assume first of all that the genus is zero. We could represent the curve Γ, $P(y,u) = 0$, parametrically in the form

$$y = R(y) \quad , \quad u = S(Y) \quad ,$$

where R and S are rational fractions and the representation is proper. Conversely, we have $Y = T(u,y)$, where T is a rational fraction. To the curve Γ there corresponds birationally the line $U = 0$, Y variable. To the integral (33) there corresponds the integral

$$x = \int \frac{R'(Y)}{S(Y)} \, dY$$

which, by hypothesis, will be of the second kind with a single simple pole, or of the third kind with two critical logarithmic points. $R'(Y)/S(Y)$ will have one of the form

$$\frac{A}{(Y-a)^2}, \quad A, \quad \frac{A}{Y-a}, \quad \frac{A}{Y-a} - \frac{A}{Y-b} \quad ,$$

where A, a, and b are constants. We deduce that Y is expressed as a function of x, in one of the corresponding forms

$$a - \frac{A}{x+C}, \quad \frac{x+C}{A}, \quad a + Ce^{x/A}, \quad \frac{a - Cbe^{x/A}}{1 - Ce^{x/A}} \quad ,$$

where C is a constant of integration and, consequently, $y = R(Y)$ is either a rational function in x or a rational fraction in $e^{x/A}$.

Let us assume now that the curve Γ has genus 1. Following the theorems of Noether and Bertini, we can make it correspond birationally to the cubic

$$U^2 = 4Y^3 - g_2Y - g_3$$

[(no. 16 and (I, 99)]. We will then have

$$y = R(Y,U) \ , \quad u = S(Y,U) \quad ,$$

where R and S are rational fractions. The integral (33) will be transformed into

$$x = \frac{1}{S}\left[\frac{\partial R}{\partial Y} + \frac{\partial R}{\partial U} \cdot \frac{dU}{dY}\right] dY = \int T\left(Y, \sqrt{4Y^3 - g_2Y - g_3}\right) dY$$

where T is a rational fraction. This integral will be of the first kind and hence of the form

$$x = C\int \frac{dY}{\sqrt{4Y^3 - g_2Y - g_3}}$$

where C is a constant. Y will be equal to $p(x/C + C')$, where C' is a constant of integration, and we will have

$$y = R(p(x/C + C'),\ p'(x/C + C')) \quad .$$

The theorem is completely proved.

It results from the proof that we obtain

The Theorem of Briot and Bouquet. *If equation (32) has its general solution uniform, then this solution is either an elliptic function, a rational function of* x, *or a rational fraction of* $e^{\alpha x}$ *where* α *is a constant.*

140 - Applications to binomial equations

For a particular example, consider a binomial equation

$$\left(\frac{dy}{dx}\right)^m = R(y) \quad ,$$

where m is a positive integer, and $R(y)$ is a rational fraction. It will have its general integral uniform if $R(y) = u^m$ defines a curve of genus 0 or 1 and if the other conditions are satisfied.

The ramification points of the surface defined by $R(y) = u^m$ are the zeros and poles of $R(y)$ whose order of multiplicity is not a multiple of m. If a_j is a point of ramification and if n_j is the remainder on division by m of the order of multiplicity of the zero or the pole which it provides, then the order of ramification r_j of a_j is the quotient by m by the greatest common divisor d_j of m and n_j. There are a_j ramification points identified with a_j (I, 174 and 175). Moreover, m and the n_j are the first in their set by virtue of the fact that $R(y) = u^m$ will be assumed to be indecomposable. Following the Riemann formula, we have, for q the number of the a_j,

$$\sum_1^q d_j(r_j - 1) = mq - \sum_1^q d_j = 2p + 2(m-1)$$

and since $2d_j \le m$,

$$m(q-2) = 2p - 2 + \sum_1^q d_j \le 2p - 2 + \frac{1}{2}mq \quad . \tag{34}$$

Thus, $m(q-4) \le 4(p-1)$, hence

$$q < 4 \quad \text{if} \quad p = 0 \, , \qquad q \le 4 \quad \text{if} \quad p = 1 \quad .$$

THE CASE OF GENUS ZERO

For $m > 1$, we have $q = 2$ or $q = 3$. If $q = 2$, we have $\sum d_j = 2$, hence $d_1 = d_2 = 1$, $r_1 = r_2 = m$. If $q = 3$, we have $\sum d_j = m + 2$ and, on dividing by m,

$$\frac{1}{r_1} + \frac{1}{r_2} + \frac{1}{r_3} = 1 + \frac{2}{m} > 1 \quad .$$

We have a finite number of solutions (cf. I, 222). We have

$$r_1 = 2, \quad r_2 = 2, \quad r_3 = \frac{m}{2}, \quad m \text{ an even integer} > 2 \quad ;$$

$$r_1 = 2, \quad r_2 = 3, \quad r_3 = 3 \, ; \quad m = 12 \quad ;$$

$$r_1 = 2, \quad r_2 = 3, \quad r_3 = 4 \, ; \quad m = 24 \quad ;$$

$$r_1 = 2, \quad r_2 = 3, \quad r_3 = 5 \, ; \quad m = 60 \quad .$$

To these cases, it is necessary to include the case $m = 1$.

THE CASE OF GENUS ONE

Following (34), it is necessary to have $q > 2$. If $q = 3$, we have $\sum d_j = m$ (cf. I, 220)

$$\frac{1}{r_1} + \frac{1}{r_2} + \frac{1}{r_3} = 1 \quad ,$$

and $n_j = d_j \mu_j$ and the n_j and m are first in their set; hence

$$r_1 = 3, \quad r_2 = 3, \quad r_3 = 3\,; \quad m = 3 \quad ;$$

$$r_1 = 2, \quad r_2 = 4, \quad r_3 = 4\,; \quad m = 4 \quad ;$$

$$r_1 = 2, \quad r_2 = 3, \quad r_3 = 6\,; \quad m = 6 \quad .$$

If $q = 4$, we have $\sum d_j = 2m$, hence

$$\frac{1}{r_1} + \frac{1}{r_2} + \frac{1}{r_3} + \frac{1}{r_4} = 2$$

and $r_1 = r_2 = r_3 = r_4 = 2$; m is equal to 2.

EQUATIONS CORRESPONDING TO THE CASE OF GENUS ZERO

Amongst the possible forms of $R(y)$, we must take those for which the integral

$$y = \int [R(y)]^{-1/m}\, dy \tag{35}$$

is of the second kind, with a single simple pole or of the third kind with two critical logarithmic points.

For $m = 1$, we obtain for $R(y)$ the following expressions, where A, a and b are constants:

$$A, \quad A(y-a), \quad A(y-a)(y-b), \quad A(y-a)^2 \quad .$$

The case $q = 3$ does not yield solutions, for x has a singularity of the same nature on each sheet or at each point of ramification corresponding to a like pole or zero of $R(y)$; there are at least two poles or three logarithmic points.

Let us assume that $q = 2$. We have $r_1 = r_2 = m$. If $m \geq 3$, we cannot have logarithmic point since there would be at least three. Hence we can have a simple pole at a ramification point. For dy/dx we will have:

$$A(y-a)^{1+1/m}, \quad A(y-a)^{1-1/m}, \quad A(y-a)^{1+1/m}(y-b)^{1-1/m} \quad .$$

These forms agree for $m = 2$. For $m = 2$, we can have two critical logarithmic points. The forms corresponding to dy/dx are, where c is again a constant,

$$A\sqrt{(y-a)(y-b)} \quad , \quad A(y-c)\sqrt{(y-a)(y-b)} \; , \quad A(y-c)\sqrt{y-a} \quad .$$

EQUATIONS CORRESPONDING TO GENUS ONE

The integral (35) must be of the first kind, where $R(y)$ has the determined form as above. For $m = 2$, we obtain for $R(y)$, the forms

$$A(y-a)(y-b)(y-c)(y-d), \quad A(y-a)(y-b)(y-c) \quad .$$

For $m = 3$, r_1, r_2 and r_3 are equal to 3, and we obtain for $R(y)$ the expressions

$$A(y-a)^2(y-b)^2(y-c)^2 \ , \quad A(y-a)^2(y-b)^2 \quad .$$

For $m = 4$, $r_1 = 2$, $r_2 = r_3 = 4$, we have the forms for $R(y)$:

$$A(y-a)^2(y-b)^3(y-c)^3 \ , \quad A(y-a)^2(y-b)^3 \ , \ A(y-a)^3(y-b)^3 \quad .$$

For $m = 6$, $r_1 = 2$, $r_2 = 3$, $r_3 = 6$, we have the following four forms for $R(y)$:

$$A(y-a)^3(y-b)^4(y-c)^5 \ , \quad A(y-a)^3(y-b)^4,$$

$$A(y-a)^3(y-c)^5 \ , \quad A(y-a)^4(y-c)^5 \quad .$$

Following the theorems of Hermite and Briot and Bouquet, the solutions of the corresponding equations are elliptic functions. For $m = 2$, we see that we can directly obtain the inverse functions of the elliptic integrals of the first kind that can be reduced to canonical forms. For $m = 3,4,6$ the expressions obtained reduce in each group, to the last in which we can take $a = 0$ and $b = 1$. We recover the functions giving the conformal representation of the half-plane on a triangle (I, 220).

141 - The theorem of Poincaré

The theorem of Painlevé of no. 134 was proved in a particular case. Painlevé considered the most general case. In particular, if we consider a differential equation of the form

$$F(y, \frac{dy}{dx}, x) = 0 \tag{36}$$

where F is a polynomial with respect to y and dy/dx is analytic with respect to x. The singular points of the solutions are on one hand, fixed, and on the other, they are poles and mobile critical algebraic points. In 1884, Fuchs indicated the conditions for which there are none of the latter and he discussed the case where the of the relation $F = 0$ (where x is regarded as a constant and y and dy/dx are regarded as independent variables) is equal to zero or one. Poincaré obtained a complete solution of the problem (1885) by proving:

If the only mobile singularities of equation (36) are poles, then the general solution is an algebraic function of the coefficients of y *and* dy/dx *in (36) when the genus is greater than one. The general solution is expressed in terms of elliptic functions if this genus is one; if the genus is zero, we obtain a Riccati equation.*

III. EQUATIONS OF ORDER GREATER THAN ONE - THE THEORY OF PAINLEVÉ - THE EQUATIONS AND FUNCTIONS OF PAINLEVÉ

142 - Second order equations with fixed critical points. Painlevé's method

Following the terminology of Painlevé, we call *the critical points of an analytic function,* the singular points (whether or not isolated) such that in an arbitrarily restricted neighbourhood of one of these points ζ, the branch of the functions, singular at the point ζ, is not uniform. Putting this another way, there exists a closed curve γ (which cannot encircle ζ) arbitrarily near to ζ, such that the branch $f(z)$ in question, extended from z_o to z_o along γ, takes at least two distinct values at the point z_o.

Let us consider a differential system

$$\frac{dy}{dx} = f(x,y,z,\alpha) \ , \qquad \frac{dz}{dx} = g(x,y,z,\alpha) \quad , \tag{37}$$

where the functions f *and* g *are analytic and uniform in* x, y, z, *and* α, *when the point* x,y,z *belongs to a domain* Δ *and when* $|\alpha| < \eta$. Let us consider the solution $y = Y(x)$, $z = Z(z)$ which corresponds to $\alpha = 0$ and which takes the values y_o, z for $x = x_o$. Let us continue this solution with the point x,y,z remaining Δ. If x', y', z' is a point of Δ obtained in this extension $(y' = Y(x'),\ z' = Z(x'))$, we can apply to this point Poincare's theorem of no. 46, tenable for one such system (see no. 47). The solution of the system in (37), which for $x = x'$, takes the values y'' and z'', and corresponds to α, is $y = y(x,x',y'',y'',\alpha)$, $z = z(x,x',y'',z'',\alpha)$. It is analytic with respect to x, y'', z'', α provided that $|x - x'|$, $|y'' - y'|$, $|z'' - z'|$ and $|\alpha|$ are sufficiently small. To each x', y', z' there corresponds a circle $|x - x'| < \rho(x',y',z')$ in which this property holds. If the set of the x', y', z' is a continuum, then we can cover it with a finite number of these circles by virtue of the theorem of Borel-Lebesque. By utilising these circles in a finite number, we obtain the following result of Poincaré.

Let $y = Y(x)$, $z = Z(x)$ *be the solution of the system in (37) corresponding to* $\alpha = 0$ *and to the initial conditions* x_o, y_o, z_o *extended in a continuum completely contained in* Δ. *Provided that* $|\alpha|$ *is sufficiently small,* $|\alpha| < \eta$, *then the solution* $y(x,\alpha)$, $z(x,\alpha)$ *which corresponds to* x *and with the same initial conditions, is expandable in a power series at* α

$$y(x,\alpha) = Y(x) + \sum_1^\infty \alpha^p Y_p(x) \ , \quad z(x,\alpha) = Z(x) + \sum_1^\infty \alpha^p Z_p(x) \quad , \tag{38}$$

whose coefficients are analytic functions holomorphic about every value x *in question.*

From this theorem we deduce the lemmas which are at the heart of the Painlevé method:

Lemma I. *If for each* α *such that* $0 < |\alpha| < \eta$, *the solution (38) of the system (37), is a uniform function of* x, *then it is also uniform for* $\alpha = 0$ *and the coefficients of the expansions (38).*

Let us assume, in effect, that one of the $Y_p(x)$, for example, is not uniform (for $p = 0$, we consider $Y(x)$) when x takes the values which it may possibly attain. There would exist one x for which $Y_p(x)$ would have at least two distinct values. For this x, $y(x,z)$ would also have at least two distinct values for all the $|\alpha| \le \eta' < \eta$ except for a finite number of α; this would not be a uniform function.

Lemma II. *If there exists a simple closed curve* γ *along which the expansions (38) are valid and if, when* x *describes* γ, *one of the coefficients is not uniform, then the corresponding branch of* $y(x,\alpha)$ *of or* $z(x,\alpha)$, *extended in the interior of* γ, *possesses at least one critical point or at least one critical group.*

In order to prove this and to explain the term critical group, let us assume, as we may, that there is a coefficient of $y(x,\alpha)$ which is not uniform. Then following Lemma I, $y(x,\alpha)$ is not uniform on γ. Then following Lemma I, $y(x,\alpha)$ is not uniform on γ. The Riemann surface on which $y(x,\alpha)$ is uniform, restricted to the interior of γ (i.e., obtained by extending $y(x,\alpha)$, with x remaining in γ), has several sheets. Its boundary may contain critical points that may or may not be isolated. Contrary to this, we shall say that it contains *a critical group.*

Here is a simple example of such a group for a function. Let $G(x)$ be a holomorphic function for $|x| > 1$ and admitting the circumference $|x| = 1$ as a cutting. (One such function may be derived from the functions previously constructed (I, 200) by changing z to $1/x$). The function $F(x) = G(x) + \sqrt{x}$ is a function defined for $|x| > 1$, which has two branches and admits $|x| = 1$ as a singular line. These points $|x| = 1$, are not critical points; for each of them, the singularity is comparable with that of $G(x)$. But they do form a critical group.

THE GENERAL METHOD OF PAINLEVÉ

We consider a second order equation

$$\frac{d^2y}{dx^2} = R(x,y, \frac{dy}{dx}) \quad , \tag{39}$$

where R *is an irreducible rational fraction in* y *and* dy/dx*, whose coefficients are analytic in* x. We look for those conditions for which the general integral is uniform or more generally does not possess any critical points or mobile critical groups. By setting $dy = zdx$, we replace the equation by a system which must have the same property. In various ways, we can introduce a parameter α such that for $0 < |\alpha| < \eta$, the solutions are again uniform or with critical points or fixed critical groups, and such that for $\alpha = 0$, the system is integrable. By applying the Lemmas I and II, we discover certain necessary conditions. We then wish to see if these conditions are sufficient.

THE FIRST APPLICATION OF THE METHOD

Let us consider the system

$$\frac{dy}{dx} = f(x,y,z) \quad , \qquad \frac{dz}{dx} = g(x,y,z) \tag{40}$$

where f and g are rational fractions in z with analytic coefficients in x and y. Let $z_1 = h(x_1,y_1)$ be a pole of the fractions $f(x_1,y_1,z)$, $g(x_1,y_1,z)$. z_1 is a zero of the denominator of f, for example. This denominator, a polynomial in z, could be decomposed into a product of indecomposable polynomials, $Q_1Q_2Q_3$ say. We assume that x_1, y_1 does not annihilate the discriminants of these polynomials. Following the theorem on the implicit functions, $h(x,y)$ is then analytic in the neighbourhood of x_1, y_1. We can set $z = h(x,y) + Z$ and the system (40) is then written as

$$Z^m \frac{dy}{dx} = F(x,y,z) \; , \qquad Z^n \frac{dZ}{dx} = G(x,y,z) \tag{41}$$

where m and n are integers, one of which is positive, and F and G are analytic and uniform about the point $x_1,y_1,0$ and different from zero at this point.

Lemma III. *In order that the critical points and critical groups of the solutions of the system (40) are fixed, it is necessary that* $m \geq n + 1$.

Let us assume in effect that $m < n + 1$, implying that $n > 0$. Let us consider the system obtained by replacing x, y, Z in (41) by

$$x_1 + \alpha^{n+1}x \qquad y_1 + \alpha^{n+1-m} \; , \qquad\qquad \alpha z$$

respectively. For $\alpha \neq 0$, the new system, along with the original, will have its critical points and groups fixed. In this new system

$$\begin{aligned} z^m \frac{dy}{dx} &= F(x_1 + \alpha^{n+1}x,\ y_1 + \alpha^{n+1-m}y, \alpha z) \quad , \\ z^n \frac{dz}{dx} &= G(x_1 + \alpha^{n+1}x,\ y_1 + \alpha^{n+1-m}y, \alpha z) \quad , \end{aligned} \tag{42}$$

we take $x_o = 0,\ y_o = 0,\ z_o \neq 0$ as initial conditions, where $|z_o|$ is sufficiently small that, about the point $0,0,z_o$ the system solved is analytic in $x,\ y,\ z$ and α. For $\alpha = 0$, the equation

$$z^n \frac{dz}{dx} = \mu \ , \qquad \mu = G(x_1, y_1, 0) \neq 0 \quad ,$$

has as its solution

$$z = [(n+1)\ x + z_o^{\ n+1}]^{1/n+1} \quad ,$$

which admits the point $\xi_o = -z_o^{\ n+1}/(n+1)\mu$ as critical point. If x describes an annulus of centre ξ_o whose inner radius is less than $|\xi_o|$ and whose outer radius is greater than $|\xi_o|$, then we can apply Lemma II provided that $|z_o|$ is sufficiently small. For $|\alpha| < \eta$, η depending on the radius ρ of the circle inside the annulus, then $z(x,\alpha)$ will be analytic in the annulus, but will be multiform: the circle $|x - \xi_o| < \rho$ will contain a ctitical point or a critical group of $z(x,\alpha)$. When we vary the argument of z_o, its modulus remains constant; this critical group will be displaced. The solutions of the system in (42) will not have their critical points and groups fixed. The lemma is thus proved.

143 - An application to the equation $y'' = R(x,y,y')$

If we write equation (39) under the form

$$\frac{dy}{dx} = z \ , \qquad \frac{dz}{dx} = R(x,y,z) \quad ,$$

we can apply Lemma III to the poles of R regarded as a function of z. As m is zero (or equal to -1), n becomes negative in order that the lemma applies; $R(x,y,z)$ must not then have any poles. $R(x,y,z)$ is a polynomial in z. Let q be its degree. Let us write equation (39) in the form

$$\frac{dy}{dx} = \frac{1}{z} \ , \qquad \frac{dz}{dx} = -z^2 R(x,y,\frac{1}{z})$$

and apply Lemma III. Here we have $m = 1$ and $n = q - 2$. It is necessary to have $q \leq 2$. Thus: *In order that the critical points or groups of the solutions to equation (39) are fixed, it is necessary that* $R(x,y,z)$ *is a polynomial in* z*, of degree two at most.* Equation (39) must have the form

$$\frac{d^2y}{dx^2} = A(x,y)\left(\frac{dy}{dx}\right)^2 + B(x,y)\,\frac{dy}{dx} + C(x,y) \tag{43}$$

where A, B and C are rational fractions in y and analytic in x.

Equation (43) is equivalent to the system

$$z\,\frac{dy}{dx} = 1, \quad -\frac{dz}{dx} = A(x,y) + B(x,y)z + C(x,y)z^2 \quad .$$

In these equations, let us replace x, y, z by $x_1 + \alpha x, y, \alpha z$ respectively, where x_1 is finite and α is a parameter. We obtain the system

$$z\,\frac{dy}{dx} = 1\ , \quad -\frac{dz}{dx} = A(x_1 + \alpha x, y) + zB(x_1 + \alpha x, y) + \alpha^2 z^2 C(x_1 + \alpha x, y) \tag{44}$$

whose solutions have for $\alpha \neq 0$, their critical points or groups fixed at the same time as those of (43). *To apply Lemma II, we first of all observe the system obtained by setting* $\alpha = 0$,

$$z\,\frac{dy}{dx} = 1\ , \quad -\frac{dz}{dx} = a(y)\ , \qquad a(y) = A(x_1, y) \quad . \tag{45}$$

Let us assume that the rational fraction $a(y)$ admits a pole γ of order p. By changing y to $y + \gamma$, we can assume that $\gamma = 0$, and write the system in the form

$$y'' = y'^2 a(y)\ , \qquad a(y) = \frac{k}{y^p} + \dots\ , \tag{46}$$

where the Laurent expansion of $a(y)$ is valid for $|y|$ sufficiently small. Let us replace the equation obtained by the new system

$$\frac{dy}{dx} = z\ , \qquad \frac{dz}{dx} = \frac{z^2}{y^p}\ (k + \dots\ ) \quad , \tag{47}$$

and introduce a system with a parameter β by setting $x = X$, $y = \beta Y$, $z = \beta^p Z$:

$$\frac{dY}{dX} = \beta^{p-1} Z\ , \qquad \frac{dZ}{dx} = \frac{Z^2}{Y^2}\ [k + \beta(\dots)] \quad .$$

The solution which takes the values Y_o and Z_o for $X = X_o$, is expanded in the form

$$Y = Y_o - \frac{1}{k}\beta^{p-1}Y_o^p \log \frac{Y_o^p - kZ_o(X-X_o)}{Y_o^p} + \ldots ,$$

$$Z = \frac{Z_o Y_o^p}{Y_o^p - kZ_o(X-X_o)} + (\ldots) \quad ,$$

when $p > 1$. This result is valid in an annulus whose centre is the point $X = X_o + Y_o^p/kZ_o$, whose inner radius ε is arbitrarily small and whose outer radius r is finite, provided that $|\beta|$ is sufficiently small (provided that X only turns a finite number of times in the annulus). This solution has critical points or groups about the centre, in the circle centred at this point and of radius ε. It is the same as the solution corresponding to the system in (47) since $X = x$. Hence, following Lemma II, the solutions to the system in (44) will have mobile critical points or groups when $p > 1$. For, even if $B(x_1,y)$ or $C(x_1,y)$ admit γ as a pole, we can choose $x_o = X_o$ and y_o in order for the conditions to be satisfied, for which x remains close to x_o, and when x_o varies, the critical points are displaced.

Remark. Direct integration of equation (46) shows that the solution admits transcendent critical points.

THE CASE OF $p = 1$

For $p = 1$, by integrating (46), we obtain

$$y' = \mu y^k (1 + \ldots) , \qquad x - x_o = \int_{y_o}^{y} \frac{1}{\mu} y^{-k} (1 + \ldots)\, dy \quad ,$$

where μ is a constant. If $k - 1$ is not of the form $1/q$, where q is an integer, then the function $y = Y(x)$ thus defined, admits a critical algebraic point and, by applying Lemma II, we can conclude the existence of mobile critical points or groups for equation (43).

For equation (43) not to have any mobile critical points or groups, it is necessary therefore that $p = 1$ and $k = 1$ or $k = 1 + 1/q$ (where q is an integer), for all poles of $a(y)$. By changing y to $1/y$, we also see that $a(y)$ must be zero at infinity and must have its residue rational at this point. As a consequence, we have

$$a(y) = \sum \frac{D_j}{y - d_j} \ ; \quad \log y' = \int a(y)dy + \log \mu \quad ,$$

where the D_j are rational. On denoting the lowest common denominator of the D_j reduced to their simplest form, by m, we see that we shall obtain

$$\left(\frac{dy}{dx}\right)^m = S(y) \quad , \qquad (48)$$

where $S(y)$ is a rational fraction. If the integral of this equation is not uniform, then it will admit critical points to which we can apply the above argument (we admit here, the general theorem of Painlevé indicated in no. 141). Consequently, *the general integral of equation (48) will become uniform and the equation will become one of the types envisaged in no. 140. Conversely, if this condition is realised, i.e., if we take*

$$a(y) = \frac{1}{m}\,\frac{S'(y)}{S(y)} \quad ,$$

then the general integral of the system in (45) is uniform. The possible values of $a(y)$ are then deduced from the results of no. 140. *The most simple of these values is* $a(y) \equiv 0$ *which results from* $S(y) = \text{const.}$

PROPERTIES OF $B(x,y)$ AND $C(x,y)$

The poles of $B(x,y)$ *and* $C(x,y)$, *where* x *is regarded as a parameter, coincides with those of* $A(x,y)$ *and are all simple.*

We have seen that the poles of $A(x,y)$ are simple. Let $y = \phi(x)$ be a pole of one of the functions A,B,C of order j for A ($j = 0$ or 1), k for B and s for C, and let us assume that s or k are greater than j; $\phi(x)$ is an analytic function. By replacing y by $\phi(x) + y$, we can assume that $\phi(x) \equiv 0$ and write equation (43), for y close to 0, in the form

$$y'' \quad \frac{y'}{y}\left[\left(1 + \frac{1}{q}\right) + y(\ldots)\right] + \frac{y'}{y^k}\left[H(x) + y(\ldots)\right]$$

$$+ \frac{1}{y^s}\left[K(x) + y(\ldots)\right]$$

where q is an integer which can equal -1 if $j = 0$. We set

$$y = \alpha Y \;, \quad x = x_1 + \alpha^k X \;, \quad \text{if} \quad s \le 2k - 1 \quad ;$$

$$y = \alpha Y \;, \quad x = x_1 + \alpha^{\frac{1+s}{2}} X \;, \quad \text{if} \quad s \ge 2k - 1 \quad .$$

The equation has the transformation

$$y'' = \frac{y'^2}{y}\left(1 + \frac{1}{q}\right) + H(x_1)\,\frac{y'}{y^k} + \alpha(\ldots), \quad \text{if} \quad s < 2k - 1 \quad ;$$

$$y'' = \frac{y'^2}{y}\left(1 + \frac{1}{q}\right) + \frac{K(x_1)}{y^s} + \sqrt{\alpha}\,(\ldots) \;, \quad \text{if} \quad s > 2k - 1 \quad ;$$

$$y'' = \frac{y'^2}{y}\left(1 + \frac{1}{q}\right) + H(x_1)\,\frac{y'}{y^k} + \frac{K(x_1)}{y^s} + \alpha(\ldots) \;, \quad \text{if} \quad s = 2k - 1 \quad .$$

For $\alpha = 0$, the three equations obtained, have the same form

$$y'' = \frac{y'^2}{y}\left(1 + \frac{1}{q}\right) + H_1 \frac{y'}{y^k} + K_1 \frac{1}{y^{2k-1}} \quad ,$$

$$H_1 = H(x_1) \ , \quad K_1 = K(x_1) \quad ,$$

(if $H_1 \neq 0$, k is taken to be a positive integer; when $H_1 = 0$, $2k - 1$ is also taken to be a positive integer). If $k > 1$, equation (50) admits the particular integral defined by $C dx = y^{k-1} dy$, where C is a constant defined by the equation

$$C^2(1 - k) = \left(1 + \frac{1}{q}\right) C^2 + H_1 C + K_1 \quad .$$

This particular integral admits a mobile critical algebraic point, which permits an application of Lemma II and a proof that equation (49) would have critical points or groups. If $k = 1$ and $q = -1$, equation (50) could be replaced by the system

$$\frac{dy}{dx} = \frac{1}{u} \ , \quad \frac{du}{dx} = -\frac{u(H_1 + K_1\ )}{y} \quad .$$

We introduce a parameter β by setting $x = \beta X$, $y = \beta Y$, $u = \beta U$, which yields

$$\frac{dY}{dX} = \frac{1}{U} \ , \quad \frac{dU}{dX} = -\beta \frac{U(H_1 + K_1 \beta U)}{Y} \quad .$$

We can again expand the solution starting from that corresponding to $\beta = 0$. We obtain the integral corresponding to the initial conditions X_o, Y_o, U_o:

$$Y = Y_o + \frac{X - U_o}{U_o} + \beta(\dots)$$

and

$$U = U_o - \beta H_1 U_o^2 \log\left(1 + \frac{X - X_o}{Y_o U_o}\right) + \dots \quad \text{if} \quad H_1 \neq 0 \quad ;$$

$$U = U_o - \beta^2 K_1 U_o^3 \log\left(1 + \frac{X - X_o}{Y_o U_o}\right) + \dots \quad \text{if} \quad H_1 = 0 \quad .$$

This is valid for X outside a circle of radius ε and of centre $X_o - Y_o U_o$ and a concentric circle of radius r, where X only turns a finite number of times in this annulus, and for $|\beta|$ sufficiently small. For equation (50) we deduce the existence of a mobile critical point or group, which again permits applying Lemma II to equation (49) and to infer the existence of a critical point or group for this equation. The condition stated for the poles of $B(x,y)$ and $C(x,y)$ is thus proved.

CONSEQUENCES

By applying the above proposition to the equation transformed by the substitution $y = 1/Y$ and by utilising the fact that the form of $A(x,y)$, hence $A(x, \frac{1}{Y})$, is known; we limit the degrees of the numerators of $B(x,y)$ and $C(x,y)$. (The denominators are the denominators of $A(x,y)$).

THE CASE OF $A(x,y) \equiv 0$

The transformation in $1/Y$ is

$$Y'' = 2\frac{Y'^2}{Y} + Y''B(x, \frac{1}{Y}) - Y^2C(x, \frac{1}{Y})$$

where $B(x,y)$ and $C(x,y)$ are here, polynomials in y, since $A(x,y) \equiv 0$ does not have any poles. It follows that $B(x,y)$ and $C(x,y)$ have at most degrees one and three respectively in y.

144 - Reduction of the equation corresponding to $A(x,y) \equiv 0$

The most simple of equations (39) with fixed critical points or groups is of type (43) with $A(x,y) \equiv 0$, $B(x,y)$ of the first degree in y, and $C(x,y)$ of the third degree in y. It is of the form

$$y'' = y' \,[[a(x)y + b(x)]] + A(x)y^3 + B(x)y^2 + C(x)y + D(x) \quad , \quad (51)$$

where the functions $a(x),\ldots,D(x)$ are analytic functions of x.

Preliminary Remarks. In everything so far, we did not assume that the coefficients of $R(x,y,y')$ are uniform functions of x. In the proofs, *we have considered a domain where those coefficients were uniform* and we have proved that, when the conditions obtained are not realised, then there existed solutions admitting mobile critical points or groups in this domain. If the coefficients are multiform, we can imagine that the point x is displaced on a Riemann surface common to these various coefficients, a surface on which these coefficients are uniform. It is on this surface that the critical points or groups of the solutions are fixed or mobile. *It is also on such a surface that we may consider the uniformity or the nonuniformity of the solutions.*

THE FIRST REDUCTION OF EQUATION (51)

If $a(x)$ is not identically zero, then the transformation $y = -2Y/a(x)$ reduces to the case where $a(x)$ has constant value -2 and changes $A(x)$ to $4A(x)/a(x)^2$. If $A(x)$ is not identically zero, then the transformation $y = Y/\sqrt{-A(x)}$ reduces to the case where $A(x)$ has constant value -1. Hence: *if* $A(x) \equiv 0$, *we can restrict to assuming* $a(x) \equiv 0$ *or* $a(x) \equiv -2$, *if* $A(x)$ *is not identically zero, we can assume* $A(x) \equiv -1$.

Let us consider this last case, $A(x) \equiv -1$. Let us set $y = Y/\alpha$ and $x = x_o + \alpha X$. Equation (51) becomes

$$Y'' = a(x_o)YY' - Y^3 + \alpha(\dots)$$

which can be replaced by a system in order that Lemma II applies. If we take

$$\frac{dY}{dX} = \frac{Y^2}{u}, \quad \frac{du}{dX} = (2 - a(x_o)u + u^2)Y + \alpha(\dots) \quad ,$$

then the coefficients of the terms in α contain Y in the denominator. For $\alpha = 0$, we obtain the system

$$\frac{dY}{dX} = \frac{Y^2}{u}, \quad \frac{du}{dX} = (u-\gamma)(u-\delta)Y \ ; \ a(x_o) = \gamma + \delta, \ \ \gamma\delta = 2 \quad , \tag{52}$$

and it again suffices that this system admits a mobile critical group inside a circle of sufficiently small radius and that Y is zero about this circle, in order to affirm that equation (51) does not have its critical points or groups fixed. In studying the system (52), it is convenient to introduce a parameter and apply Lemma II again. Let us put $u = \gamma + \beta U$, with X and Y interchanged. We obtain

$$\frac{dY}{dX} = \frac{Y^2}{\gamma + \beta U}, \qquad \frac{dU}{dX} = U(\gamma - \delta + \beta U)Y \quad . \tag{53}$$

Let us firstly assume $\delta - \gamma \neq 0$. For $\beta = 0$, we have the integral corresponding to the initial conditions X_o, Y_o, U_o:

$$Y = \frac{\gamma Y_o}{\gamma - Y_o(X - X_o)}, \quad U = U_o\left[\frac{-Y_o(X - X_o)^k}{\gamma}\right]^k \quad ;$$

$$k = 2 - \gamma^2 = \gamma(\delta - \gamma) \neq 0 \quad .$$

Moreover, in order that (52) has no mobile critical points or groups, it is necessary that $2 - \gamma^2$ is a positive or negative integer. Similarly, $2 - \delta^2$ will be a positive or negative integer. It follows that

$$4 = \gamma^2\delta^2 = (2-p)(2-q) \ , \quad \gamma^2 = 2-p \ , \quad \delta^2 = 2-q \ ,$$

where p and q are positive or negative integers. The possible values of p and q are 1 and -2, 3 and 6, 4 and 4 (to within a permutation of p and q) and those of γ and δ (we have $\gamma\delta = 2$) are deduced, and then those of $a(x_o)$ which are ± 3, $\pm i$, 0. Changing y to $-y$ in (51) reduces the case $a(x) = -3$ to $a(x) = 3$, and that of $a(x) = -i$ to $a(x) = i$.

Let us now suppose that $\gamma = \delta$. For $\beta = 0$, the system (53) expanded in powers of β, gives an integral, corresponding to $Y = Y_o$ and $U = U_o$ for $X = X_o$:

$$Y = \frac{\gamma Y_o}{\gamma - Y_o(X - X_o)} + \beta(\ldots), \quad U = U_o - \beta U_o^2 \gamma \log \frac{\gamma - Y_o(X - X_o)}{\gamma} + \ldots$$

In this case, the system (53) always has mobile critical points or groups. Consequently, *the equations in (51) whose solutions have their critical points or groups fixed, reduce via an algebraic transformation,* $y = -2Y/a(x)$ *or* $y = Y/\sqrt{-A(x)}$ *to the case where we have:*

$$a(x) \equiv 0 , \quad A(x) \equiv 0 ; \quad a(x) \equiv -2 , \quad A(x) \equiv 0 ;$$

$$a(x) \equiv 0 , \quad A(x) \equiv -1 ; \quad a(x) \equiv 3 , \quad A(x) \equiv -1 \quad ;$$

$$a(x) \equiv i , \quad A(x) \equiv -1 \quad .$$

THE CASE OF THE EQUATION CORRESPONDING TO $a(x) \equiv 0$, $A(x) \equiv 0$.

The equation of the form

$$y'' = b(x)y' + B(x)y^2 + C(x)y + D(x) \tag{54}$$

reduces to the canonical forms via a transformation

$$y = \lambda(x)Y + \mu(X) , \qquad X = v(x) \quad .$$

The equation becomes

$$\frac{d^2Y}{dX^2} = L(X)\frac{dY}{dX} + M(X)Y^2 + N(X)Y + P(X)$$

with

$$L(X) = \frac{1}{v'(x)}\left[-\frac{v''(x)}{v'(x)} - 2\frac{\lambda'(x)}{\lambda(x)} + b(x)\right], \quad M(X) = \frac{B(x)\lambda(x)}{v'^2(x)}$$

$$N(X) = \frac{1}{v'^2(x)}\left[C(x) + 2B(x)\mu(x) + \frac{1}{\lambda}[b(v)\lambda'(x) - \lambda''(x)]\right],$$

$$X = v(x) \quad .$$

If $B(x) \neq 0$, then we can choose λ, μ and v in such a way so as to force $L(X)$ and $N(X)$ identically zero, and to have for $M(X)$ a constant, that we can take to be equal to 6. The functions γ, λ, v will be given by

$$-\frac{v''}{v'} - 2\frac{\lambda'}{\lambda} + b(x) = 0, \quad \frac{B\lambda}{v'^2} = 6$$

$$\mu = \frac{1}{2B}\left[\frac{\lambda''}{\lambda} - b\frac{\lambda'}{\lambda} - C\right] \quad ;$$

μ is given by the last identity once λ is known. By integrating the first equation and then taking account of the second, we successively obtain

$$v'\lambda^2 = e^{\int b dx} , \qquad B^5 = 6e^{2\int b dx} ,$$

whence we deduce λ, and then v by a quadrature.

If $B(x) \equiv 0$, then equation (54) is a linear equation, and it is known that the critical points or groups are fixed. Consequently: *Every equation (54) which is not linear, reduces to the form*

$$y'' = 6y^2 + P(x) \tag{55}$$

by means of an analytic transformation of the indicated type, a transformation which preserves the fixity of the critical points or groups, or preserves the uniformity of the solutions on a Riemann surface suitably transformed.

Theorem. *Equation (55) can only have its critical points or groups fixed when* $P(x)$ *is a polynomial of the first degree in* x.

The transformation

$$y = \frac{Y}{\alpha^2}, \qquad x = x_1 + \alpha X ,$$

replaces the equation by

$$Y'' = 6Y^2 + \alpha^4 P(x_1) + \alpha^5 X P'(x_1) + \alpha^6 \frac{X^2}{2} P''(x_1) + \ldots$$

whose solution can be expanded with respect to α. If we write

$$Y = Y_1 + \alpha^4 Y_4 + \alpha^5 Y_5 + \alpha^6 Y_6 + \ldots ,$$

we firstly obtain, for $\alpha = 0$, $Y_1'' = 6Y_1^2$ or

$$Y_1'^2 = 4Y_1^3 - g_3$$

where g is a constant, and consequently

$$Y_1 = p(X + c) ,$$

where c is an arbitrary constant, and p the elliptic function corresponding to the invariants $g_2 = 0$ and g_3. We will assume that, for the value c, pc is finite and we will take pc and $p'c$ as the initial values of Y_1 and Y_1' respectively, corresponding to $X = 0$. The functions Y_4, Y_5, Y_6 are given by the equations of the Poincaré variations, which here are

$$Y_4'' - 12Y_4Y_1 = P(x_1) \ , \quad Y_5'' - 12Y_5Y_1 = XP'(x_1) \quad ,$$

$$Y_6 - 12Y_6Y_1 = \frac{X^2}{2} P''(x_1) \quad ,$$

as can be easily verified. We see that Y_4, Y_5, Y_6 satisfy a linear equation whose second member alone varies. The equation without second member

$$Z'' - 12Zp(X + c) = 0$$

admits $p'(X + c)$ and $Xp'(X + c) + 2p(X + c)$ as particular solutions, which again can be easily verified. The solutions of the equations with second members are obtained by the method of variation of the constants. If we set

$$Z = G_1p'(X + c) + C_2 \ [Xp'(X + c) + 2p(X + c)]$$

and if τX^q denotes the second member (equal to $P(x_1)$, $XP'(x_1)$, $\frac{1}{2} X^2P''(x_1)$), we see that

$$3g_3C_1' = \tau X^q \ [X_p'(X + c) + 2p(X + c)] \quad ,$$

$$3g_3C_2' = -\tau \ X^q p'(X + c) \quad .$$

The values of C_1 and C_2 are obtained by integrating from 0 to X. For $q = 0$ and $q = 1$, hence for Y_4 and Y_5, we see that C_1 and C_2 are meromorphic: the residues of C_1' and C_2' are, in effect, zero; moreover, integration by parts is easily carried out and only involves p and ζ. For $q = 2$, and consequently for Y_6, the residues of the second of the second members of the equations giving C_1' and C_2' corresponding to the pole $-c$ are $2c\tau$ and 2τ respectively. It follows that, about $-c$, Y_6 is the sum of meromorphic function and

$$\frac{P''(x_1)}{3g_3} \log \ (X + c) \ [(X + c)p'(X + c) + 2p(X + c)] \quad .$$

The reasoning at once shows that Y can only have critical points or groups when $P''(x_1) = 0$. This occurs for the x_1 belonging to a domain, however small, and implies that $P''(x)$ is a polynomial of degree at most one. The theorem is thus proved.

CONSEQUENCES

The nonlinear equations in (54) whose solutions have their critical points or groups fixed, reduce to the form

$$y'' = 6y^2 + mx + n \quad ,$$

where m and n are constants. If $m = 0$, we recover the equation of the elliptic functions 1, since by integrating once, we obtain

$$y'^2 = 4y^3 + 2ny + \text{const.}$$

If $m \neq 0$, we can write the equation in the form

$$y'' = 6y^2 + x \tag{56}$$

by replacing x by $\lambda x + \mu$ and y by vy where λ, μ and v are suitably chosen constants.

145 - Painlevé's equation $y'' = 6y^2 + x$

Painlevé proved that the reduced equation (56) has an integral that is generally meromorphic. *We can easily see that the solutions cannot admit critical algebraic points.* Firstly, if a is such a point, y cannot be finite at this point, for we would then obtain a finite value for y'', hence also for y', and following Cauchy's existence theorem, the solution takes a finite value for $x = a$ (finite) and whose derivative takes a finite value, is unique and holomorphic since the second member of (56) is analytic about all finite values x,y. It is therefore necessary that we have, about a, an expansion of the form

$$y = A_q(x-a)^q + \dots + A_s(x-a)^s + \dots$$

valid for $|x-a|$ sufficiently small, commencing from a term with a negative exponent q and containing terms with fractional exponents, for which s denotes the smallest of these exponents. We would have

$$y'' = A_q q(q-1)(x-a)^{q-2} + \dots + A_s s(s-1)(x-a)^{s-2} + \dots \quad ,$$

and on bringing in equation (56) and equating terms of the same degree in both members, we initially obtain $q - 2 = 2q$, hence $q = -2$, and $Aq = 1$. On the other hand, the first term with fractional exponent in the first member, would be of degree $s - 2$, in the second member, of degree $q + s = s - 2$ and we would have

$$A_s s(s-1) = 12A_s \quad , \qquad A_s \neq 0$$

which is impossible since the equation $s(s-1) - 12 = 0$ has integral roots. We also see that, *if a solution admits the point* a *as a pole, then it is a double pole, and we have about this point*

$$y = \frac{1}{(x-a)^2} - \frac{a}{10}(x-a)^2 + \dots$$

THE PROOF OF PAINLEVÉ'S THEOREM

We want to show that the solution $y = \phi(x)$ of equation (56), defined by the initial conditions $\phi(x_o) = y_o$, $\phi'(x_o) = y'_o$, where x_o, y_o, y'_o are arbitrary finite numbers, is a meromorphic function at every point at a finite distance. From Cauchy's theorem, we know that $\phi(x)$ is holomorphic in a circle of centre x_o. If we continue $\phi(x)$ radially, and if on the radius in question a pole a is encountered, then we can extend the continuation beyond this pole. We must prove that every point ξ of the plane could be attained, hence that, *if we have shown that* $\phi(x)$ *is meromorphic at all points of the segment* (x_o,ξ) *other than* ξ, *then this function is indeed meromorphic at the point* ξ. To establish this, it is necessary to prove that we can find a point x' of the segment (x_o,ξ) such that $\phi(x)$ is certainly meromorphic in a circle of centre x' and containing ξ in its interior.

We shall denote the segment (x_o,ξ) less its extremity ξ, by L.

First Hypothesis. *Let us assume that* $|\phi(x)|$ *is bounded on* L, $|\phi(x)| < M$ *say.*

As we have

$$y'' = 6y^2 + x, \quad y'y'' = 6y^2y' + xy' ,$$

$$y'^2 = 4y^3 + 2\int_{x_o}^{x} xy'dx + \text{const}$$

hence

$$y'^2 = 4y^3 + 2xy - 2\int_{x_o}^{x} ydx + y_o'^2 - 4y_o^3 - 2x_oy_o , \tag{57}$$

we see that $|\phi'(x)|$ is also bounded on L by a number M' only depending M, x_o, ξ, y_o, y'_o. On L we will have

$$|\phi(x)| < M , \quad |\phi'(x)| < M' .$$

Let us take a point x' on L, such that $|x' - \xi| < 1$. We can apply the existence theorem (nos. 47, 48) to $\phi(x)$ at the point x', where the initial conditions are $y = \phi(x')$, $y' = \phi'(x')$. Equation (56) is equivalent to the system

$$\frac{dy}{dx} = [z - \phi'(x')] + \phi'(x') , \quad \frac{dz}{dx} = 6[(y - \phi(x')) + \phi(x')^2] + (x - x') + x' ;$$

the second members are expandable as power series in $x - x'$, $y - \phi(x')$,

$z - \phi'(x')$, majored by

$$\frac{6(M+1)^2 + M' + 2 + |\xi|}{(1-X)(1-Y)(1-Z)} \quad .$$

$\phi(x)$ is therefore holomorphic in a circle of centre x' whose radius η only depends on M, M' and $|\xi|$. This circle contains the point ξ in its interior as long as $|x' - \xi| < \eta$ $\phi\ x$ *is again holomorphic at the point* ξ.

PRELIMINARY REMARK

Let us suppose that $y = \phi(x)$ *is non-zero on* L. If $\phi(x)$ admits a pole a, then we have about this pole, whose principal part is $(x-a)^{-2}$,

$$\int y dx = \frac{-1}{x-a} + \ldots, \quad \frac{y'}{y} = \frac{-2}{x-a} + \ldots \quad .$$

The function

$$y'^2 - 4y^3 - 2xy + \frac{y'}{y}$$

is holomorphic on L by virtue of equation (57). We shall set

$$u = y'^2 - 4y^3 - 2xy + \frac{y'}{y} + x \quad . \tag{58}$$

When we solve with respect to y', we obtain

$$y' = \frac{1}{2y} \pm \left[4y^3 + 2xy + u - x + \frac{1}{4y^2}\right]^{1/2} \quad .$$

On the other hand, by taking account of (56),

$$\frac{du}{dx} = 4y + \frac{x}{y} - \frac{y'^2}{y^2} + 1 \quad .$$

When we put

$$z = \frac{1}{\sqrt{y}} \ , \qquad y = \frac{1}{z^2} \quad ,$$

having made a suitable choice of a branch of z, we can write the above identities as

$$\begin{cases} \dfrac{dz}{dx} = \dfrac{z^5}{4} + \left[1 + \dfrac{xz^4}{2} + \dfrac{u-x}{4} z^6 + \dfrac{z^{10}}{16}\right]^{1/2} \\ \dfrac{du}{dx} = \dfrac{4}{z^2}\left[1 - \left(\dfrac{dz}{dx}\right)^2\right] + xz^2 + 1 \quad . \end{cases} \tag{59}$$

It is a solvable differential system when in the second line, we replace

dz/dx by its value. If we assume the initial conditions $x = x'$, $(|x' - \xi| < 1)$, $|u'| < M''$, $|z| < \varepsilon(M'')$, then the second members are analytic in $x - x'$, $u - u'$, $z - z'$ provided that $\varepsilon(M'')$ is sufficiently small, and are majored by an expression depending only on M'' and $\varepsilon(M'')$. The solution defined by these conditions is therefore holomorphic in a circle of centre x' and radius η. In this circle y will be meromorphic, and this circle will contain the point ξ if $|x' - \xi| < \eta$.

Second Hypothesis. *We assume that* $\phi(x)$ *is not bounded on* L, *but that* $|\phi(x)| \geq \rho > 0$. The function u is holomorphic on L.

Let us assume first of all that $|u|$ is bounded on L; we therefore have $|u| < M''$. If $|\phi(x)|$ does not tend towards infinity when x tends towards ξ, then there exists values x' arbitrarily close to ξ for which $|\phi(x')| < M$. For these x', equation (58) shows that $|\phi'(x')|$ is bounded, $|\phi'(x')| < M'$. The argument contained in the first hypothesis remains valid: $\phi(x)$ *is holomorphic at the point* ξ. If $|\phi(x)|$ tends towards infinity, $z = 1/\sqrt{|\phi(x)|}$ is of modulus less than $\varepsilon(M'')$ as long as x is sufficiently close to ξ; the argument made in the case of the system in (59) applies: $\phi(x)$ *is meromorphic at the point* ξ.

Let us now assume that $|u|$ is not bounded on L. This implies that the function

$$v(x) = \frac{du}{udx} = \frac{4y^3 - y'^2 + xy + y^2}{y(y'^2 y - 4y^4 - 2xy^2 + y' + xy)}$$

does not remain bounded; for, if $v(x)$ was bounded, then $\log u = \int v dx$ would be bounded. Therefore, there exist x' arbitrarily close to ξ such that $|v(x')|$ exceeds every given number. Take such a sequence of x', for which $|v(x')| \to \infty$. If for a sequence of these x', $|\phi(x')|$ and $|\phi'(x')|$ are bounded, $|\phi(x')| < M$ and $|\phi'(x')| < M'$ respectively say, then the argument of the first hypothesis proves that $\phi(x)$ *is again holomorphic at the point* ξ. The hypothesis $|\phi(x')| < M$, $\lim |\phi'(x')| = \infty$ is inadmissible, since, for these x', the expression for $v(x)$ shows that $|v(x')||\phi(x')|^2$ would remain bounded, and since $|\phi(x')| > \rho$, $v(x')$ would remain bounded. It therefore remains to look at the case where $\lim |\phi(x')| = \infty$. The hypothesis $|\phi'(x')| < M'$ would lead again to a contradiction; we also have $\lim |\phi'(x')| = \infty$. But the expression for $v(x)$ shows that $\phi'(x')^2/\phi(x')^3$ cannot exceed a fixed number, for then $|v(x')\phi(x')^2|$ would be bounded. As we can write

$$v(x) = \frac{y' + xy - xy^2 - uy + y^3}{uy^3},$$

we see that $u(x')$ tends towards zero. We then have $|u(x')| < M''$ and

$|z'| = 1/\sqrt{|\phi(x')|} < \varepsilon(M'')$ as long as x' is sufficiently close to ξ. Following the property of the system (59), $\phi(x)$ *is meromorphic at the point* ξ.

Third Hypothesis. $|\phi(x)|$ *and* $1/|\phi(x)|$ *are not bounded on* L. Consequently, if ρ is given to be positive and arbitrarily small, then there exist intervals in which $|\phi(x)| < \rho$, and others in which $|\phi(x)| > \rho$. We can assume that there is an infinity of such intervals. Since if $|\phi(x)|$ or $1/|\phi(x)|$ was bounded on the segment obtained by suprressing a segment (x_o,x_1) of L, where $\phi(x_1)$ is finite, then we could replace L by the segment (x_1,ξ) less ξ, and everything that came before, would still hold true. *We assume, then, that there exists an infinite sequence of segments* (X,X') *of* L, *where* X *and* X' *tend towards* ξ, *such that* $|\phi(x)| \le \rho$ *when* x *belongs to such a segment, whilst outside of these segments,* $|\phi(x)| > \rho$. We are going to show that, in this case, $\phi(x)$ is holomorphic or meromorphic at the point ξ, which will complete the proof of Painlevé's theorem. But we remark that for $\phi(x)$ holomorphic, or meromorphic at the point ξ, the analytic curve $|\phi(x)| = \rho$ can only intersect L at a finite number of points close to ξ. The hypothesis made is never realised.

Let us assume first of all that there exists an infinite sequence of points X for which $|\phi'(X)| < M'$. The reasoning of the first hypothesis shows that $\phi(x)$ is holomorphic at the point ξ. Let us now assume that, when X tends towards ξ, $|\phi'(X)|$ tends towards infinity. We intend to show that there exists a curve Γ with extremities X, X' on which $|\phi(x)| = \rho$, as long as X is sufficiently near to ξ, and that in the arguments of the second hypothesis, we can replace each segment (X,X') by the corresponding curve Γ. In equation (56), let us take x as the unknown function and y as the variable. The equation becomes

$$\frac{d^2x}{dy^2} = -\left(\frac{dx}{dy}\right)^3 [6y^2 + x] \quad ;$$

this is equivalent to the system

$$\frac{dx}{dy} = z, \qquad \frac{dz}{dy} = -z^3\,(6y^2 + x) \quad . \tag{60}$$

A$y = \phi(x)$ corresponds, in the neighbourhood of the point X, to the solution of the system in (60) satisfying the initial conditions $y = \phi(X)$, $z = 1/\ '(X)$, $x = X$. It defines a function $x = \psi(y)$ which is holomorphic about this point in a circle whose radius η only depends on $|\xi|$ when we assume that $|\phi(X)| < 1$, $|\phi'(X)| > 1$ and $|X - \xi| < 1$. We will assume, as indeed we can, that 2ρ had been taken to be less than η; the circle of radius η and centre $\phi(X)$ will contain the circle $|y| \le \rho$ within its interior. The solution is an analytic function of the initial conditions, as can be seen. In particular, we will have

$$z = z_o - z_o^3 ([6\phi(X)^2 + X] (y - \phi(X)) + \ldots \quad ,$$

$$z_o = \frac{1}{\phi'(X)} \quad ;$$

hence for $|z_o|$ sufficiently small, we will have

$$\left|\frac{dx}{dy}\right| = |z_o| (1 + \varepsilon), \quad |\varepsilon| < \frac{1}{2} \quad .$$

In the segment (X,X') there corresponds a curve c, $y = \phi(x)$, contained within the circle $|y| \le \rho$, and whose length is taken to be s. By integrating along this curve c, we obtain:

$$d = |X' - X| = \int |dx| = \int_c \left|\frac{dx}{dy}\right| |dy|$$

$$= |z_o| \int_c (1 + \varepsilon)|dy| = |z_o| s(1 + \varepsilon'), \quad |\varepsilon'| < \frac{1}{2} \quad ,$$

where d denotes the length of the segment (the first integral having been taken from $|X - x_o|$ to $|X' - x_o|$). The curve c is a simple curve since $x = \psi(y)$ is uniform, in the circle $|y| \le \rho$. It joins two points of the circumference $|y| = \rho$ which corresponds to X and X', and divides the interior of this circumference into two or more domains which the function $x = \psi(y)$ represents on the x-plane. The extremities of c determine on the circumference two non-zero arcs (since $\psi(y)$ is uniform); let γ be the smallest of these arcs and let Γ be the curve corresponding to the plane of the x defined by $x = \psi(y)$. The function $x = \psi(y)$ represents the domain of the plane of y bounded by c and γ on a domain Δ of the plane of the x bounded by Γ and the segment (X,X'). This domain Δ can be partially recovered, but when we continue $\phi(x)$ along Γ whilst X goes up to X', we recover the same value at the point X' as that obtained along the segment (X,X'). We can then replace the segment (X,X'') by the arc Γ on which $|\phi(x)| = \rho$. The length of Γ is given by

$$\sigma = \int_\Gamma |dx| = \int_\gamma \left|\frac{dx}{dy}\right| |dy| = |z_o| (1 + \varepsilon'')\delta \quad , \quad |\varepsilon''| < \frac{1}{2} \quad ,$$

where δ is the length of γ. This length is at most equal to the product of the chord subtended and $\pi/2$; therefore, we have

$$\sigma < \frac{3}{2} |z_o|\delta \le \frac{3\pi}{4} |z_o|s < \frac{3\pi}{2} d \quad .$$

The total length of the path obtained by replacing the segments (X,X') by the curves Γ, hence remains finite. All of the arguments outlined in the second hypothesis can be applied to this path on which $|\phi(x)| \geq \rho$, since it tends towards ξ.

Thus we have proved:

<u>The Theorem of Painlevé</u>. *The solutions of the equation*

$$y'' = 6y^2 + x$$

are meromorphic functions only admitting double poles, and for which the principal part relative to a pole a *is* $1/(x-a)^2$.

If $y = \phi(x)$ is a solution, then

$$\Phi(x) = - \int \phi(x)dx$$

is a meromorphic function whose poles are all simple and with residues equal to 1, ($\Phi(x)$ is defined to within a constant). Consequently,

$$\rho^{\Psi(x)} \quad , \; \Psi(x) = \int \Phi(x)dx$$

is an entire function. It is an entire function of order 5/2. The properties of the functions $\phi(x)$ were studied by Boutroux.

146 - Other equations of Painlevé. Irreducibility

The method used in the case $A(x,y) = 0$ and in the hypothesis $a(x) \equiv 0$, $A(x) \equiv 0$, applies to the other cases. It provided Painlevé and Gambier with all the reduced equations which can be reduced to the equations (39) whose critical points or groups are fixed. Moreover, equation (56) considered above, we obtain the equations

$$y'' = 2y^3 + xy + \alpha, \qquad y'' = \frac{y'^2}{2y} + \frac{3}{2}y^3 + 4xy^2 + 2(x^2 - \alpha)y + \frac{\beta}{y}$$

(where α and β are constants) whose general integral is meromorphic, and then two equations whose solutions admit transcendental critical points at the origin and at infinity, but whose solution becomes meromorphic at every point at a finite distance when we take $\log x$ as the variable; finally, an equation for which the origin, the point 1 and the point at infinity, are critical transcendental points of the solutions. We do not encounter the critical groups in the case of equations (39) satisfying the indicated conditions.

Painlevé proved that his method could be applied to the study of third order equations with fixed critical points. But the singularities of the solutions could be quite complicated. These third order equations were studied by Chazy and Garnier.

IRREDICIBILITY

Are the functions defined by the Painlevé equations new functions? Putting it another way, are they reducible by certain transformations of known functions? Can the equation itself, which defines these functions, be reduced to a more simple equation? To answer these questions, which must firstly be made more precise, Painlevé studied those whose general integral depends on constants of integration. He arrived at the conclusion that the functions defined by their differential equations and those of Gambier are entirely new transcendental functions.

A general theory of the reducibility of the general integral of a differential equation was established by Drach (Thesis 1898).

IV. SINGULAR INTEGRALS

147 - Singular integrals of algebraic equations of the first order

We already mentioned in no. 48 what was meant by the singular integral of an algebraic differential equation when placed at the point of view of the Cuachy existence theorem. It is an integral that cannot be obtained, for any value of the variable, by commencing from the existence theorem. In the direct study of the solutions of first order differential equations which can be integrated by quadrature (Chapter V), we defined the singular integrals as integrals which cannot be obtained by giving a particular value to the constant of integration, and which are represented by the envelopes of the general curves (nos. 63, 65, 59, 70).

Let us consider a first order algebraic differential equation,

$$P(x,y,y') = 0 \quad , \tag{61}$$

whose first member is an indecomposable polynomial in x,y,y' (the results would extend to the case where P would be a polynomial in y', and would be analytic in x and y). We have seen in no. 48 that if $D(x,y)$ is the discriminant of P considered as a function of y', i.e., the result of eliminating y' from equations (61) and

$$\frac{\partial P}{\partial y'}(x,y,y') = 0 \quad , \tag{62}$$

then the existence theorem provides all the solutions of equation (61), except the solutions satisfying the equation

$$D(x,y) = 0 \quad . \tag{63}$$

In no. 70, we arrived at the conclusion that, if the integral curves have an envelope, then the equation of this envelope is given by this same equation

(63). This fact also arises from the case where, given a point x_o, y_o for which the equation in y', $P(x_o, y_o, y') = 0$, has all its roots simple, then the integral curves passing through this point, are not tangent to each other. We remark that if one of the roots was infinite, we would consider x as the unknown function and y as the variable.

We are going to investigate the differential equation and its solutions when we are placed at a point of the curve (63). We shall confine ourselves to the most simple case where the equation (61) admits only a double root (in y'). We shall assume that this double root is finite; if it was infinite on all of an arc of the curve (63), then we can permute the roles of x and y.

A SIMPLIFIED EQUATION

If x_o, y_o is the point in question of the curve (63) and y' the value at y', then to simplify matters, we can assume that $x_o = y_o = y'_o = 0$. We obtain

$$P(x,y,y') \equiv Q_o(x,y) + y'Q_1(x,y) + y'^2Q_2(x,y) + \dots = 0 \tag{64}$$

where Q_o, Q_1, Q_2 are polynomials, and we have the conditions

$$Q_o(0,0) = 0 \ , \quad Q_1(0,0) = 0 \ , \quad |Q_2(x,y)| > \frac{1}{2}\alpha$$

For a given ε, however small, it follows that we can find a number r sufficiently small such that, for $|x| \le r$, $|y| \le r$, we have

$$|Q_o(x,y)| < \varepsilon, \quad |Q_1(x,y)| < \varepsilon, \quad |Q_2(x,y)| > \frac{1}{2}\alpha \quad .$$

When $|x| < R$, $|y| < R$, where R is fixed and greater than r, and if m is the degree of P with respect to y', then we have

$$|Q_j(x,y)| < H \ , \quad j = 3,\dots,m \quad ,$$

where H is a constant. If we write equation (64) under the form

$$y'^2Q_2(x,y) + [Q_o(x,y) + y'Q_1(x,y) + y'^3Q_3(x,y) + \dots] = 0 \quad ,$$

where we regard x and y as constants and y' as variable, then Rouche's theorem shows us that the equation admits two roots in the circle $|y'| < \rho$ provided that

$$4\varepsilon + 2\rho^3 mH < \rho^2\alpha \ , \quad (\rho < r < 1) \quad .$$

It suffices to choose ρ such that $2\rho mH < \alpha$, and then ε sufficiently small so that this condition is realised. Hence, for $|x| \le r$, $|y| \le r$, equation (64) has two roots and two alone, less in modulus than ρ. Let y_1' and y_2' be these roots. We have

$$(y_1')^k + (y_2')^k = \frac{1}{2\pi i}\int z^k \frac{P'(x,y,z)}{P(x,y,z)}\,dz\ , \quad k = 1,2,$$

where the integral is taken on the circumference $|z| = \rho$ and where P is the derivative with respect to z. We can also write

$$P(x,y,z) = z^2Q_2(0,0) + [Q_o(x,y) + [Q_1(x,y) + z^2\,Q_2(x,y) - Q_2(0,0)] + \ldots]$$

We can take ρ, then r, sufficiently small in order that the sum obtained by replacing each term in the polynomials Q_o, Q_1, $Q_2(x,y) - Q_2(0,0)$, $Q_3,\ldots,Q_m$ within the bracket, by its modulus, and z by ρ, is less than

$$\frac{1}{2}\,|Q_2(0,0)|\rho^2\quad .$$

Under these conditions, we can expand $1/P(x,y,z)$ as a power series in x and y that is absolutely and uniformly convergent for $|x| < r$, $|y| < r$ and $|z| = \rho$. It follows that the functions

$$(y_1')^k + (y_2')^k\ , \qquad k = 1,2,$$

are expandable as power series in x and y, and hence also

$$y_1' + y_2'\ , \qquad y_1'\,y_2'\quad .$$

The solutions y_1' *and* y_2' *of equation (64) that tend towards zero when* x *and* y *tend towards zero, are solutions of an equation of the form*

$$\begin{aligned} &y'^2 - 2F(x,y)y' + G(x,y) = 0\ ,\\ &F(0,0) = G(0,0) = 0\ , \end{aligned} \tag{65}$$

where F *and* G *are analytic functions of* x *and* y *about the origin.*

This proposition enters as a special case of the general theorem of Weierstrass that was proved in no. 51 for a function of two variables.

The study of equation (61) in the neighbourhood of the point in question is therefore reduced to the study of equation (65) about the origin. By solving equation (65), we obtain

$$\begin{aligned} &y' = F(x,y) \pm \sqrt{\Delta(x,y)}\\ &\Delta(x,y) = [F(x,y)]^2 - G(x,y)\quad . \end{aligned} \tag{66}$$

In the neighbourhood of the origin, the condition $\Delta(x,y) = 0$ implies that equation (65), hence also (61), has a double root. The curve $\Delta(x,y) = 0$ is a branch of $D(x,y) = 0$ (the equation (61) in y' could also have several double roots). Let us assume first of all that the curve $\Delta(x,y) = 0$ is a simple curve, and that Δ is not an exact power, which will always be the case when the polynomial $D(x,y)$ is indecomposable. We will then assume, as we may, that the origin is an ordinary double point of the curve. Two cases are possible: 1) the curve $\Delta(x,y) = 0$ is not an integral curve; 2) the curve $\Delta(x,y) = 0$ is an integral curve.

THE FIRST CASE

$\Delta(x,y) = 0$ *is not an integral curve.* In this case, we can assume that at the origin, the derivative y_1' of the function defined by $\Delta(x,y) = 0$ is not equal to 0, the value of y' given by equation (66). In effect, the points for which $y_1' = y'$, are isolated poles, and we shall assume that the origin is not one of those points. Since the origin is an ordinary point of $\Delta(x,y) = 0$, the solution of this equation for x near to zero, is holomorphic. Let $y = y_1(x)$ be this function. From no. 39, we obtain

$$\Delta(x,y) \equiv [y - y_1(x)]\, L(x,y) \qquad L(0,0) \neq 0 \quad ,$$

where $L(x,y)$ is analytic. Moreover, $y_1(0) = 0$ and $y_1'(0) \neq 0$. Let us set

$$y = y_1(x) + z \quad ;$$

equation (66) will take the form

$$z' = -y_1'(x) + F(x,y_1(x) + z) \pm \sqrt{zL(x,y_1(x)) + z)} \quad ,$$

where the functions F and L are expandable as power series in x and z about the origin, where F is zero for $x = z = 0$, as long as $L(0,0) \neq 0$. By writing y_1' in place of $y_1'(0)$, we will then have

$$z' = -y_1' + M(x,z) \pm \sqrt{z}\, N(\ ,\ ) , \quad M(0,0) \neq 0 \quad N(0,0) \neq 0 \quad , \tag{67}$$

where $M(x,z)$ and $N(x,z)$ are analytic at the origin. If we set $z = u^2$, we obtain

$$2uu' = -y_1' + M(x,u^2) + uN(x,u^2), \quad y_1' \neq 0, \quad N(0,0) \neq 0 \tag{68}$$

and an analogous equation is obtained by replacing u by $-u$. We are now in the case of no. 129; equation (68) has a unique solution equal to zero for $x = 0$. This solution has a critical algebraic point at the origin and an expansion of the form

$$u = \sqrt{-y_1'x} + \alpha_1 x + \alpha_2 x\sqrt{x} + \dots , \quad \alpha_1 \neq 0 \quad .$$

We do not change $z = u^2$ by replacing u by $-u$. Equation (66) has, therefore, in this case, a unique solution equal to zero for $x = 0$, given by

$$y = y_1(x) + u^2 = 2\alpha_1 x\sqrt{-y'x} + \beta_2 x^2 + \dots \quad ,$$

(the term in x disappears, as expected). Thus: *In this case, for every ordinary point of the curve* $\Delta(x,y) = 0$, *there passes a unique integral curve which admits, at this point, a turning point of the first kind.*

SECOND CASE

The curve $\Delta(x,y) = 0$ is an integral curve. Here, we have $y_1' = y'$ along $\Delta = 0$. When we proceed as above, we again arrive at equation (67), but when we have $z = 0$, the equation is identically satisfied, since $y_1'(x) = y'(x)$. The expression

$$-y_1' + M(x,z)$$

contains z as a factor. We have

$$z' = zM_1(x,z) \pm \sqrt{z}\, N(x,z), \quad N(0,0) \neq 0 \quad .$$

When we set $z = u^2$, we obtain the equation

$$2u' = uM_1(x,u^2) + N(x,u^2)$$

(and the equation obtained by changing u to $-u$), which admits a unique solution, zero for $x = 0$, that is holomorphic at this point. This solution has an expansion of the form

$$u = \frac{N(0,0)}{2} x + \dots$$

and we have

$$y = y_1(x) + u^2 = \gamma_2 x^2 + \dots \quad .$$

In the second case, for each ordinary point of $\Delta(x,y) = 0$, *there passes another integral curve which is tangent to the curve* $\Delta(x,y) = 0$ *at this point. The curve* $\Delta(x,y) = 0$ *is an envelope of integral curves.*

The hypothesis: $\Delta(x,y)$ is not an exact power, arises, as we have seen, from the hypothesis: $D(x,y)$ is indecomposable. It is clearly the case in general. Similarly, equation (61), in general, will only have a double root when $D(x,y)$ is zero. Also, we have seen in no. 70, that $D(x,y) = 0$, does not in general define a solution of equation (61). We can they say that, *In general, the curve* $D(x,y) = 0$ *is not an integral curve; it is then the locus of turning points of the integral curves. When* $D(x,y) = 0$ *is an integral curve, it is in general the envelope of the integral curves.*

THE CASE WHERE $D(x,y)$ DECOMPOSES

The case in question is when $\Delta(x,y)$ is an exact power. About the origin, we would have,

$$\Delta(x,y) \equiv [y - y_1(x)]^q \, L(x,y) , \qquad L(0,0) \neq 0 \quad ,$$

where q is an integer greater than one. (It is clear that $L(x,y)$ could also be regarded as a power q). By always setting $y = y_1(x) + z$, we obtain, for z, an equation analogous to (67), but one for which $\sqrt{z}$ will be replaced by $z^{q/2}$. If q is odd, we will obtain analogous results to those relative to $q = 1$: the curve $\Delta = 0$ will be a locus of turning points of the first or second kind when $\Delta = 0$ is not an integral curve, and if it is integral, then it will be the envelope of the general integral. If q is even, then the Cauchy theorem will apply directly: if $\Delta = 0$ is not an integral curve, then through its points there will pass two integral curves tangential to each other. If $\Delta = 0$ is an integral curve, it will be a *double* integral curve.

148 - Examples and complements

I. The differential equation of the circles of constant radius R whose centre describes Ox is

$$y^2y'^2 + y^2 - R^2 = 0 \quad .$$

The discriminant decomposes; its factors $y \pm R$ equated to zero, provide the integrals which are the singular integrals. The factor y^2 gives a locus of points through which there passes two tangential integral curves.

The equation of the orthogonal trajectories to these circles is

$$y^2 + y'^2(y^2 - R^2) = 0 \quad .$$

The lines $y = \pm R$ give the loci of the turning points of the integral curves (tractices); the double line $y^2 = 0$, is a double integral.

II. For the Lagrange equation $y = xy'^2 + y'^3$ considered in no. 69, the discriminant is $y(27y - 4x^3)$; $y = 0$ is a solution; it is a singular integral. The curve $27y = 4x^3$ is the locus of turning points of the integral curves. This locus is obtained by putting $C = -p^3 + 3p^2 - 3p$ in the equations of no. 69.

Remark

I. Following what was said in no. 70, there is only a singular integral if the equations in z

$$P(x,y,z) = 0 \quad \frac{\partial P}{\partial z}(x,y,z) = 0 \quad ,$$
$$\frac{\partial P}{\partial x}(x,y,z) + z\frac{\partial P}{\partial y}(x,y,z) = 0 \quad , \tag{69}$$

are compatible. *Let us assume this to be the case.* Then $D(x,y) = 0$, the discriminant of $P(x,y,z)$ (regarded as a polynomial in z) defines one or several curves along which the equations are compatible. Let us consider one of the curves Γ. *If along* Γ, *the derivative* $\frac{\partial P}{\partial y}$ *is not constantly zero, then* Γ *is an integral curve.* In effect, by letting y' denote the derivative of y with respect to x along Γ, we have, by differentiating the first equation (69) and taking account of the second,

$$\frac{\partial P}{\partial x}(x,y,z) + \frac{\partial P}{\partial y}(x,y,z)y' = 0 \quad ,$$

which, with the third equation (69) brought in, yields

$$\frac{\partial P}{\partial y}(x,y,z)\,[z - y'] = 0 \quad .$$

The hypothesis implies $y' = z$, and Γ is an integral curve.

If $\frac{\partial P}{\partial y} = 0$ *constantly along* Γ, *then we also have* $\frac{\partial P}{\partial x} = 0$, but nothing proves that Γ is an integral curve. In this case Γ could be an integral, but it is necessary to prove this directly.

By always assuming that along Γ, $P(x,y,z) = 0$ only has a double root, we intend proving the following proposition due to Darboux:

<u>Proposition</u>. *If* Γ *is an integral curve, and if along* Γ, $\frac{\partial P}{\partial y}$ *is not constantly zero, then* Γ *is a singular integral. On the other hand, if* $\frac{\partial P}{\partial y}$ *is constantly zero along* Γ, *then* Γ *is an ordinary or singular integral.*

In effect, along Γ, we will have

$$P(x,y,y') \equiv [(y' - F(x,y)^2 - \Delta(x,y)]\,H(x,y,y')$$

in the neighbourhood of a point, and for which is non-zero along Γ. Γ is defined by $\Delta = 0$. We have seen that Γ is a singular integral if Δ is an odd power, and a double integral if Δ is an even power. Along Γ, $y' - F(x,y) = 0$. Along Γ, the first factor of the second member is zero; the value of the derivative with respect to y is therefore

$$\frac{\partial P}{\partial y} = H\left[2(F - y')\frac{\partial F}{\partial y} - \frac{\partial \Delta}{\partial y}\right] = -H\frac{\partial \Delta}{\partial y} \quad .$$

If Δ is a power greater than one, then $\frac{\partial P}{\partial y}$ is zero along Γ since $\Delta = 0$ along Γ. Hence, if $\frac{\partial P}{\partial y}$ is not constantly zero, Δ is not a power and Γ is a singular integral.

Remark II.

It is well known that, if the degree of $P(x,y,y')$ with respect to y' is greater than two, the curve Γ, $D(x,y) = 0$, will be the locus of points for which $P = 0$, an equation in y', has in general a double root. It may happen that the double root is distinct from the value of the angular coefficient of the tangent to Γ; then Γ is the locus of turning points. But then one of the simple roots, at any point of Γ, may be equal to the angular coefficient of Γ at this point. Then Γ is also an integral.

Example. Let us consider the curves $y^3 = C(x - C)^2$ which, for $C \neq 0$, are cubics having a turning point at $x = C$, on Ox, where the tangent at this point is parallel to Oy. The differential equation of these curves is

$$27yy'^3 - 12xy' + 8y = 0 \quad ;$$

it is homogeneous. The discriminant, to within a numerical factor, is $y(27y^3 - 4x^3)$. The lines obtained by equating the expression in brackets, to zero, are the singular integrals (one is real). But for $y = 0$, the double root is infinite; we have a locus of turning points. The simple root is zero; $y = 0$ is also an integral corresponding to $C = 0$.

Remark III.

From the real point of view, a singular integral separates the plane into regions, for example, two, if this curve is a simple closed curve. The number of real roots of the equation in y', $P(x,y,y') = 0$ will vary, in general, by two units when Γ is crossed. If Γ, defined by $D(x,y) = 0$, is a locus of turning points, then we obtain an analogous property.

But from the complex point of view, the number of roots of $P = 0$ is always the same. The integral curves which appear as separated in the real geometry as ellipses of hyperbolas obtained in the example of no. 63 are not so in the complex domain.

149 - Singular integrals of differential systems

We can extend a system of first order algebraic equations, as we have done for a single equation. We shall confine ourselves to the case of two equations

$$P(x,y,z,y',z') = 0 \ , \qquad Q(x,y,z,y',z') = 0 \quad , \tag{70}$$

where y and z are unknown functions, y' and z' their derivatives with respect to x, and P and Q are polynomials with respect to the five variables which occur. If, at a point x,y,z,y',z', the functional determinant of P and Q with respect to y',z' is non-zero, then we can solve the system in y',z' and obtain a system analytically solvable about the point in question; the existence theorem applies. It applies in so far that the system of equation in (70) and

$$\frac{D(P,Q)}{D(y',z')} = 0 \tag{71}$$

are compatible. By eliminating y' and z' from equations (70) and (71), we obtain an equation $R(x,y,z) = 0$ which defines a surface on which the Cauchy theorem cannot apply. To study the solutions in the neighbourhood of this surface, we shall proceed in an indirect way. Let us consider the general case where P and Q contain y' and z', and let us eliminate z' from these equations. y' will be defined by

$$D(x,y,z,y') = 0 \quad , \tag{72}$$

where D is the resultant, and z' is then given by $S(x,y,z,y',z') = 0$. We shall further consider the general case where S will be of the first degree in z'. We shall have

$$z' = R(x,y,z,y') \quad , \tag{73}$$

where R is a rational fraction. At every point where equation (72) only has simple roots in y', the Cauchy theorem applies. When we assume that x,y,z,y' corresponds to a *double* root in y', we may proceed as in no. 147 and replace equation (72) in the neighbourhood of this point, which can be taken to be the origin, by an equation of the second degree with analytic coefficients. We can also assume that the double root is zero at the origin. We will obtain

$$y' = F \pm \sqrt{\Delta} \quad , \qquad z' = G \pm H\sqrt{\Delta} \tag{74}$$

where F, G, H and Δ are analytic [if z' provided by (73) was infinite on $\Delta = 0$, then we can permute z and x]. We shall assume that the origin is an ordinary point of the surface Σ defined by $\Delta(x,y,z) = 0$, a surface which is a portion of the surface obtained by equating to zero, the discriminant $D(x,y,z,y')$ regarded as a polynomial in y' alone. By choosing suitable axes, we can always assume that, in the neighbourhood of the origin, the equation $\Delta = 0$ admits $y = \phi(x,z)$ as a solution, where ϕ is analytic. We have evidently assumed that Δ is not a power.

We shall set

$$y = \phi(x,z) + Y^2 \quad . \tag{75}$$

There are two cases to be distinguished on account of whether the direction defined (in the space x,y,z) by the parameters $x'_o = 1$, y'_o, z'_o and where z'_o is given by the equations in (75) for $x = y = z = 0$ (we also assumed $y'_o = 0$), is or is not contained in the tangent plane of Σ at the origin. In the first case, the number

$$A = \phi'_x(0,0) + \phi'_z(0,0)G(0,0,0) \tag{76}$$

is zero; in the second case, it is non-zero.

THE FIRST CASE

We have $A \neq 0$. The transformation (75) yields the system

$$\begin{aligned} &2YY' + \phi'_x + \phi'_z z' = F(x, \phi + Y^2, z) + Y\psi(x, y + Y^2, z) \quad F(0,0,0) = 0 \\ &z' = G(x, \phi + Y^2, z) + Y\Psi(x, \phi + Y^2, z) \quad ; \end{aligned} \tag{77}$$

and on taking account of the second equation, the first is written

$$2YY' = -\phi'_x - \phi'_z G + F + Y(\psi - \Psi\phi'_z) \quad . \tag{78}$$

[We would also have the system obtained by replacing Y by $-Y$ in the equations (77), (78).] For $x = y = Y = 0$, the second member of (78) is equal to $-A \neq 0$. If we take x and z as unknown functions, where Y is the variable, then the system (77), (78), becomes a solvable analytic system of the form

$$\frac{dx}{dY} = YM(x,Y,z), \quad 2\alpha = M(0,0,0) = -\frac{2}{A} \quad ,$$

$$\frac{dz}{dY} = YN(x,Y,z) \ , \quad 2\beta = N(0,0,0) = -\frac{2}{A}G(0,0,0) \quad .$$

About the origin, there is a unique solution which is expanded in the form

$$x = \alpha Y^2 + \dots , \quad z = \beta Y^2 + \dots ,$$

$$y = \phi(x,z) + Y^2 = \gamma Y^3 + \dots$$

since

$$\alpha\phi'_x(0,0) + \beta\phi'_z(0,0) + 1 = 0 \quad .$$

Consequently, *If* $A \neq 0$, *the system admits a unique solution which passes through the origin. This integral curve admits at the origin, a turning point whose tangent, with direction parameters* x'_o, y'_o, z'_o, *is not in the tangent plane to* Σ *at the origin.*

THE SECOND CASE

If $A = 0$, but if

$$\phi'_x(x,z) + \phi'_z(x,z)G(x,\phi(x,z),z) - F(x,\phi(x,z),z) \tag{79}$$

is not identically zero about the origin, then we have a case of indetermination of the second member of (78) assumed to be solvable with respect to Y'. We shall put this case aside; this example only occurs on particular curves of Σ.

Let us suppose therefore that (79) is identically zero for x and z sufficiently small. This means that the direction of the direction parameters $1,y',z'$ defined by (75), is in the tangent plane to Σ when we assume that the point (x,y,z) is on Σ, $(\Delta = 0)$. Under these conditions, the second member of (78) contains Y as a factor. When we assume that Y is not identically zero, then we can divide both members of (78) by Y, and the system (77), (78) is then a solvable analytic system. *There exists a unique integral curve passing through the origin, which at this point, is tangent to* Σ, *but which is not situated on* Σ *if the second member of (78) does not contain* Y^2 *as a factor. The direction parameters of the tangent to this curve at the origin, are* $1,\ 0,\ G(0,0,0)$.

The system (77), (78) admits, on the other hand, the solution defined by $Y = 0$ and

$$z' = G(x,\phi(x,z),z) \quad . \tag{80}$$

As this equation (80) admits a unique zero solution for $x = 0$ and for which $z_0' = G(0,0,0)$, we see that we obtain, *an integral curve situated on* Σ *and tangent to the origin to the integral curve discovered beforehand.*

On taking account of these, results, we see that: *If* Σ *is a simple surface on which the equation (72) in* y', *has a double root, and if* z' *is the corresponding value provided by equation (73), where* y' *is this double root, then on denoting by* $T(M)$ *the direction of the direction parameters* $1,y',z'$ *relative to the point* $M(x,y,z)$ *of* Σ, *we find that there are two possible cases:* 1) $T(M)$ *is not constantly in the tangent plane to* Σ *at* M. *Then, in general, there passes through* M *a unique integral curve tangent to* $T(M)$ *and admitting a turning point at* M. 2) $T(M)$ *is constantly in the tangent plane to* Σ *at* M. *Then, through each point of* Σ, *there passes, in general, two integral curves: one tangent to* Σ *at* M *and the other tangent to the first at* M. Σ *is a locus of integral curves depending on one parameter; these curves are singular integrals. It is necessary, in all cases, that the exceptional case as mentioned, does present itself.*

It is clear that, in certain cases, the singular integrals cannot have an envelope, which will again be a singular integral.

The surface Σ, on which the Cauchy theorem does not apply, necessarily belongs to the set of surfaces defined by eliminating y' and z' from equations (70) and (71).

150 - An application and examples

When we consider an algebraic equation of the second order, $F(x,y,y',y'') = 0$, equivalent to the system

$$y' = z, \qquad F(x,y,z,z') = 0 \quad ,$$

then the above conditions prevail. The singular integrals will be the integral curves of the first order equation in y' deduced from $F(x,y,y'y'') = 0$, by equating to zero the discriminant of this equation in y'', $\Phi(x,y,y') = 0$ say, if along these curves, we have

$$\frac{\partial\Phi}{\partial x} + \frac{\partial\Phi}{\partial y}y' + \frac{\partial\Phi}{\partial y'}\, y'' = 0 \quad .$$

Examples I. When we have the system

$$x + ayy' = 0, \qquad z'^2 - (x^2 + y^2 - 1) = 0 \quad ,$$

where a is a constant, we see that the surface on which the equation in z' has a double root, is the cylinder Σ, $x^2 + y^2 = 1$. If $a \neq 1$, the direction T(M), with parameter directions $1, -x/ay, 0$, is not tangent to the cylinder Σ at M; Σ is a locus of turning points.

If $a = 1$, T(M) is tangent to Σ, but this gives rise to the exceptional case. If we set $x^2 + y^2 - 1 = Y^2$, then we obtain the system $z' = Y$, $YY' = 0$, whose only solution is zero. There are no singular integrals.

Examples II. The system

$$y'^2 = 4y, \qquad z' = z^2(x + y') \quad ,$$

whose integral curves are unicursal curves, admit singular integrals which are the cubics $y = 0$, $(x^2 + C)z + 2 = 0$, where C is an arbitrary constant.

Examples III. A second order algebraic equation of the second degree in y'' can be written as

$$[y'' + f\; x,y,y']^2 + g(x,y,y') = 0 \quad ,$$

where f and g are rational fractions. There will be singular integrals if the condition

$$\frac{\partial g}{\partial x} + \frac{\partial g}{\partial y}\, y' + \frac{\partial g}{\partial y}\, y'' = 0$$

is a consequence of $g(x,y,y') = 0$ and the given equation, hence if

$$\frac{\partial g}{\partial x} + \frac{\partial g}{\partial y}\, y' - \frac{\partial g}{\partial y'}\, f = 0$$

is a consequence of $g = 0$. In particular, we could take the first member of this equation, to be equal to the product of g and an arbitrary rational fraction of x,y,y', $-h(x,y,y')$ say, and we see that the equation

$$\left(\frac{\partial g}{\partial x} + gh\right)^2 + g\left(\frac{\partial g}{\partial y'}\right) = 0$$

will admit, in general, the integrals of $g = 0$, as singular integrals.

Chapter IX

DIFFERENTIAL EQUATIONS OVER THE REALS

The first proof of the existence of solutions of differential equations as given by Cauchy, published as late as 1884, and completed by Lipschitz, does not assume that those given are analytic. This was the starting point for research in differential equations over the reals. In 1890, Picard introduced a new technique, an iterative method generally known as the method of successive approximations, which applies under analogous conditions and which was introduced into a vast number of proofs of existence theorems. In the same era, Peano and Arzela proved the existence of solutions of an equation $y' = f(x,y)$, under the sole hypothesis of the continuity of the function $f(x,y)$.

In what follows, we shall first of all introduce Picard's method by deriving analogous results to those deduced from the calculus of limits (the Poincaré theorems). We shall then discuss the theory of the Jacobi multiplier and the invariant integrals introduced by Poincaré. Arzela's theorem, proved by Montel by the introduction of the functions of approximation due to Cauchy, will serve to establish the theorem of Cauchy-Lipschitz which seems to bring in the practical methods of approximate integration. Finally, we shall discuss some aspects of the theory of one-parameter groups and its application to differential equations (Lie, 1888) by keeping to the mainstream of ideas, i.e., by not assuming that they are analytic and by underling the local nature of the proofs so obtained. For the extension of this theory to every space, we will refer to the works of E. Cartan.

I. EXISTENCE THEOREMS. THE METHOD OF SUCCESSIVE APPROXIMATIONS DUE TO PICARD

151 - The existence of solutions by means of the Lipschitz condition

Within the neighbourhood of a point, we seek to determine the solution of a system of first order equations, solved with respect to the derivatives of the unknown functions. To simplify matters, we shall confine ourselves to two equations. Everything will apply, under similar conditions, to a system of n equations. Consider then, a system of two equations

$$\frac{dy}{dx} = f(x,y,z) , \qquad \frac{dz}{dx} = g(x,y,z) \quad ; \tag{1}$$

we shall assume that the variable and the unknowns are real. We shall assume that the functions f and g are continuous in a parallelopiped

$x_o \leq x \leq x_o + a$, $|y - y_o| \leq b$, $|z - z_o| \leq c$, *and that the functions satisfy a Lipschitz condition*

$$
\begin{aligned}
&|f(x,y,z)| - f(x,y',z')| < \mathrm{A}|y - y'| + \mathrm{B}|z - z'| \\
&|g(x,y,z) - g(x',y',z')| < \mathrm{A}|y - y'| + \mathrm{B}|z - z'|
\end{aligned}
\tag{2}
$$

in this parallelopiped $(x, y, z; x, y', z',$ *are taken arbitrarily in the parallelopiped;* A *and* B *are two positive constants).* Nothing changes when we replace the condition $x_o \leq x \leq x_o + a$ by $-a + x_o \leq x \leq x_o$.

The Theorem of Cauchy-Lipschitz. *By means of the conditions indicated, the system (1) admits a system of solutions* $y = \phi(x)$, $z = \psi(x)$ *such that for* $x = x_o$, $\phi(x_o) = y_o$, *and* $\psi(x_o) = z_o$, *where the functions* ϕ *and* ψ *are continuous for* $x_o \leq x \leq x_o + h$, *where* h *denotes the minimum of the numbers* a, b/M, c/M, *and where* M *is the limit of* $|f|$ *and* $|g|$ *in the parallelopiped in question.*

Here, we shall prove this theorem by Picard's method (1890), which applies to many similar cases (see I, 120-122). Firstly, we can replace the system in (1) by the system of integral equations

$$
\begin{aligned}
y(x) &= y_o + \int_{x_o}^{x} f(t,y(t),z(t))\, dt \quad , \\
z(x) &= z_o + \int_{x_o}^{x} g(t,y(t),z(t))\, dt \quad ,
\end{aligned}
\tag{3}
$$

which is clearly equivalent. If, in the second members of these equations, we replace $y(t)$ and $z(t)$ by y_o and z_o, we obtain the functions $y_1(x)$, $z_1(x)$ in the first member,

$$
\begin{aligned}
y_1(x) &= y_o + \int_{x_o}^{x} f(t,y_o,z_o)\, dt \\
z_1(x) &= z_o + \int_{x_o}^{x} g(t,y_o,z_o)\, dt \quad ,
\end{aligned}
\tag{4}
$$

which are continuous for $x_o \leq x \leq x_o + a$. If the point $x,y_1(x),z_1(x)$ is in the parallelopiped in question, then we can replace the functions $y(t)$ and $z(t)$ in the second member of (3) by $y_1(t)$ and $z_1(t)$ respectively; the first members will become the functions $y_2(x)$ and $z_2(x)$ respectively. Generally, with $y_1(x)$ and $z_1(x)$ defined by the equations (4), we can set

$$
\begin{aligned}
y_n(x) &= y_o + \int_{x_o}^{x} f(t,y_{n-1}(t),z_{n-1}(t))\, dt \quad , \\
z_n(x) &= z_o + \int_{x_o}^{x} g(t,y_{n-1}(t),z_{n-1}(t))\, dt \ , \quad n = 2,3,\ldots
\end{aligned}
\tag{5}
$$

provided that $y_{n-1}(x)$, $z_{n-1}(x)$ are continuous on the segment (x_o, x) and belong respectively to the segments $(y_o - b, y_o + b)$, $(z_o - c, z_o + c)$. Moreover $y_n(x)$ and $z_n(x)$ will be continuous and differentiable. Now, by assuming $x_o \leq x \leq x_o + h$, we have

$$|y_n(x) - y_o| = |\int_{x_o}^{x} f(t, y_{n-1}(t), z_{n-1}(t))\, dt| \leq hM \leq b \quad ,$$

and the analogous condition $|z_o(x) - z_o| \leq c$ is also satisfied provided we have $|y_{n-1}(x) - y_o| \leq b$, $|z_{n-1}(x) - z_o| \leq c$. Thus we see, by recurrence, that the functions $y_n(x)$ and $z_n(x)$ have values belonging to the segments in question. This calculation in terms of successive approximations, can be continued indefinitely.

We are now going to show that when n increases indefinitely, $y_n(x)$ and $z_n(x)$ converge uniformly towards the limiting functions $\phi(x)$ and $\psi(x)$ provided that x belongs to the segment $(x_o, x_o + h)$. Firstly, on account of the equations (4), we have

$$\begin{aligned} |y_1(x) - y_o| &\leq M(x - x_o) \\ |z_1(x) - z_o| &\leq M(x - x_o) \quad . \end{aligned} \tag{6}$$

On the other hand, by subtracting the equations (5) corresponding to the values n and $n - 1$ and taking account of the Lipschitz conditions, we obtain

$$\begin{aligned} |y_n - y_{n-1}| &= |\int_{x_o}^{x} [f(t, y_{n-1}, z_{n-1}) - f(t, y_{n-2}, z_{n-2})]dt \\ &< A\int_{x_o}^{x} |y_{n-1} - y_{n-2}|dt + B\int_{x_o}^{x} |z_{n-1} - z_{n-2}|dt \end{aligned}$$

and the same inequality for $|z_n - z_{n-1}|$.

When we take $n = 2$, we have, on account of (6),

$$|y_2 - y_1| < (AM + BM) \int_{x_o}^{x} (t - x_o)dt = (A + B)M \frac{(x - x_o)^2}{2}$$

and the same inequality for $|z_2 - z_1|$. Taking $n = 3$, we obtain, in a similar manner

$$\begin{aligned} |y_3 - y_2| &< (A + B)^2 M \frac{(x - x_o)^3}{6} \quad , \\ |z_3 - z_2| &< (A + B)^2 M \frac{(x - x_o)^3}{6} \quad . \end{aligned}$$

By recurrence, we deduce that

$$|y_n - y_{n-1}| < M(A+B)^{n-1} \frac{(x-x_o)^n}{n!} ,$$

$$|z_n - z_{n-1}| < M(A+B)^{n-1} \frac{(x-x_o)^n}{n!}$$

The series

$$y_1 + (y_2 - y_1) + \ldots + (y_n - y_{n-1}) + \ldots ,$$

$$z_1 + (z_2 - z_1) + \ldots + (z_n - z_{n-1}) + \ldots$$

are therefore absolutely and uniformly convergent for $x_o \le x \le x_o + h$. As their terms are continuous functions of x on this segment, their sums $\phi(x)$ and $\psi(x)$ are also continuous. Moreover, we have

$$\phi(x) = \lim_{n=\infty} y_n(x) , \qquad \psi(x) = \lim_{n=\infty} z_n\ x .$$

Following (5) and the theorem on the convergence of the integrals (I, 69) we will have

$$\psi(x) = \lim y_n(x) = y_o + \lim \int_{x_o}^{x} f(t, y_{n-1}(t), z_{n-1}(t))\, dt$$

$$= y_o + \int_{x_o}^{x} f(t, \phi(t), \psi(t))\, dt ,$$

and

$$\psi(x) = z_o \int_{x_o}^{x} g(t, \phi(t), \psi(t))\, dt .$$

These equations prove the theorem since they imply that $\phi(x)$ and $\psi(x)$ are differentiable, taking the values y_o and z_o for $x = x_o$, and satisfying the system in (1).

A PARTICULAR CASE

The theorem applies when, in the parallelopiped in question, the functions $f(x,y,z)$ and $g(x,y,z)$ admit first partial derivatives with respect to y and z, that are continuous in the parallelopiped. The formula of finite increases, implies the conditions (2).

Remark. When we assume $x_o \le x \le x_o + h'$, $h' < h$, we deduce the equations (5)

$$|y_n(x) - y_o| \le h'M , \qquad |z_n(x) - z_o| \le h'M$$

hence

$$|\phi(x) - y_o| \leq b\,\frac{h'}{h} \qquad |\psi(x) - z_o| \leq c\,\frac{h'}{h} \quad .$$

152 - The uniqueness of the solution. The dependence on the initial conditions

The solution obtained is the only solution of the system, taking for x_o, *the values* y_o *and* z_o, *and defined on the segment* $(x_o, x_o + h)$.

Let us assume, in effect, that there exists another system of solutions, e.g.,

$$y = \eta(x); \quad z = \zeta(x); \quad \eta(x_o) = y_o, \quad \zeta(x_o) = z_o \quad .$$

These functions would satisfy the system of integral equations in (3). By subtracting the equations so obtained from the equations in (5), we would obtain

$$\eta(x) - y_n(x) = \int_{x_o}^{x} [f(t,\eta(t),\zeta(t)) - f(t,y_{n-1}(t),z_{n-1}(t))]\,dt \quad ,$$

and the analogous equation for $\zeta(x) - z_n(x)$. When x is sufficiently near to x_o such that

$$|\eta(t) - y_o| \leq b \qquad |\zeta(t) - z_o| \leq c \quad , \tag{8}$$

for $x_o \leq t \leq x$, then we can apply the Lipschitz inequality to the second member of (7), which yields

$$|\eta(x) - y_n(x)| < A \int_{x_o}^{x} |\eta(t) - y_{n-1}(t)|\,dt$$

$$+ B \int_{x_o}^{x} |\zeta\ t\ - z_{n-1}(t)|dt \quad ,$$

and the analogous inequality. By proceeding by recurrence as above, we deduce that, for $n \geq 1$,

$$|\eta(x) - y_n(x)| < (Ab + Bc)(A + B)^{n-1}\,\frac{(x - x_o)^n}{n!} \quad ,$$

and we have the same inequality for $|\zeta(x) - z_n(x)|$. When n increases indefinitely, the second member of this inequality tends towards zero and $y_n(x)$ tends towards $\phi(x)$. It follows that

$$\eta(x) = \phi(x) \ , \qquad \zeta(x) = \psi(x) \quad ,$$

as long the inequalities (8) hold. Now, if $x_o \leq x \leq x_o + h'$ with $h' < h$, we have $|\phi(x) - y_o| < b$, $|\psi(x) - z_o| < c$ and, consequently, the inequalities

(8) occur for $x_o \le x \le x_o + h$. (For, by assuming that x' is the first value above which the inequalities do not occur, then we would have $\eta(x') = \phi(x')$, $\zeta(x') = \psi(x')$ with $|\phi(x') - y_o| < b$ and $|\zeta(x') - y_o| < c$, whence we obtain a contradiction.) The proposition is thus proved completely.

THE DEPENDENCE ON THE INITIAL CONDITIONS

The continuity of the solutions with respect to the initial conditions or parameters appearing in the equation, occurs by means of certain hypotheses. (Compare nos. 46 and 47.)

Let us assume first of all that given the system in (1), f and g are continuous and satisfy the Lipschitz conditions in a parallelopiped P which can be taken to be $|x| \le \alpha$, $|y| \le \beta$, $|z| \le \gamma$. Taking M to be the bound of $|f|$ and $|g|$ in the parallelopiped P, A and B to be the numbers appearing in the inequalities (2), and h the least of the numbers $\alpha/2$, $\beta/2M$, $\gamma/2M$, we can apply the existence theorem with the initial conditions x_o, y_o, z_o provided that $|x_o| < \alpha/2$, $|y_o| < \beta/2$, $|z_o| < \gamma/2$. We obtain a unique solution for $|x - x_o| < h$. This solution is given by the functions of x which depend on the initial conditions,

$$y = \phi(x,x_o,y_o,z_o), \qquad z = \psi(x_1 x_o,y_o,z_o)$$

say. When we assume $|x_o| < h/2$, these functions will be defined for $|x| < h/2$. *These are continuous function of* x, x_o, y_o, z_o. For they are the limits of the functions

$$y_n(x,x_o,y_o,z_o), \qquad z_n(x,x_o,y_o,z_o) \quad , \tag{9}$$

and the bounds previously obtained for $|y_n - y_{n-1}|$, $|z_n - z_{n-1}|$ remain true for all these functions for any x_o, y_o, z_o such that $|x_o| < h/2$, $|y_o| < \beta/2$ $|z_o| < \gamma/2$. The convergence is therefore uniform in x, x_o, y_o, z_o. On the other hand, the functions of the sequences in (9) are calculated termwise by the relations (4) and (5). Now, it is clear that for f and g continuous in the parallelopiped P in question, y_1 and z_1 will be continuous with respect to x, x_o, y_o, z_o in the domain

$$|x| < h/2, \quad |x_o| < h/2, \quad |y_o| < \beta/2, \quad |z_o| < \gamma/2 \quad . \tag{10}$$

Consequently we deduce, termwise, that y_n and z_n will be continuous in this same domain. With respect to uniform convergence, the bounded functions will be continuous.

THE CASE WHERE f AND g ARE DIFFERENTIABLES

Let us now suppose that f and g admit in P, the first partial derivatives with respect to y and z and that these partial derivatives are continuous. Under these conditions, the functions y_n and z_n admit partial derivatives with respect to y_o. We have, first of all,

$$\frac{\partial y_1}{\partial y_o} = 1 + \int_{x_o}^{x} \frac{\partial f}{\partial y}(t,y_o,z_o)dt\ , \qquad \frac{\partial z_1}{\partial y_o} = \int_{x_o}^{x} \frac{\partial g}{\partial y}(t,y_o,z_o)dt$$

and, by recurrence, we see that the equations in (5) are differentiable with respect to y_o. We have

$$\begin{cases} \dfrac{\partial y_n}{\partial y_o} = 1 + \displaystyle\int_{x_o}^{x} \left[\frac{\partial f}{\partial y}(t,y_{n-1},z_{n-1})\frac{\partial y_{n-1}}{\partial y_o} + \frac{f}{z}(t,y_{n-1},z_{n-1})\frac{\partial z_{n-1}}{\partial y_o}\right] dt \\ \\ \dfrac{\partial z_n}{\partial y_o} = \displaystyle\int_{x_o}^{x} \left[\frac{\partial g}{\partial y}\ t,y_{n-1},z_{n-1}\ \frac{\partial y_{n-1}}{\partial y_o} + \frac{\partial g}{\partial z}(t,y_{n-1},z_{n-1})\frac{\partial z_{n-1}}{\partial y_o}\right]dt \end{cases} .$$

We are going to show that when n increases indefinitely, these derivatives converge uniformly. If this is the case, the limiting functions $Y(x)$ and $Z(x)$ will satisfy the identities

$$Y(x) = 1 + \int_{x_o}^{x} \left[\frac{\partial f}{\partial y}(t,\phi(t),\psi(t))Y(t) + \frac{\partial f}{\partial z}((t,\phi(t),\psi(t))Z(t)\right] dt$$

$$Z(x) = \int_{x_o}^{x} \left[\frac{\partial g}{\partial y}(t,\phi(t),\psi(t))Y(t) + \frac{\partial g}{\partial t}(t,\phi(t),\psi(t))Z(t)\right] dt$$

where $\phi(x)$ and $\psi(x)$ are the solutions of the system in (1), that take the values y_o, z_o for x_o. This system of equations considered *a priori* is a differential system (in an integral form) in Y and Z. It is linear. Since $\frac{\partial f}{\partial y}$, $\frac{\partial f}{\partial z}$, $\frac{\partial g}{\partial y}$ and $\frac{\partial g}{\partial z}$ are continuous along with $\phi(t)$, $\psi(t)$ (for any y_o in the interval in question), the Lipschitz condition in Y and Z is satisfied as a result of the linearity in Y and Z. The solution admits a unique continuous solution such that for $x = x_o$, $Y(x_o) = 1$, $Z(x_o) = 0$. We are going to show that $\frac{\partial y_n}{\partial y_o}$ and $\frac{\partial z_n}{\partial y_o}$ tend uniformly (in y_o) towards those functions $Y(x)$, $Z(x)$. By subtracting the equations (11) from those satisfied by Y and Z, we have

$$Y - \frac{y_n}{y_o} = \int_{x_o}^{x} N(t)dt \quad ,$$

with

$$N(t) = Y\left[\frac{\partial f}{\partial y}(t,\phi,\psi) - \frac{\partial f}{\partial g}(t,y_{n-1},z_{n-1})\right]$$

$$+ Z\left[\frac{\partial f}{\partial g}(t,\phi,\psi) - \frac{\partial f}{\partial g}(t,y_{n-1},z_{n-1})\right]$$

$$+ \left(Y - \frac{\partial y_{n-1}}{\partial y_o}\right)\frac{\partial f}{\partial y}(t,y_{n-1},z_{n-1}) + \left(Z - \frac{\partial z_{n-1}}{\partial y_o}\right)\frac{\partial f}{\partial z}(t,y_{n-1},z_{n-1}) \ ,$$

and an analogous result for z_n. Z, Y, $\frac{\partial f}{\partial y}$, $\frac{\partial f}{\partial z}$, $\frac{\partial g}{\partial y}$, $\frac{\partial g}{\partial z}$ have their absolute values less than a number H; the coefficients of Y and Z in the first line are of modulus less than ε given that n is sufficiently large. It follows that when we set

$$\delta_n(x) = \left|Y - \frac{\partial y_n}{\partial y_o}\right| + \left|Z - \frac{\partial z_n}{\partial y_o}\right| \ ,$$

we have (for $x > x_o$)

$$\delta_n(x) < \int_{x_o}^{x} 2H(\delta_{n-1}(t) + 2\varepsilon)\, dt \ .$$

By applying this inequality, starting from $n = p + 1$ and denoting the limit of $\delta_n(x)$ by Δp, we see that by recurrence

$$\delta_n(x) < 2\varepsilon\left[K + \frac{K^2}{2} + \dots + \frac{K^{n-p-1}}{(n-p-1)!}\right] + (2\varepsilon + \Delta p)\frac{K^{n-p}}{(n-p)!} \ ,$$

$$K = 2H(x - x_o) \ .$$

It follows that, with the complementary term tending towards zero with $1/n$, we have, for n sufficiently large

$$\delta_n(x) < 4\varepsilon e^{2Hh} \ ,$$

which proves the convergence, uniform in y_o, of the $\frac{\partial y_n}{\partial y_o}$ and $\frac{\partial z_n}{\partial y_o}$, and further shows that the functions $\phi(x,x_o,y_o,z_o)$, $\psi(x,x_o,y_o,z_o)$ are differentiable with respect to y_o and z_o; the derivatives, Y and Z, for example, are continuous. Similarly, we see that ϕ and ψ are differentiable with respect to x_o. Thus: *When the functions* f *and* g *have continuous first partial derivatives with respect to* y *and* z *in* P, *then the solutions* $\phi(x,x_o,y_o,z_o)$, $\psi(x,x_o,y_o,z_o)$ *admit derivatives with respect to* x_o, y_o *and* z_o *in the domain defined by (10), that are continuous with respect to* x. *(We always assume that* f *and* g *are continuous in* P.*)*

153 - First integrals. The variable equation

Following the proposition which we have just proved, the existence theorem of the first integrals of no. 47 extend to the actual case. We could, in effect, apply the same arguments. If the point x_o, y_o, z_o satisfies the conditions in (10), then the system of solutions

$$y = \phi(x,x_o,y_o,z_o) \;, \quad z = \psi(x,x_o,y_o,z_o) \tag{12}$$

take the values y_1 and z_1 and the point x_1, and the system $y = \phi(x,x_1,y_1,z_1)$, $z = \psi(x,x_1,y_1,z_1)$ is defined in an interval containing x_o; at this it takes the values y_o, z_o. The system in (12) then can be also written as

$$y_o = \phi(x_o,x,y,z) \;, \quad z_o = \psi(x_o,x,y,z) \;.$$

We can take $x_o = 0$ and, on denoting by $F(x,y,z)$ the function $\phi(0,x,y,z)$ and $G(x,y,z)$ the function $\psi(0,x,y,z)$, we see that the solutions are defined by the system of implicit equations

$$y_o = F(x,y,z) \qquad z_o = G(x,y,z) \;, \tag{13}$$

as long as $|x|$, $|y|$, $|z|$ are sufficiently small. To each system of numbers y_o, z_o sufficiently small in absolute value, their corresponds a unique solution in the neighbourhood of the origin, given by this system. The functions F and G have first partial derivatives with respect to x,y,z. Moreover, the identities in (11) show, by recurrence, that the partial derivatives of the y_n and z_n with respect to y_o and z_o are continuous with respect to x_o, y_o, and z_o, and since there is uniform convergence, it is the same for the partial derivatives of ϕ and ψ. Hence the first partial derivatives of F and G are continuous. Finally, from the equation

$$\frac{\partial\phi}{\partial y_o} = 1 + \int_{x_o}^{x} \left[\frac{\partial f}{\partial g}(t,\phi,\psi)\frac{\partial\phi}{\partial y_o} + \frac{\partial f}{\partial z}(t,\phi,\psi)\,\frac{\partial\psi}{\partial y_o}\right] dt$$

and from the analogous equations, it results that

$$F'_y(0,0,0) = 1, \quad F'_z(0,0,0) = 0, \quad G'_y(0,0,0) = 0, \quad G'_z(0,0,0) = 1 \;.$$

The functional determinant

$$\frac{D(F,G)}{D(y,z)}$$

is continuous and equal to 1 at the origin. The system in (13) is normally solvable in y and z about the origin.

The functions $F(x,y,z)$, $G(x,y,z)$ are known as *the first integrals.* They are constant values when y and z are replaced by a system of solutions. The general result obtained for a system of n equations will be the following:

If we consider a differential system

$$\frac{dy_j}{dx} = f_j(x,y_1,y_1,\ldots,y_n) \ ; \qquad j = 1,2,\ldots,n$$

where the functions f_j *are continuous in a parallelopiped having the point* $x_o, (y_1)_o, \ldots (y_n)_o$, *and if the* f_j *have first partial derivatives with respect to the* y_k *which are continuous in this parallelopiped, then there exists a parallelopiped inside the first, in which the solutions are obtained by equating to constants, the functions*

$$F_j(x_1 y_1, y_2, \ldots, y_n) \ , \qquad j = 1,2,\ldots,n \quad .$$

These functions F_j *admit first partial derivatives with respect to the* y_k *and* x, *that are continuous. They have a functional determinant with respect to* $y\ ,y\ ,\ldots,y_n$, *which is non-zero in this parallelopiped.*

THE CASE WHERE THE EQUATIONS CONTAIN PARAMETERS

The theorem of Poincaré relating to the analytic differential equations containing parameters (no. 47), also extends to real continuous systems. When we consider the new system (1), assume that the functions f and g depend on a parameter u, and if $f(x,y,z,u)$, $g(x,y,z,u)$ are these functions which will continue to enjoy the same properties as before, for any u belonging to a segment S, then the expressions for the functions $y_n(x)$, $z_n(x)$ and the uniform convergence towards $\phi(x)$ and $\psi(x)$ shows that: *If* f *and* g *are continuous functions of* u, *then the solutions* $\phi(x)$ *and* $\psi(x)$ *which depend on* u *are also continuous functions of* u. *If* f *and* g *have first order partial derivatives with respect to* y, z, u *that are continuous with respect to* x, y, z, u, *then* $\phi(x)$ *and* $\psi(x)$ *have derivatives with respect to* u.

Remark I. It results from this, that an equation $y' - f(x,y) = 0$ admits an integrating factor provided that $f(x,y)$ admits a partial derivative with respect to y and is continuous along with this derivative.

Remark II. In the same way, we can prove the existence of derivatives of order greater than the solutions, with respect to the variable x, or with respect to the initial elements x_o, y_o, z_o, or with respect to the parameters appearing in the second members of the equations, by means of the hypotheses on the existence and the continuity of the derivatives corresponding to those second members. In particular, if f and g have derivatives continuous of all orders in a parallelopiped, then ϕ and ψ will also have derivatives of all orders.

EQUATIONS WITH VARIATIONS

As in the analytic case, the derivatives of the solutions with respect to the parameters or with respect to the initial conditions, satisfy linear equations. These equations are obtained by differentiating the equations in (1). For example, if there is a parameter u, and f and g are replaced by $f(x,y,z,u)$ and $g(x,y,z,u)$ which satisfy the above conditions, then we shall have (for u constant)

$$\frac{d}{dx}\frac{\partial\phi}{\partial u} = \frac{\partial f}{\partial y}\frac{\partial\phi}{\partial u} + \frac{\partial f}{\partial u}\frac{\partial\psi}{\partial u} + \frac{\partial f}{\partial u} \quad ,$$

$$\frac{d}{dx}\frac{\partial\psi}{\partial u} = \frac{\partial g}{\partial y}\frac{\partial\phi}{\partial u} + \frac{\partial g}{\partial z}\frac{\partial\psi}{\partial u} + \frac{\partial g}{\partial u} \quad .$$

154 - Extending the solutions. The case of linear systems

Having obtained a system of solutions on the segment $(x_o, x_o + h)$ defined in the statement of the Cauchy-Lipschitz theorem, we can start to apply the existence theorem from $x_o + h$ and the values obtained for y and z at this point. We can make this extension in steps, as long as the existence theorem applies.

SOLUTIONS NEAR TO A KNOWN SOLUTION

Let us assume that we have extended a solution $y = \phi(x)$, $z = \psi(x)$ satisfying the initial, given conditions x_o, y_o, z_o. We obtain a segment (x_o, X_o) for example (we would obtain similar results when we operate on the left of x_o), at each point x for which the values of $x, y = \phi(x)$, $z = \psi(x)$ satisfy the conditions of the existence theorem in a parallelopiped P having this point as its centre (at x_o and X_o, we have only a semi-parallelopiped). Let P' be the parallelopiped concentric with P, with P' homothetic to P in the ratio 1/4. Following the Borel-Lebesque theorem, we can cover the arc of the curve Γ described by the point $x, y = \phi(x)$, $z = \psi(x)$, with a finite number of the parallelopipeds P'. To each of them, we can apply what has been said concerning the dependence of the initial conditions. We deduce that, when we take x_o, y_1, z_1 in a small sphere having the point x_o, y_o, z_o as its centre and a small enough radius, the integral curve passing through the point x_o, y_1, z_1 will remain close to Γ. Conversely, every integral curve staying sufficiently close to Γ, will be part of this sheaf.

Moreover, if along Γ, the conditions for which the solution is differentiable with respect to the variables y_1 and z_1, are satisfied, then the sheaf will be differentiable.

SOLUTIONS DEPENDING ON PARAMETERS

In the same way, we see that, if the functions f and g depend on parameters and satisfy conditions for which the solutions ϕ and ψ are continuous, or differentiable with respect to these parameters, all along Γ, then we will obtain a family of curves close to Γ, continuous or differentiable (ϕ and ψ differentiable) with respect to these parameters.

PARTICULAR CASES

The conditions for a direct application of the existence theorem can be weakened in certain cases. If, for $x_o \le x \le x_o + a$, f and g are defined, continuous for any finite y and z, and satisfy the Lipschitz conditions (2) for any y, z, y' and z' (x satisfying the preceding condition), then it is no longer necessary to verify that the solutions approaching $y_n(x)$ and $z_n(x)$, remain close to y_o and z_o. *The eixstence theorem applies all along the segment* $(x_o, x_o + a)$. In the inequalities (6), we will take for M, the limit of $|f(x,y_o,z_o)|$ and $|g(x,y_o,z_o)|$, and the proof of the convergence will follow.

This will result, in particular, when f and g admit partial derivatives with respect to y and z which remain bounded for any y and z (where f and g are continuous for any y and z and for $x_o \le x \le x_o + a$). For example, the equation

$$\frac{dy}{dx} = h(x) + \frac{h}{y^2+1} \quad ,$$

where $h(x)$ is continuous for any x, has all its solutions continuous for any x, the initial conditions are arbitrary. We can take for $h(x)$, a continuous function without differentiating (I, 77).

THE CASE OF LINEAR SYSTEMS

What has been said will especially apply to a linear system

$$\frac{dy_j}{dx} = \sum_{k=1}^{n} a_{j,k}(x) y_k + b_j(x), \qquad j = 1,2,\ldots,n \quad , \tag{14}$$

whose coefficients are continuous. *The solutions are continuous as long as the coefficients are continuous.* If the coefficients depend on parameters and if they have partial derivatives up to order p, continuous with respect to x and the parameters, then the solutions admit derivatives with respect to this parameter, up to order p. The existence of the fundamental systems of solutions of linear systems, is then established as in no. 89.

Remark I. We have assumed that the variable x and the unknown functions are real. But some of the results hold good when, for x always real, $f(x,y,z)$ and $g(x,y,z)$ in (1), are complex functions defined for x real and y and z complex within a domain. The numbers b and c in the statement of the Cauchy-Lipschitz theorem will be replaced by the maximum moduli of $y - y_o$ and $z - z_o$. The solutions will be differentiable complex functions of the real variable x. But, in matters involving the derivatives with respect to y and z, it will be necessary to assume that f and g are analytic with respect to y and z. This will clearly be the case when the system is linear.

Remark II. Let us assume that the functions $f(x,y,z)$ $g(x,y,z)$ depend on a real parameter u and are analytic functions of this parameter. In order to see if the solutions are analytic functions of u, it would help to replace, if possible, the real variable u by a complex variable which we will continue to call u. But it is not only necessary for f and g to be defined for u complex, but that f and g are analytic functions of y and z. This last instance will occur when the system is linear. The form of the relations giving the successive approximate solutions [formulae (5)], shows that in this case of a linear system, e.g. (14), when the $a_{j,k}(x)$ and $b_j(x)$ are holomorphic functions of u in a domain, then the approximate solutions will be holomorphic since they will have derivatives. As the convergence is uniform, the limiting functions will be holomorphic (I, 186). The solutions will then be holomorphic functions of u.

155 - The case where the functions are analytic

If we assume that the functions $f(x,y,z)$ and $g(x,y,z)$ appearing in (1) are analytic functions for $|x-x_0| \leq a$, $|y-y_0| \leq b$, $|z-z_0| \leq c$, then they satisfy a Lipschitz condition. Since, for example, we can write

$$f(x,y,z) - f(x,y',z') = [f(x,y,z) - f(x,y,z')] + [f(x,y,z') - f(x,y',z')] \quad ,$$

and the increases in the holomorphic functions appear in the second member. We have, for example,

$$|f(x,y,z) - f(x,y,z')| = 1 \int_{z'}^{z} \frac{\partial f}{\partial z}(x,y,t)dt| < |z-z'|A \quad ,$$

when A is the maximum of $\left|\frac{\partial f}{\partial z}\right|$ in the region in question. We can then carry out the calculations as no. (5). We see by recurrence that the approximate functions $y_n(x)$, $z_n(x)$ are holomorphic as integrals of holomorphic functions. The limiting functions are holomorphic. We recover the result provided by calculating the limits (no. 47), but with different conditions of application.

II. FIRST INTEGRALS - INVARIANT INTEGRAL MULTIPLIERS

156 - First Integrals

Let us consider a system of differential equations solved with respect to the derivatives of unknown functions that we shall call $x_2,\ldots,x_n$, where the variable is denoted by x_1. We can write this system in the form

$$\frac{dx_1}{X_1} = \frac{dx_2}{X_2} = \ldots = \frac{dx_n}{X_n} \quad , \tag{15}$$

where $X_1,X_2,\ldots,X_n$ are given functions of $x_1,x_2,\ldots,x_n$. Under this form, the n quantities $x_1,x_2,\ldots,x_n$ play the same role and we can make an arbitrary choice of the independent variable. We shall assume that $X_1,X_2,\ldots,X_n$ have the first partial derivatives continuous in a domain Δ of the space $x_1,x_2,\ldots,x_n$ and that X_1, for example, *is non-zero at any point of this domain*. The existence theorem theorefore applies in the neighbourhood of each point of Δ. The solutions can be taken to be of the form

$$F_j(x_1,x_2,\ldots,x_n) = C_j \;, \qquad j = 1,2,\ldots,n-1 \quad , \tag{16}$$

where the C_j are constants and the F_j functions admitting continuous first partial derivatives. Conversely, when we replace the $x_2,\ldots,x_n$ as functions of x_1, the point $x_1,x_2,\ldots,x_n$ describes an integral curve; the functions F_j take constant values. Finally, the functional determinant of the F_j with respect to $x_2,\ldots,x_n$ is non-zero.

If Γ is an arbitrary integral curve, with the F_j constant on this curve, then we shall have

$$\frac{\partial F_j}{\partial x_1} + \frac{\partial F}{\partial x_2}\frac{dx_2}{dx_1} + \ldots + \frac{\partial F_j}{\partial x_n}\frac{dx_n}{dx_1} = 0 \quad , \qquad j = 1,2,\ldots,n-1$$

or, on account of (15),

$$X_1\frac{\partial F}{\partial x_1} + X_2\frac{\partial F}{\partial x_2} + \ldots + X_n\frac{\partial F}{\partial x_n} = 0 \quad , \tag{17}$$

$$F = F_1,\ldots,F_{n-1} \quad .$$

Since Γ is arbitrary, the equation (17) is an identity, hence: *The first integrals* F_j *satisfy equation (17), which is an equation with linear partial derivatives and homogeneous in* F. Conversely, if

$$F(x_1,x_2,\ldots,x_n)$$

is a solution of equation (17) defined in Δ and if we replace $x_2,\ldots,x_n$ in this function, by a system of solutions of equations (15) ($x_2,\ldots,$ are functions of x_1), then the derivative with respect to x_1 of the function obtained will be zero since (17) is satisfied. $F(x_1,\ldots,x_n)$ will be constant on every integral curve of the system in (15). (We assume that F has its first partial derivatives continuous.) Thus *every solution of equation (17) provides a first integral of the system in (15).*

Following the existence theorem, there exist n-1 first integrals of the system (15) defined in the neighbourhood of every point of Δ. Their functional determinant is non-zero in Δ. If F is a solution of equation (17) defined in the neighbourhood of this same point, then the first member of (17) is zero for $F,F_1,\ldots,F_{n-1}$. It follows that, in Δ,

$$\frac{D(F,F_1,F_2,\ldots,F_{n-1})}{D(x_1,x_2,\ldots,x_n)} \equiv 0 \quad . \tag{18}$$

But the determinant

$$\frac{D(F_1,F_2,\ldots,F_{n-1})}{D(x_2,x_3,\ldots,x_n)} \tag{19}$$

is non-zero at any point in the minor domain in question. The condition (18) then implies that F is a function of $F_1,F_2,\ldots,F_{n-1}$, (I, 126). We shall have

$$F = \Phi(F_1,F_2,\ldots,F_{n-1}) \quad . \tag{20}$$

What has just been said holds good when $F_1,F_2,\ldots,F_{n-1}$ are any first integrals whose functional determinant is non-zero. On the other hand, the function Φ, of the type obtained in (I, 126), has continuous partial derivatives with respect to $F_1,F_2,\ldots,F_{n-1}$ provided that F has continuous first derivatives. Every function F defined by (20), with Φ having continuous partial derivatives, is a solution of equation (17) since we have

$$\frac{\partial F}{\partial x_k} = \sum_{j=1}^{n-1} \frac{\partial F_j}{\partial x_k}\frac{\partial \Phi}{\partial F_j} \quad , \qquad k = 1,2,\ldots,n$$

and, on multiplying this equation by X_k and adding, we obtain zero.

Definitions. *Every function* $F(x\ ,\ldots,x_n)$ *admitting continuous first partial derivatives in the domain* Δ *and which retains a constant value when the point* $x_1,x_2,\ldots,x_n$ *describes an arbitrary integral curve* Γ *of the system is known as a first integral of the system in (15).*

We say that a system of $n-1$ *first integrals is a fundamental system in* Δ, *if one of the functional determinants of these integrals with respect to* $n-1$ *of the variables* $x, x, \ldots, x_n$, *is non-zero at any point of* Δ.

The first integrals are the solutions of the associated equation (17).

PROPERTY

If $F_1, F_2, \ldots, F_{n-1}$ form a fundamental system and if we denote the values of these functions at a point M_o at a point M_o of Δ, by $C_1, C_2, \ldots, C_{n-1}$ then the system of equations

$$F_1(x_1, x_2, \ldots, x_n) = C_1, \ldots, F_{n-1}(x_1, x_2, \ldots, x_n) = C_{n-1}$$

is solvable with respect to $n-1$ of the variables $x_1, \ldots, x_n$ in the neighbourhood of M_o, and has a *unique* solution represented by a curve passing through M_o. This curve is then the integral curve of the system in (15) which passes through M_o.

Consequently, we have the following:

Theorem. *When we know a fundamental system* $F_1, \ldots, F_{n-1}$, *of first integrals of the system (15) defined in* Δ, *we obtain all integral curves passing through the points of* Δ, *by equating these first integrals to the values that they take at an arbitrary point of* Δ *and solving the system thus obtained. The general integral, with continuous first derivatives in* Δ, *of the equation with associated partial derivatives (17) is given by the equation* $F = \Phi(F_1, \ldots, F_{-1})$, *where* Φ *is an arbitrary function with continuous first partial derivatives.*

157 - Integral combinations. Examples

A first integral $F(x_1, \ldots, x_n)$ satisfies equation (17) and conversely, the solutions of (17) provide first integrals. We will obtain such functions by seeking to determine the factors $\lambda_j(x_1, x_2, \ldots, x_n)$, $j = 1, 2, \ldots, n$, such that

$$\lambda_1 dx_1 + \ldots + \lambda_n dx_n \tag{21}$$

is a total differential and that

$$\lambda_1 X_1 + \lambda_2 X_2 + \ldots + \lambda_n X_n \equiv 0 \quad . \tag{22}$$

The λ_j will then be partial derivatives of a function F which will be a first integral. To obtain F, we will have to integrate the differential (21), and hence make quadratures. *We say that the expression in (21) is an integral combination of the equations in (15) when it is a total differential and when (22) is satisfied.*

We will then need to form a ratio equal to the ratios in (15):

$$\frac{dx_1}{X_1} = \dots = \frac{dx_n}{X_n} = \frac{\lambda_1 dx_1 + \dots + \lambda_n dx_n}{\lambda_1 X_1 + \dots + \lambda_n X_n}$$

whose numerator is the total differential of a function F and whose denominator is identically zero. There do not exist precise rules permitting the determination of such factors λ_j.

Reduction. *When we have found* p *distinct first integrals, we can lower the order of the system by* p *units.* We say that p first integrals $F_1, F_2, \dots, F_p$ are distinct if one of the functional determinants of these functions with respect to the p variables $x_1, x_2, \dots, x_n$, are non-zero in the domain Δ in which we operate. If we assume, as we may, that

$$\frac{D(F_1, \dots, F_p)}{D(x_1, \dots, x_p)} \neq 0 \quad ,$$

we can, in effect, change the variables $x_1, x_2, \dots, x_p$ by replacing them by

$$y_j = F_j(x_1, x_2, \dots, x_p, \dots, x_n) \ , \qquad j = 1, 2, \dots, p \quad , \tag{23}$$

and preserve the variables $x_{p+1}, \dots, x_n$. We can solve the system in (23) with respect to $x_1, x_2, \dots, x_p$ in a restricted domain the very least, and take the values obtained into the system in (15). X_j becomes a function of $y_1, \dots, y_p, x_{p+1}, \dots, x_n$ that can be called Y_j; we will obtain the new system

$$\frac{dx_{p+1}}{Y_{p+1}} = \dots = \frac{dx_n}{Y_n} \ , \qquad dy_1 = dy_2 = \dots = dy_p = 0 \quad .$$

There remains a systme of $n - p - 1$ equations in $x_{p+1}, \dots, x_n$, where the y_j are arbitrary constants.

Remark. If we proceed in the same way by starting from a fundamental system of first integrals, then the system in (15) takes the form

$$dy_1 = dy_2 = \dots = dy_{n-1} = 0 \quad .$$

Examples I. The differential system

$$\frac{dy}{dx} = zu \ , \quad \frac{dz}{dx} = uy \ , \quad \frac{du}{dx} = yz \quad ,$$

which can also be written as

$$\frac{dy}{zu} = \frac{dz}{uy} = \frac{du}{yz} = dx \quad ,$$

admits the integral combinations $ydy - zdz$ and $ydy - usu$. We have the first integrals $y^2 - z^2$, $y^2 - u^2$; for constants a and b, we can set

$$y^2 - z^2 = a, \quad y^2 - u^2 = b \quad , \tag{24}$$

[a and b play the same role of the y_j in (23)]. There remains to consider the equation

$$\frac{dy}{\sqrt{(y^2-a)(y^2-b)}} = dx \quad ;$$

y, in general, will be an elliptic function of x. z and u will then be given by the equations (24). By permutating the roles of y, z, u, we see that z and u are also elliptic functions of x.

Example II. The determination of a skew (gauche?) curve for which the intrinsic equations are known reduces to the integration of the differential system

$$\frac{dx}{dt} = ry - qz \; , \quad \frac{dy}{dt} = pz - rx \; , \quad \frac{dz}{dt} = qx - py \quad ,$$

where p, q and r are given continuous functions of t (see no. 207). The system does not change when we replace x,y,z by kx, ky, kz respectively, where k is a constant. We have, on the other hand, the integrable combination

$$xdx + dyd + zdz \quad ,$$

which yields the first integral $x^2 + y^2 + z^2$. We can then restrict our consideration to the solutions for which $x^2 + y^2 + z^2 = 1$, at the very least when x, y and z are assumed to be real. This amounts to saying that the integral curves are on the sphere of radius 1, centred at the origin. We can determine them by means of the parametric representation of the sphere, displaying the rectilinear generators. We will set

$$\frac{x+iy}{1-z} = \frac{1+z}{x-iy} = \lambda, \quad \frac{x+iy}{1+z} = \frac{1-z}{x-iy} = -\mu \; , \quad i = \sqrt{-1}$$

Along an integral curve, λ and μ will be complex functions of the real variable t, and conversely, we will obtain

$$x = \frac{1 - \lambda\mu}{\lambda - \mu} \; , \quad y = i\,\frac{1 + \lambda\mu}{\lambda - \mu} \; , \quad z = \frac{\lambda + \mu}{\lambda - \mu} \quad .$$

By carrying out the transformation, we see that λ and μ are solutions of the Riccati equation

$$\frac{du}{dt} = -iru + \frac{q-ip}{2} + \frac{q+ip}{2} u^2 \quad .$$

Here, we have the case of an equation with complex coefficients, being complex functions of the real variable t. The solutions exist provided that p, q and r are continuous; all the intermediate calculations do not involve any other hypothesis apart from the continuity. We must take for λ and μ, the solutions for which $-\lambda$ and $1/\mu$ are imaginary conjugates.

Remark. We shall occasionally introduce into the system (15) an auxiliary variable t, by writing

$$\frac{dx_1}{X_1} = \frac{dx_2}{X_2} = \dots = \frac{dx_n}{X_n} = dt \quad . \tag{25}$$

We then have a system of n equations. When the system (15) is integrated, we obtain t by a quadrature, since we will then have

$$dt = \frac{dx_1}{X_1(x_1, x_2(x_1), \dots, x_n(x_1))} \quad .$$

The order of the system is then not actually increased, and the integral curves in the space $x_1, x_2, \dots, x_n$ are obtained in a parametric form.

When a system is given, *a priori*, in the form (25), $X_1, \dots, X_n$, do not depend on t; we can replace it by the system (15), and when integrating, it will remain to make a quadrature.

158 - The Jacobi multiplier

Let us consider a new system (15), where X_j satisfy the conditions indicated in no. 156. Let $F_1, \dots, F_{n-1}$ be a fundamental system of first integrals. Having satisfied equation (17) by $F_1, \dots, F_{n-1}$, the X_j are proportional to the minors of the first line of the functional determinant

$$\frac{D(F, F_1, \dots, F_{n-1})}{D(x_1, x_2, \dots, x_n)} \quad .$$

If we put

$$D_j = (-1)^{j-1} \frac{D(F_1, \dots, F_j, \dots, F_{n-1})}{D(x_1, \dots, x_{j-1}, x_{j+1}, \dots, x_n)} , \qquad j = 1, 2, \dots ,$$

we have

$$D_j = MX_j \quad , \qquad j = 1,2,\dots,n \quad , \tag{26}$$

where M is a function of $x_1,x_2,\dots,x_n$, which is continuous and has first derivatives, if the hypotheses imply that the F_j have second partial derivatives. *This will certainly be the case when the* F_j *have continuous second partial derivatives. The function* M *defined by (26) is a Jacobi multiplier.* It has analogous properties to those of the integrating factor of the ordinary equations of the first order (no. 68). Following the above definition, *we see that a multiplier corresponds to every fundamental system of first integrals.* These multipliers satisfy the linear equation with partial derivatives

$$\sum_1 \frac{\partial(MX_j)}{\partial x_j} = 0 \quad . \tag{27}$$

We see this by replacing the MX_j by the D_j. In the derivations of the determinants obtained, we introduce the second derivatives of each F_k taken once with respect to x_j and once with respect to x_p with $p \neq j$. One such term arises from one part of D_j and from one part of D_p. Its coefficient is, if $p > j$, for example,

$$(-1)^{j-1}(-k)^{k+p-1} \quad \frac{D(F_1,F_2,\dots,F_{k-1},F_{k+1},\dots,F_{n-1})}{D(x_1,\dots,x_{j-1},x_{j+1},\dots,x_{p-1},\dots,x_{p+1},\dots,x_n)}$$

arising from D_j and the same determinant, and the same determinant multiplied by $(-1)^{p-1}(-1)^{j+k}$ arising from D_p. We obtain zero. The equation (27) is thus satisfied.

If M_1 is a solution of equation (27) and if we set $M = M_1$, we see that z satisfies the equation

$$M_1 \sum_1^n X_j \frac{\partial z}{\partial x_j} = 0 \quad ,$$

i.e., equation (17). The general solution of equation (27) is then

$$M\Phi(F_1,F_2,\dots,F_{n-1}) \quad . \tag{28}$$

Every function of this form is a multiplier. For, when we consider a multiplier M_1 corresponding to a fundamental system of first integrals $G_1,G_2,\dots,G_{n-1}$, we have

$$M_1 = \frac{1}{X_1}\frac{D(G_1,\ldots,G_{n-1})}{D(x_2,\ldots,x_n)} = \frac{1}{X_1}\frac{D(F_1,\ldots,F_{n-1})}{D(x_2,\ldots,x_n)}\frac{D(G_1,\ldots,G_{n-1})}{D(F_1,\ldots,F_{n-1})}$$

$$= M\,\frac{D(G_1,\ldots,G_{n-1})}{D(F_1,\ldots,F_{n-1})} \quad ,$$

and it suffices to choose $G_1,\ldots,G_{n-1}$ is such a way that

$$\frac{D(G_1,\ldots,G_{n-1})}{D(F_1,\ldots,F_{n-1})} = \Phi(F_1,\ldots,F_{n-1}) \quad .$$

For example, $G_2,\ldots,G_{n-1}$ could be given, and G_1 will be determined, as a function of the F_j, by a linear equation with partial derivatives that has solutions (no. 263), as we will see later. Consequently:

The multipliers are the solutions of equation (27). Their general form is given by the expression in (28), i.e., by

$$M = \frac{1}{X_1}\,\frac{D(F_1,\ldots,F_{n-1})}{D(x_2,\ldots,x_n)}\,\Phi(F_1,\ldots,F_{n-1}) \quad . \tag{29}$$

CHANGE OF VARIABLE

If we make a change of variable on $x_1,\ldots,x_n$, preserving x_1, we obtain a new system

$$\frac{dx_1}{Y_1} = \frac{dy_2}{Y_2} = \cdots = \frac{dy_n}{Y_n} \quad ,$$

where the first integrals $F_1,\ldots,F_{n-1}$ transform to the first integrals $G_1,\ldots,G_{n-1}$. Let us consider the multiplier

$$N = \frac{1}{Y_1}\,\frac{D(G_1,\ldots,G_{n-1})}{D(y_2,\ldots,y_n)}\,\Phi(G_1,\ldots,G_{n-1}) \quad ,$$

where Φ is the same function as in (29). With respect to the formulae of the transformation, we shall have $G_j = F_j$ and we may assume that $Y_1 = X_1$. We then have

$$N = M\,\frac{D(G_1,\ldots,G_{n-1})}{D(y_2,\ldots,y_n)} : \frac{D(F_1,\ldots,F_{n-1})}{D(x_2,\ldots,x_n)} = M\,\frac{D(y_2,\ldots,y_n)}{D(y_2,\ldots,y_n)} \quad .$$

A multiplier will be known after the transformation if it is known for the original system.

An application. *When* $n-2$ *distinct first integrals are known along with a multiplier* M, *then the integration is achieved by quadratures.*

For, if $F_1, F_2, \ldots, F_{n-2}$ are the first $n-2$ integrals, then we can make a change of variable of the form

$$x_1 = x_1, \; x_2 = x_2, \; F_1 = y_1, \ldots, F_{n-2} = y_{n-2} \quad ,$$

which reduces to the system

$$\frac{dx_1}{Y_1} = \frac{dx_2}{Y_2} \; , \qquad dy_1 = dy_2 = \ldots = dy_{n-2} = 0 \quad . \tag{30}$$

We know a multiplier N for the system, transformed from M. We therefore have

$$\frac{\partial NY_1}{\partial x_1} + \frac{\partial NY_2}{\partial x_2} = 0 \quad ,$$

which expresses the fact that N is an integrating factor of the first equation (30). The system in (30) is integrated by quadratures.

A PARTICULAR CASE

In certain mechanical systems, we have $\sum \frac{\partial X_j}{\partial x_j} \equiv 0$; we have a multipler $M = 1$.

159 - The invariant integrals of Poincaré

Let us consider a system of three equations

$$\frac{dx}{X} = \frac{dy}{Y} = \frac{dz}{Z} = dt \quad , \tag{31}$$

where X, Y, Z are functions of x, y, z only. We can consider x, y, z as the rectangular coordinates of a point P and t as the time variable. The equations (31) then define the motion of a point P when the velocity field is known. The trajectories are determined by the first two equations (31). On adding them to the last ratio, we can determine the course which they take. If P_o is the position of the frame at the instance t_o and if D_o is a domain, then the motions whose initial positions P_o are in D_o, will occupy at the instance t, the positions $P(t)$ situated in a domain $D(t)$. We shall assume D_o and $t-t_o$ sufficiently small in order to obtain on the segment (t_o, t), a sheaf of solutions, when observing the hypotheses of no. 154.

A function $f(P)$ of the position of the point P will define an *invariant volume integral* if the integral

$$\iiint f(\mathrm{P})dv \tag{32}$$

extended to the domain $D(t)$ is independent of t for arbitrary D_o. *The integral (32) will then be an invariant volume integral.* Likewise, we can assume that P_o is displaced on a surface Σ_o or on a line L_o and consider a surface integral or a curvilinear integral

$$\iint \overrightarrow{V(P)}\, \vec{n}\, d\sigma \,, \qquad \int \overrightarrow{V(P)}\, \overrightarrow{dP}$$

$\vec{n}$ where n is the unit vector of the normal to the surface $\Sigma(t)$ deduced from Σ_o. The first integral is calculated on $\Sigma(t)$, the second on $L(t)$ transformed from L_o and $\overrightarrow{V(P)}$ is a given vector at each point of the space. If, for any Σ_o or L_o, these integrals do not depend on t, then they define *the invariant integrals of the surface and the line* respectively.

These definitions extend to systems with n unknown variables $x_1,x_2,\ldots,x_n$, for which the independent variable is t. But we shall restrict our attention to the system (31).

<u>The Poincaré Theorem</u>. *In order that the function* $f(P) = f(x,y,z)$ *defines an invariant volume integral, it is necessary and sufficient that* $f(x,y,z)$ *is a Jacobi multiplier of the system (31).*

In order that (32) is an invariant integral, it is necessary and sufficient that its derivative with respect to t should be zero. As the domain $D(t)$ corresponds pointwise to D_o, then by change of variable,

$$x = \phi(t,x_o,y_o,z_o) \,, \qquad y = \psi(t,x_o,y_o,z_o)$$

$$z = \chi(t,x_o,y_o,z_o) \,,$$

defined by the solutions of (31), we can reduce the integral in (32) to an integral calculated in D_o, e.g.,

$$\iiint_{D_o} f(\phi,\psi,\chi) \left|\frac{D(x,y,z)}{D(x_o,y_o,z_o)}\right| dv_o \tag{33}$$

(We always assume that the system in (31) has properties ensuring the existence of the necessary partial derivatives.) We obtain the derivative by differentiating under the sign of integration. The derivative of the factor $f(\phi,\psi,\chi)$ is

$$\frac{\partial f}{\partial x}\frac{\partial \phi}{\partial t} + \frac{\partial f}{\partial y}\frac{\partial \psi}{\partial t} + \frac{\partial f}{\partial z}\frac{\partial \chi}{\partial t} = X\frac{\partial f}{\partial x} + Y\frac{\partial f}{\partial y} + Z\frac{\partial f}{\partial z} \,.$$

The derivative of the functional determinant is the sum of three determinants whose first is

$$\begin{vmatrix} \dfrac{\partial^2\phi}{\partial x_o \partial t} & \dfrac{\partial\psi}{\partial x_o} & \dfrac{\partial\chi}{\partial x_o} \\ \dfrac{\partial^2\phi}{\partial y_o \partial t} & \dfrac{\partial\psi}{\partial y_o} & \dfrac{\partial\chi}{\partial y_o} \\ \dfrac{\partial^2\phi}{\partial z_o \partial t} & \dfrac{\partial\psi}{\partial z_o} & \dfrac{\partial\chi}{\partial z_o} \end{vmatrix} \quad . \tag{34}$$

On taking account of (31), the first column has as its elements,

$$\frac{\partial X}{\partial x_o} = \frac{\partial X}{\partial x}\frac{\partial \phi}{\partial x_o} + \frac{\partial X}{\partial y}\frac{\partial \psi}{\partial x_o} + \frac{\partial X}{\partial z}\frac{\partial \chi}{\partial x_o} \qquad \frac{\partial X}{\partial y_o} = \ldots, \ \frac{\partial X}{\partial z_o} = \ldots \quad .$$

The determinant (34) is therefore the product of the determinant appearing in (33) is the product of this determinant and

$$\frac{\partial X}{\partial x} + \frac{\partial Y}{\partial y} + \frac{\partial Z}{\partial z}$$

and that the derivative of (33) is

$$\iiint_{D_o} \left|\frac{D(x,y,z)}{D(x_o,y_o,z_o)}\right| \left(\frac{\partial Xf}{\partial x} + \frac{\partial Yf}{\partial y} + \frac{\partial Zf}{\partial z}\right) dv_o \quad .$$

This integral can only be zero, whatever D_o, when the differential coefficient is zero, which implies that $f(x,y,z)$ is a multiplier since we must have

$$\frac{\partial Xf}{\partial x} + \frac{\partial Yf}{\partial y} + \frac{\partial Zf}{\partial z} \equiv 0 \quad .$$

Example. In order that $f = 1$ defines an invariant integral, i.e., in order that the volume is invariant, it is necessary to have

$$\frac{\partial X}{\partial x} + \frac{\partial Y}{\partial y} + \frac{\partial Z}{\partial z} \equiv 0 \quad ;$$

the velocity vector of the field must be a vortex (I, 156).

In *Methodes nouvelles de la Mécanique céleste* (t. III, 1899), Poincaré studied the invariant integrals in the general case of systems of order n and demonstrated their usefulness. On this subject, we may also refer to E. Cartan's *Leçons sur les invariants integraux* (1922).

III. THE METHOD OF CAUCHY - THE THEORY OF ARZELA. APPROXIMATE INTEGRATION

160 - Arzela's theorem on equicontinuous functions

Let us consider a family of function $\phi(x)$ defined on a segment $a \le x \le b$ and continuous on this segment. *These functions are said to be equicontinuous on this segment if, for all given* ε *positive, we can assign a number* η *such that the condition*

$$|x' - x''| < \eta , \quad a \le x' \le b , \quad a \le x'' \le b ,$$

implies

$$|\phi(x') - \phi(x'')| < \varepsilon \tag{35}$$

for all functions of the family.

Example. The functions $\phi(x)$ admitting at every point of the segment (a,b) a right ahd left hand derivative whose absolute value is at most equal to a number M, provide an equicontinuous family.

In effect, on account of the formula of finite increments (I, 38), we have

$$|\phi(x') - \phi(x'')| \le M|x' - x''| \quad ;$$

it will hence suffice to take $\eta M < \varepsilon$. We have analogous propositions to those given previously for the holomorphic functions (I, 212).

Lemma. *If the functions of the sequence* $\phi_n(x)$ *are equicontinuous on* (a,b) *and if they converge in dense points on* (a,b), *then they converge uniformly on* (a,b).

Since if η is the number corresponding to an arbitrarily small ε appearing in (35), let us partition the segment (a,b) into intervals of length less than η. On each of these partial segments s_p, we can find a point x_p on which the $\phi_n(x)$ converge. As there is a finite number of segments s_p, we can find a number N such that for all points x_p, we have

$$|\phi_n(x_p) - \phi_m(x_p)| < \varepsilon \quad \text{if} \quad n > N, m > N \quad .$$

If x is any point of (a,b), it belongs to a segment s_p; therefore if $n > N$ and $m > N$ we have

$$|\phi_n(x) - \phi_m(x)| \le |\phi_n(x) - \phi_n(x_p)| + |\phi_n(x_p) - \phi_m(x_p)|$$
$$+ |\phi_m(x_p) - \phi_m(x)| < \varepsilon$$

which proves the lemma.

<u>Arzela's Theorem</u>. *If the functions $\phi(x)$ are equicontinuous and bounded in their set, on a segment (a,b), of every sequence of functions $\phi(x)$, then we can extract another sequence of functions which converges uniformly on (a,b).*

By hypothesis, for all functions of the family, we have $|\phi(x)| < M$ on (a,b) where M is fixed.

Consider a sequence of functions of the family, functions that we shall name $\phi(x,n)$, $n = 1,2,\ldots,$ and let $x_1,x_2,\ldots,x_p,\ldots,$ be a sequence of dense points on (a,b). We can extract from the sequence of bounded numbers $\phi(x_1,n)$, a convergence sequence which converges towards a finite number. Let $\phi(x_1,n_p^1)$, $p = 1,2,\ldots,$ be this sequence. We shall consider the sequence of functions $\phi(x,n_p^1)$. We can extract a sequence $\phi(x,n_p^2)$, $p = 1,2,\ldots,$ which converges at the point x_2. This sequence will therefore converge at the points x_1 and x_2. We will extract from it a new sequence $\phi(x,n_p)$ which will converge at the points x_1,x_2,x_3 and so on. From the sequence $\phi(x,n_p^q)$ extracted from $\phi(x,n_p^{q-1})$ and which will converge at the points $x_1,x_2,\ldots,x_q$, we will extract a sequence $\phi(x,n_p^{q+1})$ which will moreover converge at the point x_{q+1}. The diagonal sequence

$$\phi(x,n_1^1),\ \phi(x,n_2^2),\ \ldots,\ \phi(x,n_q^q),\ \ldots$$

whose function of rank q will occupy the rank q at least in the sequence $\phi(x,n_q^p)$, will converge at all points $x_1,x_2,\ldots$. Following the lemma, it will converge uniformly on (a,b).

<u>Corollary</u>. *If the convergent sequences extracted from a sequence of functions $\phi_n(x)$, equicontinuous and bounded in their set on (a,b), all have the same limiting function $\phi(x)$, then the $\phi_n(x)$ converge uniformly towards $\phi(x)$ on (a,b).*

In effect, if the $\phi_n(x)$ do not converge uniformly on (a,b) towards $\phi(x)$, then there will exist an infinite sequence of values n' of n and corresponding values x_n' of x for which

$$|\phi_{n'}(x_n') - \phi(x_n')| > \alpha > 0 \quad .$$

Let us consider the sequence $\phi_n'(x)$; we can extract from it, a uniformly convergent sequence on (a,b) whose bounding function will be $\phi(x)$. Therefore, for these n', for as long as they are sufficiently large, and for any x on (a,b), we will have

$$|\phi_n'(x) - \phi(x)| < \alpha$$

contradicting the above inequality.

161 - The method of Cauchy. Arzela's theorem

Let us consider a differential equation

$$\frac{dy}{dx} = f(x,y) \tag{36}$$

where the function $f(x,y)$ is continuous for $x_o \le x \le x_o + a$ and $|y - y_o| \le b$. In this rectangle, $|f(x,y)|$ has a bound M. Let us seek a solution of equation (36) which takes the value y_o for x_o. We shall assume that x satisfies the condition

$$x_o \le x \le x_o + h \quad , \tag{37}$$

where h is the minimum of a and b/M. Let us consider points x_p on the segment $(x_o, x_o + h)$ such that

$$x_o < x_1 < x_2 < \ldots < x_p < x_{p+1} < \ldots < x_n < x_o + h = x_{n+1} \quad ,$$

and define the numbers y_p by the equations

$$\begin{aligned} y_1 &= y_o + (x_1 - x_o)f(x_o,y_o),\ldots,y_p \\ &= y_{p-1} + (x_p - x_{p-1})f(x_{p-1},y_{p-1}) \quad . \end{aligned} \tag{38}$$

Following (37), we have

$$|y_1 - y_o| \le (x_1 - x_o)\mathrm{M} \le b \,, \quad |y_{p+1} - y_p| \le (x_{p+1} - x_p)\mathrm{M} \,,$$

$$|y_{p+1} - y_o| \le \mathrm{M}(x_{p+1} - x_p + x_p - x_{p-1} + \ldots - x_o) \le b \quad .$$

To the division $x_1, x_2, \ldots, x_n$, let us assign the function $y(x; x_1, \ldots, x_n)$ which is linear for $x_p \le x \le x_{p+1}$, $p = 0,1,\ldots,n$ such that $y(x_p; x_1, \ldots, x_n) = y_p$, $p = 0,1,\ldots,n+1$.

This function which is represented geometrically by a polygonal line has everywhere a right and left-sided derivative (except at the extremities x_o, x_{n+1}, where it will have a derivative to the right at x_o and one to the left at x_{n+1}). These derivatives are bounded, for they are equal to the gradients of the successive sides of the representative polygonal line, and we have

$$\left|\frac{y_p - y_{p-}}{x_p - x_{p-}}\right| = |f(x_{p-1}, y_{p-1})| \le \mathrm{M} \quad .$$

The family of functions $y(x; x_1, x_2, \ldots, x_n)$ is hence formed from functions bounded in their set on the segment $(x_o, x_o + h)$ and their derivatives are bounded in their set; the family is equicontinuous.

Let us consider the family of these functions and apply Arzela's theorem. From every sequence of functions of the family, we can extract a uniformly convergent sequence. *Let* $\phi(x)$ *be a limiting function corresponding to a sequence for which* n *tends towards infinity and the maximum of* $x_{p+1} - x_p$ *tends towards zero. Let us show that* $\phi(x)$ *is a solution of equation (36)* (which for $x = x_o$, takes the value y_o). The function $\phi(x)$ is the limit of a certain uniformly convergent sequence $\phi(x;x_1,\dots,x_n)$. Let (x_p, x_{p+1}) be the segment, relating to this function, which contains x. For as long as n is sufficiently large, we have $x_{p+1} - x_p < \eta$, consequently (on account of equicontinuity),

$$|\phi(x;x_1,x_2,\dots,x_n) - \phi(x_p;x_1,\dots,x_n)| < \varepsilon \quad ,$$

and also (on account of uniform convergence)

$$|\phi(x) - \phi(x;x_1,\dots,x_n)| < \varepsilon \quad ,$$

hence

$$|\phi(x_p;x_1,\dots,x_n) - \phi(x)| < 2\varepsilon \quad .$$

Now on taking derivatives to the right, and for $x_{p+1} > x$, we have

$$\phi'(x;x_1,\dots,x_n) - f(x,\phi(x))$$

tends uniformly towards zero when n increases indefinitely and that the maximum $x_{p+1} - x_p$ tends towards zero. Thus on integrating, with $\phi(x)$ and hence $f(x,\phi(x))$ continuous,

$$\phi(x;x_1,\dots,x_n) - y_o - \int_{x_o}^{x} f(t,\phi(t))dt$$

tends uniformly towards zero. We therefore have

$$\phi(x) = y_o + \int_{x_o}^{x} f(t,\phi(t))dt \quad .$$

$\phi(x)$ is differentiable and satisfies equation (36). We have thus obtained Arzela's theorem (1895):

Arzela's Theorem. *With the sole condition that* $f(x,y)$ *is continuous in the rectangle* $x_o \le x \le x_o + a$, $|y - y_o| \le b$, *the equation*

$$\frac{dy}{dx} = f(x,y) \tag{36}$$

admits at least one solution which takes the value y_o *for* $x = x_o$, *and which is defined for* $x_o \le x \le x_o + h$, *where* h *is the minimum of* a *and* b/M *(when* $|f(x,y)| \le M$ *in the rectangle in question).*

The proof which has just been given and which uses the approximate polygonal line method (of Cauchy) is due to Montel (Thesis 1907). It easily extends to first order differential systems.

Clearly, we have analogous results when we replace the condition

$$x_o \le x \le x_o + a \quad \text{by} \quad x_o - a \le x \le x_o \quad .$$

We may have an infinity of solutions when the Lipschitz condition does not occur. This theorem of Arzela applies, for example, to the equation

$$\frac{dy}{dx} = 3\sqrt{y}$$

with x_o arbitrary and $y_o = 0$. We first of all obtain the singular integral $y = 0$ and the general analytic integral

$$y_{x_o}(x) = \left[\frac{2}{3}(x - x_o)\right]^{3/2}$$

but then we also obtain the integrals defined by

$$y = 0 \text{ when } x_o \le x \le x_1, \qquad y = y_{x_1}(x) \quad \text{when} \quad x \ge x_1$$

(cf. the considerations of no. 63).

162 - The theorem of Cauchy-Lipschitz

When we include the Lipschitz condition in the hypotheses in the statement of Arzela's theorem, we should recover the theorem of Cauchy-Lipschitz of no. 151. We see this by showing directly, as did Jordan, that the Lipschitz condition implies the uniqueness of the solution of equation (36), taking the value y_o for x_o. Let us assume then, that in the imagined rectangle, we also have $|f(x,y) - f(x,y')| < A|y - y'|$; where x,y and x,y' are two points of the rectangle. Let us assume that we have two solutions y and Y of the equation, satisfying the initial conditions x_o, y_o. We would have

$$\left|\frac{d}{dx}(y - Y)\right| = |f(x,y) - f(x,Y)| < A|y - Y| \quad .$$

Let us assume that $y - Y$ is non-zero for all x such that $x_o \le x \le x_1$ and let m be the maximum of $|y - Y|$ for these x. By integrating the above inequality, we would have

$$|y - Y| < Am(x_1 - x_o) \quad ,$$

hence $1 < A(x_1 - x_o)$. It is necessary to assume that $x_1 - x_o > 1/A$ in order that there should not be a contradiction; we have $y = Y$ as long as $|x_1 - x_o| < 1/A$. But we can recommence this argument starting from x_1. *The Lipschitz condition implies the uniqueness.*

CONSEQUENCE

When the Lipschitz condition is satisfied, all functions of the family $\phi(x;x_1,\ldots,x_n)$ have a single limiting function when we restrict our attention to those functions for which $n \to \infty$ and the maximum of the $x_{p+1} - x_p$ tend towards zero. Consequently, on account of Corollary no. 160:

Theorem. *If the Lipschitz condition is also satisfied by the conditions stated in Arzela's theorem, then the polygonal functions of Cauchy $\phi(x;x\ ,\ldots,x_n)$ tend uniformly towards the integral taking the value y_o for x_o when the number of sides increases indefinitely and when the maximum of $x_{p+1} - x_p$ tend towards zero.*

Remark. Let us suppose that we have noted that equation (36) in which $f(x,y)$ is continuous in a certain domain Δ, admits a solution $y = \phi(x)$ represented by a curve which is contained in this domain Δ and further, assume that the Lipschitz condition is satisfied in this domain. In the domain Δ, $|f(x,y)|$ will be bounded by a number M; we can then apply the above theorem by partitioning the interval of variation of x, into a finite number of segments. The approximation by the Cauchy polygonal functions will apply all along the integral curve. We ascertain, in effect, that the convergence of these functions exists if, instead of taking the point x_o,y_o as the point of departure of the polygonal line, we take a neighbouring point x_o',y_o' that can be made to tend towards the point x_o,y_o.

163 - Practical methods of approximation. Newton's formula. Adam's method

Following above discussion, when we consider a differential equation

$$\frac{dy}{dx} = f(x,y) \tag{36}$$

satisfying the Cauchy-Lipschitz conditions, we can obtain values as near as we wish to the solution $y = \phi(x)$, at any point x, by substituting for it, the (Cauchy) polygonal approximate solution, provided small enough intervals are taken. It will suffice to calculate the values of $\phi(x;x_1,\ldots,x_n)$ at the points x_p and for the intermediate x's, to make a linear interpolation.

We can encode the calculation into a geometrical representation.

In practice, we strive to simplify the calculation in order to reduce the number of points of the division, as was the case in the calculation of definite integrals (I, 108-112).

THE METHOD OF EXTRAPOLATION BY DIFFERENCES

We propose to calculate the approximate values of the solution $y = \phi(x)$, satisfying the initial given conditions at the point in arithmetic progression of ratio r. We can always assume that r is positive and $x_o = 0$; a change of variable immediately reduces matters to this case.

Let us denote the approximate solution by $y(x)$ and set $F(x) = f(x,y(x))$. To calculate $y(nr)$, $n = 1,2,\ldots,$ we take in the Cauchy method $y(nr) = y((n-1)r) + rF((n-1)r)$.

If we take

$$y(nr) = y((n-1)r) + \int_{(n-1)r}^{nr} F(t)dt \quad ,$$

we will have an exact integral provided that $y(x) \equiv \phi(x)$ in this interval.

We replace $F(t)$ by a polynomial calculated in terms of the values of $F(t)$ at the points $(n-1)r,\ldots,(n-q-1)r$, where q is a fixed number. We can take a polynomial of degree q. This polynomial of degree q, taking the known values at $q+1$ points in arithmetic progression, is determined by the calculus of differences.

NEWTON'S FORMULA

Let us set

$$\Delta_1(a) = F(a) - F(a-r), \quad \Delta_2(a) = \Delta_1(a) - \Delta_1(a-r),$$

$$\Delta_q(a) = \Delta_{q-1}(a) - \Delta_{q-1}(a-r), \quad q = 3,4\ldots$$

The polynomial of degree q which takes the values of $F(t)$ at the points $a, a-r,\ldots,a-qr,$ is given by Newton's Formula; we shall call it $P(t,a,r,q)$. We have

$$P(t,a,r,q) = F(a) + \sum_{p=1}^{q} \frac{u(u+1)\ldots(u+p-1)\Delta_p(a)}{p!} \; , \quad u = \frac{t-a}{r} \quad . \tag{39}$$

To establish this formula, we observe that, necessarily, we have

$$P(t,a,r,q) = F(a) + \sum_{p=1}^{q} R_p(u)\Delta_p(a) \; , \quad u = \frac{t-a}{r} \quad , \tag{40}$$

where the $R_p(u)$ are polynomials (zero for $n = 0$) independent of q. Since, when we take for $F(t)$ a polynomial of arbitrary degree $p (p \le q)$, then $\Delta_k(a)$ are zero for $k < p$ and arbitrary for $k \le p$, which determine $R_1(u)$ then $R_2(u)$, etc. To obtain $R_q(u)$, we note that it is of degree q, following what has just been said, and it must be zero for $u = 0,-1,\ldots,-q+1$, since the values of the second member of (40) for $u = 0,-1,\ldots,-q+1$ do not depend on $\Delta_q(a)$. We therefore have

$$R_q(u) = u(u+1)\dots(u+q-1)\alpha_q \quad ,$$

where α_q is a constant, and on applying this formula to this same polynomial, we see that $\alpha_q q! = 1$.

Newton's formula is thus established.

APPLICATION

In applying this method, we must first of all calculate with much accuracy the values of

$$F(0) = f(0,y_o)\ ,F(r),\dots,F(qr) \quad ,$$

by using the method of successive approximations (due to Picard) or in the analytic case, the expansion in a power series, where the limit of the remainder is provided by the method of the majorant functions. Let us assume that, generally, we have calculated

$$F(nr - qr),\ F(nr - (q-1)r),\dots,F(nr) \quad .$$

We then take

$$y(nr+r) = y(nr) + \int_{nr}^{nr+r} P(t,nr,r,q)dt \quad ,$$

which will provide $F(nr+r)$. By making the change of variable $t = nr + ru$, we shall have

$$y(nr+r) = y(nr) + rF(nr)$$
$$+ \sum_{p=1}^{q} \frac{r}{p!} \Delta_p(nr) \int_o^1 u(u+1)\dots(u+p-1)du \quad .$$

ADAM'S METHOD

This corresponds to the case where $q = 5$ is taken. Starting from formula (41), we then obtain

$$y(nr+r) = y(nr) + r\left[F(nr) + \frac{\Delta_1(nr)}{2} + \frac{5\Delta_2(nr)}{12}\right.$$
$$\left. + \frac{3\Delta_3(nr)}{8} + \frac{251\Delta_4(nr)}{720} + \frac{95\Delta_5(nr)}{288}\right] \quad .$$

Remark. We can equally consider

$$y((n+1)r) = y((n-1)r) + \int_{(n-1)r}^{(n+1)r} F(t)dt$$

and replace $F(t)$ by $P(t,nr,r,q)$. For $q = 5$, we obtain the following formula due to Nyström:

$$y(nr+r) = y(nr-r) + r\left[2F(nr) + \frac{1}{3}(\Delta_2(nr) + \Delta_3(nr) + \Delta_4(nr) + \Delta_5(nr))\right] - \frac{r}{90}(\Delta_4(nr) + 2\Delta_5(nr)) ,$$

which is more suitable for calculations.

164 - The case of equations of greater order

The same methods apply to solvable differential systems of the first order. We can just as well make a direct attack on the second order differential systems of the form

$$\frac{d^2y_j}{dx^2} = f_j(x,y_1,\ldots,y_n) , \qquad j = 1,2,\ldots,n$$

encountered in mechanics. Let us consider the case of a single equation

$$\frac{d^2y}{dx^2} = f(x,y) \quad . \tag{42}$$

We shall always denote by $y(x)$, the approximate solution and we set $F(x) = f[x,y(x)]$. By integrating between $a = nr$ and x, we shall have

$$y'(x) - y'(a) = \int_a^x F(t)dt \quad ,$$

then, on re-integrating from a to $a+r$, we obtain

$$y(a+r) - y(a) - ry'(a) = \int_a^{a+r} dx \int_a^x F(t)dt \quad .$$

When we change r to $-r$, we likewise obtain

$$y(a-r) - y(a) + ry'(a) = \int_a^{a-r} dx \int_a^x F(t)dt \quad ,$$

and, in addition,

$$y(a+r) = 2y(a) - y(a-r) + \int_a^{a+r} dx \int_a^x F(t)dt + \int_a^{a-r} dx \int_a^x F(t)dt \quad .$$

We replace $F(t)$ as above by the polynomial (39). Taking $q = 5$, we first of all calculate

$$\int_a^x dx \int_a^x F(t)dt$$

$$= r^2 \int_0^u du \int_0^u \left[F(t) + \sum_{p=1}^{5} \frac{u(u+1)\ldots(u+p-1)\Delta_p(u)}{p!} \right] du \quad ,$$

and we add the results corresponding to $u = 1$ and to $u = -1$. We obtain

$$y(nr + r) = 2y(nr) - y(nr - r)$$
$$+ r^2\left[F(nr) + \frac{\Delta_2(nr)}{12} + \frac{\Delta_3(nr)}{12} + \frac{19\Delta_4(nr)}{240} + \frac{9\Delta_5(nr)}{120}\right] \quad .$$

It is convenient to write the coefficient of r^2 in the form

$$F(nr) + \frac{1}{12}\left[\Delta_2 + \Delta_3 + \Delta_4 + \Delta_5 - \frac{1}{20}\Delta_4 - \frac{1}{10}\Delta_5\right] \quad .$$

This formula, due to Störmer, was applied to his work on the trajectories of electrified material points in a magnetic field.

The calculation of the trajectories of a ballistic has given rise to a number of methods of approximate calculation of the solutions of differential equations introduced therein. In the calculation of the perturbations due to small modifications of the initial conditions, the variational equations of Poincaré, are utilised.

IV. LIE THEORY

165 - One parameter pointwise continuous groups

We shall consider a point transformation depending on a parameter u. To simplify matters we shall assume that the points in question are in three dimensional space when we present their coordinates. At the point $P(x,y,z)$ we assign the point $Q(x',y',z')$ whose position depends on that of P and the parameter u. We have

$$x' = f(x,y,z,u), \qquad y' = g(x,y,z,u), \qquad z' = h(x,y,z,u) \qquad (43)$$

which we can abbreviate, by taking the points to be extremities of vectors fixed at the origin of the coordinates:

$$\vec{Q} = T(\vec{P},u) \quad . \qquad (44)$$

We shall assume that the transformation T is continuous with respect to $\vec{P}$ and u, and is clearly uniform in the domain containing P. The functions f, g, h are functions of x, y, z, u. We shall assume, moreover, that these functions admit partial derivatives that will be introduced and the latter will have those properties desirable for applying the arguments which will follow. In the applications f, g, h will be analytic, in general, but it is to be understood that we shall consider only the real elements.

Let us suppose that the transformation T has been applied to any arbitrary point P, with a parameter value u; this gives a point Q. Then to Q, the transformation T is applied with a parameter value v, giving a point M. With the notation (44), we can write

$$\vec{M} = T(\vec{Q},u) \ , \qquad \vec{Q} = T(\vec{P},u) \quad ,$$

or

$$\vec{M} = T[T(\vec{P},u),v] \quad .$$

If x'', y'', z'' are coordinates of M, we have on account of (43),

$$x'' = f[f(x,y,z,u), g(x,y,z,u), h(x,y,z,u), v]$$

and analogous equilities. We have made the product to the two transformations in question (I, 31).

We shall assume that the transformation (43) (44) is invertible, i.e., that the transformation is bijective (under the above conditions); we can pass from Q to P in a unique way. This inverse transforamtion is clearly continuous and depends on u continuously.

Definition. *We say that the transformations* T *defined by (43), (44) form a group when:*

1) *The product of two transformations of the family belong to the family;*
2) *The inverse transformation belongs to the family.*

By utilising the abbreviated notation, there will exist numbers w and u' such that, for any P.

$$T[T(\vec{P},u),v] \equiv T(\vec{P},w)$$
$$\vec{P} = \vec{T}(\vec{Q},u') \quad \text{if} \quad \vec{Q} = \vec{T}(P,u) \quad ; \qquad (45)$$

w is a function of u and v, u' is a function of u.

When we compose the transformation with parameter u with its inverse, with parameter $u' = \phi(u)$, we obtain a transformation whose parameter is a continuous function of u and $\phi(u)$, hence of u' and one which leaves the points unchanged. It therefore corresponds to a fixed value u_o of u. We shall call it the *identity transformation.*

If we obtain the same transformation for two values u and v of the parameter, the product of the inverse of the first and the second will be the identity transformation. We can take v as a function of u, such that the property indicated occurs for all v and $\psi(u)$ of an interval. We shall then say that the representation of the group is improper. We shall always assume that the parametrisation is proper. *A fixed transformation will be provided by a single value of the parameter* u.

Examples. The translations defined by vectors parallel to a given vector form a continuous one-parameter group. For example, the transformations $x' = x + u$, $y' = y$, $z' = z$ define a group. The representation is proper; it will become improper when we replace u by u^2. The rotations about a given axis define a continuous, one-parameter group. For example, by taking $x' = x \cos u - y \sin u$, $y' = x \sin u + y \cos u$, $z' = z$, with $0 \leq u < 2\pi$, we will obtain a proper representation.

SIMILAR GROUPS

If we make an invertible pointwise transformation of the domain in question onto another domain for which the points P',Q',M'... correspond bijectively to the points P,Q,M..., then this transformation changes a continuous one-parameter group to another continuous one-parameter group. For, if $\vec{P}' = \overrightarrow{S(P)}$ is this transformation and if $\vec{P} = S_{-1}(\vec{P}')$ is its inverse, then the transformation (44) is changed to

$$\vec{Q}' = S(T[S_{-1}(\vec{P}'),u])$$

and we see that conditions (45) imply

$$ST[S_{-1}ST[S_{-1}(\vec{P}'),u]v] \equiv ST[T(S_{-1}(\vec{P}'),u),v] = ST[S_{-1}(\vec{P}'),w] \quad ,$$

$$\vec{P}' = ST[S_{-1}(\vec{Q}'),u'] \text{ if } \vec{Q}' = ST[S_{-1}(\vec{P}'),u] \quad .$$

Similarly, a bijective change of variable on u, $u = \Phi(U)$ say, transforms a continuous group to a continuous group.

Two continuous one-parameter groups having the property that one may be transformed to the other via a bijective transformation of the parameter are called *similar groups*.

INVARIANT CURVES

When u varies, the locus of points Q such that

$$\vec{Q} = T(\vec{P},u) \quad ,$$

where P is given, is a curve Γ passing through P (for $u = u_o$). Following the definition of the group, Γ is also the locus of transformations of all its points Q [the first equality (45)]. The curve Γ is therefore *invariant under the transformations of the group*. Conversely, the curves Γ invariant under the transformations of the group are the loci of transformations of their points, and hence are obtained as we have just stated.

These curves are also called the group *trajectories*.

The differential equations of these curves are easily formed. Through a point $Q(x',y',z')$, there passes one and only one curve Γ which can be considered as the locus of transformations of the point P identified with Q. The neighbouring points of Q on Γ, correspond to the values of u close to u_o, a value providing the identity transformation. We therefore have

$$\frac{dx'}{\frac{\partial f}{\partial u}(x',y',z',y_o)} = \frac{dy'}{\frac{\partial g}{\partial u}(x',y',z',u_o)} = \frac{dz'}{\frac{\partial h}{\partial u}(x',y',z',u_o)} \tag{46}$$

which yields the differential equation of the invariant curves.

THE DIFFERENTIAL EQUATION OF THE GROUP

The group is defined by the equations (43) which, for $P(x,y,z)$ fixed, define the group trajectories. On account of the group properties, we have identically

$$T(\vec{Q},v) = T(\vec{P},u) \quad . \tag{47}$$

Let us assume that, in this identity, w is fixed, then v is a function of u, $v = \phi(u)$ and if a trajectory is displaced, we have on differentiating with respect to u,

$$\frac{\partial f}{\partial x}(x',y',z',v)\frac{dx'}{du} + \frac{\partial f}{\partial y}(x',y',z',v)\frac{dy'}{du}$$

$$+ \frac{\partial f}{\partial z}(x',y',z',v)\frac{dz'}{du} + \frac{\partial f}{\partial u}(x',y',z',v)\,\phi'(u) = 0$$

and analogous equations. We can therefore form a ratio equal to the ratios (46), a ratio of the form

$$\frac{\phi'(u)du}{U(x',y',z',v)} \quad .$$

But, dx'/du, etc., are independent of the value given to w, hence to v. Finally, we obtain

$$\frac{dx'}{X(x',y',z')} = \frac{dy'}{Y(x',y',z')} = \frac{dz}{Z(x',y',z')} = \phi'(u)du \quad ,$$

and the equations (43) are the integrals of this system which reduce to x, y, z respectively for $u = u_o$.

A GROUP CORRESPONDING TO A DIFFERENTIAL SYSTEM

Conversely, let us consider a linear system of the form

$$\frac{dx'}{X(x',y',z')} = \frac{dy'}{Y(x',y',z')} = \frac{dz'}{Z(x',y',z')} = dt \quad . \tag{49}$$

When we integrate at first the system formed by the first three ratios, its integral is of the form $F(x',y',z') = C_1$, $G(x',y',z') = C_2$, where F and G are two first integrals. We then have to make a quadrature, which yields $t + C_3 = H(x',y',z')$; C_1,C_2,C_3 are arbitrary constants. The integrals taking the values x, y, z for $t = 0$ are then given by

$$F(x',y',z') = F(x,y,z) \ , \qquad G(x',y',z') = G(x,y,z) \ ,$$

$$H(x',y',z') = H(x,y,z) + t \quad ,$$

a solvable system in x',y',z' or x,y,z in a certain domain. We thus define a continuous one-parameter group. For the inverse transformation (the passage from x',y',z' to x,y,z) is achieved by setting $t = -t$. When we make the transformation corresponding to the value t, and then that corresponding to t', then the product is the transformation which corresponds to $t + t'$. (Everything is assumed to take place within a restricted domain and in accordance with the conditions stated in nos. 151-153.) The integrals of the system in (49) therefore define a continuous group.

REDUCTION TO THE CANONICAL FORM

The system in (48) reduces to the system in (49) via the change of parameter $t = \phi(u)$. *We shall say that the group is considered in the canonical form.*

If, in the system (49), we take

$$\xi' = F(x',y',z'), \quad \eta' = G(x',y',z'), \quad \zeta' = H(x',y',z')$$

as new variables, then the integrals become

$$\xi' = \xi, \qquad \eta' = \eta, \quad \zeta' = \zeta + t \quad .$$

The corresponding group is the group of translations.

Consequently:

Theorem. *By means of the conditions of continuity and differentiability which are always raised in the analytic case, every continuous one-parameter group is similar to a group of translations in a restricted domain at the very least.*

Corollary. *The group in question is commutative (or abelian): the product of two transformations of the group does not depend on their order.*

Example. Let us consider the group of the plane

$$x' = xe^{u\alpha}, \quad y' = ye^{u\alpha}, \quad \alpha = 2 + \cos\,(\mathrm{arc} + g\,\frac{y}{x}) \quad ,$$

whose trajectories are the half-lines stemming from the origin. We have

$$\frac{dx'}{du} = x\alpha e^{u\alpha} = x'\alpha\,, \qquad \frac{dy'}{du} = y'\alpha$$

and hence the system

$$\frac{dx'}{x'\left(\cos(\mathrm{arc} + g\,\frac{y'}{x'}) + 2\right)} \quad \frac{dy'}{y'\left(2 + \cos(\mathrm{arc} + g\,\frac{y'}{x'})\right)} = du$$

whose first integrals are

$$\frac{y'}{x'}\,, \qquad \frac{\log x'}{2 + \cos(\mathrm{arc} + g\,\frac{y'}{x'})} - u \quad .$$

The transformation

$$y' - \xi' x' = 0\,, \qquad x' = \rho^{\beta\eta'} \quad ,$$

$$\beta = 2 + \cos(\mathrm{arc} + g\,\xi') \quad ,$$

reduces to the group of translations.

166 - Infinitesimal transformations of the group

Let us consider a function $\Phi(P) = \Phi(x,y,z)$ of the position of the point P. Let us take the group in *the canonical form* with the differential equations of the group in the form (49), t is the parameter and the identity transformation corresponds to $t = 0$. When we replace, in $\Phi(P)$, the point P by its transformation Q, we obtain a function $\Phi(Q)$ which for $t = 0$ reduces to $\Phi(P)$. The differential of this function $\Phi(Q)$ for $t = 0$, a differential written as $\delta\Phi$, with δt an increment in t, defines the infinitesimal transformation of $\Phi(P)$. We clearly have

$$\delta\Phi(x,y,z) = \left(\frac{\partial\Phi}{\partial x}\frac{dx'}{dt} + \frac{\partial\Phi}{\partial y}\frac{dy'}{dt} + \frac{\partial\Phi}{\partial z}\frac{dz'}{dt}\right)\,\delta t \quad ,$$

the coefficients $dx'/dt,\ldots,$ taken for $t = 0$, hence equal $X(x,y,z),\ldots$ Thus

$$\delta\Phi = \left(\frac{\partial\Phi}{\partial x}X + \frac{\partial\Phi}{\partial y}Y + \frac{\partial\Phi}{\partial z}Z\right)(x,y,z)\,\delta t \quad .$$

In particular, *the infinitesimal transformation of the group* is defined by the infinitesimal transformations of the coordinates.

$$\delta x = X(x,y,z)\delta t \ , \quad \delta y = Y(x,y,z)\delta t, \quad \delta z = Z(x,y,z)\delta t \quad .$$

Once given, this infinitesimal transformation defines the group since it reduces to the given one of the canonical system (49). The infinitesimal transformation of the function $\Phi(x,y,z)$ is written

$$\delta\Phi = \frac{\partial\Phi}{\partial x}\delta x + \frac{\partial\Phi}{\partial y}\delta y + \frac{\partial\Phi}{\partial z}\delta z \quad .$$

In order that the function Φ should be *invariant* under the group transformations, i.e., to have a value not depending on t, it is necessary to have the infinitesimal transformation zero, hence that Φ should be a solution of the partial differential equation

$$\Pi(\Phi) \equiv \frac{\partial\Phi}{\partial x} X + \frac{\partial\Phi}{\partial y} Y + \frac{\partial\Phi}{\partial z} Z = 0 \quad . \tag{50}$$

This necessary condition is sufficient. We know, in fact, that the solutions of equation (50) are of the form $\Phi(F,G)$, where F and G are both first integrals of the system of the first two equations (49) [we suppress the fourth member of (49)]. Now, F and G are invariant.

SUCCESSIVE DIFFERENTIALS OF $\Phi(P)$

We can clearly define and calculate the successive differentials of $\Phi(x,y,z)$ about $t = 0$, provided that the partial derivatives of the functions X, Y, Z occuring in the canonical equation (49), exist. We will first of all obtain

$$\delta^2\Phi = \left(\frac{\partial}{\partial x} X + \frac{\partial}{\partial y} Y + \frac{\partial}{\partial z} Z\right)^{(2)} \Phi(x,y,z)\delta t^2$$

$$+ \left(\frac{\partial\Phi}{\partial x}\delta X + \frac{\partial\Phi}{\partial y}\delta Y + \frac{\partial\Phi}{\partial z}\delta Z\right)\delta t \quad ,$$

etc.

167 - The extended group

Let us assume that the point P describes a curve C along which y and z will be assumed to be functions of x and differentiable. To the curve C there will correspond a curve $C(u)$ along which y' and z' will be functions of x'. The derivatives of y' and z' with respect to x' are given as functions of x, y, z and the derivatives of y and z with respect to x, by the formulae deduced from the equations obtained on differentiating the formulae of the transformation (43). We will have

$$\frac{dy'}{dx'} = k\left(x,y,z,\ \frac{dy}{dx},\ \frac{dz}{dx},\ u\right)$$
$$\frac{dz'}{dx'} = \ell\left(x,y,z,\ \frac{dy}{dx},\ \frac{dz}{dx},\ u\right) \quad . \tag{51}$$

If we add these formulae to the formulae in (43) we obtain another one-parameter transformation. It is the extended transformation of (43). These one-parameter transformations again form a group. For when the product of these transformations is made, corresponding to the values u and v of the parameter, we obtain a transformation which, in coordinate terms, belongs to the group (43) and corresponds to a value $w = \phi(u,v)$ of the parameter. When we consider the derivatives corresponding to points of the product transformation, (x'',y'',z''), their expression is then the same when calculated from $x,y,z,\ \frac{dy}{dx},\ \frac{dz}{dx}$, by making the product of the transformations u,v or by considering the transformation $\phi(u,v)$ alone. The same applies to the inverse transformation.

Similarly, when we assume P to describe a surface and when z is assumed to be a function of x and y, with p and q its partial derivatives, and p' and q' the derivatives of z' with respect to x' and y' following the transformation, we shall have

$$p' = m(x,y,z,p,q,u), \qquad q' = n(x,y,z,p,q,u) \quad .$$

The set of transformations defined by the relationships in (43) and these last two will form a group. It is the extended group.

Likewise, we can extend the group by considering the second derivatives etc. The infinitesimal transformation of the extended group (in the canonical form) introduces the differentials of X, Y, Z.

<u>Example</u>. Let us restrict our attention to the case of a plane transformation defined by the infinitesimal transformation

$$\delta x = X(x,y)\,\delta t, \qquad \delta y = Y(x,y)\,\delta t \quad .$$

For t close to zero, we have

$$x' = x + Xt + \ldots, \qquad y' = y + Yt + \ldots \quad ,$$

whence we deduce

$$\frac{dy'}{dx'} = \frac{\frac{dy}{dx} + t\left(\frac{\partial Y}{\partial x} + \frac{dy}{dx}\frac{\partial Y}{\partial y}\right) + \cdots}{1 + t\left(\frac{\partial X}{\partial x} \quad \frac{dy}{dx}\frac{\partial X}{\partial y}\right) \quad \ldots} \quad .$$

The coefficient of t in the second member provides the infinitesimal transformation of the extended element

$$\delta \frac{dy}{dx} = \left[\frac{\partial Y}{\partial x} + \left(\frac{\partial Y}{\partial y} - \frac{\partial X}{\partial x}\right)\frac{dy}{dx} - \frac{\partial X}{\partial y}\left(\frac{y}{x}\right)^2\right]\delta t \quad .$$

The extended group is again canonical.

168 - An application to first order differential equations

Let us consider a sheaf of surface $\Phi(P)$ = const. If we apply the transformations of a group taken in the canonical form, to these surfaces, we obtain the surfaces defined by $\Phi(x',y',z')$ = const.* [We assume that the equations of the group are $x' = f(x,y,z,t),\ldots.$] For each value of t, we have a sheaf. The *sheaf* in question will be invariant if, for each t, $\Phi(x',y',z')$ = const. defines a surface $\Phi(x,y,z)$ = const. For each t, the system $\Phi(x,y,z) = \alpha$, $\Phi(f(x,y,z,t)g,h) = \beta$, has its functional determinants with respect to any two of the variables x,y,z identically zero (otherwise the two equations will only be satisfied on a curve). Consequently, $\Phi(x',y',z')$ is a function of $\Phi(x,y,z)$ and t alone. This necessary condition is evidently sufficient. For x,y,z fixed, we have

$$\frac{d\Phi\ x',y',z'}{dt} = \frac{\partial\Phi}{\partial x}(x',y',z')\frac{dx'}{dt} + \frac{\partial\Phi}{\partial y}\frac{dy'}{dt} + \frac{\partial\Phi}{\partial z}\frac{dz'}{dt}$$

$$= \Pi(\Phi(x',y',z')) \quad ,$$

where $\Pi(\phi)$ is the function defined by the equation (50). But, on the other hand, this derivative is a function of $\Phi(x,y,z)$ and t, hence also of $\Phi(x',y',z')$ and t alone, and since it does not depend on t, but only on x',y',z', we have

$$\Pi[\Phi(x',y',z')] = \lambda(\Phi(x',y',z')) \quad ,$$

where λ is a determined function. Thus, in order for a sheaf of surfaces $\Phi(x,y,z)$ = const. to be invariant, it is necessary to have

$$\Pi(\Phi) \equiv \lambda(\Phi) \quad .$$

Conversely, if this condition is realised, we will have the equality in (52) and consequently,

$$\frac{d\Phi(x',y',z')}{dt} = \lambda[\Phi(x',y',z')] \quad ,$$

*The inverse transformation of $x' = f(x,y,z,t)\ldots$ is $x = f(x',y',z',-t)$, ...

hence,

$$\frac{d\Phi}{\lambda(\Phi)} = t + \text{const} \quad ,$$

and on integrating

$$\Phi(x',y',z') = \mu[\Phi(x,y,z\ ,t] \quad .$$

Consequently:

The necessary and sufficient condition for the sheaf of surfaces $\Phi(x,y,z) = \text{const.}$ *to be invariant with respect to the transformations of the canonical group defined by (49) is that*

$$\Pi(\Phi) \equiv \frac{\partial\Phi}{\partial x} X + \frac{\partial\Phi}{\partial y} Y + \frac{\partial\Phi}{\partial z} Z = \lambda(\Phi) \quad ,$$

where $\lambda(\Phi)$ *is an arbitrary function of* Φ. (In the case of a group of transformations of the plane, the surfaces are replaced by curves.)

A DIFFERENTIAL EQUATION ADMITTING THE TRANSFORMATIONS OF A GROUP

Consider a differential equation of the form

$$P(x,y)dx + Q(x,y)dy = 0 \quad . \tag{53}$$

We say that this equation admits the transformations of the one-parameter group defined by

$$\frac{dx'}{X(x',y')} = \frac{dy'}{Y(x',y')} = dt \ ; \quad x' = f(x,y,t) \ , \quad y' = g(x,y,t) \tag{54}$$

when the sheaf of the integral curves is invariant with respect to the transformations of the group.

To find the condition for this to be the case, we subject the equation to the transformation of the extended group and claim that it is not modified. This necessitates that the group is known to be in the finite form $x' = f(x,y,t)$, $y' = g(x,y,t)$. Let us assume that only the infinitesimal transformation X,Y is known. Then the infinitesimal transformation of the extended group is following the calculation of no. 167,

$$X,\ Y,\ \left(\frac{\partial Y}{\partial x} + \frac{\partial Y}{\partial y} z - \frac{\partial X}{\partial x} z - \frac{\partial X}{\partial y} z^2\right) \quad , \qquad z = \frac{dy}{dx} \quad .$$

The equation (53) is written

$$\Phi(x,y,z) \equiv P(x,y) + Q(x,y)z = 0 \quad , \qquad z = \frac{dy}{dx} \quad . \tag{55}$$

The transformations of the extended group transform $\Phi(x,y,z) = 0$ to $\Phi(x',y',z') = 0$, which must define the same surface as the first equation, for any t. It follows that the derivative of $\Phi(x',y',z')$ with respect to t (with x,y,z constants) will be zero, hence $\Pi([\Phi(x',y',z'))$ will be zero when $\Phi(x',y',z')$ will be zero, and as above, we see that this condition is sufficient.

We have

$$\Pi(\Phi) \equiv X \frac{\partial\Phi}{\partial x} + Y \frac{\partial\Phi}{\partial y} + \left(\frac{\partial Y}{\partial x} + \frac{\partial Y}{\partial y} z - \frac{\partial X}{\partial x} z - \frac{\partial X}{\partial y} z^2\right) \frac{\partial\Phi}{\partial z} \quad ,$$

hence the equation

$$X \left(\frac{\partial P}{\partial x} + z \frac{\partial Q}{\partial x}\right) + Y \left(\frac{\partial P}{\partial y} + z \frac{\partial Q}{\partial y}\right) + Q \left(\frac{\partial Y}{\partial x} + \frac{\partial Y}{\partial y} z - z \frac{\partial X}{\partial x} - z^2 \frac{\partial X}{\partial y}\right) = 0$$

must be a result of (55) and this suffices. Hence, it is necessary and sufficient that

$$X \left[Q \frac{\partial P}{\partial x} - P \frac{\partial Q}{\partial x}\right] + Y \left[Q \frac{\partial P}{\partial y} - P \frac{\partial Q}{\partial y}\right] + Q \left[Q \frac{\partial Y}{\partial x} - P \frac{\partial Y}{\partial y}\right] + P \left[Q \frac{\partial X}{\partial x} - P \frac{\partial X}{\partial y}\right] \equiv 0$$

is satisfied. If we assume that $PX + QY \neq 0$, then we can write this condition under the form

$$- \frac{\partial}{\partial x} \frac{Q}{PX + QY} + \frac{\partial}{\partial y} \frac{P}{PX + QY} = 0 \quad , \tag{56}$$

and we obtain the theorem:

<u>Theorem</u>. *In order that equation (53) admits transformations of the group whose infinitesimal transformation is defined by equations (54), it is necessary and sufficient that* $\frac{1}{PX + QY}$ *is an integrating factor, or else that this equation (53) does not differ from equation (54)* $(PX + QY \equiv 0)$.

<u>The theory of elementary integration</u>. It follows from the above theorem that knowing when a group does not change the equation implies knowing an integrating factor and consequently we can assert in this case that the equation is integrated by quadratures.

Conversely, if the integrating factor M is known, then it suffices to take for X and Y those functions such that $PX + QY = 1/M$, in order to derive a one-parameter group of transformations for which the equation is invariant. This shows that every equation (53) admits the transformations of an infinity of groups.

The elementary methods of integration of first order equations amounts to knowing a group admitted by the equation. Thus the equations of the form (53) which do not contain the variable x admit the group of parallel translations to Ox, and these are the only ones. The homogeneous equations (no. 65) admit the group of homothéties with respect to the origin and these are the only ones admitting this group.

Remark. This group of homothéties has for its canonical equations

$$\frac{dx'}{x'} = \frac{dy'}{y'} = dt\ , \qquad x' = xe^t\ , \qquad y' = ye^t \quad ;$$

the equations which admit it will be obtained by the determination of a pair of functions P and $Q \equiv 1$ satisfying equation (56).* But we may also proceed as in no. 165. The equation of the group is taken to be of the form

$$\frac{y'}{x'} = \frac{y}{x}\ , \qquad \log y' = \log y + t\ ;$$

the change of variable

$$\xi = \frac{y}{x}\ , \qquad \eta = \log y$$

which reduces the group to one of translations, will reduce the equation to an equation invariant with respect to the modified group

$$\xi' = \xi, \quad \eta' = \eta + t, \quad (X = 0, \quad Y = 1) \quad ,$$

to which there corresponds the equation

$$d\eta = P(\xi)d\xi$$

[We see this at once, having regarded (56).] On returning to the original equation, we indeed find a homogeneous equation.

We can apply this procedure of reduction when the group is known in its finite form.

169 - Equations of higher order. Groups with several parameters

In the same manner, we may seek to determine the condition in order that a system of first order equations admits the transformations of a one-parameter group. For a system of two equations

$$\frac{dx}{P} = \frac{dy}{Q} = \frac{dz}{R} \quad ,$$

to which there corresponds, as we have seen (in no. 156), the partial differential equation

---

*The integration of the linear partial differential equation thus obtained is achieved by the technique which will be outlined in Chapter XV.

$$\text{Ⓗ}(\psi) \equiv P\frac{\partial\psi}{\partial x} + Q\frac{\partial\psi}{\partial y} + R\frac{\partial\psi}{\partial z} = 0 \quad , \tag{57}$$

one proves that the system will be invariant under the transformations of the canonical group (59) when, for ψ an integral of equation (57),

$$\Pi(\psi) \equiv X\frac{\partial\psi}{\partial x} + Y\frac{\partial\psi}{\partial y} + Z\frac{\partial\psi}{\partial z}$$

is equally a solution. This is again a necessary and sufficient condition. For this to be so, it is necessary and sufficient that

$$\Pi(\text{Ⓗ}(\psi)) - \text{Ⓗ}(\Pi(\psi)) \equiv v\,\text{Ⓗ}(\psi) \quad ,$$

where v is a function of x,y,z. For a second order equation reducing to a system of two first order equations, it would then be possible to see if it is invariant under the transformations of an extended group. This will not be the case in general. But, when there exists a group admitted by the equation, we can benefit by simplifying it. This is independent of all the theory in no. 72.

GROUPS WITH r PARAMETERS

We know of such groups, in geometrical terms: the group of displacements in the plane (a three-parameter group), the group of spatial displacements (a six-parameter group). In their realisations, these groups admit a number of infinitesimal transformations equal to the number of parameters. This is a general fact. The general study of the r-parameter groups, their infinitesimal transformation and their structure was the prime objective of the works of Lie, and later, E. Cartan. Concerning these matters, we refer to Lie's *Theorie der Transformationgruppen* and the relevant works of E. Cartan.

Chapter X

FIRST AND SECOND ORDER EQUATIONS IN THE REALS

The study of real integral curves of first order differential equations of the form $y' = R(x,y)$, where $R(x,y)$ is a rational fraction, was undertaken by Poincaré in his earlier works (1880 to 1885). This necessitated a global, rather than a local study within the neighbourhood of a point. Poincaré extended his techniques and results to differential systems, non-solvable equations and to equations of higher order. His findings were taken up, in particular by Bendixon around 1900, and more recently by Birkhoff. An important issue, left in suspense by Poincaré, was settled by Denjoy in 1932. In what follows, we shall confine ourselves to an account of the local nature of the form of the solutions of a first order equation, solvable in the neighbourhood of a singularity of the most simple type, and then, to bring forth the most simple results obtained by Poincaré in the case of equations of the form indicated above.

As a consequence, we shall consider a question of quite a different nature. The problems arising in the theory of thermal conductivity led Sturm and Liouville to study the real zeros of the real solutions of second order linear equations. In 1836 Sturm announced the very simple properties of those zeros which play an important role in the series expansions which arise when solving second order partical differential equations. We shall make some remarks on this type of problem.

I. THE STUDY OF THE SOLUTIONS OF FIRST ORDER EQUATIONS IN THE NEIGHBOURHOOD OF SINGULAR POINTS OF THE EQUATION

170 - A simple example

Let us consider the homogeneous differential equation

$$\frac{dy}{dx} = \frac{Cx + Dy}{Ax + By} \quad , \qquad AD - BC \neq 0 \tag{1}$$

where A, B, C, D are real constants, and investigate the nature of the integral curves in the neighbourhood of the origin. We can write equation (1) in the parametric form

$$\frac{dx}{Ax + By} = \frac{dy}{Cx + Dy} = dt \quad , \tag{2}$$

and apply the known results (no. 85). The characteristic equation is

$$(A - r)(D - r) - BC = 0 \quad , \tag{3}$$

and we obtain the following cases:

FIRST CASE

Equation (3) has its roots r, ρ *real and of the same sign.* If $r \neq \rho$, then the equations of the integral curves are

$$x = \lambda\alpha e^{rt} + \mu\beta e^{\rho t} , \quad y = \lambda\gamma e^{t} + \mu\delta e^{\rho t} , \quad \alpha\delta - \beta\delta \neq 0 \tag{4}$$

where α, β, γ, δ are fixed constants and λ and μ are arbitrary constants. If $r = \rho$, we have, for $B \neq 0$,

$$x = (2\lambda Bt + 2B\mu)e^{rt} , \quad y = (\lambda(D-A)t + 2\lambda + \mu(D-A)e^{rt} , \qquad r = \frac{A+D}{2} \tag{5}$$

and if $B = 0$, hence $D = A$, $r = A$,

$$x = \lambda e^{At} , \quad y = (C\lambda t + \mu)e^{At} \quad . \tag{6}$$

If we take r and ρ negative, we see that in the three cases (4) (5) and (6), the integral curves tend towards the origin when t tends towards $+\infty$. The tangent to the origin is the same for all integral curves save one alone, except in the special case $B = C = 0$, $A = D$, where the integral curves are the lines passing through the origin. When r and ρ are positive, we can make t tend towards $-\infty$. We say that the origin is a NODE.

Remark. The solution does not change when t is changed to $t + v$, i.e., by multiplying μ by a similar positive number. We can then take $\mu = 1$ or $\mu = -1$. When λ and μ are changed into $-\lambda$ and $-\mu$ respectively, we obtain two curves symmetric with respect to the origin. *We regard them as extensions of each other, or even as constituting a single integral curve.*

SECOND CASE

If the roots r and ρ of equation (3) *are real and of opposite sign,* then only the lines obtained in equations (4) for $\lambda = 0$, $\mu = \pm 1$ and for $\lambda = \pm 1$ and $\mu = 0$, which are distinct since $\alpha\delta - \beta\delta \neq 0$, pass through the origin; they are the arrangement given in figure 42 and are asymptotic to the two integral curves reduced to lines.

In this case we say that the point 0 is a saddle point.

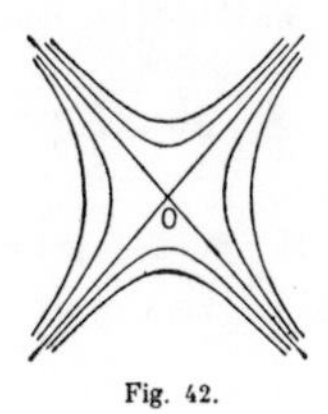

Fig. 42.

THIRD CASE

The roots of equation (3) are *imaginary*, hence are imaginary conjugates. These roots are of the form $a \pm bi$, $(i = \sqrt{-1})$. If a is zero, then the integral curves (4) are of the form

$$x = \lambda'(\alpha' \cos bt + \beta \sin bt) \qquad y = \lambda'(\gamma' \cos bt + \delta' \sin bt) \quad .$$

These are homothetic ellipses, and hence are *closed curves about the origin;* we say that the origin is a CENTRE.

If, for example, a is positive (otherwise t is changed to $-t$), we see that the points M of the curve are obtained, starting from the points P of the preceding ellipse, by taking $\overrightarrow{OM} = e^{at}\,\overrightarrow{OP}$; when t passes from $+\infty$ and $-\infty$, the factor e^{at} passes from $+\infty$ to 0. The origin is an asymptotic point. The integral curve encircles the origin in a spiral-like fashion, and likewise at the point at infinity. (In the case where the ellipse is a circle, we have a logarithmic spiral.)

We say that the origin is a FOCUS.

Remark. A change of variables with constant real coefficients

$$X = mx + ny \,, \qquad Y = px + qy \,, \qquad mq - np \neq 0 \quad , \tag{7}$$

reduces equation (1) to a canonical form if m, n, p, q are suitably chosen. It is such a transformation, analogous to that made in no. 93, that we are going to use in the general case.

171 - A study of the general case when the roots of the equation in r are real and of the same sign

We shall consider an equation of the general form

$$\frac{dy}{dx} = \frac{Cx + Dy + g(x,y)}{Ax + By + f(x,y)} \quad ; \tag{8}$$

A, B, C, D are real constants, $AD - BC \neq 0$, and f and g real functions of x,y, continuous and satisfying a Lipschitz condition in x and y in a circle with centre the origin. We finally assume that

$$f(x,y) = o(|x| + |y|) \ , \quad g(x,y) = o(|x| + |y|) \ ,$$

[this signifies that the ratio of f to $|x| + |y|$ tends to zero when $|x| + |y|$ tends to zero (no. 52)]. We can write equation (8) in the form

$$\frac{dx}{dt} = Ax + By + f(x,y), \quad \frac{dy}{dt} = (x + Dy + g(x,y) \ . \tag{9}$$

Through every point of a circle with centre the origin and with sufficiently small radius, and distinct from the origin 0, there passes one and only one integral curve since the conditions of the Cauchy-Lipschitz theorem are satisfied for the system (9) and yield a solution which does not reduce to $x = y = 0$.

The change of variable (7) replaces the system (9) by an analogous sytem with the reservation that m, n, p, q are real:

$$\frac{dX}{dt} = A'X + B'Y + F(X,Y) \ , \quad \frac{dY}{dt} = C'X + D'Y + G(X,Y) \ .$$

We have

$$A'(mx + ny) + B'(px + qy) \equiv m(Ax + By) + n(Cx + Dy)$$

$$C'(mx + ny) + D'(px + qy) \equiv p(Ax + By) + q(Cx + Dy) \ .$$

We can have $B' = 0$, $C' = 0$ if A' and D' are distinct roots of equation (3)

$$(A - r)(D - r) - BC = 0 \ .$$

Hence, if equation (3) has real distinct roots, r and ρ say, then the system (9) takes the canonical form

$$\frac{d}{dt} = rX + F(X,Y) \ , \quad \frac{dY}{dt} = \rho Y + G(X,Y) \ , \tag{10}$$

with the help of a *real* substitution (7). If equation (3) has a double root, r, we can only annihilate B', and we obtain

$$\frac{dX}{dt} = rX + F(X,Y) \ , \quad \frac{dY}{dt} = r'X + rY + G(X,Y) \ . \tag{11}$$

Let us now return to the first case and assume that r and ρ are of the same sign. We can, by changing (if needs be) t to $-t$, assume that r and ρ are negative. If we multiply the two equations (10) by X and Y

respectively, and add them, we see that if we set $R^2 = X^2 + Y^2$, and if $\rho < r$, we have

$$\frac{1}{2}\frac{dR^2}{dt} = rX^2 + \rho Y^2 + o(R^2) \quad ,$$

hence

$$2\rho R^2 + o(R^2) < \frac{dR^2}{dt} < 2rR^2 + o(R^2) \quad .$$

Consequently, provided r is taken to be sufficiently small,

$$3\rho R^2 < \frac{dR^2}{dt} < rR^2 \quad ,$$

or

$$3\rho dt < \frac{dR^2}{R^2} < r\, dt \ , \qquad (dt > 0) \quad .$$

We will then have, by integrating from t_o to $t > t_o$,

$$R_o^2 e^{3\rho(t-t_o)} < R^2 < R_o^2\, e^{r(t-t_o)} \quad .$$

This inequality shows that, when t increases, the point of the integral curve approaches the origin; hence, the integral curve is extendable and that the origin is only attained for t infinite.

We obtain the same result if $\rho = r < 0$. For, if the following combination of equations (11), is formed

$$\frac{1}{2}\frac{d}{dt}(X^2 + kY^2) = rX^2 + kr'XY + rkY^2 + o(X^2 + Y^2) \quad , \tag{12}$$

where k is a positive constant and, if $kr'^2 < 4r^2$, then the form constituted by the first three terms of the second member is a definite form, then the ratios of the forms

$$X^2 + kY^2,\ -(rX^2 + kr'XY + rkY^2),\ \ X^2 + Y^2$$

taken pairwise, are fixed between two positive numbers. It follows that for R sufficiently small, we have

$$(X^2 + kY^2)e^{s'(t-t_o)} < X^2 + kY^2 < (X^2 + kY^2)e^{s(t-t_o)} \quad ;$$

where s and s' are two negative constants. We have a similar inequality for R^2. Thus:

If the roots of equation (3) are real and of the same sign, then the arcs of the integral curves which penetrate a circle of centre 0 *and small enough radius, end at the origin.*

On the other hand, let us multiply the second equation (10) by X, the first by $-Y$, add, and then divide by $X^2 + Y^2$. We shall have

$$\frac{XdY - YdX}{(X^2+Y^2)} = (\rho - r)\frac{XY}{X^2+Y^2} + o(1) \quad ,$$

where $o(1)$ tends to zero when t increases indefinitely, or, by setting $Y = X + g\theta$,

$$\frac{d\theta}{dt} = (\rho - r)\sin\theta\cos\theta + o(1) \quad . \tag{13}$$

If we consider small angles of opening ζ having the origin as vertex and OX and OY as bisectors, then equation (13) shows that, outside of these angles

$$\frac{d2\theta}{\sin 2\theta} = (\rho - r + \varepsilon')\, dt \ ,$$

where $|\varepsilon'|$ is less than $\frac{1}{2}|\rho - r|$ provided that t is sufficiently large. On integrating, we have

$$tg\theta = tg(\theta_1)e^{(\rho-r+\varepsilon)(t-t_1)} \ , \qquad |\varepsilon| < \frac{1}{2}|\rho - r| \quad .$$

If t increases indefinitely, then the second member tends to zero or infinity following the sign of $\rho - r$. It follows that, if the point M of the integral curve is outside one of these angles of opening ζ, in question, then it enters into one of these angles, starting from a value of t and remains there. Hence: *If* $\rho \neq r$, *the integral curves are all tangent to the origin, at* OX *or at* OY *say.*

If $\rho = r$, the same calculation starting from equations (11) yields

$$\frac{d\theta}{dt} = r'\cos^2\theta + o(1) \quad . \tag{14}$$

If r *is non-zero, we can apply the above argument.* But if $r' = 0$, it is necessary to establish an extra hypothesis to ensure that θ has a limit. Let us suppose that

$$|f(x,y)| < \phi(|x| + |y|) \ , \qquad |g(x,y)| < \phi(|x| + |y|) \quad ,$$

where $\phi(u)$ tends towards zero with u. In the expression (14), we will have

$$|o(1)| < \frac{\phi(KR)}{R}K'$$

where K and K' are constants. It will suffice to have the integral

$$\int_o^{R_o} \frac{\phi(KR)}{R} dt$$

convergent. With respect to the exponential form of R, *we see that this condition will be satisfied provided that*

$$\phi(|x| + |y|) = (|x| + |y|)^{1-\gamma} \qquad (15)$$

where γ *is an arbitrarily small fixed positive number.*
Summarising:
If the roots of equation (3) are real and of the same sign, the integral curves end at the origin and are tangent there, to two lines when these roots are distinct; if the roots are confounded, then the integral curves can be tangent to the variable lines. Moreover, it is necessary in this case, to make the extra hypothesis which has been mentioned. The origin is a node.

Remark. When $\rho = r$, $r' = 0$, we can easily construct examples of the case where θ increases indefinitely, hence where the origin is a focus; likewise, when the Lipschitz condition is satisfied at the origin. This is the case for the system

$$\frac{dX}{dt} = X - Y \frac{1}{\log(X^2+Y^2)}, \qquad \frac{dY}{dt} = Y + X \frac{1}{\log(X^2+Y^2)} .$$

This system is analytic about the origin, except at 0. But following the condition in (15), if suffices that the functions F(X,Y) and G(X,Y) are algebraic at the origin in order that the exceptional case does not present itself.

172 - The case where the roots are imaginary

Let us now suppose that the roots of the equation in r, are imaginary, hence are imaginary conjugates. The change of variables (7) is then with complex coefficients; p and q are imaginary conjugates of m and n, X and Y are imaginary conjugates. If we set $X = \xi + i\eta$, $Y = \xi - i\eta$, where ξ and η are real, we shall obtain

$$mx + ny = \xi + i\eta \quad , \qquad px + qy = \xi - i\eta \quad ;$$

x and y will then be given as functions of ξ and η, by the formulae with real coefficients. We thus see, by setting

$$r = v + i\tau \ , \qquad \rho = v - i\tau \quad ,$$

that a real transformation brings the system in (9) to the system

$$\frac{d\xi}{dt} = v\xi - \tau\eta + H(\xi,\eta) \ , \qquad \frac{d\eta}{dt} = \tau\xi + \eta v + K(\xi,\eta) \ , \tag{16}$$

where the functions H and K of ξ and η are real and always satisfy the same conditions as f and g.

Let us firstly assume $v \neq 0$. From the system in (16), we deduce

$$\begin{aligned} \xi \frac{d\xi}{dt} + \eta \frac{d\eta}{dt} &= v(\xi^2 + \eta^2) + o(\xi^2 + \eta^2) \ , \\ \frac{d\eta}{dt} - \eta \frac{d\xi}{dt} &= \tau(\xi^2 + \eta^2) + o(\xi^2 + \eta^2) \ , \end{aligned} \tag{17}$$

and the first equation falls within the conditions of no. 171. The integral curves tend towards the point O, and can only attain it for t intinite. By setting $\eta = \xi + g\theta$, the second equation yields

$$\frac{d\theta}{dt} = \tau + \varepsilon, \quad \tau \neq 0 \ , \quad \varepsilon = \frac{o(R^2)}{R^2} \ ,$$

hence,

$$\theta = \theta_o + (\tau + \varepsilon)(t - t_o) \ . \tag{18}$$

The integral curves are spirals. The origin is a focus.

THE CASE WHERE $v = 0$

When we take $v = 0$, then all that has just been said on the subject of θ, remains true. *The integral curves encircle the origin indefinitely,* where the latter can only be a focus, a centre or a singularity of another type. The first equation (17) here reduces to

$$\frac{dR^2}{dt} = 2o(R^2) \ , \qquad R^2 = \xi^2 + \eta^2 \ . \tag{19}$$

R could tend towards zero, or could indeed have a finite limit or none at all, but it is non-zero for t finite. In effect, the equation in (19) shows that for as long as R is sufficiently small, we have

$$\frac{dR^2}{R^2 dt} > -\sigma \ , \qquad \sigma > 0 \ ,$$

hence, if $dt > 0$,

$$\frac{dR^2}{R^2} > -\sigma \, dt \ , \quad R^2 > R_o^2 e^{-\sigma(t-t_o)} \ , \quad t > t_o \ ,$$

and if $dt < 0$, we likewise have

$$\frac{dR^2}{R^2} > \sigma\, dt\ , \qquad R_o^2 < R^2 e^{\sigma(t-t_o)} \quad , \qquad t < t_o \quad .$$

When t increases or decreases indefinitely from t_o, equation (18) shows that $|\theta|$ tends towards infinity, as long as R is sufficiently small in order that the equation applies. Moreover, $d\theta/dt$ has a constant sign, hence θ always varies in the same sense when t increases or decreases: θ increases or decreases from 2π when t varies approximately from $2\pi/\tau$ whilst R^2 varies by an infinitesimally small amount. If we start from a point (R_o, θ_o) inside a circle of centre 0 and of sufficiently small radius, we will obtain, by varying t from $2\pi/\tau + o(1)$, an arc M_oM_1 of the solution, inside this circle, on which θ will vary from 2π. *The point* M_1 *could be confounded with* M_o; *then we would have a closed integral curve about the point* 0. *If* M_o *and* M_0 *are not confounded,* we could have $OM_o > OM_1$ or $OM_o < OM_1$ (fig. 43).

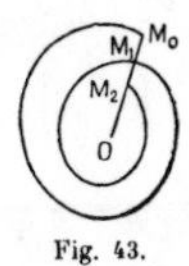

Fig. 43.

In this second case, we interchange the roles of M_o and M_1. We can then assume that, for t increasing, M describes an arc M_oM_1 with $OM_1 < OM$. With t continuing to increase, θ continues to vary in the same sense, and we obtain a second arc M_1M_2 which cannot cut the first and which is not closed since through every point of the circle there only passes one solution, etc. We obtain an integral curve in the form of a spiral which turns indefinitely about 0; it intersects every radius $\eta = \xi + t\theta$ at successive points approaching the origin. On such a radius, the successive points of intersection of the integral curve have a limiting position. If this position is the point 0 for one of the radii, then (19) shows that it is the same for the others. Hence, *either the integral curve admits the origin as an asymptotic point, or it is asymptotic to a continuous simple closed curve about the origin and is intersected at a single point by the radii* $\theta =$ const. In effect, let $R_1 \geq 0$, be the limit of the values of R at the successive points $M_o, M_1, \ldots, M_p, \ldots$, where the integral curve intersects the radius $\theta = \theta_o$ and let t_p be the value of t providing the point M_p. The points of intersection of the integral curve with another radius $\theta = \theta_1$, corresponding to values t'_p such that $t_p < t'_p < t_{p+1}$. Following (18), as

$t_{p+1} - t_p$ tends to $2\pi/|\tau|$, we see that, on account of (19), we have for the point M'_p corresponding to t'_p,

$$\overline{OM'}_p^2 = \overline{OM}_p^2\, e^v\,, \qquad v = 2\int_{t_p}^{t'_p} o(R^2)dt \quad .^*$$

$|v|$ is uniformly bounded, since R is uniformly small along the arc M_oM_1, hence, *a fortiori*, along $M_pM'_p$. The limit of the values R for the points M'_p is therefore positive with R_1. Following this same calculation, for two neighbouring values θ_2 and θ_3, the values of t'_p and t'_{p+1} are near to each other on account of (18), and hence v is near to zero if $\theta_3 - \theta_2$ is sufficiently small. The limiting values of R for θ_2 and θ_3 are close and the limiting curve is indeed continuous.

All the integral curves originating from a point situated in the domain bounded by the arcs M_oM_1, M_1M_2 and by the segments M_oM_1 and M_1M_2 will behave as the integral curve which has just been discussed, since they do not intersect it. They will be asymptotic to the origin if the integral curve in question admits this point as an asymptotic point, or even asymptotes to the limiting curve. *In the last case, the origin is a focus and the discussion ends.*

A STUDY OF THE ASYMPTOTIC CURVE

Let us take a fixed integral curve $M_o, M_1, \ldots, M_p, \ldots$ asymptotic to the limiting curve Γ. Along the arc M_pM_{p+1}, R is a function of θ, $\psi_p(\theta)$ say. When p increases indefinitely, $\psi_p(\theta)$ converges towards the limiting function $\psi(\theta)$ and $R = \psi(\theta)$ is the equation of Γ. The derivative of $\psi_p(\theta)$ with respect to θ is given by the equations (16) and (17). We have

$$\frac{d(\psi_p(\theta))^2}{d\theta} = 2(\xi^2+\eta^2)\,\frac{\xi H(\xi,\eta) + \eta K(\xi,\eta)}{\tau(\xi^2+\eta^2) + \xi K(\xi,\eta) - \eta H(\xi,\eta)} \quad . \tag{20}$$

The second member is a function of ξ,η which is bounded and continuous, hence uniformly continuous, in an annulus contained in the circle with which we are concerned, with an arbitrarily small inner radius. This is a uniformly continuous function of θ in this annulus. Hence the functions of the first member of (20) are bounded in their set and are equicontinuous. Arzela's theorem (no. 160) applies. There certainly exists a sequence of functions which converge uniformly towards a limiting function. If we take for p, the indices of these functions, then we have, uniformly,

---

* $\overline{OM} = R$ is in general a pseudo-distance.

$$\lim \frac{d\psi_p(\theta)^2}{d\theta} = \chi(\theta) \quad ,$$

where $\chi(\theta)$ is a continuous function. Since, on the other hand, $\psi_p(\theta)$ tends towards a limit function $\psi(\theta)$, this function $\psi(\theta)$ is differentiable and we have

$$2\psi(\theta)\psi'(\theta) = \chi(\theta) \quad .$$

The curve Γ *has a tangent which vaires continuously.* Moreover, by passing to the limit in (20), we obtain

$$\frac{d\psi(\theta)^2}{d\theta} = 2\psi(\theta)^2 \frac{\xi H + \eta K}{\tau\psi^2 + \xi K - \eta H} , \quad \xi = \psi \cos \theta,\ \eta = \psi \sin \theta ,$$

which proves that Γ is a solution of the equation obtained by transforming equation (16) (after elimination of t) into pseudo-polar coordinates (the axes are not rectangular, in general). Thus: *The curve* Γ *is an integral curve.**

INSPECTION OF VARIOUS POSSIBLE CASES

We can reiterate the same arguments starting from a new point M_o inside Γ. Two cases are possible.

FIRST CASE

We can find a point M_o such that the integral curve starting from this point ends at the origin. This will be the same (on account of what was said previously) for all integral curves starting from points situated in the domain bounded by the segment and by the arc M_oM_1. *The origin is therefore a focus.*

SECOND CASE

No integral curve inside Γ ends at the origin. There are various possibilities. It could happen that all the integral curves are closed curves encircling the origin for as long as they are sufficiently near. It could also happen that there exists an infinity of similar closed integral curves to Γ and inside Γ, where the other integral curves are asymptotic to these closed curves or to an infinity between themselves. (We see as before that if we follow a nonclosed integral curve, inside Γ, in the sense of R increasing, then it is asymptotic to a closed integral curve.)

In all these cases, *we shall say that the point* O *is a centre.*

---

* We can achieve the same result by other means (see no. 178, The study of spirals).

Example. Let us consider the equation in polar coordinates

$$\frac{dR^2}{d\theta} = \omega(R)R^4 \sin\frac{1}{R^2} \quad ,$$

where $\omega(R)$ has a derivative; likewise for $R = 0$, $R = 0$. The expression $R^4 \sin 1/R^2$ is defined and is differentiable for $R \neq 0$; if we take it equal to zero for $R = 0$, it has a bounded derivative, and hence satisfies a Lipschitz condition for $0 \leq R \leq 1$. In general, we have

$$\int_{R_o}^{R_1} \frac{dR^2}{R^4 \sin 1/R^2} = \omega(\theta_1 + \text{const.}),$$

$$\log tg\frac{1}{2R^2} - \log tg\frac{1}{2R_1^2} = \omega(\theta_1 + \text{const.}) \quad ,$$

where ω is a mean value of $\omega(R)$ between R_o and R_1. It follows that the integral curves are the circles

$$R = \frac{1}{\sqrt{k\pi}} \quad , \qquad k = 1,2,3,\ldots$$

and the asymptotic curves to these circles if the function $\omega(R)$ is, for example, constant. But we may also assume that $\omega(R)$ is equal to zero in an infinity of intervals $\sqrt{k\pi}$, $\sqrt{(k+1)\pi}$ and constant in another sequence of analogous intervals. We shall have intervals of values of R for which the integral curves will be circles and others in which we shall have spirals asymptotic to two circles.

Remark I. In this second case, those given are no longer analytic.

Remark II. By taking, for example:

$$\frac{dx}{dt} = y + x(x^2 + y^2) \sin\frac{1}{x^2 + y^2} \, ,$$

$$\frac{dy}{dt} = -x + y(x^2 + y^2)^2 \sin\frac{1}{x^2 + y^2} \, ,$$

where the sine terms are zero for $x = y = 0$, and the Lipschitz condition in x and y is satisfied for $x^2 + y^2 < 1$ since the second members have continuous first partial derivatives.

173 - The case where the roots of the equation in r are real and of opposite signs

In this case, the real substitution allows us to return to the equations in (10), which we shall write by displaying the signs:

$$\frac{dX}{dt} = rX + F(X,Y) \; , \qquad \frac{dY}{dt} = -\rho Y + G(X,Y) \quad . \tag{21}$$

We shall assume, as we may, that r and ρ are positive. F and G are continuous and satisfy a Lipschitz condition. By always setting $X^2 + Y^2 = R^2$, $Y = X \, tg \, \theta$, the system can always be written as

$$\begin{aligned} \frac{1}{2}\frac{dR^2}{dt} &= (r\cos^2\theta - \rho\sin^2\theta)R^2 + o(R^2) \\ \frac{d\theta}{dt} &= -(\rho + r)\sin\theta\cos\theta + \frac{o(R^2)}{R^2} \quad . \end{aligned} \tag{22}$$

Here again, if we take matters outside of angles of opening ζ having the axes as bisectors, for ζ arbitrarily small but fixed, then we have for R sufficiently small

$$\frac{d\theta}{dt} = -(\rho + r + \varepsilon)\sin\theta\cos\theta, \qquad |\varepsilon| < \frac{\rho + r}{2} \quad ;$$

where θ always varies in the same sense, and as in no. 171, we have

$$tg\theta = tg(\theta_1)e^{-(\rho+r+\varepsilon)(t-t_1)} \quad .$$

An integral curve which passes through a point M, *outside of two angles in question, will enter into one of these angles and will remain there, according to a value of* t, *or will even leave the circle with centre* 0, *in which it is situated.* We have, in effect, following the first equation in (22),

$$R^2 = R_1^2 e^{v} \quad , \qquad |v| < 2\left(\rho + r + \frac{o(R^2)}{R^2}\right)|t - t_1| \quad ;$$

the integral curve cannot end at the origin for a finite value of t.

Thus *the integral curves which end at the origin, if there are any, admit a tangent which can only be* OX *or* OY.

To pursue matters, *we shall now establish the hypothesis that the functions* $f(x,y)$ *and* $g(x,y)$ *are analytic in the circle in question,* hence expandable in a power series in x and y, converging in a circle of centre 0. This will be the same for F(X,Y) and G(X,Y). As these functions are zero at the origin and equal to $(|X| + |Y|)$, their expansions commence from second degree terms.

To study the system in (21), it remains to consider the integral curves in the angles $2|\theta| \le \zeta$, $2|\pi/2 - \theta| \le \zeta, \ldots$. We shall restrict our attention to the first of these angles; we would obtain a similar result for the other cases. Let us set $Y = Xu$. We obtain

$$\frac{dX}{dt} = rX + X^2F_1(X,u) \quad ,$$

$$u\frac{dX}{dt} + X\frac{du}{dt} = -\rho Xu + X^2G_1(X,u) \quad ,$$

where F and G are analytic. We then derive the differential equation

$$X\frac{du}{dX} = \frac{-(\rho+r)u + XH(X,u)}{r + XF_1(X,u)}$$

$$= -\frac{\rho+r}{r}u + XK(X,u) \quad , \tag{23}$$

where $K(X,u)$ is an analytic function. This equation is of the type studied by Briot and Bouquet (no. 130) and, for the coefficient of u negative, it admits a holomorphic solution at the origin.

$$u = cX + \ldots$$

which is clearly real and defines two integral curves passing through the origin, one corresponding to $X > 0$ and the other to $X < 0$. Each of them is an extension of the other. Let us denote this solution by u_1, assume $X > 0$ and set $u = u_1 + U$. Equation (23) becomes (see no. 130)

$$X\frac{dU}{dX} = -\frac{\rho+r}{r}UK_1(X,U)$$

where $K_1(X,U)$ is analytic and equal to 1 at the origin. It follows that for X and $|U|$ sufficiently small, we have

$$\frac{dU}{U} = -\sigma\frac{dX}{X} \quad ,$$

where σ is taken to be between two positive numbers. Since U is non-zero, we shall have on integrating

$$|U| > |U_o|\left(\frac{X_o}{X}\right)^{\beta}, \qquad \beta > 0 \ , \quad X < X_o$$

$|U|$ does not remain finite when X tends towards zero. In the angular domain corresponding to U bounded, there does not exist any other integral curve ending at the origin other than that corresponding to the solution u_1 of equation (23). Consequently: *By means of the extra hypothesis that* $f(x,y)$ *and* $g(x,y)$ *are analytic at the origin, there only exist four integral curves ending at the origin. They are respectively tangent, there to two distinct lines, and extend pairwise. The point* 0 *is a col.*

174 - The case of rational equations. Poincaré's theorems

The results obtained in the qualitative study of solutions about a singularity of the type in question, especially apply to the case where $f(x,y)$ and $g(x,y)$ are polynomials. Following the hypotheses made, these polynomials have their terms of lowest degree, those of the second degree at least. *In the case where the roots of equation (3) are real and of the same sign, the exceptional case does not present itself; the origin is a node. In the case where the roots are real and of the opposite sign, the original is a col. (no. 173). If the roots are imaginary, without being purely imaginary, then the origin is a focus.*

The dubious case is that where the roots of equation (3) are purely imaginary. We can have (no. 172) either a focus or a centre. Now, when we have a centre, the circumstances, as we have seen, can be quite diverse in the general case where the functions $f(x,y)$ and $g(x,y)$ are only continuous and satisfy the Lipschitz conditions. Poincaré has shown that, in the case where $f(x,y)$ and $g(x,y)$ are polynomials, the origin is a focus, in general, and has shown how one goes about checking that this is not the case. We shall confine ourselves to only a brief description of his method.

PRELIMINARY REMARK

Poincaré called an integral curve *stable* if, for M_o one of its points and C_o a circle of arbitrarily small radius having this point as centre, after leaving this circle, it happens to cross it again. As a consequence of this definition and when the origin is a centre, the only case where the integral curves are stable is when all these curves are closed curves. Poincaré showed that either the origin is a focus, or the integral curves are stable, i.e., *that, either the origin is an asymptotic point, or all the integral curves are closed (in the neighbourhood of the origin).* In this last case, we shall call the origin *a true centre.*

Poincaré's method. We consider equations (16) in the case $v = 0$, by assuming that H and K are polynomials in ξ and η contain neither independent terms nor terms of the first degree. By changing t to $-t/\tau$, we can assume $t = 1$. By putting the new variables x,y in place of ξ,η, we obtain the system

$$\frac{dx}{dt} = y + X_2 + \dots + X_n \quad ;$$

$$\frac{dy}{dt} = - x + Y_2 + \dots + Y_n \quad ,$$

where the X_j and Y_j are homogeneous polynomials in x and y of degree equal to their index. To the system in (24) we can associate the partial differential equation

$$\frac{\partial\Phi}{\partial x}\frac{dx}{dt}+\frac{\partial\Phi}{\partial y}\frac{dy}{dt}=0 \quad , \tag{25}$$

where $\frac{dx}{dt}$ and $\frac{dy}{dt}$ must be replaced by the second members of the equations in (24). The integrals of equation (25) provide the first integrals of the system. Let us seek a solution of (25) under the form

$$\Phi = x^2 + y^2 + \Phi_3 + \Phi_4 + \ldots$$

where the Φ_j are homogeneous polynomials of degree equal to their index. The condition

$$(y + X_2 + \ldots)\left(2x + \frac{\partial\Phi_3}{\partial x} + \ldots\right) + (-x + Y_2 + \ldots)\left(2y + \frac{\partial\Phi_3}{\partial y} + \ldots\right) = 0$$

shows that once $\Phi_3,\ldots,\Phi_{j-1}$ are determined, we will have to determine Φ_j, the equation

$$y\frac{\partial\Phi_j}{\partial x} - x\frac{\partial\Phi_j}{\partial y} = \Psi_j \quad , \tag{26}$$

where Ψ_j is a polynomial depending on the X_j and Y_j and on the Φ_k already calculated. The determination of Φ_j can be obtained by setting $x = R\cos\theta$, $y = R\sin\theta$. The first member of (26) is equal to

$$-\frac{\partial\Phi_j}{\partial\theta} \quad .$$

Hence if we set

$$\Phi_j = R^j \sum(a_n \cos n\theta + b_n \sin n\theta)$$

$$\Psi_j = R^j \sum(c_n \cos n\theta + d_n \sin \theta) \quad ,$$

where the indices n have the parity of j, we see that the a_n and b_n are easily calculated provided $c_o = 0$, which is necessarily the case if j is odd.

Thus, Φ_3 can be calculated at once, but Φ_4 can only be calculable if the term c_o of Φ_4 is zero. If this condition is realised, then there will remain an arbitrary coefficient in Φ_4. We can calculate Φ_5, but the calculation of Φ_6 will only be possible if the term c_o of Ψ_6 is zero. Then we can continue; and so on. But in the general case, we will stop at the calculation of Φ_4. Let us assume generally that, for any choice of coefficients which remain arbitrary in the above polynomials, we terminate the calculation at Φ_{2n}. When we replace Ψ_{2n} by $\Psi_{2n} - c_o R^{2n}$, we can calculate a polynomial, that we shall again call Φ_{2n}, by the equation

$$\frac{\partial\Phi_{2n}}{\partial\theta} = \Psi_{2n} - c_o R^{2n} \quad .$$

The polynomial

$$\Omega = R^2 + \Phi_3 + \ldots + \Phi_{2n} \tag{27}$$

thus determined is such that, for x and y satisfying the system in (24), the expression

$$\frac{d\Omega}{dt} = \frac{\partial\Omega}{\partial x}\frac{dx}{dt} + \frac{\partial\Omega}{\partial y}\frac{dy}{dt}$$

is a polynomial in x and y whose term of lowest degree is reduced to $-x_o R^{2n}$. Let us assume for example that c_o is positive. All that we have to say will be true for c_o negative, by changing t to $-t$. Let us assume that R is so small that

$$\Omega = R^2 + o(R^2) \ , \qquad \frac{d\Omega}{dt} = -c_o R^{2n} + o(R^{2n}) < 0 \quad .$$

When t increases, Ω decreases and following the equations stated, we have

$$\frac{d\Omega}{dt} = -c_o \Omega^n (1 + \varepsilon'), \qquad |\varepsilon'| < \frac{1}{2} \quad ,$$

or

$$\frac{d\Omega}{\Omega^n} = -c_o(1 + \varepsilon')dt \ , \qquad n > 1 \quad ,$$

and on integrating from t_o to $t > t_o$,

$$\frac{1}{\Omega^{n-1}} > \frac{1}{\Omega^{n-1}} + \frac{1}{2}c_o(n-1)(t - t_o) \quad .$$

Hence Ω tends towards zero when t increases indefinitely. The integral curve tends towards the origin and since θ increases indefinitely (no. 172) the origin is an asymptotic point. We thus obtain the first theorem due to Poincaré.

Theorem. *If there is no termination in calculating the polynomials* Φ_n, *which is generally the case, then the point* 0 *is a focus.*

We see that the method and the result both hold if $f(x,y)$ and $g(x,y)$ instead of being polyonials, are power series in x and y, commencing from terms of the second degree.

The second theorem of Poincaré. When we can continue indefinitely the calculation, we obtain a series

$$\Phi = x^2 + y^2 + \Phi_3 + \ldots + \Phi_n + \ldots$$

which formally satisfies the partial differential equation (25). Poincaré showed that this equation converges uniformly in the neighbourhood of the origin. It therefore defines a first integral. The curves $\Phi = \text{const.}$ are integral curves that are actually closed and which encircle the origin. *The origin is a true centre,* since there passes one of these closed integral curves through each neighbouring point of the origin. Hence:

In order that the origin is to be a true centre, i.e., for the motion to be stable (where t denotes time), then it is necessary and sufficient to have no termination in the calculation of the Φ_n.

Remark. If, when changing y to $-y$, without changing x, then the Y_j do not change whilst the X_j change to $-X_j$. The integral curves emanating from a point of OX are (obliquely) symmetric with respect to OX: they will intersect themselves on OX and hence will be closed. The origin will be a ture centre.

175 - Bendixon's method

When we replace the system in (24) by an equation in x and y and make the transformation $x = R\cos\theta$, $y = R\sin\theta$, we obtain, on account of the equations already made in no. 172,

$$\frac{dR}{d\theta} = R^2\Phi_3 + R^3\Phi_4 + \ldots \qquad (28)$$

where the Φ_j are polynomials in $\cos\theta$ and $\sin\theta$, with degrees at least equal to their (respective) indices, and for which the series is convergent for $R < R'_o$ and θ real. Moreover, on account of its origin, this series converges for $R < R_1 < R'_o$ if R_1 is taken to be sufficiently small, when we replace each sine and cosine by its series expansion, in which each term is replaced by its absolute value, with the condition $|\theta| < 3\pi$ for example. The second member of equation (28) is then expandable in a convergent series in θ and R for θ and R complex, $|\theta| < 3\pi$, $|R| < R_1$. Let us consider the solution which for $\theta = \theta_o$ real, belonging to the segment $0, 2\pi$, takes a value R of sufficiently small modulus. Following the theorem of Poincaré and Weierstrass on the initial conditions (Theorem I, no. 46), this solution is expandable as a power series in R_o and $\theta - \theta_o$ for $|R_o|$ and $|\theta - \theta_o|$ less than values independent of θ_o. It follows that, for R_o real, positive and sufficiently small, that the solution $R(\theta)$ of equation (28) can be expanded in the form

$$R(\theta) = R_o\psi_1(\theta) + R_o^2\psi_2(\theta) + \ldots$$

where the functions $\psi_j(\theta)$ are analytic for θ belonging to the segment $(0,2\pi)$ and the series is absolutely and uniformly convergent.

The functions $\psi_1(\theta),\psi_2(\theta),\ldots$ are calculated iteratively by taking this expansion into equation (28) and equating the powers of R_o; $\psi_1(\theta)$ is identically 1. The $\psi_j(\theta)$ are zero for $\theta = 0$ and $j > 1$ and are given by

$$\psi_2'(\theta) = \Phi_3(\theta), \quad \psi_3'(\theta) = 2\psi_2(\theta)\Phi_3(\theta) + \Phi_4(\theta),\ldots$$

When the functions $\psi_j(\theta)$ are all periodic with period 2π, the solution R is periodic as long as R is sufficiently small and the integral curves are closed, and we have a true centre. Let us assume that one of the function $\psi_j(\theta)$ is not of period 2π, whilst $\psi_2(\theta),\ldots,\psi_{j-1}(\theta)$ admits this period. We will have

$$R(2\ ) - R_o = R_o^j[\psi_j(2\pi) - \psi_j(0)] + R_o^{j+1}(\ ) + \ldots$$

Hence $R(2\pi) - R_o$ will equal a non-zero number provided that R_o is sufficiently small. The integral curve will not be closed and it will encircle 0 indefinitely. Furthermore, by making a suitable choice of orientation, we see that the integral curve commencing from R_o intersects $0x$ at the point r_1 given by $r_1 < R_o - dR_o^j$, where d is a fixed positive number. It will cut $0x$ at successive points r_p and we will have

$$r_p < r_{p-1} - dr_{p-1}^j \quad .$$

This sequence of points tends towards 0. Since if they have a limit point with abscissa a, we would have $a < r_p < a + \varepsilon$, for ε sufficiently small, commencing from a value p. Hence

$$r_{p+1} < a + \varepsilon - da^j \quad ,$$

which yields a contradiction if $\varepsilon < da^j$. Thus, if one of the function $\psi_j(\theta)$ does not have period 2π, the point 0 is a focus. By this method, it is also proved that: *When the characteristic equation (3) has purely imaginary roots and when $f(x,y)$ and $g(x,y)$ are polynomials (or more generally, analytic functions in a circle with centre the origin), then the origin can only be a focus or a true centre.*

We also have another criterion for showing that the origin is a centre: *In order for the origin to be a true centre, it is necessary and sufficient that the $\psi_n(\theta)$ are always periodic, of period 2π.*

II. THE GENERAL THEORY OF POINCARÉ

176 - The general problem. Projection onto the sphere. Characteristics

Given a differential equation

$$\frac{dx}{X(x,y)} = \frac{dy}{Y(x,y)} \tag{29}$$

where X and Y are real polynomials of the variables x and y, we propose to study, in the general case, where three polynomials are of the same degree m, the form of the integral curves (real curves), as far as infinity. The singularities of the equation at a finite distance correspond to the values of x and y which annihilate X and Y simultaneously (since if X alone is annihilated at one point, then the equation is regular when we regard x as the unknown function and y as the variable; a unique integral curve passes through this point and its tangent is parallel to Oy). Assume the general case where the common points of the algebraic curves $X = 0$, $Y = 0$ (real points), are simple points on each of the curves and where the curves are not tangent to each other at these points. By taking such a point to be the origin, we will have

$$X = Bx + By + \ldots, \quad Y = Cx + Dy + \ldots, \quad AD - BC \neq 0 ,$$

and we will have a node, a col. a focus or eventually a true centre at this point (this last case is exceptional preceding results).

Through every point at a finite distance, distinct from these singular points, there passes one and only one integral curve, where y and x are regular analytic functions of the other variable, about this point.

PROJECTION ONTO THE SPHERE

To study integral curves at infinity, from the point of view of real geometry, i.e., by distinguishing the various points at infinity, we project the plane onto a sphere, where the centre of projection is the centre of the sphere, taken outside of the plane Oxy. Let Ω be this centre of the sphere Σ, taken, for example, to be on the perpendicular to the plane Oxy through O. At a point M of the plane there corresponds two points on the sphere. If Σ_1 and Σ_2 are the two hemispheres separated by the great circle parallel to the plane Oxy, then one of the two images M_1 of M is on Σ_1 and the other, M_2, is on Σ_2. The points at infinity of the plane are projected onto the great circle separating Σ_1 and Σ_2. This great circle will be called the equator E.

We shall call every simple curve of the sphere which is *closed* and which is formed by analytic arcs (no. 41), *a spherical cycle*. A spherical cycle divides the sphere into two domains, one of which (arbitrary) will be considered as the interior, the other as the exterior. (The stenographic projection onto a plane with respect to a point which on the cycle effectively brings us back to Jordan's theorem.)

CHARACTERISTICS

We shall call the image curves on the sphere, of the integral curves of equation (29), *the characteristics*. Like every spherical image of a plane curve, a characteristic is symmetric with respect to Ω. The points of the characteristic situated on the equator E are the images of the points at infinity, of the integral curve. If a characteristic is not closed, then one of its points splits it into two parts that we call *semi-characteristics*. To study a point at infinity of an integral curve, we set

$$x = \frac{1}{\xi}, \quad y = \frac{\eta}{\xi} \quad , \tag{30}$$

if the spherical image of the point does not lie on the great circle corresponding to $x = 0$, The transformed point will be at a finite distance $\xi = 0$, η finite. Equation (29) is transformed into

$$\frac{d\xi}{X_1(\xi,\eta)} = \eta \frac{d\xi - \xi d\eta}{Y_1(\xi,\eta)} \quad , \qquad X_1(\xi,\eta) = \xi^m X\left(\frac{1}{\xi}, \frac{\eta}{\xi}\right)$$

$$Y_1(\xi,\eta) = \xi^m Y\left(\frac{1}{\xi}, \frac{\eta}{\xi}\right) \quad ,$$

or

$$\frac{d\eta}{\eta X_1 - Y_1} = \frac{d\xi}{\xi X_1} \quad . \tag{31}$$

When the point at infinity is on $x = 0$, we set

$$x = \frac{\xi}{\eta}, \quad y = \frac{1}{\eta} \quad .$$

177 - Singular points of the system of characteristics. Poincaré's relationships between the number of nodes, cols. and foci

By a *system of characteristics* we mean the set of image curves of the integral curves. The singular points of the system are the images of the singular points of the integral curves. If the curves $X = 0$, $Y = 0$ have common points at a finite distance, then on Σ_1 and Σ_2 there corresponds to them, singular points of the system of characteristics that we call nodes, cols. (saddle points), foci or centres, simultaneous with the points of the plane provided them.

If the curves $X = 0$, $Y = 0$ intersect at infinity, at a point which may be taken to be not on $x = 0$, then at such a point the numbers ξ and η

defined by (30), will have the values 0 and η_o which will not be the same for the two curves. The functions X_1 and Y_1 will be zero for these values $0,\eta_o$; this point will be a singular point of equation (31). Under these conditions, the curves

$$\eta X_1(\xi,\eta) - Y_1(\xi,\eta) = 0 , \qquad \xi X_1(\xi,\eta) = 0 , \tag{32}$$

have the point $\xi = 0$, $\eta = \eta_o$, in common, but the terms of lowest degree in the second equation are of the second degree when the point $0,\eta_o$ is taken to be the origin. We have not studied the singularities in this case. *We shall therefore assume that the curves* $X = 0$, $Y = 0$ *have no common points at infinity* (this is the case with the real points). In general, this is the case if X and Y are any polynomials. Let us denote by $X_2(x,y)$ and $Y_2(x,y)$ the terms of degree m in X and Y respectively. On account of the hypothesis that has just been made, the equations $X_2(1,) = 0$ and $Y_2(1,\eta) = 0$ do not have any common real root. Now we have $X_1(0,\eta) = X_2(1,\eta)$, $Y_1(0,\eta) = Y_2(1,\eta)$. The equations $X_1(0,\eta) = 0$ and $Y_1(0,\eta) = 0$ therefore do not have any common (real) root. The equations (32) will be satisfied simultaneously and uniquely if $\xi = 0$ and $X_1(0,\eta) - Y_1(0,\eta) = 0$, $X_1(0,\eta) \neq 0$. Now $X_1(0,\eta)$ and $Y_1(0,\eta)$ are of degree m [since $X_2(1,\eta)$ and $Y_2(1,\eta)$ are generally of degree m at the axes Oxy], hence, if m is even, the equation $\eta X_1(0,\eta) - Y_1(0,\eta) = 0$ has at least one real root (and for this root, $X_1(0,\eta)$ is non-zero). Thus: *If* m *is even, there exists at least one singular point of the system of characteristics situated on the equator.*

Moreover, in the general case, the roots of $\eta X_1(0,\eta) - Y_1(0,\eta) = 0$ will be simple, and the singularity will be of the type considered. *When* m *is odd, the curves* $X = 0$, $Y = 0$, *of the same degree* m *and without common real points at infinity, have at least a common point at a finite distance.*

For the total number of common points is the odd number m^2 and there is at least one common real point. We thus obtain the first theorem due to Poincaré.

Theorem 1. *By means of the above hypotheses, the system of characteristics has an even number of singular points and has at least two. These points are nodes, cols. (saddle points?), foci (or centres).*

The number of these singular points is in fact even due to the symmetry with respect to the centre Ω of the sphere. *The equator is a characteristic.* Equation (31) is effectively satisfied for $\xi = 0$. *The* singular points situated on the equator can *only then be nodes or cols (saddle point)* since, in the case of foci or centres, there is no characteristic tangent to a great circle at such a point (on the sphere, the tangents are replaced by tangential great circles).

THE INDEX OF A CYCLE

Let us consider a spherical cycle Γ situated on the hemisphere Σ_1 and has only one of its regions confounded with the equator. We shall say that this cycle travels in the direct sense when its projection onto the plane Oxy is taken in the direct sense. At each point M_1 of the cycle, X and Y have determined values; when M_1 approaches the equator, the transformation (30) is made and we take

$$\frac{Y}{X} = \frac{Y_1}{X_1} \quad .$$

We shall assume that the spherical cycle in question does not coincide, on any of its parts, with the image cycle of the curve $X = 0$. When the point M_1 is taken to be on the cycle, Y/X varies continuously and has a well-determined value at each point, *if the cycle does not pass through any singular point of the system of characteristics. Let* p *be the number of points where* Y/X *becomes infinite and passes from the* + *sign to the* - *sign, and* q *the number of points where this quotient passes from* $-\infty$ *to* $+\infty$*. We shall call the number*

$$I = \frac{p - q}{2} \quad ,$$

the index of the cycle.

It is clear that when *the interior of the cycle* (the exterior being the part of the sphere which is taken to be Σ_2) is separated into two domains by an analytic curve joining two points of the cycle Γ and satisfying the same conditions as Γ (it contains no portions of $X = 0$ and does not pass through any singular point), then the two domains resulting, will be bounded by two distinct cycles Γ_1 and Γ_2 with indices I_1 and I_2, and we shall have

$$I = I_1 + I_2 \quad .$$

Theorem. *Let* N, F *and* C *be the number of nodes, foci and cols (saddle points?) repsectively, inside a cycle with index* I. *We have*

$$I = N + F - C \quad .$$

To prove this, we decompose the interior of Γ into very small domains bounded by the cycles and prove the proposition for the cycles so obtained; it will be proved for Γ following the above formula. As the singular points of the system of characteristics are simple points common to the curves Γ' and Γ'' defined by $X = 0$ and $Y = 0$ (it applies to singular points which are not on the equator E), we can assume that these points are within the cycles γ not containing any singular points of Γ' and Γ'' and intersecting Γ' and

Γ'' at only two points respectively. We can assume that the remainder of the region inside Γ is partitioned into domains, the boundary of which only intersects Γ' or Γ'', but not both these curves. Under these conditions:

1) A cycle γ which does not intersect Γ', gives $I = 0$.

2) A cycle γ which intersects Γ' but not Γ'' gives $I = 0$. For on γ, Y has a constant sign. Every branch of Γ' whose crossing changes the sign of X, crosses γ an even number of times, and it has as many paths from + to - as it has from - to +.

3) Let γ be a small cycle containing a singular point of the system of characteristics. We can assume that this point is $x = 0$, $y = 0$, and that γ has as its image in the plane Oxy, a small simple curve encircling the origin. We have

$$X = Ax + By + o(|x| + |y|) , \quad Y = Cx + Dy + o(|x| + |y|) ,$$

$$AD - BC \neq 0 .$$

For $X = 0$, we have $Ax + By$ near to zero and consequently Y has the sign of

$$\frac{CB - AD}{B} x .$$

For $x > 0$, y increases when Γ' is crossed; X passes from the sign of -B to the sign of +B, hence Y/X passes from the sign of AD - BC to that of BC - AD. We obtain the same result when Γ' is crossed to the left of Oy. Hence I is equal to 1 if $AD - BC > 0$ and to -1 if $AD - BC < 0$. Now AD - BC is the product of the roots of equation (3) of no. 170. If $AD - BC > 0$, the roots are of the same sign or are imaginary; we obtain a node, a focus or a centre. If AD - BC 0, we obtain a col (saddle point). We therefore have $I = N + F - C$ in this case where γ contains a singular point.

<u>Remark</u>. The index is equal to the variation of

$$\frac{1}{2\pi} \operatorname{arc} tg \frac{Y}{X}$$

when γ is described. (Cf. I, 53).

INDEX OF THE EQUATOR

The index of the equator corresponds to the variation of

$$\frac{Y_1(0,\eta)}{X_1(0,\eta)} = \frac{Y_2(1,\eta)}{X_2(1,\eta)} .$$

Via a rotation of the axes Oxy, we can assume that there are no points at infinity of Γ' on Oy. The preceding expression only becomes infinite for η

finite. Equation (31) provides the cols (saddle points) and nodes situated on the equator; they correspond to the roots of the equation

$$X_3(\eta) = \eta X_1(0,\eta) - Y_1(0,\eta) = 0 \quad .$$

If η_o is one such root, then we have in the neighbourhood of the point $0,\eta_o$,

$$\frac{d\xi}{\xi X_1(0,\eta) + \ldots} = \frac{d\eta}{\alpha\xi + \beta(\eta - \eta_o) + \ldots}$$

we will have a node if $X_1(0,\eta_o)\beta > 0$ and a col (saddle point) if this quantity is negative. Now, we have

$$\beta = \left(\frac{dX_3(\eta)}{d\eta}\right)_o = X_3'(\eta_o)$$

and

$$\frac{Y_1(0,\eta)}{X_1(0,\eta)} = \frac{\eta X_1(0,\eta) - X_3(\eta)}{X_1(0,\eta)} = \eta - \frac{X_3(\eta)}{X_1(0,\eta)} \quad .$$

When η is finite, the infinites of Y_1/X_1 are those of $-X_3/X_1$. For η infinite, Y_1/X_1 is finite whilst $-X_3/X_1$ is infinite and passes from the sign - to +. Matters are then reduced to determining the variation of

$$\frac{1}{2\pi} \text{ arc } tg \left(\frac{-X_3(\eta)}{X_1(0,\eta)}\right) \quad ,$$

equal to that of

$$\frac{1}{2\pi} \text{ arc } tg \left(\frac{X_1(0,\eta)}{X_3(\eta)}\right) \quad ,$$

when η varies from $-\infty$ to $+\infty$. By adding 1 to this variation, we will obtain the index sought after. When $X_3(\eta_o) = 0$, the quotient of $X_1(0,\eta)$ by $X_3(\eta)$ passes from the sign + to the sign -, if $X_1(0,\eta)X_3'(\eta_o) < 0$ and from the sign - to the sign + in the opposite case. It follows that *the index of the equator is*

$$1 + C_E - N_E \quad ,$$

where $2C_E$ *and* $2N_E$ *are the numbers of cols (saddle points) and nodes respectively, situated on the equator.*

Poincaré's relationship. When we consider a cycle of Σ_1 constituted by a circle parallel to the equator, then when it tends towards the equator, the index of this cycle tends towards that of the equator. Consequently, if N, C, F are the numbers of nodes, cols (saddle points) and foci (or centres) respectively on a hemisphere and C_E and N_E the numbers of cols (saddle points) and nodes respectively on a semiequator, then we have

$$N + F - C = 1 + C_E - N_E$$

or

$$(N + N_E) + F = 1 + (C + C_E) \quad .$$

178 - The form of characteristics. Theory of contact. Bounding cycles

When we extend a semi-characteristic originating from an ordinary point A of Σ several possibilities may occur. The semi-characteristic could encircle a focus, which will never be attained. If the characteristic ends at a col (saddle point), we can extend it beyond this col (saddle point) by following, not the branch of the characteristic which has the same tangent as the semi-characteristic in question, but one of the other two. In all cases, when the col (saddle point) is on the equator, we will extend it, on remaining in Σ. If the characteristic ends at a node, we will regard it as stopping at this point. *The foci and nodes are therefore points of termination, whilst a col (saddle point) can never be regarded as such.* The semi-characteristic extended could return to the point A, or meet the second semi-characteristic originating from A; the characteristic forms a cycle which could pass through several cols (saddle points). The closed characteristics passing through the cols are finite in number (if there are any), since there is only a finite number of cols.

Theorem 1. *An extended semi-characteristic* C *originating from an ordinary point* A, *which does not end at a node and which is intersected at only a finite number of points by every cycle, reduces to a cycle.*

The hypothesis implies that C does not end at a focus (otherwise a great circle passing through this point would intersect it at any infinity of points). C is extendable indefinitely. If we tessellate the hemisphere Σ_1 by meridians and parallels, C only intersects these curves at a finite number of points, and hence remains to commence from a certain point in one of its domains Δ thus obtained. Let us subdivide Δ by some other meridians nad parallels. C will remain to commence from one of its points inside one of these new domains; and so on. C will end at a certain point A' bounded by these domains of subdivision; A' is an ordinary point and consequently, can only be A, otherwise C would be extendable beyond A'. It is indeed a cycle.

ARCS SUBTENDING AND ARCS SUPERINTENDING A CHARACTERISTIC ARC

If A and B are two points of a characteristic C and if they are joined by a curve arc Γ formed by analytic arcs and only having the points A and in common with C, we form a spherical cycle which divides Δ into two domains Δ and Δ'. When the arcs of C are extended beyond A, on the one hand, and beyond B on the other, such that they are in the same domain Δ or Δ', then we say that *the arc* Γ *subtends the arc* AB *of the characteristic.* If the characteristic extended beyond A is in Δ whilst that extended beyond B is in Δ', then we shall say that *the arc* Γ *super-intends the arc* AB *of the characteristic.*

THE NUMBER OF CONTACTS OF AN ARC Γ AND OF A CHARACTERISTIC

The number of points of contact of an arc Γ, formed by a finite number of analytic arcs, and an arbitrary characteristic, is finite. Since, if $F(x,y) = 0$ is the equation of the projection onto the plane Oxy of an analytic arc Γ' of Δ, then the points of contact satisfy, besides $F(x,y) = 0$, the equation

$$X \frac{\partial F}{\partial x} + Y \frac{\partial F}{\partial y} = 0 \quad . \tag{33}$$

They are given by the zeros of an analytic function of x (obtained by extracting y as a function of x, from $F(x,y) = 0$ and bringing it into the second equation) and hence are finite in number. The property holds for each of the constituent arcs of Γ and thus holds for Γ, but there might be generalised contact at an angular point of Γ if, in the neighbourhood of this point M, the curve Γ remains on the same side of the characteristic through M. All that has just been said clearly assumes that Γ is not taken to be an arc of a characteristic.

By taking into account the order of multiplicity of the points of contact, we show that *the number of points of contact with the characteristic of a cycle without angular points, not passing through a singular point, is an even number.* The proof is based on the same considerations as above, and the result extends to the general case.

Theorem II. *If an arc* AB *of a characteristic, which does not pass through a singular point, is subtended by a curve-arc* Γ, *formed by a finite number of analytic arcs, then the number of contacts of* Γ *with the system of characteristics is odd.*

If $F(x,y) = 0$ is the equation of Γ, then the characteristic is given by $x = f(t)$, $y = g(t)$ with $x' = X$, $y' = Y$ and t increasing from t_o to t_1 on the arc AB. The function $F(f(t),g(t))$ changes sign when t crosses the values t_o and t_1, if at t_o, it passes from + to -; it will pass from - to + at the point t_1 since Γ subtends the arc of the characteristic. The derivative of F with respect to t, which is the first member of (33), has opposite signs for t_o and t_1. Hence on Γ

$$X \frac{\partial F}{\partial x} + Y \frac{\partial F}{\partial y}$$

passes from the sign + to the sign -, or conversely, when it passes from A to B. The number of contacts is odd.

We have assumed that Γ is a single analytic arc (without singularities). Generally, F = 0 is replaced by several successive equations $F_1 = 0$, $F_2 = 0, \ldots$; we assume that $F_1, F_2, \ldots,$ have the same sign in the neighbourhood of Γ in the domain bounded by Γ and AB.

The result extends to the case where the arc AB *passes through a col (saddle point).* Since, on appealing to what was said in the local study of the integral curves in the neighbourhood of such a point (no. 173), we see that we can replace AB by an arc A'B' of characteristic as near as we wish to AB, and no longer passing through the col (saddle point). This is due to the fact that, at a col (saddle point), the characteristic is extended with an angular at the col (saddle point) (fig. 44).

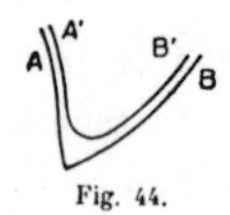

Fig. 44.

Similarly, by taking account of this last result, we have:

<u>Theorem III</u>. *If an arc* AB *of a characteristic is super-intended by a curve arc* Γ *constituted by a finite number of analytic arcs, then the number of contacts of* Γ *with the system of characteristics, is even.*

ARCS WITHOUT CONTACT

If A is an ordinary point on Σ_1, a small circle of centre A is crossed by a sheaf of characteristics given by a first integral equated to a constant. These characteristics admit orthogonal trajectories. The arcs of these trajectories are clearly curves without contact. Generally, a curve intersecting the characteristic originating from A at a non-zero angle, will define an arc without contact, in the neighbourhood of A.

THE FORM OF CHARACTERISTICS NOT ENDING AT A NODE

Let us consider an arc AB without contact (fig. 45). Let us assume that a semi-characteristic C originating from a point M_o of AB intersects AB again at an initial point M_1. The arc M_oM_1 super-intends C since it is without contact, otherwise, if M_oM_1 were to subtend C, there would at least be one contact (Theorem II).

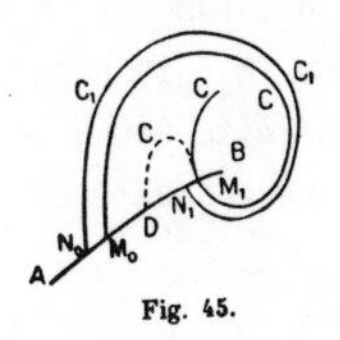

Fig. 45.

If we consider the semi-characteristic beyond M_1, then it no longer intersects the arc M_oM_1. For, if it were to intersect M_oM_1 at D, then the arc M_1CD would be subtended by M_1D.

Let C_1 be a characteristic near to C starting from the point N_o of AB, outside of M_oM_1. It intersects AB at N_1 between M_o and M_1. Since N_1 must be near to M_1 and the arc $N_oC_1N_1$ crosses neither M_oM_1 nor C, hence it does not enter into one of the domains bounded by M_oCM_1 and by the arc M_oM_1 of AB. It follows that the semi-characteristic C remains within a domain bounded by the cycle formed by $N_oC_1N_1$ and by the arc $N_1M_oN_o$ of AB. Under these conditions, we say that C is a spiral.

Definition. *By a spiral we mean every semi-characteristic not ending at a node and which remains within a domain bounded by a cycle passing through its origin.*

The use of the word "spiral" is indeed justified.

Theorem IV. *Every semi-characteristic which does not end at a node and which does not reduce to a cycle is a spiral.*

Following Theorem I, there exists a cycle which intersects this semi-characteristic C at an infinity of points. This cycle Γ has only a finite number of contacts with the characteristics. We can divide it into arcs only having contacts with their extremities. One of these arcs A'B' will intersect C at an infinity of points. One arc M_oCM_1 of C will intersect A'B' at points M_oM_1 distinct from A'B'. There will then exist an arc AB of Γ without contact which will intersect C at M_o and M_1. Following what was said before, C will be a spiral.

THE STUDY OF SPIRALS

Let C be a spiral which intersects the arc A'B' which has just been introduced, at an infinity of points. We have a first point M_o which we will take to be the origin on C, then M_1 (fig. 46) and then M_2. As we have seen above, M_2 does not lie between M_1 and M_o; it is in a domain Δ bounded by the arc M_oM_1 of A'B', and by M_oCM_1. This is the domain where C remains beyond M_1, hence M_1 is between M_o and M_2. It follows that C intersects A'B' in an infinite sequence of points $M_oM_1M_2\ldots$ which are successively encountered when one passes from B' to A' for example (fig. 46).

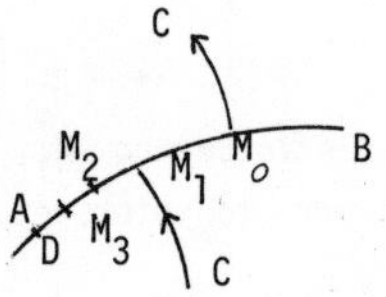

Fig. 46

These points have a limit point D. If D is a singular point of the system of characteristics, then as a result of the local study, it is a focus. The spiral C admits the point D as an asymptotic point and in this case its title is justified.

Let us assume that D is an ordinary point. Through D there passes a characteristic γ. Let us denote by C_p ($p = 1,2,\ldots$), the semi-characteristic originating from M_p and direct it towards M_{p+1}. The C_p correspond to solutions of equation (29) taken with the initial conditions as near as is wished to those of γ. It follows that if we extend γ and the C_p, then the points of γ will not cease to be the limit of points of C_p as long as singular points of the differential equation are not encountered along γ. But following the remark already applied in the proofs of Theorems II and III, the property holds if γ passes through a col (saddle point), provided that γ is suitably extended. Can γ end at a focus or a node? On account of the property of these points, the neighbouring integral curves equally end there, which is contrary to hypothesis. It follows that

<u>Theorem V</u>. *A spiral which does not end at a focus is asymptotic to a characteristic which reduces to a cycle.*

In effect, γ neither ends at a foci nor at a node. This could only be a cycle or a spiral (Theorem IV). It is impossible for it to be a spiral. For an analytic arc without contact would intersect it at two points μ and μ' which will both be limiting points of intersection of the C_p, hence of C, which is impossible since these points have only one limit point.

All of the semi-characteristics originating from the points of the arc A'B' taken between M and M_1, extended in Δ, will equally be asymptotic to γ.

We see that the situation in no. 172 is re-encountered, *but no longer from a local point of view.*

LIMITING CYCLES

The closed characteristics to which the spirals are asymptotic are called *limiting cycles.*

Poincaré showed that inside and outside a limiting cycle there is at least a focus or a node and that, if none of the limiting cycles do not pass through a col (saddle point), then the number of these limiting cycles is finite. He then gave examples to complete the discussion of the problem of determining the general form of characteristics.

179. Extension of the method and the results

Poincaré extended his method to the case of equations of the form $F(x,y,y') = 0$ where F is a polynomial with respect to x,y,y', where y' is the derivative of y with respect to x. This study concerns the genus of a (real) surface, deduced from $F(x,y,z) = 0$ by a birational transformation. This genus, for a closed surface without points at infinity, is that defined in no. 9 for surfaces topologically equivalent to Riemann surfaces: a sphere has genus 0, a "cushion"-like surface with p holes, has genus p. In particular, a torus of revolution is of genus 1. In the general case, if p is the genus, then the relationship between the number of nodes, foci and cols (saddle points) becomes $N + F - C = 2 - 2p$. It is similar to the formula of Lhuilier for polyhedra (9). When the genus is 1, there can be no singular point. This is the case in particular when the surface in question is a torus. In such a case, the characteristics will be cycles or spirals. But the problem of the limiting curves is no longer posed in the same way. In certain straightforward cases, Poincaré showed that they do not exist; every characteristic in spiral form happens to pass into the neighbourhood of one of its singular points, an infinite number of times; the characteristics are stable (see no. 174, preliminary remark). In the case of the torus, Poincaré asked if there could also exist limiting curves and consequently instable characteristics. Denjoy in 1932 proved that this is not so, and for the equations studied by Poincaré, the characteristics are always stable. For these matters, and for the extension of Poincaré's methods to the case of greater order, we refer to Vol. I of the works of Poincaré, to the paper of Denjoy (Journal de Mathématiques, 1932), and to the works of Bendixon and Birkhoff.

III. STURM-LIOUVILLE EQUATIONS

180 - The Sturm-Liouville problem with boundary values; the reduced equation

We shall consider a linear, homogeneous second order equation in the general form

$$y'' + p(x)y' + q(x)y = 0 \quad , \tag{34}$$

whose coefficients are real functions of the real variable x, continuous on a segment (a,b). Its general solution exists on account of the discussion in no. 154. It is given in terms of two particular solutions forming a fundamental system. Suppose we seek a solution $y(x)$ defined by the limiting conditions

$$\begin{gathered} \alpha y(a) + \gamma y'(a) = A \ , \quad \beta y(b) + \delta y'(b) = B \ , \\ (\alpha^2 + \gamma^2)(\beta^2 + \delta^2) \neq 0 \quad , \end{gathered} \tag{35}$$

where α, β, γ, δ, A and B are given real numbers. Given the form of the general solution of the equation, which depends linearly and homogeneously on the two constants of integration λ and μ, then matters are reduced to solving a system of two equations in λ and μ. The conditions anticipated will arise. In particular, if the problem only admits the solution yielding $y \equiv 0$ when $A = B = 0$, it will have a unique solution when $A^2 + B^2 \neq 0$. The case $A = B = 0$ is known as the Sturm-Liouville case.

We can also study the same problem when equation (34) has a second member; there will be no essential difference since it will suffice to supplement the solution of the corresponding equation without second member, with a particular solution of the complete equation, which will amount to changing the values of A and B in the relationships in (35).

Examples. For the equation

$$y'' - \omega^2 y = 0 \quad ,$$

the problem with limits $y(a) = y(b) = 0$ has no solution. Since, having the general solution

$$y = \lambda e^{\omega x} + \mu e^{-\omega x}$$

we would obtain the system

$$\lambda e^{\omega a} + \mu e^{-\omega a} = 0 \ , \qquad \lambda e^{\omega b} + \mu e^{-\omega b} = 0 \quad ,$$

which is impossible, since the determinant is non-zero $\neq 0$.

For the equation

$$y'' + \omega^2 y = 0 \quad ,$$

the same problem leads to considering the determinant of the system which is here, $\sin(b-a)\omega$. The problem is possible if $(b-a)\omega = k\pi$, where k is an integer.

THE REDUCED EQUATION

Equation (34) could also be written in the form

$$\frac{d}{dx}\left[\mathrm{P}(x)\,\frac{dy}{dx}\right] + \mathrm{Q}(x)y = 0 \tag{36}$$

by setting

$$\mathrm{Q}(x) = q(x)\mathrm{P}(x) \ , \quad \mathrm{P}(x) = e^{\mathrm{R}(x)} \ , \quad \mathrm{R}(x) = \int_a^x p(t)dt \quad .$$

The problems with boundary values are the same for both equations.

It was under the form (36) that the equation was studied by Sturm and Liouville.

We could make a second reduction. Let us set

$$\xi = \int_a^x \frac{dt}{\sqrt{\mathrm{P}(t)}} \quad , \qquad y = \frac{1}{4\sqrt{\mathrm{P}(x)}}\ z(\xi) \quad .$$

Equation (36) transforms identically to

$$\frac{1}{4\sqrt{\mathrm{P}(x)}}\left[\frac{d^2z}{d\xi^2} + z\mathrm{Q}_1(\xi)\right] = 0$$

with

$$\mathrm{Q}_1(\xi) = \mathrm{Q}(x) - \frac{1}{4}\,4\sqrt{\mathrm{P}(x)}\left(\frac{\mathrm{P}'(x)}{4\sqrt{\mathrm{P}(x)}}\right)' \quad .$$

We recover the reduced form

$$\frac{d^2z}{d\xi^2} + z\mathrm{Q}(\xi) = 0 \quad . \tag{37}$$

We note that when we commence from (34), where $p(x)$ and $q(x)$ are continuous on (a,b), the function $\mathrm{P}(x)$ introduced in equation (36) is positive, P and

Q are continuous on (a,b); the passage to the new variables z and ξ is well defined. The variable ξ will vary from 0 to a value Λ, x will be a continuous function. But we must assume that $P(x)$ has a continuous second derivative, hence that $p(x)$ has a continuous first derivative.

181 - Theorems of Sturm

These theorems relate to the zeros of solutions of second order equations taken to be of the form (36). We assume $P(x)$ and $Q(x)$ to be continuous on a segment (a,b). Every solution $y(x)$ (we leave aside once and for all the identically zero solutions) can only be zero a finite number of times on (a,b) if c were an accumulation point of the zeros of $y(x)$, we would have $y(c) = 0$ since $y(x)$ is continuous. On account of Rolle's theorem, $y'(x)$ would also have zeros with accumulation point c; we would also have, since $y'(x)$ is continuous, $y'(c) = 0$. Following the theorem of existence of the solutions, the only solution corresponding to $y(c) = y'(c) = 0$ is the solution $y(x) = 0$.

Theorem I. *If a solution $y_1(x)$ of equation (36) is zero at least twice on (a,b) and if c and d are two consecutive zeros of $y_1(x)$, then every solution $y_2(x)$ which is linearly distinct from $y_1(x)$, is zero once and only once between c and d.*

To establish the theorem, it suffices to show that $y_2(x)$ is zero at least once between c and d. It will happen that this solution is zero only once, for if it had two zeros c' and d', $y_1(x)$ would be zero between c' and d' contrary to hypothesis.

Now, suppose that $y_2(x)$ is non-zero between c and d. This solution is not zero at c nor at d, since the Wronskian

$$y_1'(x)y_2(x) - y_1(x)y_2'(x)$$

is non-zero on (a,b) (Liouville's theorem, no. 77). Hence the quotient $y_1(x)/y_2(x)$ is continuous and is zero at c and d; its derivative must be zero between c and d, and this is impossible since the numerator of this derivative is the Wronskian. The theorem is proved.

We could therefore say that *the zeros of a solution separate those of the other solutions* (which are not proportional to it). This applies as long as a solution has at least two zeros on (a,b). We say that such a solution is *oscillatory*.

Theorem II. *Consider the two equations*

$$\frac{d}{dx}\left(P(x)\,\frac{dy}{dx}\right) + Q(x)y = 0 \ , \quad \frac{d}{dx}\left(P(x)\,\frac{dz}{dx}\right) + Q_1(x)z = 0 \quad ,$$

where the functions $P(x)$, $Q(x)$, $Q_1(x)$ *are continuous on* (a,b), *and assume that on* (a,b) *we have* $Q(x) < Q_1(x)$, $P(x) > 0$. *If a solution* $y(x)$ *of the first equation is zero at two consecutive points* c, d, *then every solution* $z(x)$ *of the second equation is zero at least once between* c *and* d.

Let us multiply both members of the first equation $z = z(x)$, and those of the second by $-y = -y(x)$ and add. We obtain

$$\frac{d}{dx}\left[P\left(z\,\frac{dy}{dz} - y\,\frac{dz}{dx}\right)\right] = (Q_1 - Q)yz \quad , \tag{38}$$

and, on integrating from c to d

$$\left[P\left(z\,\frac{dy}{dx} - y\,\frac{dz}{dx}\right)\right]_c^d = \int_c^d (Q_1 - Q)yz\, dx \quad .$$

If the theorem is not true, then y and z would have a constant sign between c and d. We may assume that this is the $+$ sign, by changing, if needs be, the signs of y and z; this does not change the above equation. The second member of the equation will be *positive* and the first member would have the sign of $P(d)z(d)y'(d) - P(c)z(c)y'(c)$.

Now $y'(d)$ is then negative and $y'(c)$ positive; the preceding expression would be negative or zero and we would have a contradiction. The theorem is therefore true.

We see that everything remains true if, at isolated points, we have $Q(x) = Q_1(x)$ or $P(x) = 0$. We could say, in short, that when the coefficient of $P(x)$ in equation (36), is positive, *we increase the oscillation of the solutions by increasing* $Q(x)$.

APPLICATION

Let us consider the equation

$$\frac{d^2y}{dx^2} + Q(x)y = 0 \quad , \tag{39}$$

where $Q(x)$ is continuous for $a \le x \le b$, and assume that

$$0 < \omega^2 < Q(x) < \Omega^2 \quad .$$

We can compare the zeros of the solutions to those of the zeros of the equations

$$\frac{d^2z}{dx^2} + \alpha^2 z = 0 \; , \qquad \alpha = \omega \; , \qquad \alpha = \Omega \quad .$$

These solutions $\sin\alpha(x - x_o)$ have $x_o = \frac{k\pi}{\alpha}$ for zeros, where k is an integer and $\alpha = \omega$ or Ω.

It follows that the *difference* δ *of two successive zeros of a solution of equation* (39) is at least π/Ω in order that this interval contains a zero of every solution of the equation at z for $\alpha = \Omega$, and that it is at most π/ω. We have

$$\frac{\pi}{\Omega} \leq \delta \leq \frac{\pi}{\omega} \quad .$$

This arises from the limits of the number of zeros (these all being simple on account of what was said above).

Other theorems of Sturm. Other theorems of comparison were given by Sturm. We refer, on this subject, to *Leçons sur les methodes de Sturm* by Böcher.

182 - An application to Bessel functions

The Bessel function $J_v(x)$ (no. 124) satisfies the equation

$$y'' + \frac{1}{x} y' \left(1 - \frac{v^2}{x^2}\right) y = 0 \quad ,$$

which becomes

$$z'' + \left(1 - \frac{4v^2 - 1}{4x^2}\right) z = 0 \quad ,$$

when we set $z = \sqrt{x}y$. Furthermore, we have

$$J_v(x) = \left(\frac{x}{2}\right)^v \sum_o^\infty \frac{\left(-\frac{x^2}{4}\right)}{\Gamma(q+1)\Gamma(v+q+1)} \tag{40}$$

Let us assume that v *is real and positive.* If we take an interval (a,b), where a is positive and b is arbitrarily large, then what has been said before applies: the zeros are all simple and their difference is as near as is desired to π, providing a is sufficiently large. The number of positive zeros whose value is less than X is then approximately equal to X/π. If $n(X)$ is this number of zeros, we have

$$\lim_{X\to\infty} \frac{n(X)}{X} = \frac{1}{\pi} \quad .$$

The series appearing in (40) is an even function of x. If we set

$$\frac{-x^2}{4} = u \quad ,$$

then we obtain the entire function

$$\sum \frac{u^q}{\Gamma(q+1)\Gamma(v+q+1)} \quad . \tag{41}$$

Following what has just been said, this function admits negative zeros $-u_1$, $-u_2,\ldots,-u_q,\ldots$ all of which are simple, and we have

$$\lim_{q\to\infty} \frac{u_q}{q^2} = \frac{\pi^2}{4} \quad .$$

The application of a theorem of Laguerve (I, 193) shows that all the zeros of the function (41) are real and negative. The u_q are the only zeros. We also obtain all of the zeros of the Bessel function (40).

183 - An equation containing one parameter

Let us consider the equation

$$\frac{d^2y}{dx^2} + kQ(x)y = 0 \quad , \tag{42}$$

where $Q(x)$ is real and continuous (and not identially zero) for $a \le x \le b$ and k a parameter. *We seek the integrals (not identially zero) which are zero at* a *and* b.

Following the remarks made at the end of no. 154, we may take the parameter k to be complex. The solutions of (42) will be complex functions of the real variable x. The system equivalent to (42) depends analytically on k, which is arbitrary; the solutions will then be holomorphic functions of k, for every finite value of k, and these will be entire functions of the parameter k. The general solution is of the form

$$\lambda y_1(x,k) + \mu y_2(x,k) \quad ,$$

where λ and μ are arbitrary constants and y_1 and y_2 a fundamental system of solutions. We must have

$$\lambda y_1(a,b) + \mu y_2(a,b) = 0 \quad , \quad \lambda y_1(b,k) + \mu y_2(b,k) = 0 \quad ,$$

which can only occur when

$$F(k) \equiv y_1(a,k)y_2(b,k) - y_1(b,k)y_2(a,k) = 0 \quad .$$

The function $F(k)$ is an entire function which is not identially zero since the problem is impossible if $k = 0$. If it has any, its zeros will provide the values of k for which the problem is possible. These values, which will form a sequence, will be called the *proper values* or *fundamental values*. To each there will correspond a solution, defined to within a constant, which will be called *the fundamental solution* or *the proper function*.

THE PROPER VALUES ARE REAL

In effect, if $\alpha + i\beta$ was a proper value and $y = u + iv$ the corresponding solution (α, β, u and v real), then we would have

$$\frac{d^2u}{dx^2} + (\alpha u - \beta v)Q(x) = 0 \quad ,$$

$$\frac{d^2v}{dx^2} + (\alpha u + \beta v)Q(x) = 0 \quad ,$$

from which we extract

$$v\,\frac{d^2u}{dx^2} - u\,\frac{d^2v}{dx^2} = \beta(u^2 + v^2)Q(x) \quad .$$

On integrating, we would obtain

$$0 = \left[v\,\frac{du}{dx} - u\,\frac{dv}{dx}\right]_a^b = \beta\int_a^b (u^2 + v^2)Q(x)dx \quad .$$

Similarly, by multiplying the equations in (43) by u and v respectively and adding, and then integrating from a to b, we would have

$$0 = \left[u\,\frac{du}{dx} + v\,\frac{dv}{dx}\right]_a^b = \int_a^b \left[\left(\frac{du}{dx}\right)^2 + \left(\frac{dv}{dx}\right)^2\right]dx$$

$$- \alpha\int_a^b (u^2 + v^2)Q(x)dx \quad .$$

As β is non-zero, we would then have

$$\int_a^b (u^2 + v^2)Q(x)dx = 0 \quad ,$$

hence

$$\int_a^b \left[\left(\frac{du}{dx}\right)^2 + \left(\frac{dv}{dx}\right)^2\right] dx = 0 \quad ,$$

which is actually impossible. Consequently: *The proper values are real.* The entire function $F(k)$ can only have real zeros, and henceforth we can restrict ourselves to considering the real values of k and the real solutions exclusively.

<u>Theorem</u>. *There exists an infinity of proper values.*

The proposition results from the consideration of the number of zeros of the solutions. Let $y(x,k)$ be one of the solutions. By hypothesis, there exists points of the segment (a,b) at which $Q(x)$ is non-zero; hence we can find a segment (c,d) on which $|Q(x)| > \omega^2 > 0$. Let us take k with the sign of $Q(x)$ on (c,d). We will have $kQ(x) > |k|\omega^2$ on (c,d) and on applying Theorem II of no. 181, $y(x,k)$ will be zero

$$\frac{(d-c)}{\pi}\,\omega\,\sqrt{k} - 1$$

times at least on (c,d). Hence, when $|k|$ increases indefinitely, with the sign of k suitably chosen, the number of zeros of the solutions increases indefinitely. By changing the sign of k if needs be we can assume that this occurs when k tends towards $+\infty$.

Let us consider then the particular solution $y(x,k)$ such that $y(a,k) = 0$ and $y'(a,k) = 1$ (where y' is the derivative with respect to x). For each k, $y(x,k)$ has a finite number of zeros on (a,b); all of these zeros are simple. If x_j is one of these zeros, other than a or b, we have $y(x_j,k) = 0$, $y'(x_j,k) \neq 0$; $y(x,k)$ and $y'(x,k)$ are continuous at x and k following Poincaré's theorem (on equation containing one parameter). It follows that the equation $y(x,k') = 0$ for k' near to k, admits one and only one root in an interval $x_j - \varepsilon$, $x_j + \varepsilon$, so that, if $y(b,k') \neq 0$, there exists an interval $k' - \eta$, $k' + \eta$ in which the number of zeros of $y(x,k)$ remains fixed. By applying the Borel-Lebesque theorem, we deduce from this, that if $y(b,k)$ is non-zero for $k'' \leq k' \leq k'''$, the number of zeros of $y(x,k)$, for these values of k, is constant. *The number of zeros of* $y(x,k)$ *can only change when* $y(b,k)$ *is zero*. As $y(x,o) = x - a$ only has one zero, in order that $y(x,k)$ can have n zeros, it is to necessary for $y(b,k)$ to be zero at least $n - 1$ times. Now, we have seen that n increases indefinitely with k. The theorem is established.

THE CASE WHERE $Q(x)$ IS POSITIVE ON (a,b)

In this case, for $Q(x)$ continuous, there exists two numbers ω and Ω such that $0 < \omega^2 < Q(x) < \Omega^2$. *The proper values are all positive.* For, if k is one of these values and y is the corresponding *fundamental solution*, the equation

$$\int_a^b y\left[\frac{d^2y}{dx^2} + kyQ(x)\right] dx = 0$$

which, on integrating by parts the first term, is written

$$0 = \left[y\,\frac{dy}{dx}\right]_a^b = \int_a^b \left(\frac{dy}{dx}\right) dx - k\int_a^b Q(x)y^2 dx \quad ,$$

shows that $k > 0$ when y is non-constant. Let then $k_1, k_2, \ldots, k_n, \ldots,$ $k_n > k_{n-1}$ be the sequence of proper values, and $y_1, \ldots, y_n, \ldots$ the corresponding fundamental solutions. *The number of zeros of* y_n *belonging to the interval* (a,b) *is equal to* $n - 1$. To prove this it is sufficient to establish that whenever $y(b,k)$ is zero, it effectively introduces a zero of $y(x,k)$, hence (see fig. 47)

$$\frac{\partial y}{\partial k}(b,k) \cdot \frac{\partial y}{\partial x}(b,k) > 0 \quad .$$

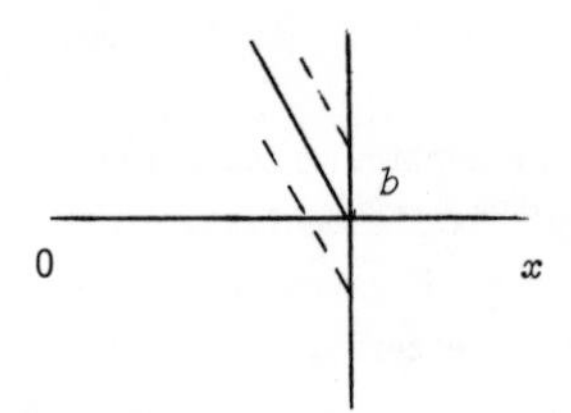

Fig. 47

Now $y(x,k)$ satisfies the equation and $\frac{\partial y}{\partial k}(y,k)$ is the solution of the equation with variations. We have

$$\frac{d^2y}{dx^2} + kQy = 0 \ , \quad \frac{d^2}{dx^2}\left(\frac{\partial y}{\partial k}\right) + Qy + kQ\left(\frac{\partial y}{\partial k}\right) = 0 \quad ,$$

whence we deduce

$$\frac{\partial y}{\partial k}\frac{d^2y}{dx^2} - y\frac{d^2}{dx^2}\left(\frac{\partial y}{\partial k}\right) = y^2Q \quad ,$$

and, on integrating from a to b, we see that

$$\left[\frac{\partial y}{\partial k}\frac{\partial y}{\partial x} - y\frac{d}{dx}\left(\frac{\partial y}{\partial k}\right)\right]_a^b > 0 \quad .$$

As y is zero for a and b, and $\frac{\partial y}{\partial k}$ zero for $x = a$, we indeed have

$$\frac{\partial y}{\partial k}(b,k) \cdot y'(b,k) > 0 \quad .$$

Consequently:

If $Q(x)$ *is positive on the segment* (a,b), *the fundamental solution of rank* n, $y_n(x)$, *has* $n - 1$ *zeros in the interval* (a,b).

The functions y_n form an orthogonal sequence on (a,b) *in the general sense* already considered in nos. 113, 118, etc. We have, in effect,

$$\frac{d^2y_n}{dx^2} + k_n Q y_n = 0 \ , \quad \frac{d^2y_m}{dx^2} + k_m Q y_m = 0$$

and on taking the combination obtained by multiplying the first equation by y_m, the second by $-y_n$, and adding, then integrating from a to b, we obtain

$$0 = \left[y_m \frac{dy_n}{dx} - y_n \frac{dy_m}{dx}\right]_a^b = (k_m - k_n) \int_a^b Q y_n y_m \, dx \quad .$$

As $k_n - k_m \neq 0$, we have

$$\int_a^b Q y_n y_m dx = 0 \quad .$$

The sequence is orthogonal with respect to $Q(x)$. On the other hand, since $Q(x) > 0$, we have

$$\int_a^b Q y_n^2 \, dx \neq 0 \quad .$$

we can *normalise* the sequence (I, 93).

If we try to expand a function $f(x)$ continuous on (a,b) as a series of fundamental solutions,

$$f(x) = c_1 y_1(x) + \ldots + c_n y_n(x) \tag{44}$$

say, then we can calculate the coefficients c_n by the formulae

$$c_n \int_a^b Q y_n dx = \int_a^b Q f y_n \, dx \quad ,$$

which yields these c_n uniquely. It will be necessary to see under what conditions the series, thus constructed, converges and has $f(x)$ as its sum. We shall not embark on this problem.

COMPLETE ORTHOGONAL SEQUENCES

We say that a sequence of continuous functions $y_n(x)$ say $(n = 1,2,\ldots)$, on a segment (a,b), is *a complete sequence*, if there exist no function $f(x)$ continuous on (a,b) such that

$$\int_a^b f(x) y_n(x) dx = 0 \ , \qquad n = 1,2,\ldots$$

For the sequence y_n above, orthogonal with respect to $Q(x)$, the condition is that the equations

$$\int_a^b Q(x)f(x)y_n(x)dx = 0\ , \qquad n = 1,2,\ldots \tag{45}$$

implies $f(x) \equiv 0$.

It can be seen that the sequence of fundamental solutions $y_n(x)$ corresponding to $Q(x) > 0$ on (a,b) is a *complete* sequence.

Remark I. It is not necessary for an orthogonal sequence to be complete for it to provide expansions of the form (44). For example, the sequence of functions $\sin nx$, $n = 1,2,\ldots,$ is orthogonal on the segment $-\pi, +\pi$, and is not complete, since the equations (45) are tenable for the functions $\cos nx$, $n = 0,1,\ldots$. However, the sequence $\sin nx$ allows an expansion of all *odd* continuous periodic functions, of period 2π satisfying the Jordan conditions (I, 87).

Remark II. It is clear that if an orthogonal sequence is not complete, hence if the equations (45) occur for one function $f(x)$, then this function cannot be represented by a series (44) which is uniformly convergent (since all the c_n would be zero). We would then try to complete the sequence by adding $f(x)$ to the functions of this sequence, in order to increase the possibilities of representation of the functions.

Remark III. The sequence of functions $\cos nx$, $n = 0,1,\ldots,$ and $\sin nx$, $n = 1,2,\ldots,$ is complete on the segment $-\pi$, $+\pi$, since a continuous function on $-\pi, +\pi$ whose Fourier coefficients are all zero is identically zero (I, 87).

Remark IV. The sequence of polynomials $1,x,x^2,\ldots$ is complete on all segments (a,b). For if $f(x)$ is continuous on (a,b), we can associate with ε, a polynomial $P(x)$ such that $|f(x) - P(x)| < \varepsilon$ on (a,b) [Theorem of Weierstrass (I, 90)]. By hypothesis

$$\int_a^b f(x)x^n dx = 0\ , \qquad n = 0,1,\ldots\ ,$$

we deduce

$$\int_a^b f(x)P(x)dx = 0\ , \qquad \int_a^b f(x)[P(x) - f(x) + f(x)]dx = 0\ ,$$

hence

$$\int_a^b f(x)^2 dx = \int_a^b f(x)[f(x) - P(x)]\, dx\ ,$$

which is impossible since the first member is a positive member whilst the second is arbitrarily small (we have $|f(x)| < M$ and $|f - P| < \varepsilon$. As a result, the sequence of Legendre polynomials which are derived from this sequence is also a complete sequence.

185 - Applications and extensions

The application of the series expansions of fundamental solutions appears in the study of the second order partial differential equations of mathematical physics.

For example, the problem of vibrating strings with variable linear density leads to the partial differential equation

$$\frac{\partial^2 z}{\partial x^2} = Q(x) \frac{\partial^2 z}{\partial t^2} , \qquad Q(x) > 0 ,$$

with the boundary values $z(a,t) = 0$, $z(b,t) = 0$, $z(x,o) = (x)$, $\frac{\partial z}{\partial t}(x,o) = \psi(x)$. We seek solutions of the form $y(x) \cos \omega t$, $y(x) \sin \omega t$, where ω is a constant and $y(x)$ is a solution of

$$\frac{d^2 y}{dx^2} + \omega^2 Q(x) y = 0 , \qquad y(a) = 0 , \qquad y(b) = 0 .$$

We would have particular solutions given by the fundamental solutions

$$y_n(x) \sin \omega_n t , \qquad y_n(x) \cos \omega_n t ,$$

and then we would see if the conditions of the problem are satisfied with a series of such functions.

EXTENSIONS

The study of solutions of the equations of type (42) satisfying boundary conditions other than those considered, leads to results quite close to those which have been established. We refer, on this subject, to *Les Leçons de Picard sur les problems aux limites dans la theorie des equations differentielles*, and concerning the question of series expansions of fundamental functions, we refer to the works of E. Schmidt.

Sturm and Liouville also studied the equations of the general type in (36) containing one parameter and satisfying boundary conditions of the general type (see the work of Böcher, as cited).

Chapter XI

THE CALCULUS OF VARIATIONS

The original problems associated with the calculus of variations were studied by Newton in 1687 and by Jean and Jacques Bernouilli in 1696 and 1697. These are problems relating to the solid of revolution of least resistance, the brachistochrone and isoperimetric problems. Euler generalised their results in 1744. But it was Lagrange who in 1760 professed the general methods of the study of integral variations and applied his results to problems in theoretical mechanics. Legendre introduced the study of the second variation in 1786; his results were re-discovered by Lagrange (*Theorie des fonctions analytiques*, Paris, am V.). The second variation was again studied by Jacobi (1837). The insufficiency of the conditions given by Legendre and Jacobi was demonstrated by Weierstrass in 1870, who professed new methods for determining the necessary and sufficient conditions for the existence of a maximum or a minimum (1879). His findings were pursued by Kneser. The method of Weierstrass was simplified by Hilbert (1900). Sufficient conditions for a minimum were established by Darboux in 1867 in the particular case of geodesics.

In what follows, we will restrict matters to the most simple problems in the calculus of variations. Having defined the extermals, discussed some examples in the case of integrals where only a single unknown function occurs, and outlining a particular example in which one can advance further by direct methods, we will prove the results of Weierstrass in the case of a single unknown function. We shall then remark on the most general cases and on the most intricate problems.

I. THE CASE OF SIMPLE INTEGRALS - EXTREMA

186 - The nature of the problem

We shall consider a simple integral

$$I = \int_a^b f(x,y,y')dx \quad , \qquad a < b \quad , \tag{1}$$

where a and b are two given numbers and $f(x,y,y')$ a function of three variables x, y, y', which is defined in a domain $a \leq x \leq b$, $c \leq y \leq d$, $c' \leq y' \leq d'$ and which possesses there first and second continuous partial derivatives. We are given the values A and B of y for $x = a$ and $x = b$. We wish to determine the function $y(a)$ such that $y(b) = A$, $(\ ) = B$, admitting a derivative with respect to x, a derivative which we shall assume to be continuous, in such a way that the integral I is a maximum or minimum. We shall assume that A and B belong to the segment (c,d) and that the values given to y and y' belong to the domain in question.

We can always reduce matters, to finding a minimum by changing, if needs be, $f(x,y,y')$ to $-f(x,y,y')$. In most cases, y' will not be subjected to any condition; $c' = -\infty$, $d' = +\infty$. It is to be understood that all elements are assumed real.

Geometrically, we are considering continuous curves $y = y(x)$ joining the points with coordinates (a,A) and (b,B), points that we shall also name A and B, and admitting a tangent which varies continuously, and amongst these curves, we seek a curve Γ for which the integral I is a minimum.

We confine ourselves to considering those curves which are not intersected at a point by a parallel to Oy (fig. 48).

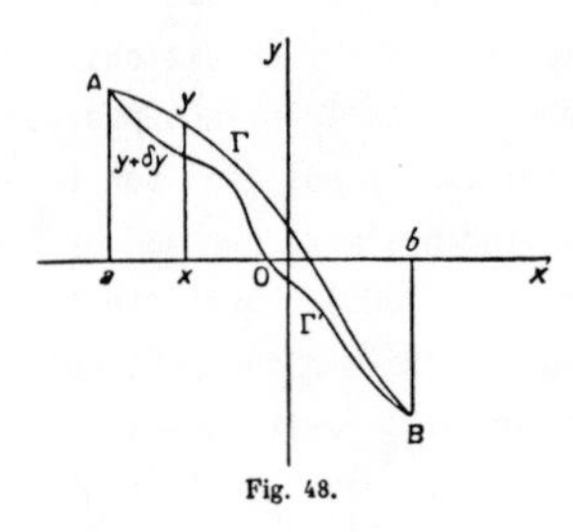

Fig. 48.

We can thus write the integral in the form

$$\mathrm{I} = \int_{\mathrm{A}}^{\mathrm{B}} f(x,y,y')dx$$

and regard it as a kind of curvilinear integral.

187 - Finding the necessary conditions

Let us imagine that we have obtained a curve Γ giving the required minimum. We are going to obtain a necessary condition for this to be so, by comparing the value of I on Γ to the value taken by I on certain families of curves. Let us take a function, which we shall call δy because we assume it is small, and having the following properties: δy is continuous on (a,b) and is zero for $x = a$ and $x = b$; δy has a continuous derivative that we shall call $\delta y'$. We therefore have

$$\delta y_a = 0 , \quad \delta y_b = 0 , \quad \frac{d}{dy}\,\delta y = \delta y' \quad . \tag{2}$$

δy is taken such that the function f is again defined for the values x, $y + \lambda\delta y$, $y' + \lambda\delta y'$, for any λ on the segment $(-1,+1)$ and where y is the ordinate of a point of Γ. Let us consider the family of curves whose ordinate is $y + \lambda\delta y$. On a curve of the family, the integral in question is equal to

$$I(\lambda,\delta y) = \int_a^b f(x,y, + \lambda\delta y, y' + \lambda\delta y')dx \quad .$$

In order for I, which corresponds to $\lambda = 0$, to provide a minimum, it is necessary for the function $I(\lambda,\delta y)$ to be a minimum for $\lambda = 0$. Now this function is differentiable. Its derivative with respect to λ must be zero for $\lambda = 0$. The derivative is

$$\int_a^b \left[\frac{\partial f}{\partial y}(x,y + \lambda\delta y, y' + \lambda\delta y')\delta y + \frac{\partial f}{\partial y'}(x,y + \lambda\delta y, y' + \lambda\delta y')\delta y'\right] dx \quad .$$

We must then have

$$\int_a^b \left[\delta y \frac{\partial f}{\partial y}(x,y,y') + \delta y' \frac{\partial f}{\partial y'}(x,y,y')\right]dx = 0 \quad , \tag{3}$$

for any function δy *satisfying the conditions in (2).* To simplify this condition, we are going to integrate by parts, the second term

$$\int_a^b \frac{d}{dx}(\delta y)\frac{\partial f}{\partial y'}(x,y,y')dx$$

(which can only be achieved by assuming that y' has a derivative), which yields

$$\left[\frac{\partial f}{\partial y'}\delta y\right]_b^a - \int_a^b \delta y \frac{d}{dx}\left(\frac{\partial f}{\partial y'}\right) dx \quad .$$

The part integrated is zero by virtue of the conditions in (2), and we see that equation (3) becomes

$$\int_a^b \delta y \left[\frac{\partial f}{\partial y} - \frac{d}{dx}\frac{\partial f}{\partial y'}\right] dx = 0 \quad . \tag{4}$$

We then have

$$\frac{d}{dx}\left(\frac{\partial f}{\partial y'}\right) = \frac{\partial^2 f}{\partial y'\partial x} + \frac{\partial^2 f}{\partial y'\partial y} y' + \frac{\delta^2 f}{\delta y'^2} y'' \quad . \tag{5}$$

188 - The first fundamental lemma in the calculus of variations. The Euler equation. Extremals

Equation (4) must hold for every function δy consider as above. As a result, the coefficient of δy must be zero by virtue of the following lemma.

Lemma. *If the function* $F(x)$ *is continuous on the segment* (a,b) *and if*

$$\int_a^b \delta y \; F(x)dx = 0$$

for every continuous differentiable function δy *with continuous derivative, zero for* $x = a$ *and* $x = b$ *and such that* $|\delta y| < \varepsilon$, *then, necessarily,* $F(x) \equiv 0$ *on* (a,b).

Let us assume in effect that $F(x)$ is non-zero at a point of the segment (a,b). As this function is continuous, it would have an absolute value greater than a positive number η throughout an interval (c,d) which we may take to be completely contained within (a,b). Let us take

$$\delta y = 0 \quad \text{for} \quad a \le x \le c \quad \text{and} \quad d \le x \le b$$

$$\delta y = k(x-c)^2\,(d-x)^2 \quad \text{for} \quad c \le x \le d \quad ;$$

this function has a continuous first derivative on (a,b). We can, in fact, dispose of k in order that $|\delta y| < \varepsilon$, where ε is given in advance. We shall have

$$\left|\int_a^b \delta y F(x)dx\right| > |k|\eta \int_e^d (x-c)^2(d-x)^2 dx$$

$$= |k|\eta \frac{(d-c)^5}{30} \neq 0 \quad .$$

The lemma is thus established.

A CONSEQUENCE

For condition (4) to be satisfied, it is necessary to have

$$\frac{\partial f}{\partial y} - \frac{d}{dx}\frac{\partial f}{\partial y'} = 0 \quad , \tag{6}$$

and this condition is seen to be sufficient. Explicitly, and with a change of signs, we have

$$\frac{\partial^2 f}{\partial y'^2} y'' + \frac{\partial f}{\partial y \partial y'} y' + \frac{\partial^2 f}{\partial y' \partial x} = \frac{\partial f}{\partial y} = 0 \quad . \tag{7}$$

The equation (6) or ((7)) is known as

Euler's Equation. *Only the curves* Γ *joining* A *and* B *and satisfying this equation are susceptible to minimising the integral* I.

EXTREMALS

An extremal (of the variational problem in question) is an integral curve of the Euler equation which is of second order if $\partial^2 f/\partial y'^2$ is not identically zero; the extremal depend on two arbitrary constants. In a number of cases, it will be possible to have an extremal passing through two given points A,B. When the problem is solved, it remains to see if the curve obtained effectively provides a minimum.

The complete solution of a variational problem therefore depends on three operations:

1) Finding the extremals by either direct integration, by reducing to quadratures, or by reducing to approximate integrations.

2) Finding the extremals passing through the given points.

3) Finding those curves joining A and B, which give the minimum.

189 - Particular cases. Change of variable

Examples. In certain cases, integrating the Euler equation reduces matters to the integration of a first order equation.

FIRST CASE

The unknown function y does not occur directly. Let us assume that the function f only depends on x and y'. We then have $\partial f/\partial y \equiv 0$ and the Euler equation (6) reduces to

$$\frac{d}{dx}\left(\frac{\partial f}{\partial y'}\right) = 0 \quad .$$

We straightaway obtain a *first integral*

$$\frac{\partial f}{\partial y'}(x,y') = \text{const.} \tag{8}$$

This reduces matters to a first order equation. (The title "first integral" indeed conforms with the previous terminology: the second order equation is equivalent to a system of two first order equations by setting $y' = z$, and we indeed have a first integral of this system.)

SECOND CASE

The variable x does not occur in f. The Euler equation is then

$$\frac{\partial f}{\partial y}(y,y') - \frac{d}{dx}\left(\frac{\partial f}{\partial y'}(y,y')\right) = 0 \quad ;$$

it leads to a second order equation which does not contain x, hence it can be reduced by taking y as the variable and y' as the unknown function. Let us set $y' = p$; it becomes

$$\frac{\partial}{\partial y} f(y,p) - p\frac{d}{dy}\left[\frac{\partial f}{\partial p}(y,p)\right] = 0 \quad ,$$

or

$$\frac{\partial f}{\partial y} + \frac{\partial f}{\partial p}\frac{p}{y} - \frac{d}{dy}\left(p\frac{\partial f}{\partial p}\right) = 0 \quad ,$$

and then finally

$$\frac{d}{dy}\left[f(y,y') - y'\frac{\partial f(y,y')}{y}\right] = 0 \quad .$$

We have the first integral

$$f(y,y') - y'\frac{\partial f}{\partial y'}(y,y') = \text{const.} \tag{9}$$

CHANGE OF VARIABLE

Let us observe the problem from a geometric point of view. We must join A to B by a curve rendering I minimum. It is clear that, for this to be so, it is necessary, if C and D are any two points of this curve Γ, to have a minimum when we restrict to the curves joining C to D (that we require to be tangent to the curve Γ at C and D; we see that in each case the lemma still holds true). If C and D are not on a parallel to Ox, then we can consider the curves joining C to D that are only intersected at one point by a parallel to Ox or to Oy. The Euler equation must not change by permuting *in the integral* I, the roles of x and y, i.e. by taking the integral

$$\int_C^D f(x,y,\frac{1}{x'})x'dy$$

and finding the corresponding extremal. It can be seen directly that the two equations only differ by the factor y' which will have to be taken into account.

In practice we could then change the variable in the integral to obtain the general extremals. But when it is a question of seeing if the extremal passing through A and B is a minimum, then it will be of importance to show the variable of the problem posed. However far we go, the independent variable only has a secondary role in determining the necessary conditions for there to be a minimum.

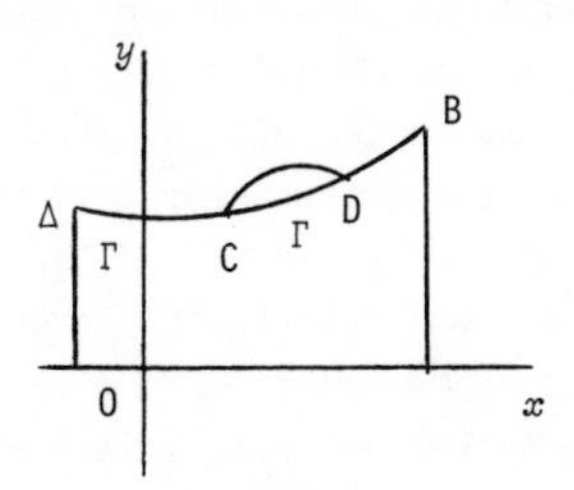

Fig. 49

Examples. Let us consider the integral

$$\int_A^B \frac{ds}{y^\alpha} = \int_a^b \frac{\sqrt{1+y'^2}}{y^\alpha}\, dx \quad ,$$

where α is a real fixed number which we shall take to be arbitrary; y *will then be taken to be positive* (for suitable $\alpha < 0$, we can assume y to be arbitrary).

Following (9), the extremals have the equation

$$\frac{\sqrt{1+y'^2}}{y^\alpha} - \frac{y'}{y^\alpha\sqrt{1+y'^2}} = \text{const} \quad ,$$

or

$$y^\alpha\sqrt{1+y'^2} = C = \text{const.}$$

If we assume $\alpha = 0$, we obtain $y' = \text{const.}$ and the extremals are straight lines. The problem posed in this case is that of the shortest path between A and B, but amongst the curves only intersecting the parallels to Oy at a single point.

Let us assume $\alpha \neq 0$. We can integrate by setting $y' = \cot gt$. We obtain

$$y^\alpha = C \sin t, \qquad dx = (tgt)\, dy \quad ,$$

and on taking $\beta = \frac{1}{\alpha}$, we have

$$\begin{cases} x = x_o + \lambda\beta \int \sin^\beta t\, dt \\ y = \lambda \sin^\beta t \quad . \end{cases} \tag{10}$$

These curves are derived from each other by translations parallel to Ox and homotheties with respect to the origin, which can be anticipated from the start, since the second of these operations only takes effect by multiplying the integral in question by a constant.

When β is positive (hence $\alpha > 0$), the reduced curve γ corresponding to $x_o = 0$, $\lambda = 1$, is complete everywhere to a finite distance. This is obtained by varying t between 0 and π. It has the form given in fig. 50 and is concave towards the base. A straight line only intersects it in at most two points. It stops at two points D and E of Ox. When we are given A and B (with positive ordinates), there exists a unique extremal, passing through these points. Since if C is the point of intersection of AB with Ox, it will be necessary to find a chord B'A'C' of γ, parallel to BAC and such that the ration of $\overline{B'A'}$ to $\overline{A'C'}$ is equal to the ratio of $\overline{BA}$ to $\overline{AC}$. The translation C'C followed by a homothety, will then provide the required extremal. Now, when the chord C'B' is displaced whilst

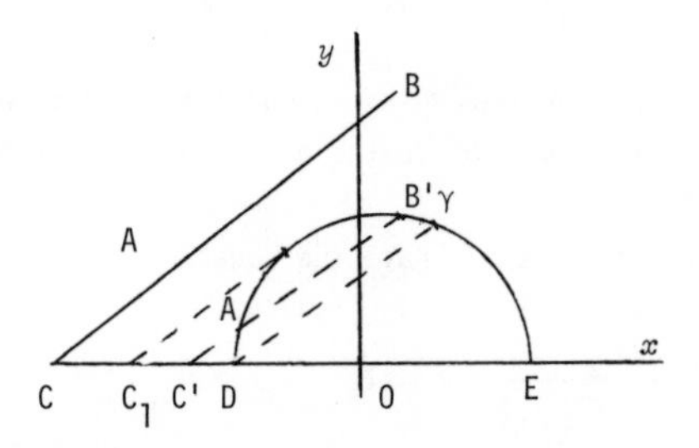

Fig. 50

remaining parallel to CB, with the point C' displaced from the position D up to the position C_1 corresponding to the tangent parallel to CB, C'A increases from 0 and A'B' decreases as far as zero; the ratio takes every value once and only once. Thus:

When α *is positive, there passes one and only one extremal through every pair of points* AB (with positive ordinates).

(If A and B were on a parallel to Ox, we would consider the ratio of $\overline{AB}$ to the distance from AB to Ox.)

The case $\alpha = 1$ corresponds to finding the shortest distance from one point to another in the Poincaré geometry. We recover the circles centred on Ox which were obtained directly (I, 166) and for which the direct proof shows that we effectively have an absolute minimum.

When $\alpha = 1/2$, the equations (10), where $\beta = 2$, define cycloids whose turning points are on Ox. The problem then becomes that of the brachistochrone (J. Bernouilli, 1696): assuming the plane Oxy vertical and Oy directed towards the base, we seek to join A to B by a curve Γ such that a moving weighing point on Γ having lost its motion at the highest point, arrives at the lowest point in a minimum of time, assuming there is no friction.

The case where α is negative. In this case, for β negative, y decreases from infinity to 1 when t increases from 0 to $\pi/2$ (we always consider the reduced curve, $\lambda = 1$, $x_0 = 0$). As for x, it remains finite if $-\beta < 1$ or $-\alpha < 1$; it becomes infinite if $-\alpha \geq 1$.

In the second case, we have a curve with parabolic branches in the direction of Oy, whilst, if $-\alpha > 1$, there are asymptotes parallel to Oy. The concavity is always directed towards the top (the sense of 0 positive) (fig. 51). Given ABC, let us now consider the parallels to ABC taken along the points A' of γ and intersecting Ox at C', the point A' being above the point of contact A_1 of the tangent to γ parallel to ABC.

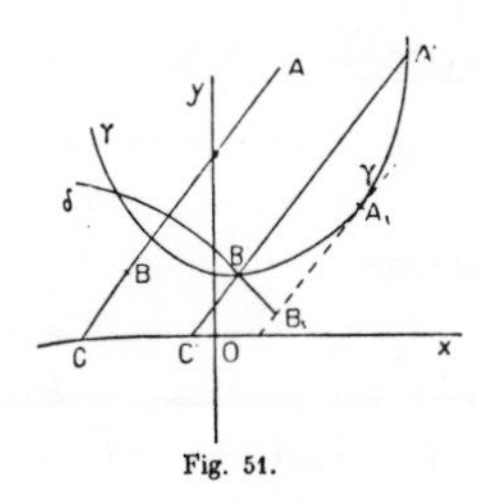

Fig. 51.

On A'C' let us take the point B' such that $\overline{B'C'}/\overline{A'C'} = \overline{BC}/\overline{AC}$. The points of intersection of the locus of δ of B' with γ, will provide chords parallel to ABC and for which the above equality will hold, and then by a translation and a homothety, the extremals passing through A and B. The locus δ of the point B' is a curve directing its concavity from the side of x negative when the gradient $1/m$ of CA is positive. The coordinates of B', x_1 and y_1, are deduced, in effect, from those of A', x and y, by the relationships

$$x_1 = x - m(1-k)y , \qquad y_1 = ky , \qquad 0 < k < 1 ,$$

where k is a constant. On differentiating with respect to t, we have

$$\frac{x_1'}{y_1'} = \frac{x'}{ky'} - m\,\frac{1-k}{k} ,$$

which proves our assertion. The curve δ has an oblique asymptotic direction, hence finite by being outside of γ; its points of departure B_1 is also outside of γ. The curves γ and δ have either two or zero common points, and one in the case where there is contact. Hence, *when α is negative, there passes either two or zero extremals through two points* A *and* B *(and one in the limiting cases).*

For $\alpha = -1$, we consider finding the minimum of the integral

$$2\pi \int_A^B y ds \quad ,$$

which represents the area spanned by the arc of the curve joining AB when revolved about Ox. The extremals are the small links having the base Ox.

For $\alpha = -1/2$, hence $\beta = -2$, the curves (10) are the paraboles for which Ox is the directrix; the problem corresponds to finding the trajectories of a point subjected to a constant force parallel to Oy, when the principle of least action is applied.

190 - The objection and lemma of DuBois Reymond

We remarked in no. 187 that the integration by parts performed in simplifying the first member of the equation in (3), necessitated the assumption that y'' exists and is continuous.

We have therefore added a hypothesis whose necessity does not appear because of those given in the problem. DuBois Reymond, who had made this objection to the methods of Lagrange and Euler, responded in the following way. Let us take equation (3) and, instead of reducing $\delta y'$ to δy, let us do the contrary. We integrate by parts the term in δy. If we set

$$\Phi(x) = \int_a^x \frac{\partial f}{\partial y}(x,y,y')dx \quad , \tag{11}$$

we have

$$\int_a^b \delta y \frac{\partial f}{\partial y}(x,y,y')dx = \Big[\delta y \ \Phi(x)\Big]_a^b - \int_a^b \delta y'\Phi(x)dx$$

and the derivative of $I(\lambda,\delta y)$ for $\lambda = 0$, takes the form

$$\int_a^b \delta y' \left[\frac{\partial f}{\partial y'} - \Phi(x)\right]dx \quad . \tag{12}$$

It is necessary for this integral to be zero for any function δy satisfying the conditions in (2). DuBois Reymond showed that it is a question of having

$$\frac{\partial f}{\partial y'} - \Phi(x) = \text{const.} \tag{13}$$

In order to prove this, we rely on this lemma:

Lemma of DuBois Reymond. *If* $F(x)$ *is continuous on the segment* (a,b), *then in order to have*

$$\int_a^b F(x)\delta y' dx = 0 \quad ,$$

for every function satisfying the conditions in (2), it is necessary and sufficient for $F(x)$ *to be constant on* (a,b).

The condition is clearly sufficient.

To prove the necessity, we observe that, if C is any constant, we also have, with respect to the conditions in (2),

$$\int_a^b [F(x) - C]\delta y' dx = 0 \quad .$$

Let us choose C in such a way that

$$\int_a^b [F(x) - C]dx = 0 \quad ,$$

and take

$$\delta y = k\int_a^b [F(x) - C]dx \quad , \qquad \delta y' = k[F(x) - C]$$

where $k \neq 0$ is a constant. The conditions in (2) are satisfied. We must have

$$k\int_a^b [F(x) - C]^2 dx = 0$$

which implies $F(x) - C \equiv 0$.

Remark. Note that we can also say that the proposition is still true if we take $|\delta y| < \varepsilon$, for some fixed ε, but taken to be arbitrarily small.

CONSEQUENCES OF THE EQUALITY (13)

It follows that, for $y(x)$ and $y'(x)$ being the values of y and y' on a curve Γ, annihilating the derivative of $I(\lambda,\delta y)$ for any δy, the equality (13) must hold, where $\Phi(x)$ is defined by the equality in (11). $y' = y(x)$ therefore satisfies the equation

$$\frac{\partial f}{\partial y'}(x,y,y') = \Phi(x) + \text{const.}$$

It can be regarded as an equation defining y' implicitly. In the second member, $\Phi(x)$ has a continuous derivative. If the derivative of the first member with respect to y' is non-zero, then

$$\frac{\partial^2 f}{\partial y'^2}(x,y,y') \neq 0 \quad . \tag{14}$$

The implicit function theorem (I, 120) allows us to assert that the solution y' of this equation is differentiable and that its derivative y'' is continuous.

Thus, *by means of the condition (14) which is assumed to be satisfied along the curve in question, we can assert that y'' exists and is continuous, with any hypothesis a priori not included with those originally made (no. 186).*

When we observe that the first member of the condition in (14) is the coefficient of y'' in the equation (7) of the extremals, then we see that this condition also plays a part in the actual definition of these curves: the existence theorem of the solutions of (7) no longer applies when (14) occurs.

II. THE STUDY OF THE CASE WHERE F DOES NOT DEPEND ON Y

191 - The sufficient condition for the absolute minimum. The figurative. Discontinuous solutions

Let us take the particular case of the integral

$$I = \int_a^b f(x,y')dx \quad . \tag{15}$$

By applying the lemma of DuBois Reymond, we see that it suffices here to assume that f has a partial derivative with respect to y' that is continuous, in order to obtain directly the equation of the extremals;

$$\int_a^b \delta y' \frac{\partial f}{\partial y} dx$$

must be zero for all functions δy satisfying the conditions in (2). We therefore have

$$\frac{\partial f}{\partial y'}(x,y') = \text{const.} \tag{16}$$

A SUFFICIENT CONDITION FOR THE MINIMUM

Let us consider an extremal joining the two points A and B. Let Γ be this curve, y the value of y on Γ and y' its derivative. We have $\frac{\partial f}{\partial y'}(x,y') = k$ along Γ. Let Γ' be another curve joining A to B, Y the value of y on Γ' and Y' its derivative. On Γ', the integral takes the value

$$I' = \int_a^b f(x, Y')dx \quad ,$$

hence

$$I' - I = \int_a^b [f(x, Y') - f(x, y')]dx \quad .$$

Now

$$\int_a^b \frac{f}{y'}(x, y')(Y' - y')dx = k \int_a^b (Y' - y')dx = 0 \quad ,$$

we can also write

$$I' - I = \int_a^b E_1(x, Y', y')dx \quad ,$$

by setting

$$E_1(x, Y', y') = f(x, Y') - f(x, y') - f(Y' - y') \frac{\partial f}{\partial y'}(x, y') \quad . \tag{17}$$

Consequently:

Along Γ, *it suffices, for all* Y' *satisfying the conditions of the statement, to have* $E_1(x, Y', y') > 0$ *when* $Y' \neq y'$, *in order for* Γ *to yield an absolute minimum.*

A GEOMETRIC INTERPRETATION

Consider three rectangular axes Ox, Oy', Oz on which the coordinates are x, y', z and consider the surface Σ whose equation is $z = f(x, y')$. The section of Σ through a plane $x = \text{const.}$, is a curve G_x whose equation, in its plane, is $z = f(x, y')$. The expression (17) can be written as

$$Z - z - (Y' - y') \frac{dz}{dy'} \quad ;$$

this represents the difference between the side from a point of G_x and the side from the point of the same ordinate Y' situated on the tangent to the points (y', z). To say that $E_1(x, Y', y')$ is positive for the value x amounts to saying that the curve G_x is above its tangent at the point M_x corresponding to (x, y') (fig. 52). *In particular, this occurs when* G_x *is concave towards positive* z. Moreover, the gradient of the tangent at the point M_x is constant when x varies. The locus of points M_x is the contact curve of Σ and the circumscribed cylinder whose generators have 0, 1, $\frac{\partial f}{\partial y'} = k$, as direction parameters. *A curve* Γ *will provide an absolute minimum if the surface* Σ *is above this cylinder* (except at the point of contact).

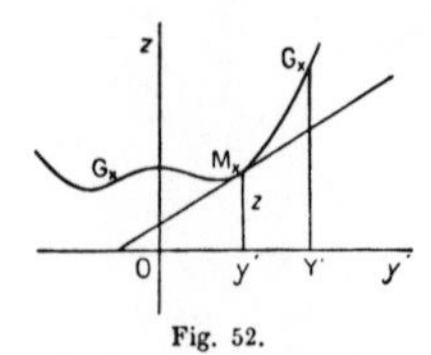

Fig. 52.

We shall say that the curve G_x (which generally depends on x), or the surface Σ, is the *figurative* of the problem (this term was used in the general case by Hadamard).

DISCONTINUOUS SOLUTIONS

Let us consider the circumscribed cylinder parallel to a direction 0, 1, k. In general, it will be tangent to G_x at several points. It is the point at which the side to the origin of the tangent is a minimum, that is to be considered. The locus of this point M , provides a circumscribed cylinder having the required property. It provides for each x, the value of y', and by taking

$$y = \int y'dx + \text{const} \quad ,$$

we obtain a sheaf of extremals giving an absolute minimum.

It could happen that for one (or several) x, the tangent with gradient k is actually a bitangent to G_x (fig. 53) and that the curve of contact of the lowest circumscribed cylinder corresponds to the point M_x when $x' > x$ and to the point M'_x when $x' < x$.

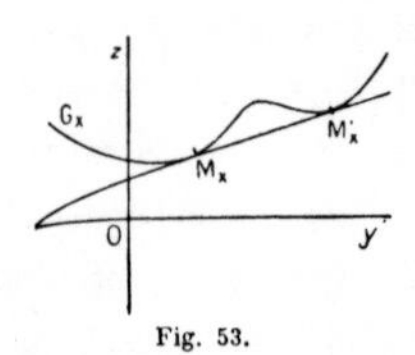

Fig. 53.

This leads to considering an extremal curve Γ formed by two distinct arcs of integral curves of equation (16), intersecting at the point with abscissa x and having distinct tangents at this point. Such a curve will *again give an absolute minimum,* for the calculation leading us to considering (17) (which is, moreover, independent of the fundamental lemmas), holds true if y' or Y' have a finite number of discontinuities. *We say that such a solution is a discontinuous solution of the variational problem.*

Example. Consider the integral

$$I = \int_a^b [xy' + (y'^2 - 1)^2]dx \quad ,$$

where a and b are taken to be arbitrary, and y' is not subjected to any condition. For the equation of extremals, we have

$$x + 4y'(y'^2 - 1) = k \quad .$$

We immediately integrate, and setting $y' = -t$, we have

$$x = k + 4t(t^2 - 1)$$

$$y = k' - t^2(3t^2 - 2) \quad .$$

These are quartics which are easily constructed. But we shall direct our attention to the surface Σ, which has the equation

$$z = xy' + (y'^2 - 1)^2 \quad .$$

The form of the curves G_x are obtained straightaway, by taking as axes Oz and the line $z = xy'$; this last line is a bitangent (fig. 54). For $x = k$, the tangent to G_x with gradient k, is a bitangent, and by following the variation of the form of G_x when x varies, we see at once that the tangent with gradient k, whose ordinate with the origin is minimum, has a point of contact whose ordinate y' jumps by two units when the value k is crossed. Discontinuous solutions will arise when the points A and B are suitably placed.

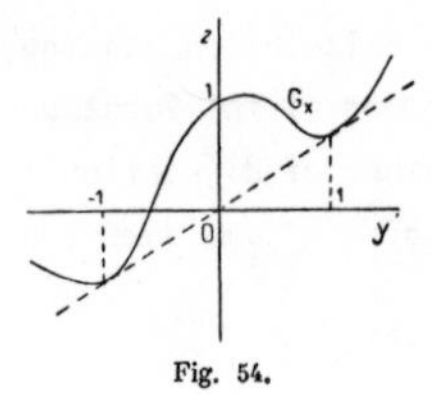

Fig. 54.

192 - Examples. The Newton problem. I.

Let us consider the integral

$$\int_a^b x^\beta \sqrt{1+y'^2}\, dx \,, \quad \beta > 0 \,, \quad 0 < a < b \quad .$$

The extremals are defined by the equation

$$\frac{x^\beta y'}{\sqrt{1+y'^2}} = \text{const.}$$

The extremals corresponding to $y' \neq 0$ were studied in no. 189. In effect, when we change the role of x and y, we recover the examples in no. 189. The curves are those of the figure (51), with the roles of Ox and Oy changed. Extremals do not always pass through two points A,B. Moreover, there could pass an extremal which is intersected at two points by a parallel to Oy. This does not answer the actual question. To these curves it is necessary to include the lines y = const.

We have

$$z = x^\beta \sqrt{1+y'^2} \quad .$$

The curves G_x are all the same to within an affinity. Their concavity is turned upwards. We can see at once that the second derivative of z with respect to y' is positive. The arcs of the curvilinear extremals that we consider here are arcs which do not contain points at which the tangent is parallel to Oy. For B on a line from A $(b > a)$, we must take the arcs of the extremals passing through B and bounded by their summit S. The locus of S is a closed curve passing through B and by its projection B' onto Oy (fig. 55), is symmetric with respect to BB'.

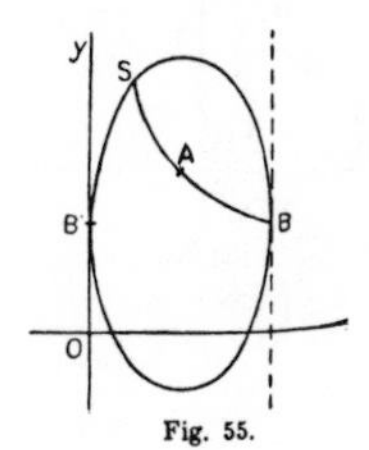

Fig. 55.

If A is inside this curve, we obtain an extremal arc AB providing an absolute minimum. (We may observe that the second solution of the Euler equation passing through A and B will have its summit between A and B, otherwise, we would have a contradiction.) Thus the problem is completely solved in this case. But if A is outside of the locus of S (and always to the left of B), it would be necessary to join A to B by a curve with angular points formed by curvilinear and rectilinear extrema. We shall not embark on this matter here.

Example II. The Newton Problem. In the study of a body of revolution offering the least resistance when displaced in a direction parallel to its axis in a liquid, we may assume, in an initial theoretical study, that each minute part of the surface offers a resistance which is proportional to its surface, and is a function of themotion and the angle of the tangent plane with the axis. By taking these effects to be independent, the resistance will be the product of a function of the motion and an integral of the form

$$\int_a^b x\phi(|y'|)ds = \int_a^b xf(|y'|)dx \quad , \qquad f(|y'|) > 0 \quad .$$

Oy is the axis of revolution and AB is a meridian arc. We assume $a < b < 0$ (fig. 56) and B_1 is taken from B above A_1, in turn taken from A. The function $f(|y'|)$ is assumed to be zero for $|y'|$ infinite, increasing with $1/|y'|$ and a maximum for y' zero.

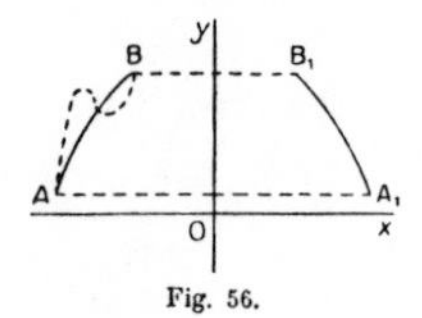

Fig. 56.

When we admit curves as those appearing along the dotted lines in figure 56, we can give the integral a value as small as is wished. The minimum will be zero. *But let us assume that we have admitted uniquely, some positive values for* y'. The problem is then posed. The extremals have as equation

$$xf'(y') = k \,, \qquad k > 0 \quad ,$$

or, by setting $y' = t$,

$$x = \frac{k}{f'(t)} \,, \qquad y = k' - k \int \frac{tf''(t)}{f'(t)^2}\, dt \quad .$$

The surface Σ is defined uniquely for $y' > 0$. The curves G_x are deduced by affinity with $z = f(y')$. We shall assume that $f''(y')$ is zero only once. The curve G_x has the form represented in figure 57, with a point of inflexion I. If $t_o = y'_o$ is the value of y' at this point, then the extremals admit the point corresponding to the turning point, and at this point, their concavity changes direction. They have the form indicated in figure 58, where the regions 1 and 2 correspond in figure 57 and 58.

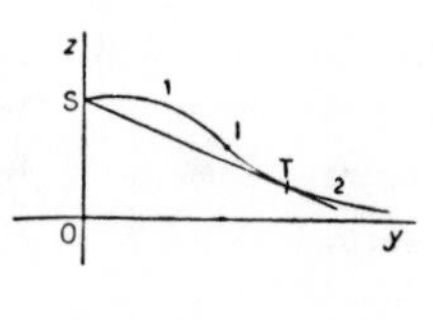

Fig. 57.

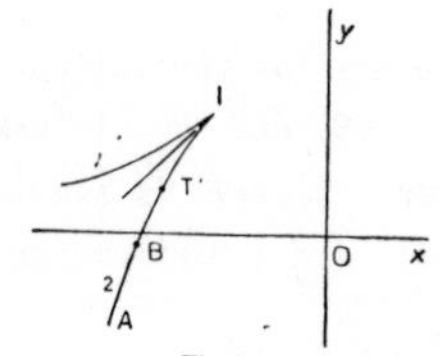

Fig. 58.

Let us apply the result relative to $E_1(x,Y',y')$. The parts of the extremals which guarantee an absolute minima will correspond to the portion of the curve G_x for each point of which G_x is above its tangent. From the point S of G_x situated on Oz we take the tangent to the curve G_x; its point of contact is T (fig. 57). Corresponding to T, we have a point T' on each extremal. It is the part of the branch 2 of the extremum situated below this point T' (fig. 58) that will answer the question. If two points A and B are joined by such a branch, then we have an absolute minimum. Here again, if A is given, B is located in a part of the plane which can be easily bounded and which does not contain the axis Oy. In order to attain the axis, we show that we must use, at the same time, the curvilinear extremals and a segment y = const, which is indeed an extremal since $f'(\infty) = 0$, but corresponding to a maximum of I.

III. EXTREMUM CONDITIONS IN THE GENERAL CASE

193 - The first variation in general

We intend studying the small variations of the integral

$$I = \int_A^B f(x,y,y')dx$$

when we replace the curve Γ joining A to B on which it was calculated by a neighbouring curve Γ' joining two neighbouring points of A and B, A' and B' say (fig. 59). We will assume that the curve Γ' depends on a parameter λ and that, for $\lambda = 0$, it reduces to Γ. The ordinate of a point of Γ' will be defined by $y = \phi(x,\lambda)$; for $\lambda = 0$, we will obtain the ordinate y of the point of Γ with abscissa x.

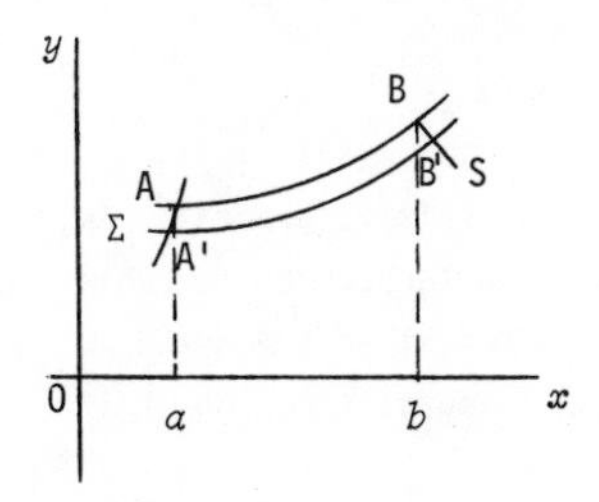

Fig. 59

We shall assume that Γ is defined somewhat beyond the points A and B $\phi(x,\lambda)$ will then be defined for $a - \varepsilon \le x \le b + \varepsilon$, $\varepsilon > 0$, and $0 \le \lambda \le 1$. We shall assume that it admits continuous first and second partial derivatives. We will assume that the extremities A' and B' of Γ', describe curves Σ and S passing through A and B and admit tangents at these two points. The coordinates a', A' of A' will then be functions of λ. If m_A is the slope (assumed to be finite in the proof) of the tangent to Σ at the point A, with coordinates (a,A), then we have

$$A' - A = m(a' - a) \quad , \tag{18}$$

where m tends towards m_A when λ tends towards zero (the point A' must then tend towards the point A). Now, with these hypotheses established, we have

$$A' - A = \phi(a',\lambda) - \phi(a,0) = (a' - a)\phi'_x(a'',\lambda') + \lambda\phi'_\lambda(a'',\lambda') \quad ,$$

where a'' and λ' tend towards a and 0 respectively, when λ tends towards zero. By substituting into (18), dividing by λ and letting λ tend to zero, we see that the abscissa has a derivative with respect to λ for $\lambda = 0$, given by

$$\left(\frac{da'}{d\lambda}\right)_o \left[\phi'_x(a,0) - m_A\right] = \phi'_\lambda(a,0) = 0 \quad .$$

This assumes $\phi'_x(a,0) \neq m_A$. When the curve Σ is tangent to Γ at A, then we would make the hypothesis that $da'/d\lambda$ exists for $\lambda = 0$.

More generally, we shall denote by δy, the derivative $\phi'_\lambda(x,0)$; $\phi'_x(x,0)$ is y' since $\phi(x,0) \equiv y$. We shall call δx_A and δy_A the derivatives of a' and A' for $\lambda = 0$, whose existence we are about to prove. The above formula will then be written as

$$y'_A\delta x_A - \delta y_A + \delta y(a) = 0 \quad . \tag{19}$$

Under this form, the result is general; δy_A and δy_A are direction parameters of the tangent to the curve Σ, which can be taken to be arbitrary. We have a similar result at the point B.

Let us consider the integral taken on Γ':

$$I(\lambda) = \int_{a'}^{b'} f(x,\phi(x,\lambda),\phi'_x(x,\lambda))dx \quad .$$

It is a function of λ which has a derivative for $\lambda = 0$. This derivative is obtained by the usual method (I, 70). We must differentiate under the sign of integration, by assuming $a' = a$, $b' = b$ are fixed, and adding the terms providing the limits, terms which exist for $\lambda = 0$. These complementary terms are

$$\delta x_B f(x,y,y')_B - \delta x_A f(x,y,y')_A \quad .$$

Differentiation under the summation sign yields for $\lambda = 0$,

$$\int_a^b \left[\delta y \, \frac{\partial f}{\partial y}\,(x,y,y') + \frac{d}{dx}\,(\delta y)\,\frac{\partial f}{\partial y'}\,(x,y,y')\right] dx \quad . \tag{20}$$

We can again integrate by parts the second term as in no. 187, but the part completely integrated is no longer zero; we obtain as the value of the integral (20).

$$\left[\frac{\partial f}{\partial y'}\,\delta y\right]_a^b + \int_a^b \delta y \left[\frac{\partial f}{\partial y} - \frac{d}{dx}\left(\frac{\partial f}{\partial y'}\right)\right] dx \quad .$$

By replacing $\delta y(a)$ by its value extracted from (19) and $\delta y(b)$ by the analogous expression, we obtain the value of the derivative of $I(\lambda)$, for $\lambda = 0$, *a derivative which we shall denote by* δI. We have

$$\delta I = \int_a^b \delta y \left[\frac{\partial f}{\partial y} - \frac{d}{dx}\left(\frac{\partial f}{\partial y'}\right)\right] dx + \left[\delta x\left(f - y'\,\frac{\partial f}{\partial y'}\right) + \delta y\,\frac{\partial f}{\partial y'}\right]_A^B \quad , \tag{21}$$

for which the second term is equal to

$$\delta x_B\left(f - y'\frac{\partial f}{\partial y'}\right)_b + \delta y_B\left(\frac{\partial f}{\partial y'}\right)_b - \delta x_A\left(f - y'\frac{\partial f}{\partial y'}\right)_a - \delta y_A\left(\frac{\partial f}{\partial y'}\right)_a \tag{22}$$

δI *is known as the first variation.* It is to be understood that δy denotes the derivative of $\phi(x,\lambda)$ with respect to λ for $\lambda = 0$; δx_A, δx_A, δx_B, δx_B are direction parameters of the tangents at the loci of A and B, related to δy by the equation (19), and the analogous equation $\phi(x,\lambda)$ must have continuous first and second partial derivatives.

We can regard the notation $\delta y,\ldots,\delta I$ as a differential notation and imagine that all these quantities, which are derivatives with respect to λ, are multiplied by $\delta\lambda$.

THE CASE WHERE Γ IS AN EXTREMAL

In this case, the coefficient of δy in the integral of the second member of (21) is zero and δI reduces to the part completely integrated. We have

$$\delta I = \left[\delta x\ f - y' \frac{\partial f}{\partial y'} + \delta y \frac{\partial f}{\partial y'}\right]_A^B \quad .$$

194 - Variational problems with variable limits. Transversals

We always assume that $f(x,y,y')$ has the properties enumerated in no. 186. Given two continuous curves S and Σ admitting tangents which vary continuously, we join a point A taken on Σ to a point B taken on S (the abscissa of A is less than that of B) by a curve Γ intersected at a single point by a parallel to Oy. We seek to determine A, B, Γ in order that the integral

$$I = \int_a^b f(x,y,y')dx$$

calculated on Γ, is either a maximum or minimum.

(Generally, we say *an extremum*.) We also assume that y and y' are continuous on Γ.

Let us assume that A, B and Γ answer the question. The curve Γ must provide the extremum when A and B are assumed fixed, and hence is amongst the curves joining two fixed points. This must be an extremal (no. 188), an integral curve of the Euler equation

$$\frac{\partial f}{\partial y} - \frac{d}{dx}\left(\frac{\partial f}{\partial y'}\right) = 0 \quad .$$

Let us assume now, that for A fixed, we vary B along S. The curve Γ must yield the extremum amongst those curves joining A to a variable point B' on S. In particular, this must be the case when we consider a family of curves $y = \phi(x,\lambda)$ joining A to a point B' of S, and reducing to Γ for $\lambda = 0$. It will therefore be necessary for the first variation δI relative to these curves, to be zero. As Γ is an extremal and A is fixed, we have

$$\delta I = \delta x_B \left(f - y' \frac{\partial f}{\partial y'}\right)_b + \delta y_B \left(\frac{\partial f}{\partial y'}\right)_b = 0 \quad . \tag{23}$$

Remark. To see that this condition is not illusory, it is clearly necessary for δx_B and δx_B not to be zero simultaneously. If Γ is not tangent to S at the point B, then we can take for the function $\phi(x,\lambda)$ of no. 193,

$\phi(x,\lambda) = y + \varepsilon(x-a)\lambda$, where y is always the ordinate of Γ and ε is fixed and is small. The conditions imposed are satisfied and as $\delta y(b) = \varepsilon(b-a) \neq 0$, then δx_B and δx_B, which are related by

$$\delta y_B - y'_B \delta x = \delta y(b) ,$$

and not zero simultaneously. When Γ is tangent to S at the point B, we can take $\phi(x,\lambda) = y + \varepsilon(x-a)\lambda^2$; here, we have $\delta y = 0$. We give ε a suitable sign in order that Γ' intersects S (fig. 60). If the contact at B is of order 1, then δx_B exists and is non-zero; the formula (23) is tenable and yields $f(x,y,y')_B = 0$. For higher order contact, we change the exponent of λ.

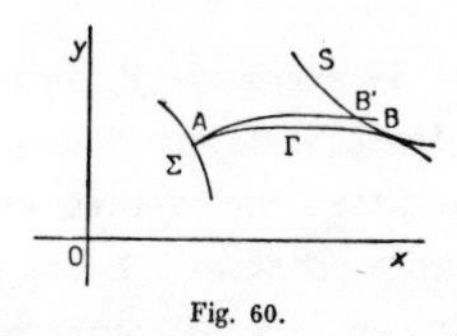

Fig. 60.

NECESSARY CONDITIONS OF THE FIRST ORDER

The same applies when we have

$$\delta x_A \left(f - y' \frac{\partial f}{\partial y'}\right)_a + \delta y_A \left(\frac{\partial f}{\partial y'}\right)_a = 0 \quad , \tag{24}$$

where δx_A and δy_A are not simultaneously zero. Hence:

For the possibility of the problem, it is necessary for there to exist extremals joining the points S *and* Σ, *and* A *and* B *are such that the direction parameters* δx_A *and* δy_B *of the tangent to* Σ *at* A, δx_B *and* δy_B *those of the tangent to* S *at* B, *satisfy the conditions (23) and (24).*

If we let Y'_A and Y'_B be the angular coefficients of the tangents to Σ and S at A and B, then the conditions (23) and (24) are also written as

$$\left[f + \frac{\partial f}{\partial y'}(Y' - y')\right]_A = 0 \ , \quad \left[f + \frac{\partial f}{\partial y'}(Y' - y')\right]_B = 0 \tag{25}$$

As there are two arbitrary constants C and C' in the solutions of the Euler equation, the conditions (25) yield two relationships between the constants. The problem is therefore possible in certain cases.

When the above problem, which we call the *problem of the first order*, is solved, then it remains to be seen if the curve (or curves) obtained, yields an extremum. This is the problem of the second order.

The problem of the first order can be indeterminate. It could also be impossible; then we would have to try to join S to Σ by pieces of extremals.

A PARTICULAR CASE

S or Σ could be reduced to a point. For example, for A fixed, we would have to determine the extremals passing through A and intersecting S at a point B for which condition (23) is satisfied.

TRANSVERSALS

Generally, we say that an extremal Γ intersects a curve S *transversally* at a point B when condition (25) is realised at this point. By applying this term (due to Kneser), we see that the problem which consists of finding a curve Γ joining a point of S to a point of Σ and making the integral I, an extremum, amounts to finding *an extremal intersecting* S *and* Σ *transversally*.

In this case where the integral I is of the form

$$\int_a^b \Phi(x,y)ds \ , \qquad ds = \sqrt{1+y'^2}\ dx \quad ,$$

the condition of transversality becomes a condition of *orthogonality*. Since, by cancelling out the factor $\Phi(x,y)$ in condition (25), which is non-zero in general, there remains

$$\sqrt{1+y'^2} + \frac{y'}{\sqrt{1+y'^2}}(Y' - y') = 0 \quad ,$$

which gives $y'Y' + 1 = 0$.

A SHEAF OF TRANSVERSALS

Let us consider a point A and the extremals passing through this point. If A is a regular point from the point of view of the solutions of the Euler equation, then there will pass through A, an extremal and a single tangent to a line with angular coefficient y' (since the Euler equation is solvable in y''). We thus obtain a sheaf of extremals originating from A (fig. 61). In the neighbourhood of A, these extremals will not have common points other than A. There will exist to the right and left of A, curves C will be transversally intersected by all these extremals. In effect, at a point M near to A, there will pass a single extremum of the sheaf, hence, at this point $M(x,y)$, the value of y' on this extremal will be a function of x,y above, and the transversality condition at M will become $Y' = \psi(x,y)$.

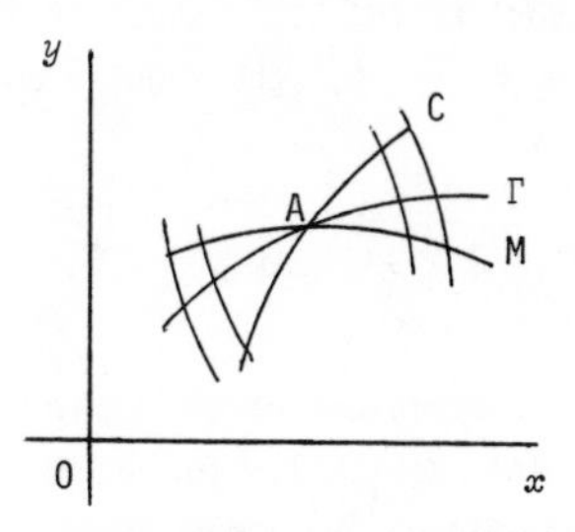

Fig. 61

The solutions of the first order differential equation $y' = \psi(x,y)$ will settle the matter. *We say that such curves constitute a sheaf of transversals associated to the point* A.

If C *is a curve of this sheaf, the value of the integral* I *computed from* A *to a variable point* M *of* C, *on the extremal joining* A *to* M, *is independent of* M.

This property arises from the hypothesis that the function $f(x,y,y')$ has a *sufficient number* of partial derivatives and that

$$\frac{\partial^2 f}{\partial y'^2}(x_o,y_o,\lambda) \qquad \frac{\partial f}{\partial y'}(x_o,y_o,\lambda) \quad ,$$

is non-zero, for any λ, where x_o, y_o are the coordinates of A. The functions $\partial^2 f/\partial y'^2$ and $\partial f/\partial y'$ will not be zero *in the neighbourhood of this point*. We can solve the Euler equation with respect to y''. By applying the existence theorem of the solutions of this equation, we see that the solutions will be of the form $y = \phi(x,\lambda)$ when we take the initial conditions $x = x_o$, $y = y_o$, $y' = \lambda$, and ϕ will be a function of x and λ admitting partial derivatives with respect to x and λ, following the theorem of Poincaré-Weierstrass. One of these solutions will pass function of λ whose derivative for each λ will be given by δI, which here reduces to

$$\delta y \ \frac{\partial f}{\partial y'} + \delta x\left(f - y' \frac{\partial f}{\partial y'}\right) \quad ,$$

where δx is the derivative of x with respect to λ along C and $\delta y = Y'\delta x$. This expression is zero, hence $I(\lambda)$ is constant. Q.E.D.

A GENERALISATION

Likewise, we could consider a sheaf of transversals associated to a given curve C.

Firstly, we will associate to C, a sheaf of extremals intersecting C transversally. At each point of C, the angular coefficient y' of the extremal is again given by

$$f(x,y,y') + (Y' - y')\frac{\partial f}{\partial y'}(x,y,y') = 0 \quad ,$$

where Y' is the angular coefficient of the curve C. We shall assume that through a neighbouring point $M(x_1,y_1)$, for which y' will have a value y_1', and this solution will be unique. We indeed have a sheaf of extremals. At the point M, y_1' is well determined and the angular coefficient of the curve C will be given by

$$Y_1' = y_1' - \frac{f(x_1,y_1,y_1')}{\frac{\partial f}{\partial y'}(x_1,y_1,y_1')} \quad .$$

This will be a function of x_1,y_1 satisfying the conditions allowing differential equations to have solutions.

If we consider then, the extremals joining A to the points M of C, taken to be of the form $y = \phi(x,\lambda)$, we see that they depend on the parameter λ and satisfy the conditions of no. 193, for each value λ_o of λ, by changing λ to $(\lambda - \lambda_o) + \lambda_o$. The integral y' can be derived as a function of x,y, along C. We will obtain a sheaf of extremals Γ and the curves which are cut transversally, will form a sheaf containing C.

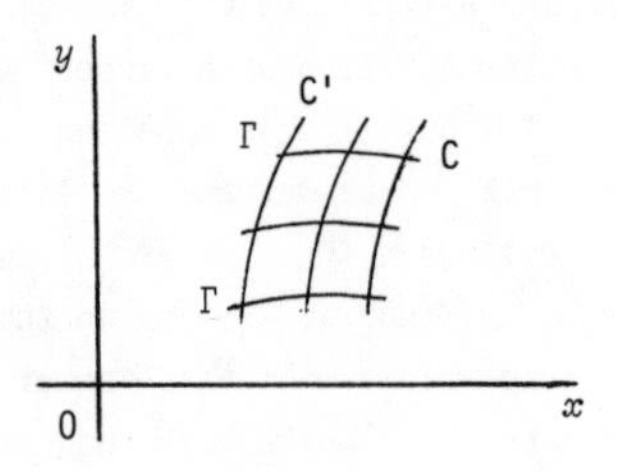

Fig. 62

All of this arises from the conditions which immediately involve the existence theorems of implicity functions and the solutions of differential equations. As was the case for the extremals from A, it can be seen that:

Taking C' *to be a curve of the sheaf of transversals associated to* C*, then the integral* I *computed between* C *and* C' *along a variable extremal, has a constant value.*

EXAMPLES

We can apply these considerations to the integrals studied in no. 189. As they have an expression of the form $\Phi(x,y)ds$, the condition of transversality is an orthogonality condition. For example, in the Poincaré geometry, the sheaf of transversals associated with a point A, will be formed by circles whose non-Euclidean centre is A. But the sheaf of transversals associated to a line C of the geometry (an orthogonal circle) is not formed by lines of the Poincaré geometry, since this will be the sheaf of the loci of points equidistant from C (this being the non-Euclidean distance): these are circles (of the usual geometry) passing through the points where C intersects Ox.

Remark. The fact that the value of I computed on an extremal between two curves of a sheaf of transversals, is constant, is established independently of the fact that this integral gives or does not give, a true extremum.

195 - Necessary conditions for the extremum. The conditions of Weierstrass and Legendre

Let Γ be an extremal joining two points AB. We seek those conditions for which the integral I, computed on Γ, can yield an extremum. We shall always assume that it is *a minimum* that is in question.

Let us take on Γ an arbitrary point C with abscissa c and take through C, a line Δ with angular coefficient p distinct from the angular coefficient $y'(c)$ of the tangent to Γ at the point C (fig. 63). Let us take another point D on Γ with abscissa d, near to C, such that $d < c$ (but $d > a$): (In all cases, if C is the point A, we would take $d > c$ and consequently modify that which follows.)

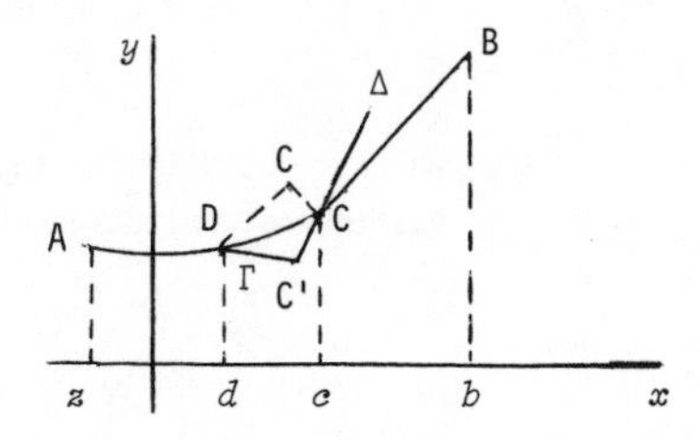

Fig. 63

Finally, let us take on Δ, a point C' near to C and with abscissa c' less than c. The point C' is either above or below Γ depending on whether p is greater or less than $y'(c)$.

We are going to compare the integral

$$\int_d^c f(x,y,y')dx$$

computed on Γ, to that integral computed on a curve Γ' joining D to C', and then on Δ from C' to C. We take a curve Γ' depending on a parameter λ such that we can apply the result of no. 193. With y' always denoting the ordinate of a point of Γ, we take as the ordinate on Γ'

$$\phi(x,\lambda) = y + \varepsilon\lambda(x-d) \quad ,$$

where ε is positive or negative depending on whether p is less than or greater than $y'(c)$ Γ' intersects Δ at a point C'. The integral computer on Γ' between D and C' is a function of λ whose derivative for $\lambda = 0$ is, since Γ is an extremal,

$$\delta x_c \left(f - y' \frac{\partial f}{\partial y'}\right)_c + \delta y_c \left(\frac{\partial f}{\partial y'}\right)_c \quad ,$$

i.e.,

$$\delta x_c \left[f - y' \frac{\partial f}{\partial y'} + p \frac{\partial f}{\partial y'}\right]_c$$

Let us consider the integral computed on Δ from C' to C. This is

$$\int_{C'}^{C} f(x,y,p)dx \quad .$$

It is a function of λ, and, on account of the mean value formula, it has for $\lambda = 0$, a derivative which is

$$-\delta x_c\, f(x,y,p)_c \quad .$$

The integral computed on DΓ'C'ΔC, which reduces to the integral computed on Γ between D and C for $\lambda = 0$, has then a derivative which is

$$-\delta x_c\, E(c,y(c),y'(c),p) \qquad (26)$$

when we set

$$E(x,y,y',p) = f(x,y,p) - f(x,y,y') - (p-y')\frac{\partial f}{\partial y'}(x,y,y') \quad . \qquad (27)$$

From this, we are going to deduce that, for there to be a minimum, it is necessary that

$$E(c,y(c),y'(c),p) \geq 0 \quad . \tag{28}$$

Let us assume, in effect, that this inequality is never satisfied. The first member will be negative. In the expression in (26), δx_c *is negative.* The relation between λ and the abscissa x of C' is given by

$$y(x) + \varepsilon\lambda(x-d) = y(c) + p(x-c) \quad ,$$

which at once gives λ as a function of x, and then the value of the derivative of λ with respect to x for $\lambda = 0$; we thus obtain

$$\delta x_c = \frac{\varepsilon(c-d)}{p-y'(c)} < 0 \quad .$$

The expression (26) is negative. The derivative, for $\lambda = 0$, of the value of I on DΓ'C'ΔC is negative; we can therefore take such a contour on which I is less than the value calculated on Γ between D and C. We can then round off the angles at D, C', C (fig. 64), in such a way so as to obtain a curve from A to B on which y has a continuous derivative, the deformations being sufficiently weak such that I, when computed on this curve, is again less than the value obtained on Γ

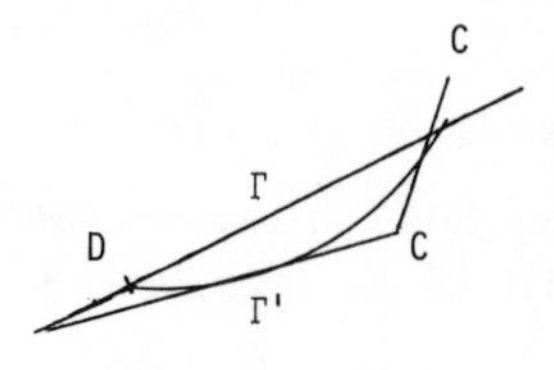

Fig. 64

Thus: It is necessary for the inequality (28) to be satisfied, at every point C of Γ and for every finite value of p (for $p = y'(c)$, the expression E is zero). Thus:

A necessary condition (due to Weierstrass). *In order for an extremal* Γ *joining two points* A, B, *to yield a minimum, it is necessary that, throughout* Γ, *the expression* $E(x,y,y',p)$ *defined by the equality in (27), is positive or zero for any finite* p.

It is to be understood that x,y,y' are the coordinates of a point of Γ and the angular coefficient of the tangent at this point.

Remarks I. We have implicitly assumed that in $f(x,y,y')$ we can assign all possible values to y'. If the variation of y' is subjected to the restrictions, then the number p appearing in the expression E (due to Weierstrass) will also be subjected to these same restrictions, and the condition will remain valid.

Remarks II. In the particular case where y does not appear in the equation, the expression E becomes the expression E_1 of no. 191, the equality in (17): for there to be any possibility of a minimum, it is necessary that $E_1(x,Y',y') \geq 0$ and, on the other hand (no. 191), if $E_1(x,Y',y') > 0$, there is an absolute minimum.

Remarks III. We see that *the condition* $E \geq 0$ *is also necessary for there to be a relative minimum,* i.e., in order that the value of I on Γ is less than its value on infinitely near curves.

Legendre conditions. We have assumed that the second derivative of f with respect to y' is continuous in the domain in question. We can then apply the Taylor formula to the expression E which becomes

$$\frac{(p-y')^2}{2} \frac{\partial^2 f}{\partial y'^2}(x,y,q) \quad ,$$

where q is taken to be between y' and p. When p tends towards y', q tends towards y'. *It is therefore necessary to have*

$$\frac{\partial^2 f}{\partial y'^2}(x,y,y') \geq 0 \quad ,$$

in order to obtain a minimum. This is LEGENDRE'S CONDITION. It does not imply the Weierstrass condition. *But if we assume that, for any finite* p, *we have*

$$\frac{\partial^2 f}{\partial y'^2}(x,y,p) \geq 0 \quad ,$$

along Γ, *then the Weierstrass condition is satisfied.*

Example. For all integrals of the form

$$\int \Phi(x,y)ds \ , \qquad \Phi(x,y) > 0 \quad ,$$

we have $\dfrac{\partial^2 f}{\partial y'^2} > 0$, and the Weierstrass condition is satisfied.

196 - A field of extremals. Hilbert's theorem. The sufficient conditions of Weierstrass and Legendre

We have seen in no. 194 that, by means of conditions relative to the differentiability of the function $f(x,y,y')$ and provided the second derivative with respect to y' is non-zero at a point, we can define a sheaf of extremals about this point, such that through each point, there passes one and only one extremal of this sheaf. We considered the extremals passing through a point A, and also the extremals which intersect a curve arc transversally.

In the first case, if Γ is one of these extremals, and Γ_1 and Γ_2 are two other extremals encircling it, then we obtain a domain Δ bounded by Γ_1 and Γ_2, and, for example, a curve of the sheaf of transversals, such that through every point of Δ there passes one and only one extremal *of the sheaf*. The angular coefficient of this extremal is a function $z(x,y)$ of the coordinates of $M(x,y)$. *This function* $z(x,y)$ *has continuous first partial derivatives.* In effect, if we set $y' = z$, then the Euler equation is transformed into a system of the form $z' = F(x,y,z)$, $y' = z$ whose solution in question is $y = \phi(x,\lambda)$, $z = \psi(x,\lambda)$, [where ψ is in fact $\phi'_x(x,\lambda)$]; this depends on the angular coefficient of the extremal of the sheaf at the point A. z has partial derivatives in x and λ and $y = \phi(x,\lambda)$ has these as well and is solvable in λ (since $\psi'_\lambda = 1$ at the point A, hence ϕ'_λ is non-zero in Δ if Δ is sufficiently close to A and Γ).

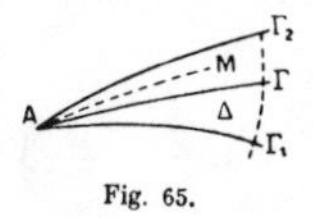

Fig. 65.

Hence z, regarded as a function of x and y, has partial derivatives with respect to x and y, that are continuous.

A FIELD OF EXTREMALS IN GENERAL

Generally, we say that we have defined a field of extremals in a domain Δ, when we have defined a sheaf of extremals depending on one parameter and such that:

1) *Through a point of* Δ *there passes one and only one extremal of the sheaf;*

2) *If* $z(x,y)$ *is taken to be the angular coefficient at* $M(x,y)$ *of the extremal of the sheaf passing through this point* M, *then this function*

$z(x,y)$ admits first partial derivatives which are continuous in Δ. $z(x,y)$ will be called the function of the field.

Following what has just been said, a sheaf of extremals passing through a point A at which $\partial^2 f/\partial y'^2$ is non-zero, defines a sheaf of extremals in a domain ending at the point A, to the right of A, and likewise, another field to the left of A.

Hilbert's Theorem. *If Δ is a sheaf of extremals attached to a function $f(x,y,y')$, $z(x,y)$ the function of the field, and W a rectifiable closed curve inside Δ, then the curvilinear integral*

$$\int_W \left[f(x,y,z) - z\,\frac{\partial f}{\partial y'}(x,y,z)\right]dx + \frac{\partial f}{\partial y'}(x,y,z)dy \tag{29}$$

is zero.

Firstly, let us assume that W is a simple curve. In order to prove the theorem, by applying the Riemann formula, it suffices to show that, in Δ,

$$\frac{\partial f}{\partial y}\left[f(x,y,z) - z\,\frac{\partial f}{\partial y'}(x,y,z)\right] \equiv \frac{\partial}{\partial x}\left[\frac{\partial f}{\partial y'}(x,y,z)\right] .$$

Now this can be written as

$$\frac{\partial f}{\partial y} - z\left(\frac{\partial^2 f}{\partial y\,\partial y'} + \frac{\partial^2 f}{\partial y'^2}\,\frac{\partial z}{\partial y}\right) \equiv \frac{\partial^2 f}{\partial x\partial y'} + \frac{\partial^2 f}{\partial y'^2}\,\frac{\partial z}{\partial x} ,$$

or

$$\frac{\partial f}{\partial y} - \frac{\partial^2 f}{\partial x\partial y'} - z\,\frac{\partial^2 f}{\partial y\,\partial y'} - \left(z\,\frac{\partial z}{\partial y}\quad \frac{\partial z}{\partial x}\right)\frac{\partial^2 f}{\partial y'^2} \equiv 0 . \tag{30}$$

The coefficient of the last term is

$$z\,\frac{\partial z}{\partial y} + \frac{\partial z}{\partial x} = \frac{dz}{dx}$$

when we assume that a displacement is made on the extremal passing through the point (x,y). The first member is nothing but the equation of the extremals written in a different form: it suffices to replace z by y' and dz/dx by y'' in order to recover the Euler equation. Hence (30) is satisfied and the theorem is proved in this case. We then know that it is again true if W is any polygonal line, since W into a finite number of simple closed polygonal lines. It follows that it will be true for all cases since the integral (29) can be considered as the limit of an integral taken on polygonal line (I, 182).

A sufficient condition due to Weierstass. Let us consider an arc AB taken on an extremal Γ and assume that this arc is *inside* a field of extremals Δ, where Γ is an extremal of the field. Let us consider a curve Γ' joining A to B, and contained in the domain Δ, where Γ' is always taken to be one of the curves on which we calculate the integral I. In order to calculate

$$\int_{\Gamma'} f(x,y,y')dx - \int_{\Gamma} f(x,y,y')dx \quad , \tag{31}$$

we can, without changing anything, remove the integral (29) taken along the closed line constituted by Γ', along A to B, then Γ from B to A. Now, on Γ, the expression under the sign of integration in (29), reduces to $f(x,y,z)dx$, since $dy = y'dx = zdx$. On subtracting (29) from (31), there only remains the term relative to Γ'. *The difference (31) is equal to*

$$\int_{\Gamma'} \left[f(x,y,y') - f(x,y,z) + (z - y') \frac{\partial f}{\partial y'} (x,y,z)\right]dx \quad ,$$

or, by introducing the function E defined by (27), it is equal to

$$\int_{\Gamma'} E(x,y,z,y')dx \quad . \tag{32}$$

Consequently:

The Weierstrass condition. *When the arc* AB *of* Γ *belongs to a field of extremals* Δ *and if, taking* $z(x,y)$ *to be the function of the field, we have, for any finite* p *and any* x,y *in* Δ,

$$E(x,y,z(x,y),p) > 0 \quad ,$$

then the integral computed on Γ *is less than the integral calculated on every acceptable curve joining* A *to* B *and remaining in the field.*

In particular, by applying the Taylor formula as at the end of no. 195, we arrive at:

The Legendre condition. *When the arc* AB *of the extremal* Γ *belongs to a field* Δ *and if, throughout* Δ, *we have*

$$\frac{\partial^2 f}{\partial y'^2} (x,y,p) > 0$$

for every finite value of p, *then* Γ *yields a minimum relative to all acceptable curves* Γ' *sitauted in* Δ *and joining* A *to* B.

197 - Conjugate foci. The Jacobi condition. Strict extremum, weak extremum. Absolute extremum

Let us return to the determination of a field of extremals arising from a sheaf of extremals originating from a point A and encircling a particular extremal Γ (no. 196). We can try to extend this field, which only exists in the neighbourhood of A, by going along Γ, and by diminishing its opening. It is necessary for y' to remain finite and to be able to solve the equation $y = \phi(x,\lambda)$ in the neighbourhood of Γ. This will be possible as long as $\partial\phi/\partial\lambda$ is non-zero.

Hence, on Γ, y' will be finite and $\frac{\partial\phi}{\partial\lambda} \neq 0$; this will again be the case in a small domain about Γ, a domain in which we will have a field of extremals. By stating this condition analytically, we obtain an equation due to Jacobi. We note that the condition $\frac{\partial\phi}{\partial\lambda} = 0$ is the condition of contact of $y = \phi(x,\lambda)$ with its envelope. Hence, *we can define a field of extremals containing the part of* Γ *situated between* A *and the first point of contact of* Γ *with the envelope of extremals passing through* A. *There will in general be a point of contact to the right of* A, *and one to the left of* A. These points are called *the conjugate foci of* A *on* Γ. When an extremal arc AB is such that this AB does not contain one of the conjugate foci of A, *then we say that it satisfies the Jacobi condition.*

THE NECESSITY OF THE JACOBI CONDITION (DARBOUX-KNESER)

Let us consider the most simple case, where the extremal Γ originating from A, has first order contact with its envelope T, and *let us assume that the Legendre condition* $\partial^2 f/\partial y'^2 > 0$ *is realised on* Γ from A to the point of contact C with T (fig. 66).

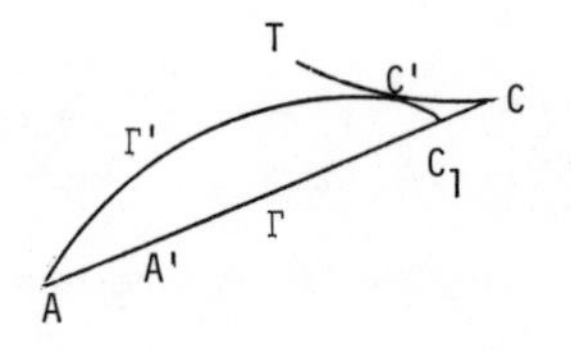

Fig. 66

This condition will then arise in the neighbourhood of Γ. Let Γ' be an extremal originating from A, neighbouring Γ, tangent to T at the point C' and intersecting Γ at C_1 between A and C. We can take Γ' sufficiently near to Γ in order that the domain Δ bounded by Γ, Γ' to the point C' then T from C to C', is a field of extremals, where the extremals of the field are the extremals originating from A and bounded at their point of contact with T. If we join a point A' of Γ close to A the point C_1, by a curve contained in Δ and on which y' remains near to the values $z(x,y)$, then the function $E(x,y,z(x,y),y')$ will be positive, hence the value of I is greater on this curve compared with that obtained on Γ. By passing to the limit, we see that the value of I on Γ' between A and C_1 is greater than that obtained on Γ between these two points. As C_1 is *on* Γ' beyond the conjugate foci of A, we see that, *by means of the conditions which were underlined (the Legendre condition satisfied), the Jacobi condition is necessary.*

STRICT AND WEAK EXTREMA

The sufficient conditions which were given in no. 196 for there to be a minimum are the conditions for which we have a *strict minimum.* The varied curves are the curves of the domain Δ whose tangent is not subjected to any condition. In the problem posed, it is clear that if y was subjected to some condition; as was the case in the Newton problem (no. 192), then the number p appearing in the Weierstrass or Legendre conditions, would be subjected to the same restriction.

But we can assume that I is compared on the neighbouring curves from one another, and on which the angular coefficient y' at points with the same abscissa, has neighbouring values. This is known as seeking a *weak extremum.*

In this case, the number y' which appears in (32) will be close to the number y' relative to the extremal Γ and we see that the following result is obtained:

Every extremal arc AB *which satisfies the Jacobi condition and the Legendre condition* $\partial^2 f/\partial y'^2 > 0$ *provides at least one weak minimum between any two of its interior points.*

ABSOLUTE EXTREMUM

If it is possible to take every rectangle $a \leq x \leq b$, $|y| < H$ in a field of extremals containing the extremal joining the points A and B, and if either the Weierstrass or Legendre condition is satisfied there, then the extremal in question will be seen to yield an absolute minimum.

198 - Examples and various remarks

For all integrals of the form

$$\iint \Phi(x,y)ds \quad , \qquad \Phi(x,y) > 0$$

the Legendre condition is always satisfied; we have $\dfrac{\partial^2 f}{\partial y'^2} > 0$.

Example I. For the integrals

$$\int \frac{ds}{y^\alpha}, \qquad \alpha > 0 , \qquad y > 0 \quad ,$$

we have seen (no. 189) that two points A and B can be joined by one and only one extremal. The extremals passing through a point, do not have any envelope, there are no conjugate foci and the Jacobi condition is always satisfied. The extremals passing through A define a field throughout the half-strip $y > 0$, $a < x < b$ (α is the abscissa of A). Hence, given an extremal Γ joining AB, it will suffice to take A' on Γ to the left of A, in order to define a half-strip containing the arc AB in its interior and which contains a field of extremals of which Γ is a part. Hence, *in all these cases, the extremal joining two points* A *and* B, *yields an absolute minimum.*

This applies in particular to the brachistochrone problem.

Zermelo's formula. Let us consider two extremals Γ, Γ' (see fig. 66) originating from A and tangential at C and C' respectively, to their envelope T. We can take the equation of the extremals originating from A, to be of the form $y = \phi(x,\lambda)$, where λ is the angular coefficient of these curves at the point A. This function $\phi(x,\lambda)$ has continuous partial derivatives as a result of suitable hypotheses on $f(x,y,y')$. The integral I taken on Γ' from A to C' is a function of λ. We can compute its derivative for the value of λ which provides Γ. It is given by the first variation. [If λ_o is the value of λ which gives Γ, then we assume that λ is replaced by $\lambda_o + (\lambda - \lambda_o)$.] Since Γ is an extremal tangent to the curve T, we have

$$\delta I = \delta x_c \left(f - y' \frac{\partial f}{\partial y'}\right)_c + \delta y_c \left(\frac{\partial f}{\partial y'}\right)_c = \delta x_c f(x,y,y')_c$$

and δx_c is non-zero when the contact is ordinary, which is assumed.

On the other hand, let us consider the value of I computed on the curve T, from C' to C. This is also a function of λ whose derivative is

$$- \delta x_c f(x,y,y') \quad .$$

It follows that, for the value λ corresponding to Γ, the derivative of

$$\int_A^{C'} f(x,y,y')dx + \int_{C'}^{C} f(x,y,y')dx$$

is zero, and since Γ is arbitrary, we see that

$$I(\Gamma') + I(C'TC) = I(\Gamma) \quad , \tag{33}$$

where the values of I are taken on Γ' from A to C', on T from C' to C, and in the second member, on Γ from A to C. *This is Zermelo's formula.*

Example II. Let us take the integrals

$$I = \int y^{\alpha} ds \,, \qquad \alpha > 0 \,, \quad y > 0 \quad .$$

We have seen in no. 189 that two points A,B cannot always be joined by an extremal, and that when they can be connected, there are two extremals solving the problem. $f''y'^2$ is always positive.

If a point A is given, the points B which can be attained by an extremal originating from A, are inside the envelope T of extremals passing through A. The extremals are convex curves, the envelope itself is convex and is tangent to Ox at the foot D of the perpendicular from A onto Ox (fig. 67). We note (as did L. Lindelöf and Wiman) that, for C the point of contact of the extremal Γ with T, the tangents to Γ at A and C, intersect on Ox. (This can be seen by taking the equation of the extremals originating from A in the form $\mu y = \psi(\mu x + v)$, where v and μ are two parameters corresponding to the translations and homotheties.) If B is inside T, there are two extremals joining A to B, but only one Γ, has its point of contact C with T outside of AB and satisfies the Jacobi condition.

Let us take A' on Γ just to the left of A; B will again be inside the interior of the envelope T' of the extremals originating from A'. These extremals, bounded at their point of contact with T' define a field in the domain Δ bounded by T' and by the parallels to 0 through A' and B' which is taken just to the right of B. Hence I, computed on Γ between A and B is a minimum relative to its value computed on all acceptable curves joining A to B remaining above T', and consequently also remaining above T since A' is arbitrarily near to A. Likewise, passing to the limit shows that, if B is on T, then I computed on Γ from A to B is *at most* equal to its value on another acceptable curve situated above T (except at B).

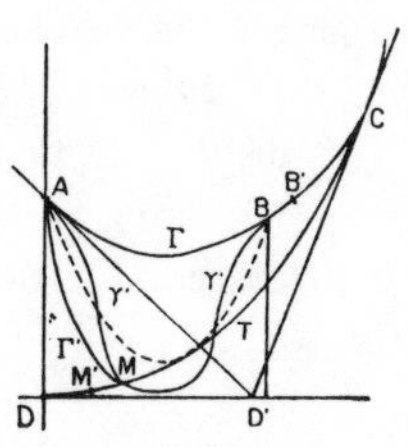

Fig. 67.

Let us consider a curve γ' joining A to B, but intersecting T at an initial point M from A, or tangent to T at M. We will not increase the value of I on γ' by replacing the part of γ' situated between A and M by the extremal Γ' joining A to M, which is tangent to T at M.

On account of Zermelo's formula, we can replace Γ' by the curve constituted by an extremal Γ'' from A, tangent to T at M' between D and M, to which we adjoin the arc M'M of T. We can pass to the limit and take M' at D. We can then replace γ' between A and M by the segment AD and by the arc DM of T; the value of I can only diminish. But we can go further by taking advantage of the form of I. I computed on a curve joining D to B is greater in value than that computed on the segment BD', where D' is the point of abscissa b of Ox. In effect, for a similar value of y, the ds corresponding to this curve is greater than that corresponding to the segment [compare (I, 166)]. Finally, the integral on the path γ' is greater than that computed on the two segments DA, D'B, to which we can add DD', on which $y = 0$.

This result is again tenable, *a fortiori,* if B is to the right of T. Summarising: *The integral* I *computed on any acceptable curve joining* A *to* B, *is greater than the integral calculated on the contour* ADD'B *if* B *is to the right of* T *or on* T. *It is greater than the minimum of the two values computed, one on this contour, and the other on the extremal* Γ *joining* A *to* B, *not touching* T *between* A *and* B, *when* B *is to the left of* T.

We see that: *There is always an absolute minimum, with the condition of admitting, in certain cases, a curve presenting angular points.*

We remark that $y' = \infty$ can be regarded as satisfying the equation of the extremals; the contour ADD'B is formed by extremals.

It remains to compare the values on ADD'B and on the curvilinear extremal when B is inside T. The result of this comparison when $\alpha = 1/2$ can be found in Hadamard's *Leçons sur le calcul de variations.*

A REMARK ON THE DISCONTINUOUS SOLUTIONS

Hilbert observed that it is not possible to have discontinuous solutions, i.e., presenting angular points, when $\partial^2 f/\partial y'^2$ *is non-zero* (and is finite). In this case, we say that the problem is *regular* (Kneser).

Let us assume, in effect, that the point $A(x_o, y_o)$ is an angular point, and let B be taken on the left of A, C to the right of A and Δ a line passing through A (fig. 68). If we vary A on Δ and replace the line BAC by BA'C depending on a parameter λ, then the derivative with respect to this parameter must be zero.

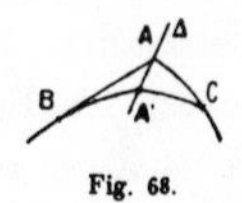

Fig. 68.

Hence, on account of the formula for the first variation, we must have for any x_o, y_o,

$$(\delta y_o - y_o'\delta x_o)\,\frac{\partial f}{\partial y'}\,(x_o,y_o,y_o') - (\delta y_o - y_1'\delta x_o)\,\frac{\partial f}{\partial y'}\,(x_o,y_o,y_1')$$

$$- \delta x_o\ [f\ x_o,y_o,y' - f\ x_o,y_o,y_o'\ ] = -0 \quad ,$$

where y_o' is the value of y' at the point to the left of A and y_1' the value of y' at the point to the right of A. In particular, it will be necessary to have

$$\frac{\partial f}{\partial y'}\,(x_o,y_o,y_o') = \frac{\partial f}{\partial y'}\,(x_o,y_o,y_1') \quad ,$$

hence

$$\frac{\partial f}{\partial y'}\,(x_o,y_o,y')$$

takes the same value at least twice; its derivative with respect to y' must be zero.

In the regular case, the only possible angular points will be *at the boundary of the domain in question.*

IV. ON CERTAIN VARIATIONAL PROBLEMS. FIRST ORDER CONDITIONS

199 - The case of several unknown functions

To simplify matters, we will only conisder two unknown functions y,z, but there could be n of them. We consider an integral

$$I = \int_a^b f(x,y,z,y'z')dx \quad ,$$

where f is a given function of the five variables above. We assume that f

has continuous partial derivatives of the first two orders, with respect to the variables in question. y and x are unknown functions of x, admitting continuous derivatives y',x' on the segment (a,b) and taking the given values for $x = a$ and $x = b$. We can assume that I is calculated on a curve Γ joining two given points of three dimensional space and only intersecting the planes parallel to Oyz at a point. We can write

$$I = \int_A^B f(x,y,z,y',z')dx \quad .$$

We wish to determine y and z, i.e., Γ, in order that I should be a maximum or a minimum. We can proceed as in no. 187, by assuming that Γ is replaced by a neighbouring curve depending on one parameter. This curve will be defined by increments $\lambda\delta y$, $\lambda\delta z$ of y and z respectively, where δy and δz are zero at the points A and B, and admitting continuous derivatives. We will state that the derivative with respect to λ is zero for $\lambda = 0$; this will yield a necessary condition. We will have

$$\int_a^b \left(\frac{\partial f}{\partial y}\,\delta y + \frac{\partial f}{\partial z}\,\delta z + \frac{\partial f}{\partial y'}\,\delta y' + \frac{\partial f}{\partial z}\,\delta z'\right) dx = 0 \quad , \tag{34}$$

where $\delta y'$ and $\delta z'$ are the derivatives of δy and δz. We integrate by parts as in no. 187, which will give us

$$\int_a^b \left\{\delta y \left[\frac{\partial f}{\partial y} - \frac{d}{dx}\frac{\partial f}{\partial y'}\right] + \delta z \left[\frac{\partial f}{\partial z} - \frac{d}{dx}\frac{\partial f}{\partial z'}\right]\right\} dx = 0 \quad .$$

This occurs in particular for $\delta z = 0$ or $\delta y = 0$; the fundamental lemma (no. 188) again shows that, for (34) to hold, it is necessary and sufficient that the differential system

$$\frac{\partial f}{\partial y} - \frac{d}{dx}\frac{\partial f}{\partial y'} = 0 \, , \qquad \frac{\partial f}{\partial z} - \frac{d}{dx}\frac{\partial f}{\partial z'} = 0 \quad , \tag{35}$$

is satisfied. These equations are the Euler equations. They are second order equations; the system is equivalent to a system of four first order equations. Explicitly, we have

$$\frac{\partial^2 f}{\partial y'^2}\, y'' + \frac{\partial^2 f}{\partial y'\partial z'}\, z'' + \frac{\partial^2 f}{\partial y'\partial y}\, y' + \frac{\partial^2 f}{\partial y'\partial z}\, z' + \frac{\partial^2 f}{\partial y'\partial x} - \frac{\partial f}{\partial y} = 0 \quad ,$$

and the analogous equation. We can solve for y'' and z'' *if the determinant*

$$\frac{\partial^2 f}{\partial y'^2}\frac{\partial^2 f}{\partial z'^2} - \left(\frac{\partial f}{\partial y'\partial z'}\right)^2$$

is non-zero. We will say that the problem is *regular* when this determinant is non-zero for the values in question. The integral curves of the system in (35) are the extremals; they depend on four parameters. We could then, in certain cases, solve the problem, which consists in sending an extremal through two points. We would then have solved the first order problem.

PARTICULAR CASES

In certain cases, we will have first integrals (of the system of the fourth order). If f does not contain y, we have $\partial f/\partial y' = \text{const.}$, hence if f contains neither y nor z, matters are reduces to a system of two equations of the first order

$$\frac{\partial f}{\partial y'} = \text{const} \quad , \qquad \frac{\partial f}{\partial z'} = \text{const} \quad .$$

Example. If we have

$$I = \int \phi(z)ds \ , \ ds^2 = dx^2 + dy^2 + dz^2 \ , \ \phi(z) > 0 \quad ,$$

we can take z as an independent variable, as we have seen in no. 189; this does not change the extremals (other than the curves on which $z = \text{const}$). We will have

$$\int \phi(z) \sqrt{1 + x'^2 + y'^2} \ dz \quad .$$

The extremals will be given by

$$\phi(z) \frac{x'}{\sqrt{1 + x'^2 + y'^2}} = k \ , \qquad \phi(z) \frac{y'}{\sqrt{1 + x'^2 + y'^2}} = k' \quad ,$$

whence we deduce

$$y'k - x'k' = 0$$

and the new first integral $yk - xk' = k''$; k, k' and k'' are constants. The extremals are in the planes parallel to Oz. Matters are reduced to a planar problem and to the integration, by taking $y = 0$ for example of a single equation

$$\phi(z) \frac{x'}{\sqrt{1 + x'^2}} = k$$

which is integrated by quadratures.

We remark that if a projection is made onto a curve Γ joining A to B on the plane parallel to Oz passing through AB, then the integral I computed on this projection λ is less than that on Γ by virtue of the reasoning which was already made [no. 198 and (I, 166)]. Matters are reduced *a priori* to a planar problem. This is the case for the brachistochrones, in particular.

THE FIRST VARIATION

The considerations of no. 193 can be extended. The first variation is

$$\delta I = \int_A^B \left\{ \delta y \left[\frac{\partial f}{\partial y} - \frac{d}{dx} \frac{\partial f}{\partial y'} \right] + \delta z \left[\frac{\partial f}{\partial z} - \frac{d}{dx} \frac{\partial f}{\partial z'} \right] \right\} dx$$
$$+ \left[\delta x \left(f - y' \frac{\partial f}{\partial y'} - z' \frac{\partial f}{\partial z'} \right) + \delta y \frac{\partial f}{\partial y'} + \delta z \frac{\partial f}{\partial z'} \right]_A^B \quad . \tag{36}$$

As in no. 194, it will allow us to deal with problems with variable limits; the points A and B are displaced respectively on curves or surfaces. The curves susceptible to providing an *extremum* will be the extremals intersecting *transversally*, the curves or surfaces in question. For example, if A is on a surface S and if $p, q, -1$ are the coefficients of the tangent plane to S, then we must have

$$\delta x_A \left(f - y' \frac{\partial f}{\partial y'} - z' \frac{\partial f}{\partial z'} \right)_A + \delta y_A \left(\frac{\partial f}{\partial y'} \right)_A + \delta z_A \left(\frac{\partial f}{\partial z'} \right)_A = 0$$

for all δx, δy, δz relative to A, such that $p\delta x + q\delta y - \delta z = 0$. We obtain

$$\frac{f - y' \frac{\partial f}{\partial y'} - z' \frac{\partial f}{\partial z'}}{p} = \frac{\frac{\partial f}{\partial y'}}{q} = - \frac{\partial f}{\partial z} \quad ,$$

which yields two relations between the parameters on which the extremal depends (since we must also state that A is on S).

Similarly, if B is on a curve, then along this curve, the direction parameters of the tangent δx, δy, δz will be related by the conditional equation

$$\delta x \left(f - y' \frac{\partial f}{\partial y'} - z' \frac{\partial f}{\partial z'} \right) + \delta y \frac{\partial f}{\partial y'} + \delta z \frac{\partial f}{\partial z'} = 0 \quad .$$

We will then have two relations between the four parameters on which the extremal depends.

THE SHEAF OF TRANSVERSALS

Let us take a surface S. Through each point M of S, there passes an extremal Γ intersecting S transversally at M. Let us assume that M describes a curve Λ on S; the corresponding extremals Γ describe a surface Σ. The curves of Σ which are cut transversally by the extremals Γ are determined by a first order differential equation; they form a sheaf of curves Λ'. *The integral* I *calculated on a curve* Γ *between the point* M *where it intersects* Λ *and the point* M' *where it intersects* Λ', *has a constant value*. This can be seen as in no. 194. Conversely, if on the extremals Γ, we take from the point M on Λ, an arc MM' such that I is constant, we obtain a curve Λ'. It follows that when we take on the extremals Γ intersecting S, an arc MM', from the point M on S, on which I remains constant, then the locus of M' is a surface whose curves are intersected transversally by the Γ; this surface is cut transversally by the Γ. Thus:

Commencing from a given surface S, *we can define a sheaf of surfaces containing* S *such that this sheaf is cut transversally by the extremals depending on two parameters* (these extremals constitute a congruence, see no. 249), I computed on one of these extremals, between two or these surfaces, is constant.

We arrive at analogous results by considering the extremals intersecting a curve transversally, or passing through a point.

It is clear that these properties occur by means of the differentiability conditions, and in the neighbourhood of a point.

A PARTICULAR CASE

Once again, if we have

$$I = \int F(x,y,z)ds \ , \quad ds^2 = dx^2 + dy^2 + dz^2 \quad , \tag{37}$$

the condition of transversality is an orthogonality condition. In particular, if $F \equiv 1$, the extremals are lines and the preceding properties concern the parallel surfaces and the canal surfaces (see no. 243).

Remark. The first variation gives, in particular, the variation of the length L of a segment which is displaced: if MM' is this segment, then the relationship (36) gives for the integral (37), with $F \equiv 1$,

$$\delta L = \left[\frac{\delta x \quad y'\delta y \quad z'\delta z}{\sqrt{1+y'^2+z'^2}}\right]_M^{M'} = \frac{\overrightarrow{MM'}\ \overrightarrow{\delta M'} \quad \overrightarrow{M'M}\ \overrightarrow{\delta M}}{|\overrightarrow{MM'}|}$$

We recover that which is obtained by differentiating $|\overrightarrow{MM'}|^2$.

Example. By assuming that a point M describes an ellipse E and that from M the tangents MM', MM" are taken to a homofocal ellipse E' inside, M and M' being the points of contact (fig. 69), then, a well-known theorem due to Poncelet, we have

$$\frac{\overrightarrow{MM'}}{|\overrightarrow{MM'}|}\,\overrightarrow{\delta M} = -\,\frac{\overrightarrow{MM''}}{|\overrightarrow{MM''}|}\,\overrightarrow{\delta M}\ .$$

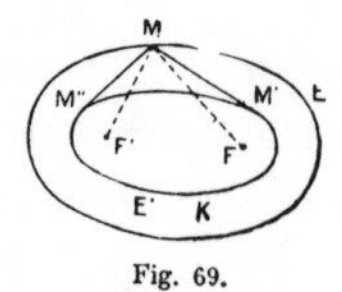

Fig. 69.

Applying the formula (38) to the segments MM' and MM", we deduce that the sum of the lengths of these segments subtracted from the length of the arc M'M" of E' is constant. Consequently, if we hold, on an ellipse E', a closed thread of length greater than the ellipse, MM'KM"M say, then the locus of M is an ellipse homogocal to E' (a theorem of Graves).

It can be shown that the theorems relating to the tangents to the ellipse extend to the ellipses of the Poincaré geometry, defined as the loci of points whose sum of distances to two fixed points, is constant. The theorem of Graves can be extended (the lengths being those of the Poincaré geometry).

For the hyperbola, we have a closely related proposition, due to Chasles, that can easily be deduced.

A SUFFICIENT CONDITION FOR THE EXTREMUM

The method of Weierstrass-Hilbert extends. We will define a field of extremals in a domain Δ of the space (x,y,z) by taking a congruence of extremals (extremals depending on two parameters) such that:

1) Through a point (x,y,z) of Δ, there passes one and only one extremal;

2) At the point (x,y,z) the values of y' and z' on the extremal associated to this point are functions $u(x,y,z)$ and $v(x,y,z)$ which admit continuous first partial derivatives. In a field of extremals, the sufficient condition of Weierstrass for the minimum will be $E(x,y,z,y',z,'u,v) > 0$, where E denotes the expression

$$f(x,y,z,y',z') - f(x,y,z,u,v) - (y'-u)\,\frac{\partial f}{\partial y'}\,(x,y,z,v)$$

$$- (z'-v)\,\frac{\partial f}{\partial z'}\,(x,y,z,u,v) \quad ,$$

where $u = u(x,y,z)$, $v = v(x,y,z)$. If it is satisfied for all finite values of y' and z', then we have a strict minimum (relative to the curves of Δ). By introducing the second derivatives from the Taylor formula, then we obtain a sufficient condition due to Legendre: it suffices that the quadratic form

$$h^2\,\frac{\partial^2 f}{\partial y'^2}\,(x,y,z,y',z') + 2hk\,\frac{\partial^2 f}{\partial y'\partial z'}\,(x,y,z,y',z')$$

$$+ k^2\,\frac{\partial^2 f}{\partial z'^2}\,(x,y,z,y',z')$$

should be positive definite for any y' and z'. Similarly, we obtain a sufficient condition for the weak minimum.

200 - Isoperimetric problems

These are problems of the following type. Let us take the case of a single unknown function y, and the integral I of no. 186. We seek to determine y, hence the curve Γ joining A to B, such that I is a minimum, but confine ourselves to considering the curves Γ for which $g(x,y,y')$ is another given function and C is a constant; we have

$$\int_a^b g(x,y,y')dx = C \quad . \tag{39}$$

We impose on g, the same hypotheses as on $f(x,y,y')$.

Let us assume that the problem is solvable. Let us consider again, the one parameter family of curves in no. 193. For these curves, $I(\lambda)$ must have a zero derivative for $\lambda = 0$. On the other hand, for the curves in question, the derivative of the first member of (39), which is constant, must be zero. Therefore, we will have (on integrating by parts)

$$J(\delta y) \equiv \int_a^b \delta y \; \left[\frac{\partial f}{\partial y} - \frac{d}{dx}\left(\frac{\partial f}{\partial y'}\right)\right] dx \tag{40}$$

for every function δy satisfying the conditions of no. 187, if (39) holds, hence if

$$K(\delta y) \equiv \int_a^b \delta y \; \left[\frac{\partial g}{\partial y} - \frac{d}{dx}\left(\frac{\partial g}{\partial y'}\right)\right] dx = 0 \quad . \tag{41}$$

Let us denote by δy_1, an initial function δy for which $K(\delta y_1) \neq 0$. Such a function exists if Γ is not an extremal of the variational problem relative to g. *For* δy *now taken to be arbitrary* (but satisfying the conditions of no. 187), let us take $\phi(x,\lambda) = y + \lambda(\delta y - k(\lambda)\delta y_1)$, $k(\lambda)$ is determined by (39) and $k = k(0)$ is given by tne equality $K(\delta y) = kK(\delta y_1)$. We must then have $J(\delta y) - kJ(\delta y_1) = 0$, i.e.,

$$J(\delta y) - \frac{J(\delta y_1)}{K(\delta y_1)} K(\delta y) = 0 \quad .$$

Consequently, *for any* δy, *we must have*

$$\int_a^b \delta y \left[\frac{\partial(f+\rho g)}{dy} - \frac{d}{dx} \frac{\partial(f+g)}{\partial y'}\right] dx = 0 \quad ,$$

$$\rho = -\frac{J(\delta y_1)}{K(\delta y_1)} \quad .$$

We have to write the conditions of the extremum for the integral along $f + \rho g$. This holds in the case where Γ is an extremal of the problem relative to g. *In order for the problem to be possible, it is necessary for the curve* Γ *to be an integral curve of the Euler equation*

$$\frac{\partial(f + \rho g)}{\partial y} - \frac{d}{dx} \frac{\partial(f + \rho g)}{\partial y'} = 0 \quad , \tag{42}$$

where ρ *is an unknown constant.* When the curve Γ is a solution of this equation, joins A to B and satisfies the condition (39), then the equation (40) holds for all functions δy for which (41) is zero. *The condition of the third order is verified.* As the integral curves of (42) depend on three constants (ρ and the two constants of integration), then the problem of the first order can be dealt with in certain cases.

Example I. More precisely, the isomperimetric problem, which gave its name to all the others, consists in finding a curve Γ joining the points A and B taken on Ox ($x = a$, $y = 0$; $x = b$, $y = 0$), having a given length L and such that the area of the domain bounded by Ox and Γ, is a maximum. We assume that Γ is only intersected at one point by a parallel to Oy and a tangent which varies continuously. We have here

$$I = \int_a^b y\,dx \,, \qquad \int_a^b \sqrt{1 + y'^2}\, dx = L \quad .$$

The equation determining the curves Γ is

$$1 + \rho \frac{d}{dx} \frac{y'}{\sqrt{1+y'^2}} = 0 \quad ,$$

which admits the first integral

$$\frac{y'}{\sqrt{1+y'^2}} = \frac{-x}{\rho} + \lambda \quad .$$

The result of integration is

$$(y - \mu\rho)^2 + (x - \lambda\rho)^2 = \rho^2 \quad ,$$

where λ, μ, ρ are constants. Γ is therefore a circular one. The problem of the first order such as that put into the equation will be possible if $L > b - a$ and if $2L < \pi(b-a)$ since we must join A to B by a circular arc intersected at a single point by a parallel to Oy. But we can re-introduce the remark made in no. 189. When we remove the condition that Γ is only intersected at a point by a parallel to Oy and if the problem is assumed to be solvable, then we can consider a secant line intersecting Γ at two consecutive points C and D. It will be necessary for the arc CD to be a solution of the isoperimetric problem in question, with the curves joining C to D and whose length is given. If the perpendiculars to the chord CD only intersect the arc CD at a point, then this arc must be that of a circle. We can deduce that the entire arc AB is an arc of a circle.

It remains to be seen if the arc of the circle effectively gives the maximum [cf. (I, 92)].

Example II. Given a curve Γ of length L joining A to B, we seek the minimum of

$$\int_a^b y ds \quad ,$$

where ds is the arc element. This amounts to finding the weighing homogeneous curve (Oy vertical) of given length, whose centre of gravity is the lowest possible, i.e., the curve of equilibrium of a homogeneous weighing thread. We have here

$$\int_a^b ds = L \quad .$$

The equation is obtained by stating that

$$\int_a^b (y + \rho) ds$$

is an extremum. The problem has been dealt with (see no. 189). Γ will be a small chain whose base is parallel to Ox.

201 - The case of double integrals. Minimal surfaces

The function $f(x,y,z,p,q)$ given and admitting continuous first and second partial derivatives in the domain of values in question, we consider the integral

$$I = \iint_\Delta f(x,y,z,p,q)dxdy \quad ,$$

computed in a domain Δ of the plane Oxy bounded by a simple closed curve C, where z is the representation $z = \phi(x,y)$ of a point of a surface S, and p and q the partial derivatives of ϕ with respect to x and y. The surface S is constrained to pass through a given curve C' whose projection onto Oxy is C. We seek to determine S in order for I to be a maximum or minimum. We assume p and q to be continuous.

We obtain the first order conditions by always proceeding in the same manner. Supposing that a surface S fulfills the requirement, we introduce a family of surfaces defined by $Z = z + \lambda\delta z$, where λ is the representation on S, λ a parameter and δz an arbitrary function, zero on C. The derivative of the corresponding integral,

$$I(\lambda) = \iint_\Delta f(x,y,z + \lambda\delta z,\ p + \lambda\delta p,\ q + \lambda\delta q)dxdy \quad ,$$

must be zero for $\lambda = 0$. δp and δq are partial derivatives of δq with respect to x and y. We then have the condition

$$\iint_\Delta \left[\frac{\partial f}{\partial z}\delta z + \frac{\partial f}{\partial p}\delta p + \frac{\partial f}{\partial q}\delta q\right] dxdy = 0 \quad , \tag{43}$$

where the partial derivatives appearing in the integral are taken at the point x,y,z,p,q of S. We transform the terms in δp and δq by applying the Riemann formula (I, 142), by assuming that the conditions are fulfilled in order for it to be applicable. For example, we have

$$\int_{C^+} \left(\frac{\partial f}{\partial z}\delta z\right) dy = \iint_\Delta \left[\frac{\partial f}{\partial p}\delta p + \delta z \frac{d}{dx}\left(\frac{\partial f}{\partial p}\right)\right] dxdy \quad .$$

The first member is zero since $\delta z = 0$ on C and we see that we can transform the term in δp appearing in (43) to a term in δz. Likewise, we deal with the term in δq and the equation (43) takes the form

$$\iint_\Delta \left[\frac{\partial f}{\partial z} - \frac{d}{dx}\frac{\partial f}{\partial p} - \frac{d}{dy}\frac{\partial f}{\partial q}\right] \delta z\ dxdy \quad .$$

This being the case for all functions δz imagined, we see that, as in no. 188, we must have

$$\frac{\partial f}{\partial z} - \frac{d}{dx}\frac{\partial f}{\partial p} - \frac{d}{dy}\frac{\partial f}{\partial q} = 0 \tag{44}$$

[and this suffices for (43) to occur].

If the problem posed is at all possible, then it can only be solved in terms of surface solutions of the partial differential equation (44). We must then see if it is possible to make such a surface pass through the given curve C'. This is quite a complicated problem which consists of passing an extremal through two points.

The equation (44) is the Lagrange equation of the problem. It is an equation of the Monge-Ampère type (see Chapter XVI). The surface solutions are known as *extremal surfaces,* or in short, *extremals*.

AN EXAMPLE. MINIMAL SURFACES

In particular, when we seek to minimise the area of the surface S passing through a closed curve C', then we have to consider the integral

$$I = \iint_\Delta \sqrt{1 + p^2 + q^2}\; dxdy \quad .$$

The Lagrange equation here is,

$$\frac{d}{dx}\frac{p}{\sqrt{1 + p^2 + q^2}} + \frac{d}{dy}\frac{q}{\sqrt{1 + p^2 + q^2}} = 0 \quad ,$$

and, on expanding, and using the Monge notations, we obtain

$$r(1 + q^2) - 2pqs + t(1 + p^2) = 0 \quad . \tag{45}$$

We shall return to these surfaces in Chapter XIV.

202 - The parametric form. Geodesics

In many variational problems of a geometric origin, we seek a curve Γ mimising a curvilinear integral attached to this curve, joining two points AB and which is not subjected to intersecting a parallel to Oy only at a point, in the case of the plane, and a parallel to Oyz in the spatial case. For example, it is a question of minimising

$$\int_A^B f(x,y,z)ds \tag{46}$$

where ds is the differential of the arc of Γ. *We then take* Γ *to be in the parametric form.* We will consider a proper parametric representation of Γ

$$x = \phi(t), \qquad y = \psi(t), \qquad z = \theta(t) \quad , \tag{47}$$

such that ϕ', ψ', θ' (which we assume to exist), are non-zero simultaneously and such that the arc AB of Γ is obtained by varying t from α to β, where α and β are numbers chosen once and for all; for example $\alpha = 0$, $\beta = 1$. The integral (46) becomes

$$\int_\alpha^\beta f(\phi(t),\psi(t),\theta(t))\sqrt{\phi'(t)^2 + \psi'(t)^2 + \theta'(t)^2}\ dt \quad . \tag{48}$$

Matters are reduced to finding the minimum of an integral of the form

$$\int_\alpha^\beta F(x,y,z,x',y',z')dt \tag{49}$$

where F *is not an arbitrary function.* In the case (48), we see that it is a *homogeneous function of degree* 1, with respect to x',y',z'.* It can be seen that for an integral (49) to be independent of the representation in (47), i.e., to define a number attached to a curve defined by (47), it is *necessary to have* F *homogeneous of degree* 1 *in* x',y',z'. The first order conditions of the extremum for the integral (49) are, following no. 199,

$$\frac{\partial F}{\partial x} - \frac{d}{dt}\frac{\partial F}{\partial x'} = 0 \ , \qquad \frac{\partial F}{\partial y} - \frac{d}{dt}\frac{\partial F}{\partial y'} = 0 \ , \qquad \frac{\partial F}{\partial z} - \frac{d}{dt}\frac{\partial F}{\partial z'} = 0 \quad . \tag{50}$$

These equations, which define the extremals, are not independent. We have, in effect, since F is homogeneous of degree 1 in x',y',z',

$$F \equiv x'\frac{\partial F}{\partial x} + y'\frac{\partial F}{\partial y'} + z'\frac{\partial F}{\partial z'} \quad ,$$

hence on differentiating with respect to t, we have

$$\sum \frac{\partial}{\partial x}x' + \sum \frac{\partial}{\partial x'}\frac{dx'}{dt} = \frac{dF}{dt} = \sum x'\frac{d}{dt}\frac{\partial}{\partial x'} + \sum \frac{\partial}{\partial x'}\frac{dx'}{dt}$$

or

$$\sum x'\frac{d}{dt}\frac{\partial F}{x'} - \sum x'\frac{\partial F}{\partial x} = 0 \quad . \tag{51}$$

On multiplying the first equation (50) by x', the second by y', the third by z' and adding, we obtain the equation (51). The equations in (50) reduce to two.

---

* We use the term "homogeneous" as previously applied (I, 116). In this same case, some authors say *positively homogeneous* (see no. 65).

Following the remarks which have already been made several times, the extremals defined by the equations in (50), are the same as those obtained by taking x, y or z as the independent variable, instead of t, which is possible since

$$F(x,y,z,x',y',z') \equiv x'F(x,y,z,1,\frac{y'}{x'},\frac{z'}{x'}) \quad .$$

But the equations in (50), where x. y. z play a symmetric role, can be more convenient, similarly, for the study of the first order problem.

We can extend the definition of the field of extremals and give a sufficient condition for there to be a minimum relative to the curves situated in this field. We introduce the function

$$E(x,y,z,x',y',z',X',Y',Z') = F(x,y,z,X',Y',Z')$$

$$- F(x,y,z,x',y',z') - \sum (X' - x') \frac{\partial F}{\partial x'} (x,y,z,x',y',z')$$

with the condition

$$x'^2 + y'^2 + z'^2 = X'^2 + Y'^2 + Z'^2 = 1 \quad .$$

At every point of the field, x',y',z' will be defined; these will be the direction cosines of the extremal of the field passing through this point (where the positive tangent is taken continuously in the direction from A to B). *It will suffice, for all other values of* X', Y', Z', *to have throughout the field,*

$$E(x,y,z,x',y',z',X',Y',Z') > 0 \tag{52}$$

in order for there to be a strict minimum.

With respect to the Euler condition for the homogeneous functions, the expression for E can be simplified; we have

$$E = F(x,y,z,X',Y',Z') - \sum X' \frac{F}{x'} (x,y,z,x',y',z') \quad . \tag{53}$$

Once again, we can express E, taken in its original form, by using the Taylor formula and obtain sufficient conditions of the Legendre type: it will suffice to have the quadratic form in X', Y', Z',

$$\left(X' \frac{\partial}{\partial x'} + Y' \frac{\partial}{\partial y'} + Z' \frac{\partial}{\partial z'}\right)^2 F(x,y,z,x',y',z')$$

positive definite, for any x,y,z in the field and x',y',z', in order for there to be a strict minimum. If the condition holds for x,y,z on Γ, where x',y',z' are the direction cosines of the tangent to Γ in the direction from A to B, then we at least obtain a weak minimum.

In the case of integrals of the form (46), where $f(x,y,z) > 0$, we see directly that the expression in E is positive. Once a part of the extremal belongs to a field, it will provide a strict minimum relative to all the curves remaining in this field.

GEODESICS

The geodesics of a surface S are the lines of S joining two points A and B of S and whose length is a minimum. If the surface is given in its parametric form, as a function of two parameters u and v, we have

$$ds^2 = \mathrm{E}du^2 + 2\mathrm{F}dudv + \mathrm{G}dv^2 \quad ,$$

where E, F, G are functions of u and v. We intend minimising

$$\int_{\mathrm{A}}^{\mathrm{B}} ds \quad .$$

When we express u and v as functions of a parameter t, as we have just done in the spatial case, then we obtain the integral

$$\int_{\alpha}^{\beta} \sqrt{\mathrm{E}u'^2 + 2\mathrm{F}u'v' + \mathrm{G}v'^2}\; dt \quad . \tag{54}$$

The geodesics will be defined by

$$\frac{\partial\sqrt{\Delta}}{\partial u} - \frac{d}{dt}\,\frac{\mathrm{E}u' + \mathrm{F}v'}{\sqrt{\Delta}} = 0 \;, \qquad \frac{\partial\sqrt{\Delta}}{\partial v} - \frac{d}{dt}\,\frac{\mathrm{F}u' + \mathrm{G}v'}{\sqrt{\Delta}} = 0 \quad ,$$

where $\sqrt{\Delta}$ is the root appearing in the integral in (54).

They provide at least one weak minimum between two of their points when they are sufficiently near, for the second member of the analogous expression to (53) has the sign of

$$\begin{aligned}
&(\mathrm{EU'}^2 + 2\mathrm{FU'V'} + \mathrm{GV'}^2)(\mathrm{E}u'^2 + 2\mathrm{F}u'v' + \mathrm{G}v'^2) \\
&\quad - (\mathrm{E}u'\mathrm{U'} + \mathrm{F}(u'\mathrm{V'} + v'\mathrm{U'}) + \mathrm{G}v'\mathrm{V'})^2 \\
&\quad = (\mathrm{EG} - \mathrm{F}^2)(\mathrm{U'}v' - \mathrm{V'}u')^2 > 0 \quad .
\end{aligned}$$

We shall discuss a geometric property of these curves in Chapter XIII.

203 - Canonical equations

Let us consider the general case where we seek the minimum of an integral

$$\int_A^B f(x,y_1,\ldots,y_n,y'_1,\ldots,y'_n)dx \quad .$$

The extremals will be given by

$$\frac{\partial f}{\partial y_j} - \frac{d}{dx}\left(\frac{\partial f}{\partial y_j}\right) = 0 \;, \quad j = 1,2,\ldots,n \quad . \tag{55}$$

We can replace this system of n equations of the second order by a system of $2n$ equations of the first order by taking the y'_j and the y_j as the unknown functions. Hamilton devised the most simple system of order $2n$, by introducing as new unknown functions, besides the y_j, the functions

$$p_j = \frac{\partial f}{\partial y_j} \;, \qquad j = 1,2,\ldots,n \quad . \tag{56}$$

We assume that we can extract, from these equations, the y'_j as functions of the p_j and y_j, which will be the case when the corresponding functional determinant is non-zero in the domain in question. This hypothesis is that made in no. 199 when we have assumed the problem to be regular. *Let us assume then, that the problem is regular.* For the sake of simplicity, Hamilton introduced, in place of the function f, the function

$$H(x,y_1,\ldots,y_n,p_1,\ldots,p_n) = \sum_1^n y'_j p_j - f(x,y_1,\ldots,y_n,y'_1,\ldots,y'_n) \quad .$$

We have

$$dH = \sum_1^n y'_j dp_j + \sum_1^n p_j dy'_j - \frac{\partial f}{\partial x}dx - \sum_1^n \frac{\partial f}{\partial y_j}dy_j - \sum_1^n \frac{\partial f}{\partial y'_j}dy'_j \quad ,$$

and, since we have the equalities in (56)

$$dH = -\frac{f}{x}dx - \sum_1^n \frac{f}{y_j}dy_j + \sum_1^n y'_j\,dp_j \quad .$$

It follows that, on account of (55),

$$\frac{\partial H}{\partial y_j} = -\frac{\partial f}{\partial y_j} = -\frac{d}{dx}\frac{\partial f}{\partial y'_j} = -\frac{dp_j}{dx} \;, \qquad j = 1,2,\ldots,n \quad ;$$

$$\frac{\partial}{\partial p_j} = y'_j = \frac{dy_j}{dx} \;, \qquad j = 1,2,\ldots,n \quad .$$

The system in (55) is thus replaced by the *canonical system.*

$$\frac{dy_j}{dx} = \frac{\partial H}{\partial p_j}, \quad \frac{dp_j}{dx} = -\frac{\partial H}{\partial y_j} \qquad j = 1,2,\ldots,n \quad ,$$

where H is the Hamiltonian function defined by the equality in (57).

These equations are regarded as a particular case of the equations arising in theoretical mechanics where the Lagrange equations are associated to the Hamiltonian principle of least action. We refer on this matter to E. Cartan's *Leçons sur les invariants integraux*.